Contents

KU-767-699

Mapping to your course

This book covers the Level 3 Diploma in Professional Cookery (VRQ) and the Level 3 NVQ Diploma in Professional Cookery. It is also suitable for the Level 3 NVQ Diploma in Professional Cookery (Patisserie and Confectionery). The table below shows where you can find the content that covers each unit.

Unit code	Unit title	Chapter of the book
	Level 3 Diploma in Professional Cookery (VRQ)	
	Supervisory Skills in the Hospitality Industry	Chapter 1
	The Principles of Food Safety Supervision for Catering	Chapter 2
	Exploring Gastronomy	Chapter 3
	Advanced Skills and Techniques in Producing Vegetable and Vegetarian Dishes	Chapter 7
	Advanced Skills and Techniques in Producing Meat Dishes	Chapter 8
	Advanced Skills and Techniques in Producing Poultry and Game Dishes	Chapter 9
	Advanced Skills and Techniques in Producing Fish and Shellfish Dishes	Chapter 10
	Produce Fermented Dough and Batter Products	Chapter 11
	Produce Petits Fours	Chapter 12
	Produce Paste Products	Chapter 13
	Produce Hot, Cold and Frozen Desserts	Chapter 14
	Produce Biscuits, Cake and Sponges	Chapter 15
	Food Product Development	Chapter 17
	Level 3 NVQ Diploma in Professional Cookery **(Level 3 NVQ Diploma in Professional Cookery (Patisserie and Confectionery), only units marked * apply)**	
*2GEN3	Maintain Food Safety when Storing, Preparing and Cooking Food	Chapter 2
*HSL4	Maintain the Health, Hygiene, Safety and Security of the Working Environment	Chapter 2
*MSC D1 HSL2	Develop Productive Working Relationships with Colleagues	Chapter 1
2PR17	Produce Healthier Dishes	Chapter 18
3FC1	Cook and Finish Complex Fish Dishes	Chapter 10
3FC2	Cook and Finish Complex Shellfish Dishes	Chapter 10
3FC3	Cook and Finish Complex Meat Dishes	Chapter 8
3FC4	Cook and Finish Complex Poultry Dishes	Chapter 9
3FC6	Cook and Finish Complex Vegetable Dishes	Chapter 7
3FP1	Prepare Fish for Complex Dishes	Chapter 10
3FP2	Prepare Shellfish for Complex Dishes	Chapter 10
3FP3	Prepare Meat for Complex Dishes	Chapter 8
3FP4	Prepare Poultry for Complex Dishes	Chapter 9
3FP5	Prepare Game for Complex Dishes	Chapter 9
3FPC1	Prepare, Cook and Finish Complex Hot Sauces	Chapter 6
3FPC10	Prepare, Finish and Present Canapés and Cocktail Products	Chapter 4
3FPC11	Prepare, Cook and Finish Dressings and Cold Sauces	Chapter 4
*3FPC12	Prepare, Cook and Finish Complex Hot Desserts	Chapter 14
*3FPC13	Prepare, Cook and Finish Complex Cold Desserts	Chapter 14
*3FPC14	Produce Sauces, Fillings and Coatings for Complex Desserts	Chapters 13 and 14
3FPC2	Prepare, Cook and Finish Complex Soups	Chapter 6

Unit code	Unit title	Chapter of the book
3FPC3	Prepare, Cook and Finish Fresh Pasta Dishes	Chapter 5
*3FPC4	Prepare, Cook and Finish Complex Bread and Dough Products	Chapter 11
*3FPC5	Prepare, Cook and Finish Complex Cakes, Sponges, Biscuits and Scones	Chapters 12 and 15
*3FPC6	Prepare, Cook and Finish Complex Pastry Products	Chapter 13
*3FPC7	Prepare, Process and Finish Complex Chocolate Products	Chapter 16
*3FPC8	Prepare, Process and Finish Marzipan, Pastillage and Sugar Products	Chapter 16
*3FPC9	Prepare, Cook and Present Complex Cold Products	Chapter 4
*PERR/10	Employment Rights and Responsibilities in the Hospitality, Leisure, Travel and Tourism Sector	Chapter 1
*HSL3	Contribute to the Control of Resources	Chapter 1
*HSL30	Ensure Food Safety Practices are Followed in the Preparation and Serving of Food and Drink	Chapter 2
*HSL9	Contribute to the Development of Recipes and Menus	Chapter 17

About the authors

Professor David Foskett MBE CMA FIH is Head of School at the London School of Tourism and Hospitality at the University of West London.

Neil Rippington is Dean of the College of Food at University College Birmingham.

Patricia Paskins is Senior Hospitality Lecturer and Work-Based Learning Co-ordinator at the London School of Tourism and Hospitality at the University of West London. She was awarded the Craft Guild of Chefs award for Chef Lecturer of the Year, 2013.

Steve Thorpe FIH is Head of the Hotel, Hair and Beauty School at City College, Norwich.

Dedication

This book is dedicated to the memory of Professor Dr William Barry, academic and entrepreneur, who was a keen supporter of students entering the hospitality industry.

Foreword

Welcome to *Practical Cookery Level 3*. At Level 3 you will learn to develop increasing precision, speed and control in your existing culinary skills, as well as developing more refined and advanced techniques. You will also develop the skills you'll need to work as a senior chef, including the supervisory aspects of the role such as managing others and controlling resources.

I fell in love with the idea of being a chef when, at the age of sixteen, I visited a three-star restaurant beneath the cliffs of Provence with my family and experienced the magic combination of extraordinary food and beautiful surroundings. It took many years to realise my dream, during which time I cooked the same dishes over and over again, perfecting the techniques and seeking the best way to harness flavour. It is this experimentation, dedication to precision and development of expertise that is at the heart of your Level 3 course.

Science has always been fundamental to the development of my restaurant and has been a constant source of inspiration for my cooking. I'm particularly interested in multisensory perception – how the brain influences our appreciation of food, how we perceive flavour and how learned preferences enhance our enjoyment of a dish. This Level 3 course will allow you to explore the concept of gastronomy further, and to understand how science (among other influences) impacts on the dining experience and our appreciation of food. You'll also get an opportunity to explore a range of techniques that help ensure precision. I have, for example, used water baths in my restaurant to cook with exceptional precision and consistency. You'll find some discussion of how to use these techniques, as well as recipes that incorporate sous vide cooking, in this book.

While modern culinary trends are important, classical skills should never be forgotten. *Practical Cookery* has always been essential for those looking to learn the fundamental foundations of classical cookery, and this new Level 3 text combines the old with the new, the modern with the classical. This is a balance that is key to my menus and which will prove useful to you as you progress in your career as a professional chef.

Heston Blumenthal

Acknowledgements

We appreciate the sponsorship received from The William Barry Trust.

We are also very grateful to Watts Farms for providing much of the fruit and vegetables shown in the photographs. Watts Farms is a family-run fresh produce business specialising in the growing, packing and supply of a wide range of seasonal produce from spinach, baby leaf salads, herbs and legumes to fruit, chillies, asparagus, brassicas and specialist interest crops.

Some recipes and text in Chapters 11, 12, 14, 15 and 16 are taken from *Professional Patisserie for Levels 2 and 3 and Professional Chefs*, and were written by Mick Burke and Chris Barker.

Nutritional analysis for this edition has been provided by Joanne Tucker at the University of West London.

Photography

Most of the photos in this book are by Andrew Callaghan of Callaghan Studios. The photography work could not have been done without the generous help of the authors and their colleagues and students at the University of West London (UWL) and City College, Norwich (CCN). The publishers would particularly like to acknowledge the following for their work.

Gary Farrelly and Ketharanathan Vasanthan organised the cookery at UWL. They were assisted in the kitchen by:
- Catherine Bent
- Manish Gobin
- Eddie Leong Chun How
- Ra-Hyun Hwang
- Chin Keat Lim
- Fern Lough
- Tarkan Nevzat
- Vikram Rathour
- Shamin Talib.

Steve Thorpe organised the photography at the Hotel School at CCN. He was assisted in the kitchen by:
- Greg Arundell
- Nick Blackmore
- Sam Brown
- Daniel Knight
- Bethany Redhead.

The authors and publishers are grateful to everyone involved for their hard work.

Picture credits

Using the QR codes

There are free videos on the website. Look out for the QR codes throughout the book.

To view the videos you will need a QR code reader for your smartphone/tablet. There are many free readers available, depending on the smartphone/tablet you are using.

Once you have downloaded a QR code reader, simply open the reader app and use it to take a photo of the code. The file will then load on your smartphone/tablet.

If you cannot read the QR code or you are using a computer, the web link next to the code will take you directly to the same place.

1 Supervisory skills in the hospitality industry

This chapter covers the following units:

NVQ:

→ Maintain the health, hygiene, safety and security of the working environment.
→ Develop productive working relationships with colleagues.

VRQ:

→ Supervisory skills in the hospitality industry.

Introduction

Supervising means working with a team or with individuals to make sure that tasks and procedures are completed in an appropriate way and within the allocated time. Supervision will inevitably also involve some management procedures such as:

→ solving day-to-day problems and making decisions
→ liaising with management effectively
→ conducting training sessions, meetings and monitoring performance
→ organisation of employees, including allocation and delegation of tasks
→ establishing, keeping and updating records
→ planning and forecasting
→ complying with legislation.

Throughout this chapter 'the supervisor' will be referred to. This term could also describe duties completed by those with the following job titles: a head chef (of a small establishment), sous chef, section chef, chef de partie, team leader, assistant catering manager, restaurant manager, head cook or assistant head cook, section manager, floor supervisor and many more.

Learning objectives

By the end of this chapter you should be able to:

→ Apply and monitor good health and safety practices.
→ Apply and monitor good health and safety training.
→ Develop best practices for working in a safe and healthy way.
→ Explain how to apply staff supervisory skills within a small team.
→ Contribute to effective use of resources.
→ Develop effective working relationships within the team.
→ Describe a range of supervisory tasks and how they apply to a team.
→ Discuss the purpose of supervision and the characteristics of leadership.
→ Identify characteristics of leadership and leadership styles.
→ Identify the characteristics of a good team and the benefits of team development.
→ Identify the training requirements in the hospitality industry.
→ Explain the different methods of training.

Supervising health and safety

The responsibility of chefs, supervisors and others concerned with health and safety is to ensure that the health and safety policies and standards of the workplace are upheld, training and instruction is given so as to prevent accidents and to help staff work efficiently and safely.

Many chefs have a supervisory role built into their job specification and responsibilities. Chef de parties, sous chefs and head chefs may all be responsible for supervising health and safety in the workplace and making others aware of the importance of health and safety procedures. It is important to supervise day-to-day work and to train employees in good practice to ensure that they achieve consistently high standards in health and safety and meet all legal requirements. As part of the supervisor role, they may also advise management and keep them informed on health and safety issues.

The management/supervision of health and safety at work involves:

● Enabling managers, supervisors and employees to implement health and safety procedures in the best possible way.
● Production and application of organisational policies.
● Measuring health and safety performance and reviewing that performance.
● Assisting with auditing of the health and safety system as often as is required.
● Planning health and safety requirements when there are changes in policy, procedure or equipment.
● Awareness of and working within current legislation.
● Provision of relevant training, mentoring and supervision for staff.
● Ensuring that risk assessments are used in working situations.
● Reviewing and developing and implementing the health and safety policy.

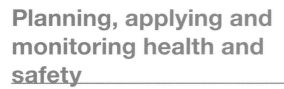

Planning, applying and monitoring health and safety

Responsibilities of supervisors

As a chef supervising others, you need to be involved in the planning of health and safety initiatives, the training of employees and the monitoring of health and safety performance standards. In order to identify hazards in the workplace, there must be risk assessment and regular safety inspections. This means observing how people actually carry out their daily work. If you have any concerns, they need to be discussed with employees and management as appropriate.

As a working chef/supervisor, you will also need to identify the health and safety training needs of the establishment, reviewing each individual's training needs and assessing how safe people are in completing the tasks that make up their job role. Individual employees may need different levels of training and training for specific parts of health and safety requirements. The supervisor will need to monitor and plan for individual training needs, ensuring that the required training takes place at appropriate times and is recorded.

Often the chef as a supervisor will be part of a committee on health and safety that establishments are required to have to ensure safety inspections are carried out and that accidents are investigated.

Chefs are responsible for overseeing and carrying out tasks in accordance with the establishment's health and safety policy. The procedures and tasks must be safe to a high standard and comply with legislation.

The supervisor must, at all times, give guidance, demonstrate good practice and ensure that all work activities are carried out in a disciplined manner. Different employees will need different levels of supervision. Competent people also need supervising to check that they do not fall into bad habits or take dangerous shortcuts. Particular attention must be paid to people who are vulnerable to a higher risk of injury such as young or inexperienced employees.

Activity

Suggest three kitchen tasks that could be considered of higher risk of causing injury. Identify the safety measures you would put in place to enable these tasks to be completed safely.

Ensuring compliance with legislation

To ensure that legislation regarding safety and security is implemented, it is necessary:

1. To be familiar with the relevant legislation.
2. To ensure that the requirements are carried out effectively.
3. To implement a system of checks to ensure that the legislation is complied with.

Employers' responsibilities to their employees

Every employer has a duty to ensure health and safety and welfare at work of all employees in so far as is reasonably practicable. In hospitality this means:

- Providing and maintaining kitchens, restaurant, accommodation and systems of work that are safe and without risk to health.
- Making sure that storage areas and transporting items such as food and equipment are safe and will not cause injury or risk to health.
- That there is information, instruction, training and supervision to ensure the health and safety of employees at work.
- Maintaining the premises and building to make sure they are safe and pose no risk to health.
- Maintaining entrances and exits to the workplace and access to work areas that are safe and without risk to health.
- Providing a clean and safe environment with good welfare facilities.
- Where necessary, providing health surveillance of employees.
- Providing and maintaining the required personal protective equipment (and clothing).

Employers must provide a written statement about:

- the general policy towards employees' health and safety at work;
- the organisation and arrangements for carrying out that policy.

An organisation with a board of directors/governors must formally and publicly accept its collective role in providing health and safety leadership in the organisation.

A system of checks, both spot-checks and regular inspections at frequent intervals, needs to be set up and the observations and recommendations resulting from these inspections should be recorded and passed to the relevant person for action. The details would include the time and date of inspection, exact areas or equipment involved and a clear description of any breach of security or fault of safety equipment. Records should be accurate and legible. It is important that any shortcomings are dealt with at once.

The type of equipment that needs to be inspected to make certain that it is available and ready for use includes security equipment, machinery safety guards, warning signs, first-aid and fire-fighting equipment. The supervisor or person responsible for these items needs to regularly check and record that they are in working condition and that, if they have been used, they are restored ready for further use.

Security systems and fire-fighting equipment are usually checked by the manufacturers or a contractor. However, it is the responsibility of the management of the establishment to ensure that this equipment is maintained and working correctly. It is advisable that all staff are given basic fire training and are trained in the use of fire extinguishers.

First-aid equipment is usually the responsibility of the designated first-aider, whose functions include replenishing first-aid boxes. However, this is another task that would be overseen by the supervisor.

Routine checks and inspections need to be carried out in any establishment to see that standards of health, safety and security are maintained for the benefit of workers, customers and anyone else on the premises. Visitors, suppliers and contractors are also entitled to expect the premises to be safe when they enter. Particular attention needs to be paid to exits and entrances, passageways and the provision of adequate lighting. Floors need to be sound, uncluttered and safe to walk on. Disposal and waste bin areas need particular care regarding cleanliness, health and safety. Toilets, staff rooms and changing rooms need to be checked and cleaned regularly.

Checks or inspections may be carried out by a person responsible for health and safety within the organisation with authority to take action to remedy faults and discrepancies, and to implement improvements. However, it is important that this person works closely with the supervisor who is the person most familiar with the area and working procedures

Occupational health

Occupational health is concerned with protecting the safety, health and welfare of people relating to their work or employment, including any illness or disease caused by the employment. The aims of occupational health programmes include provision of a safe and healthy working environment and advising on working procedures and substances that could affect health. In a kitchen, this could be frequent exposure to strong cleaning chemicals, causing respiratory difficulties or skin problems, or problems caused by frequent heavy lifting.

Occupational health departments provided by some companies may also offer staff health advice, screening, vaccination, weight loss and smoking cessation advice, pre-employment medical checks and fitness to work checks. They frequently offer advice on safe working procedures when there is a change in circumstances, for example, a kitchen employee who is pregnant.

Increasingly, occupational health provision for companies is sourced from an outside agency or health centre.

Duties of employees

All employees in an establishment must be made aware of the need for safety and security and their legal responsibilities towards themselves, their colleagues, their employers and others.

It is the responsibility of everyone in the workplace to be conscious of health, safety and security. Everyone in the working area needs to be aware of equipment and procedures for maintaining these.

Every employee has a duty to take reasonable care of his or her own health and safety and that of other people who may be affected by what he or she does or does not do in the course of carrying out work. Employees must co-operate with the employer to enable the employer to comply with the necessary requirements.

Employees must also remain vigilant and report anything they think could be a health and safety hazard or could cause injury.

Maintaining a healthy and safe working environment

Kitchens and hospitality premises in general can potentially be dangerous places. It is a legal requirement for every establishment to identify risks, introduce the necessary measures to minimise risks and possible harm, and to ensure that the workplace is as safe as possible.

Most of what is required is about common-sense, developing good practices and making everyone aware of the importance of high standards of health and safety. Making health and safety part of daily working practices, planning, feedback, handover and meetings will all help to ensure a safe workplace.

Every individual at work anywhere on the premises needs to develop a positive attitude towards identifying potentially hazardous situations so potential accidents can be minimised. On-going training is also essential to develop good practice and should include information on the hazards to look for, hygienic methods of working and the procedures to follow in the event of an accident or incident. Records of staff training, the topics covered and the dates they were completed should be kept.

Every organisation will have procedures to follow in the event of a fire, accident, flood or bomb alert; each employee needs to be familiar with and regularly reminded of these procedures.

Every establishment must have processes in place to record accidents. It is also desirable to have a procedure to record items in need of maintenance due to wear and tear or damage so that these faults can be remedied. Details of incidents such as power failure, flooding, infestation, contamination should be recorded in an incident record.

Workplace security

Workplace security is important to ensure the safety of employees and everyone else using the premises, as well as the actual premises and their contents and fittings. Security measures are necessary to protect against:
- personal injury
- theft
- fraud
- vandalism
- personal assault
- terrorism.

All employees must remain vigilant with regard to security matters and immediately report anything they think might be a breach of security. All staff must be trained in the use of the company's security measures and induction may be a good time to do this. Everyone must observe and use security measures put in place by the employer properly. These may include:
- employee ID and swipe cards
- locks and security key pads
- CCTV cameras
- safes and secure boxes or drawers
- passwords
- staffed security/reception desks.

Large amounts of cash may be handled by hospitality establishments and this could be vulnerable to theft, fraud

or misuse. Encourage regular removal of cash to safes or to finance personnel, regular collection of cash by security companies and also encourage use of debit/credit cards so less cash is handled.

Items lost, damaged or discarded should be noted and a record kept, giving details of why and how it happened and what subsequent steps have been taken.

Health and safety legislation and enforcement

It is important that supervisors are well aware of the health and safety legislation that applies to their industry and establishment and the people working within it. Legislation is updated occasionally to deal with modern working procedures and equipment but the main Acts are outlined below.

The aims of legislation are:
- To secure the health, safety and welfare of people at work.
- To provide regulations and approved codes of practice that set the standards of health, safety and welfare.
- To establish the minimum health and safety requirements for different areas of the workplace.
- To protect people other than employees who may be in the area, for example, guests, customers, contractors, suppliers and others, against any health and safety risks within the workplace.
- To control the storage and use of dangerous substances such as explosive, corrosive, highly flammable or otherwise dangerous materials.

Health and Safety at Work Act (1974)

This legislation deals with workplace health and safety in England, Wales and Scotland. Largely similar provision is covered in Northern Ireland under the Health and Safety at Work (Northern Ireland) Order 1978. The Act is largely about establishing and maintaining good health and safety practice in the workplace. It requires that employers comply with all parts of the Act, including completing risk assessments of all areas and procedures and introducing safe ways of working. Employers will need a health and safety policy including risk assessments (unless there are fewer than five employees), to keep premises and equipment in safe working order and provide/maintain personal protective equipment for employees. Employees must work in a way that does not endanger themselves and others. They must comply with health and safety procedures introduced by their employers and report any defect or problem that could affect health and safety.

There were updates to this Act in 1994.

Management of Health and Safety at Work Regulations 1999

Where an employer has five or more employees, there must be a written health and safety policy in place that is issued to every member of staff. This will outline the responsibilities of the employer and employee in relation to health and safety. It states that health and safety information must be provided for all employees and these could be in the form of training packs, leaflets, posters, DVDs and stickers, all of which are available from the Health and Safety Executive (HSE).

Personal Protective Equipment at Work Regulations 1992

These regulations require employers to assess the need for and provide suitable personal protective equipment (PPE) and clothing at work. In a commercial kitchen environment, this may include chefs' uniforms and items for cleaning tasks such as rubber gloves, goggles and masks. Employers must keep these items clean, in good condition and provide suitable storage for them. Employees must use them correctly and report any defects or shortages.

Manual Handling Operations Regulations 1992

These regulations are in place to protect employees from injury or accident when required to lift or move heavy or awkwardly shaped items. A risk assessment must be completed, employees trained in correct manual handling techniques and lifting or moving equipment provided where appropriate.

Provision and Use of Work Equipment Regulations 1998 (PUWER)

'Work equipment' covers items such as food processors, slicers, ovens, fryers and knives. These regulations place duties on employers to ensure that work equipment is suitable for its intended use, is maintained in efficient working order and is in good repair, and that adequate information, instruction and training on the use and maintenance of the equipment and any associated hazards is given to employees. Work equipment that poses a specific risk must be used only by designated people who have received relevant training. Requirements cover dangerous machinery parts, protection against certain hazards (such as falling objects, ejected components, overheating and entrapment). Also covered is the provision of certain stop and emergency cut-out controls, isolation from energy sources, stability lighting and markings and warnings.

Control of Substances Hazardous to Health 2002 (COSHH)

Under these regulations, risk assessments must be completed by employers of all hazardous chemicals and substances that employees may be exposed to at work, their safe use and disposal. Each chemical must be given a risk rating that is recorded and employees given relevant information and training on the chemicals they will be using. Some examples of chemical substances found in kitchens are:

- cleaning chemicals – alkalis and acids
- detergents, sanitisers, de-scalers, de-greasers
- chemicals associated with burnishing
- pest control chemicals, insecticides and rodenticides.

Workplace (Health, Safety and Welfare) Regulations 1992

These regulations are in place to ensure the safety and well-being of employees within their working environment. The act covers such things as suitable working premises, staff hygiene facilities, heating, ventilation and lighting.

The Reporting of Injuries, Diseases and Dangerous Occurrences Regulations 1995 (RIDDOR) (updated October 2013)

The law that requires employers, and/or the person with responsibility for health and safety within a workplace, to report and keep records of any:

- work-related fatal accidents
- work-related disease
- accidents and injury resulting in the employee being off work for three days or more
- a dangerous workplace event (including 'near-miss' occurrences)
- major injuries, loss of limbs or eyesight.

The 2013 changes simplify the reporting procedures and provide shorter lists of the incidents/diseases that need to be reported but most requirements will remain broadly unchanged.

Take it further

Accidents and incidents under RIDDOR need to be reported and this can be done by contacting the HSE by telephone on 0845 300 9923 or by completing the RIDDOR form on their website www.hse.gov.uk/riddor

Fire Precautions (Workplace) Regulations 1997

The Fire Precautions (Workplace) Regulations 1997 state that premises with over five employees must have a written fire risk assessment with details of the appropriate fire safety precautions in place. These may include:

- Provision of emergency exit routes and doors.
- 'Fire Exit' signs, emergency lighting to cover the exit routes where necessary.
- Fire-fighting equipment, fire alarms and fire detectors, where necessary.
- Fire training for employees in fire safety following the written risk assessment.
- Production of an emergency plan and enough trained people/equipment to carry out the plan.
- All equipment such as fire extinguishers, alarms systems and emergency doors to be regularly maintained and faults rectified as soon as possible.
- Employers to plan, organise, control, monitor and review the measures taken to protect employees and others from fire.

The Electricity at Work Regulations 1989

These regulations state that all pieces of electrical equipment used in the workplace should be checked every 12 months by a qualified electrician and this must be recorded. It is also recommended that a supervisor or other responsible person check these items on a regular basis. All equipment should be included in the health and safety risk assessment and staff must have training in the safe use of the equipment.

Other relevant regulations that apply to the workplace

- Lifting Operations and Lifting Equipment Regulations 1998 (LOLER)
- Noise at Work Regulations 1989
- The Health and Safety (First Aid) Regulations 1981
- Data Protection Act 1988
- Health and Safety (Display Screen Equipment) Regulations 1992
- Equality Act 2010
- Race Relations Act 1977
- Asylum and Immigration Act 1996
- Sex Discrimination Act 1975
- Human Rights Act 1988
- Licensing Act 1964
- Working Time Regulations 1998

- National Minimum Wage Act 1998
- Equality Act 2010
- Disability Discrimination Act 1995.

Activity

Choose two of the Acts listed above and explain how they may affect the staffing of a large kitchen.

Sources of support for supervisors of health and safety

Supervisors should have a good knowledge of health and safety matters and of the related legislation, but there are a number of helpful sources where they can find health and safety information, including:

- workplace HR departments and health and safety representatives
- the Health and Safety Executive (HSE) – there is a wealth of information on their website, some of it specific to kitchens
- 'Safer Food, Better Business' (available from the Food Standards Agency)
- Environmental Health Officers
- the Food Standards Agency
- fire safety officers
- product manufacturers and the literature and learning materials they produce
- various health and safety publications and chef/hospitality textbooks
- trade unions.

Delegation

Health and safety in a kitchen area can be complex and time-consuming for one supervisor to cover alone. The responsibilities could therefore be shared between one or more other supervisors. Some tasks could also be delegated to other members of the team, for example, checking the first-aid box contents, checking that the necessary health and safety signage and information are in place, mentoring a new recruit.

> **Key term**
>
> **Delegate** – to give some of the tasks you may previously have done yourself to someone else to complete.

Approved Codes of Practice

The Health and Safety Commission can provide practical guidance for employers and employees to help them to comply with the regulations and duties which apply to them. The advice is issued as a Code of Practice. If the Health and Safety Commission approves these standards, they can bear the title 'Approved Code of Practice'. They include a standard, a specification and any other form of practical advice.

Approved Codes of Practice have a special place in legislation. They are not law but a failure to observe any part of an approved code of practice may be admissible in criminal court proceedings as evidence about a related alleged contravention of health and safety legislation. If the advice in the approved code of practice has not been followed, then it is up to the defendants to prove that they have satisfactorily complied with the requirement in some other way. Examples of Approved Codes of Practice include:
- Management of Health and Safety at Work
- Workplace Health, Safety and Welfare
- Control of Substances Hazardous to Health
- Safe Use of Work Equipment
- Safe Use of Lighting Equipment
- First Aid at Work.

Guidance

The Health and Safety Commission also produces guidance on many other technical health and safety subjects. Guidance does not have the same standing in law as Approved Codes of Practice, but it can be used by employers to show that they have complied with a recognised standard. Examples of guidance include:
- Manual Handling
- Reducing Noise at Work
- Guide to COSHH Assessment.

Activity

A kitchen supervisor wishes to create a file of useful health and safety information to be available for staff. Where could the supervisor find the necessary information and what is the information that could be found from each source?

Setting health and safety standards

- **The Health and Safety Commission (HSC)** – appointed by the Secretary of State. The Commission is responsible for proposing health and safety law and standards. It consults professional bodies with an interest in health and safety such as trade unions and industry.
- **The Health and Safety Executive (HSE)** – appointed by the HSC to regulate health and safety law in industry and public areas.
- **Local authorities** – the HSC gives local authorities delegated power to regulate health and safety law in premises such as retail shops, offices, catering services, restaurants, hotels, etc.
- **Authorised officers** – health and safety law is enforced by:
  - health and safety inspectors from the HSE
  - environmental health officers (EHOs) and technical officers from local authorities
  - fire officers from the fire service.

For factories, farms and hospitals, the enforcing officer is a health and safety inspector from the HSE.

For shops, restaurants and leisure centres, the enforcing officer is the local EHO. Fire officers can visit all these premises for the purposes of enforcing the law on fire safety and fire precautions.

Enforcement

Inspections can be carried out by HSE inspectors or environmental health officers/practitioners.

Inspectors have a number of powers including the right of entry to premises, to serve legal notices requiring improvement work to be done and prohibiting work procedures, processes and the use of work equipment.

Inspectors may:
- Enter premises at any reasonable time.
- Take a police officer with them.
- Take an authorised person or equipment to help with the investigation.
- Make necessary examinations and inspections.
- Take measurements, photographs or recordings.
- Take samples of articles or substances.
- Dismantle or test any article or substance.
- Take possession of and detain any article or substance for examination or to ensure that no-one tampers with it.
- Require any person to give information to assist with any examination or investigation.

- Require documents to be produced, inspected or copied.
- Require that assistance and facilities be made available to allow the inspector's powers to be exercised.
- Seize and render harmless any article or substance which the inspector believes to be a cause of imminent danger or serious personal injury.
- Stop certain procedures.

Improvement notices

If the inspector thinks there is a contravention of health and safety legislation, an improvement notice may be served on the person responsible stating the details of the contravention. The notice also requires the person responsible to remedy the contravention within a fixed time. The person responsible is the director, manager or supervisor in charge of the premises at the time of the inspection.

The notice must state:
- That a contravention exists.
- The details of the law contravened.
- The inspector's reasons for his or her opinion.
- That the person responsible must remedy the contravention.
- The time given for the remedy to be carried out. This must not be less than 21 days.

If a person fails to comply with an improvement notice, he or she commits a criminal offence.

Prohibition notices

If an inspector believes that work activities involve a serious risk of 'personal injury', a prohibition notice may be served on the person in charge of the work activity. This will prevent further work being carried out in the premises or area deemed to be dangerous. The notice must:
- State that, in the inspector's opinion, there is a risk of serious personal injury.
- Identify the matters which create the risk.
- Give reasons why the inspector believes there to have been a contravention of health and safety law.
- Direct that the activities stated in the notice must not be carried on, by or under the control of the person served with the notice, unless the matters which are associated with the risk have been rectified.

Activity

1 Suggest three findings in a kitchen that may cause an inspector to serve an improvement notice.
2 Suggest three findings that may cause an inspector to serve a prohibition notice.

Appeals

The person who is served with an improvement or a prohibition notice can appeal against the notice to an employment tribunal. The grounds of the appeal could be based on:
- The inspector interpreted the law incorrectly or exceeded their powers.
- A contravention might be admitted but the appeal would be that the remedy was not practicable or not reasonably practicable.
- A contravention might be admitted. The appeal would be based on the fact that the incident was so insignificant that the notice should be cancelled.

If an appeal is made against an improvement notice, then the notice is suspended until the appeal is heard. If an appeal is made against a prohibition notice, the notice is suspended only if the employment tribunal suspends it. If there is no compliance with the notice, the person served with the notice can be prosecuted.

Offences

A contravention of the Health and Safety at Work Act 1974 or any of the regulations made under the Act is a criminal offence. Both an individual and a corporate body can commit an offence and be tried for it in court. It is an offence to:
- Fail to carry out a duty placed on employers, self-employed employees, owners of premises, designers, manufacturers, importers and suppliers.
- Intentionally or recklessly interfere with anything provided for safety.
- Require payment for anything that an employer must by law provide in the interests of health and safety.
- Contravene any requirement of any health and safety regulations.
- Contravene any requirement imposed by an inspector.
- Attempt to prevent a person from appearing before an inspector or from answering his/her questions.
- Contravene an improvement or prohibition notice.
- Intentionally obstruct an inspector in the exercise of his/her powers or duties.
- Intentionally make a false entry in a register, book, notice or other document which is required to be kept.
- Intentionally or recklessly make false statements.
- Pretend to be a health and safety inspector.
- Fail to comply with a court order.

If a person is found guilty of a health and safety offence, the penalty may be a substantial fine; serious cases could result in imprisonment.

Accidents in the workplace

An accident is:

- An unplanned and uncontrolled event.
- An event that causes injury, damage or loss.
- An event that could lead to a near-miss accident, or could result in no loss or damage at all.

Accidents do not just happen. They arise from uncontrolled events and often from a chain of uncontrolled events. There are a number of reasons why accidents may occur.

- **Human factors and errors** – the ability (or lack of ability) to recognise hazards and risks, lack of skills, general attitude to safety (taking short cuts, insufficient care taken), tiredness, use of alcohol or drugs.
- **Occupational factors** – being exposed to possible hazards in the workplace specific to the occupation. A chef may risk cuts or burns.
- **Environmental factors** – this refers to the working environment such as poor lighting or poor air quality. It may also refer to the time available to carry out certain jobs and the pressure of the work environment.
- **Organisational factors** – these could affect the safety of staff. They include, for example, the safety standards of the organisation, safety precautions, enforcing/encouraging of precautions and standards by the employer, the effectiveness of communication between work colleagues and the employer, the amount of training individuals have received, advice and supervision.

Causes of accidents

Accidents happen in many ways. The three main causes are:

1 Unsafe actions
2 Unsafe conditions
3 A combination of unsafe actions and unsafe conditions.

Accidents will occur in the workplace if health and safety is not taken seriously and there is no culture of safety in the workplace. There will be a low risk of accidents in an organisation that has an effective health and safety policy, a strong safety culture and a real commitment from management.

Table 1.1 Significant health and safety factors in the hospitality industry

Cause	Percentage of all accidents	Significant factors
Slips, trips, falls	30% but they account for 75% of all major injuries	88% of these are caused by slippery floors due to spillage not cleared up, or wet or oily floors. Boxes and buckets or other trip hazards left in passageways. Uneven floors are also an issue
Handling and lifting	30%	33.3% due to lifting pans, trays and other kitchen equipment 33.3% due to handling sharp objects such as knives and 33.3% due to awkward lifts from low ovens or high positions
Exposure to hazardous substances, hot surfaces, steam	16%	Often from splashes with hot liquids/oils or hot objects. Causes are poor maintenance; steam from ovens/steamer; carrying hot liquids; misuse of cleaning materials; cleaning deep fat fryers; equipment failure; carelessness; hot surfaces
Struck by moving articles including hand tools and mixers	10%	33% most probably from knives; 33% from other equipment; 25% from falling articles; 9% from assault
Walking into objects	4%	75% of cases involved walking into a fixed as opposed to a moveable object
Machinery	3%	Slicers 30%; mixers 15%; also caused by vegetable cutting machines; vegetable slicing, mincing and grating attachments; pie and tart machines; dough mixer; dough moulder; mincing machine; dishwasher
Falls	1.8%	75% falls from low height but half of the major injuries occurred on stairs
Fire and explosion	1.6%	80% during manually igniting gas fire appliances, mainly ovens
Electric shock	0.5%	25% due to poor maintenance; 25% trolley involved; 25% unsafe switching and unplugging (75% of these in wet conditions); 25% poor maintenance
Transportation	3%	50% involved forklift trucks

Examples of potential hazards in the kitchen

- unsafe equipment, using sharp or mechanical equipment without guards
- walking on slippery floors
- carrying saucepans of boiling water or hot oil
- carrying sharp knives
- using damaged equipment
- lifting heavy loads in an unsafe manner
- not wearing protective clothing
- using unsafe chemicals or cleaning chemicals without following the manufacturers' instructions
- inadequate maintenance of work equipment
- poor environmental conditions such as extreme temperatures, high humidity, poorly designed buildings
- dirty environment
- broken machine guards
- loose clothing that can get trapped in machines.

Slips, trips and falls

The highest number of accidents occurring in hospitality/catering premises is due to falling, slipping or tripping. A major reason for the high incidence of this kind of accident is that water and grease are likely to be split and the combination of these substances makes the floor surface very slippery. Any spillage must be cleaned immediately and warning notices put in place, where appropriate, highlighting the danger of the slippery surface. Ideally a member of staff should stand guard until the hazard is cleared.

Another cause of falls is the placing of articles on the floor in corridors, passageways or between stoves and tables. Persons carrying trays and containers have their vision obstructed and items on the floor may not be visible to them. The person may fall onto a hot stove and the item being carried may be hot. These falls can have severe consequences. The solution is to ensure that nothing is left on the floor that may cause a hazard. If it is necessary to have articles temporarily on the floor, then it is desirable that they are guarded to prevent accidents. Kitchen personnel should be trained to think and act in a safe manner at all times so health and safety become part of the culture of the workplace

Effects of accidents

Table 1.2 Effects of accidents on employers, employees and close friends and family

The possible effect of accidents on employers	The possible effect on employees of injury or illness	Impact on employees, close friends and family
• Lives can be lost • People injured • Money wasted • Machinery damaged • Products damaged • Reputation lost • Damage to the environment • Legal action	• Pain and suffering • Loss of earnings • Loss of quality of life • Long-term health problems • Personal injury or death	• Distress and grief • Anxiety • Loss of earnings looking after sick or injured person • Loss of quality of life

Table 1.3 Costs of accidents to employers

Direct costs	Indirect costs
• Damage and repairs to buildings, vehicles, machinery or stock • Legal costs • Fines (criminal court case) • Compensation (civil court case or agreed compensation) • Loss of product • Overtime payments • Employee medical costs • Employer's liability and public liability claims • Increased insurance premiums	• Loss of output to business • Product liability payments • Time and money spent on investigating an accident • Loss of good will between employees and management • Loss of consumer confidence • Damage to corporate reputation • Hiring and training of replacement staff • Loss of expertise

As an employer, what could be the effect and possible financial costs if a chef:

- Sustained serious cuts from a slicing machine because the safety guard was damaged?
- Tripped on a loose piece of damaged flooring while carrying a pan of hot soup?

Accident recording

All accidents should be reported to your line manager, chef or a more senior supervisor. Each accident is recorded on an accident form, which must be provided in every business. An example of an incident report form, showing all the detail required, is shown in the figure below.

Accident reports include confidential information, so completed reports must not be kept in a book or stored in a way that makes them easily accessible to anyone.

Full name and contact details of injured person:			
Occupation		Supervisor:	
Time of accident:	Date of accident:	Time of report:	Date of report:
Nature of injury or condition:			
Details of treatment:			
Extent of injury (after medical attention):			
Place of accident or dangerous occurrence:			
Injured person's evidence of what happened (include equipment/items and/or other persons):			
Witness evidence (1):		Witness evidence (2):	
Supervisor's recommendations:			
Date:		Supervisor's signature	

▲ Incident report form

Manual handling

The incorrect handling of heavy and awkward loads causes accidents, which can result in staff being off work for some time, often with back injury. It is important to lift heavy items in the correct way. The safest way to lift items is to bend at the knees rather than bending the back. Strain and damage can be reduced if two people do the lifting rather than one.

You are running a training session for kitchen staff on manual handling. Provide a statement to display in the kitchen saying why correct handling is important. Design a poster to use at the training session listing the main points of correct manual handling as they apply to where you work.

Manual handling checklist

- Assess the load. Can you get help with the lifting and would the use of a trolley be appropriate?
- Consider how easy or difficult it is to grip, how far you need to carry it?
- Check the place you are moving the load to is clear.
- With feet apart, bend from the knees alongside the load and grip it firmly.
- Straighten up, lifting the load using the knees not the spine and keeping the load close to the body.
- Place safely in the new position.
- When goods are moved on trolleys, trucks or any wheeled vehicles, they should be loaded carefully (not overloaded) and in a manner that enables the handler to see where they are going
- In store rooms, it is essential that heavy items are stacked at the bottom and that steps are used with care. (Ladder training may be needed.)
- Particular care is needed when large pots are moved containing liquid, especially hot liquid or oil. They should not be filled too full.
- A sign warning that equipment handles and lids can be hot should be given. The standard kitchen indicator is a small sprinkle of flour on the handle or lid, but better still wrap or cover it with an oven cloth.
- Extra care is needed when taking a tray from a hot oven or salamander that the tray does not burn someone else.

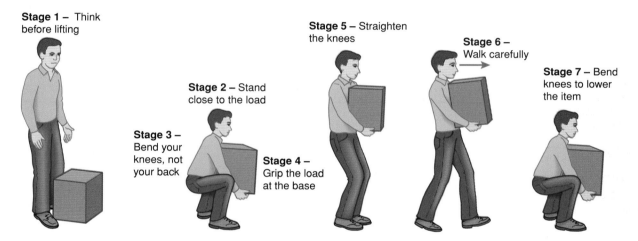

Stage 1 – Think before lifting

Stage 2 – Stand close to the load

Stage 3 – Bend your knees, not your back

Stage 4 – Grip the load at the base

Stage 5 – Straighten the knees

Stage 6 – Walk carefully

Stage 7 – Bend knees to lower the item

Fire safety

All employers have an explicit duty for the safety of their employees in the event of a fire. The Regulatory Reform Fire Safety Order 2005 places a greater focus on fire prevention. It places responsibility for the fire safety of the occupants of premises and people who might be affected by fire on a defined responsible person, usually the employer. The responsible person must:

- Make sure that the fire precautions, where reasonably practicable, ensure the safety of all employees and others in the building.
- Make an assessment of the risk of and from fire in the establishment and put suitable precautions and safety measures in place. Special consideration must be given to dangerous chemicals or substances, and the risks that these pose if a fire occurs.
- Review the preventative and protective measures.

Fire safety requires constant vigilance to reduce the risk of a fire by adopting safe working methods and providing frequent staff training and information. Risk is reduced in the event of a fire by the provision of detection and alarm systems and well practised emergency and evacuation procedures.

Fire precautions

Identified hazards must be removed or reduced as far as is reasonable. All persons must be protected from the risk of fire and the likelihood of a fire spreading.

- All escape routes must be safe and used effectively.
- Means for fighting fires must be available on the premises.
- Means of detecting a fire on the premises and giving warning in case of fire on the premises must be available.
- Arrangements must be in place for action to be taken in the event of a fire on the premises, including the instruction and training of employees.
- All precautions provided must be installed and maintained by a competent person.

Although businesses no longer need a fire certificate as they previously did, the fire and rescue authorities will continue to inspect premises and ensure adequate fire precautions are in place. They will also wish to be satisfied that the risk assessment has been completed, is comprehensive, relevant and up to date.

The fire triangle

For a fire to start, three things are needed:

1 a source of ignition (heat)
2 fuel
3 oxygen.

If any one of these is missing, a fire cannot start. Taking steps to avoid the three coming together will therefore reduce the chances of a fire occurring.

Methods of extinguishing fires concentrate on cooling or depriving the fire of oxygen such as an extinguisher that uses foam or powder to exclude oxygen.

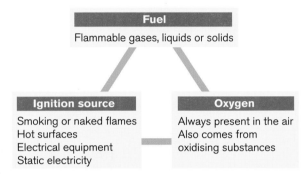

▲ The fire triangle

Once a fire starts it can spread very quickly from one source of fuel to another. As it grows, the amount of heat it gives off will increase and this can cause other fuels to self-ignite.

Fire detection and fire warning

There needs to be effective means of detecting any outbreak of fire and for warning people in your workplace quickly enough so that they can get to a safe place before the fire makes escape routes unusable.

In very small workplaces where a fire is unlikely to cut off the means of escape (open-air areas and single-storey buildings where all exits are visible and the distances to be travelled are small), it is likely that any fire will quickly be detected by the people present and a simple alarm or shout of 'Fire' may be all that is needed.

In larger workplaces, particularly multi-storey premises, an electrical fire warning system with manually operated call points is likely to be the minimum needed. In unoccupied areas, where a fire could start and develop to the extent that escape routes may become affected before it is discovered, it is likely that a form of automatic fire detection will also be necessary.

Lighting of escape routes

All escape routes, including external ones, must have sufficient lighting for people to see their way out safely. Emergency escape lighting may be needed if areas of the workplace are without natural daylight or are used at night.

Means of fighting fire

There needs to be effective fire-fighting equipment in place for employees to use, without exposing themselves to danger. The equipment must be suitable and appropriate staff will need training and instruction in its proper use.

In small premises, having one or two portable extinguishers in an appropriate location may be all that is required. In larger or more complex premises, a greater number of portable extinguishers, strategically sited throughout the premises, are likely to be the minimum requirement.

Fire-fighting equipment

Portable fire extinguishers enable suitably trained people to tackle a fire in its early stage, if they can do so without putting themselves in danger. When deciding on the types of extinguisher to provide, you should always get suitable advice from extinguisher manufacturers or fire prevention authorities. Consider the materials, surfaces and substances likely to be found in your workplace. Fires are classified in accordance with British Standard EN2 as shown in Table 1.4.

The fire-extinguishing medium in portable extinguishers is expelled by internal pressure, either permanently stored or by means of a gas cartridge. Generally, portable fire extinguishers are categorised according to the medium they contain:

1 water
2 foam
3 powder
4 carbon dioxide
5 vaporising liquids, including halons.

Some fire extinguishers can be used on more than one type of fire. For instance, AFFF extinguishers (see diagram) can be used on both Class A fires and Class B fires. Your fire equipment supplier will be able to advise you.

Health and safety ⚠

A fire blanket will exclude oxygen from the fire and it will start to diminish. When using a fire blanket, always wrap the top of the blanket round the hands first.

KNOW YOUR FIRE EXTINGUISHER COLOUR CODE

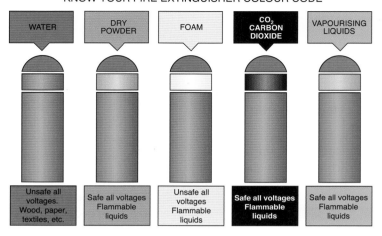

WATER	DRY POWDER	FOAM	CO₂ CARBON DIOXIDE	VAPOURISING LIQUIDS
Unsafe all voltages. Wood, paper, textiles, etc.	Safe all voltages Flammable liquids	Unsafe all voltages Flammable liquids	Safe all voltages Flammable liquids	Safe all voltages Flammable liquids

Table 1.4 British Standard fire classifications and types of fire extinguisher

Fire class	Description	Type of fire extinguisher to use
Class A	Fires involving solid materials where combustion normally takes place with the formation of glowing embers (for example, organic matter such as wood and paper).	Water, foam or multi-purpose powder extinguishers, with water and foam considered the most suitable. Do not use on cooking oil or fat fires.
Class B	Fires involving liquids or liquefiable solids (for example, paints, oils or fats but not cooking oils or fats).	Foam (including multi-purpose aqueous film-forming foam (AFFF), carbon dioxide, halon or dry powder types). Do not use on cooking oil or fat fires.
Class C	Fires involving gases	Dry powder extinguishers
Class D	Fires involving metals	
Class F	Fires involving cooking oils or fats (for example, deep fat fryer fires). These can cause very serious fires and burn fiercely once alight.	'Wet chemical' extinguisher – this has a yellow band across the front. Alternatively for small fires, a fire blanket can be used. None of the standard fire extinguishers should be used on these fires and it would be very dangerous to do so.

Health and safety ⚠

Do not attempt to extinguish fires unless you have been properly trained in procedures and the use of extinguishers/fire blankets. Never put yourself or others in additional danger.

Activity

Design an information leaflet to be used at a staff induction session to explain what to do in the event of a kitchen fire. Also list three things staff should NOT do.

Hazards in the workplace and risk reduction

The following aspects of the kitchen environment have the potential to give rise to hazards:

- equipment – liquidisers, food processors, mixers, mincers, blow torches and other equipment
- substances – cleaning chemicals, detergents, sanitisers
- work methods – carrying knives and equipment incorrectly and not following a proper procedure
- work areas – spillages not cleaned up, overcrowded work areas, insufficient work space, uncomfortable work conditions due to extreme heat or cold.

The duties of employers, as specified by the Management of Health and Safety at Work Regulations 1999, are listed below. Employers have a duty under these regulations to carry out risk assessments and COSHH assessments.

> **Key term**
>
> **Hazard** – anything that could possibly cause harm, such as chemicals, electricity, working with machinery.
>
> **Risk** – the chances, high or low, that someone could be harmed by the hazard.

Managing risk

Managing risk is not a complicated procedure. To start with, a health and safety policy must be in place for the business which is recorded where there are five or more employees. This will outline all the requirements and detail the specific procedures in place to ensure health and safety at work. Employers' planned arrangements for maintaining a safe workplace should cover the usual management functions of:

- planning
- organisation
- control.

Involve employees

Chefs and the other kitchen employees are the people most at risk of having accidents, or experiencing ill health, and they also know the most about the jobs they do so are in the best position to help managers and supervisors develop safe systems of work that are effective in practice. An actively engaged workforce is one of the foundations that support good practice in health and safety. It ensures that all those involved with a work activity are participating in assessing risks.

> **Activity**
>
> 1 What are the items of PPE you would expect to be provided for kitchen staff?
> 2 State what you would consider to be good welfare facilities for chefs employed by a large company.

Reporting hazards

All staff should be made aware that if they see a hazard in the work area that could cause an accident they should:

- Make the hazard safe, as long as it can be done so without risking personal safety.
- Report the hazard to a supervisor or manager as soon as possible, making sure no one enters the area without being aware of the danger.
- If there is a hazard that cannot be made safe, warn others, block the route past the hazard and use a sign such as a bright yellow wet floor sign.

Safety and hazard signage

Safety signs generally take the following forms:

- Prohibition signs – RED (for example, 'no smoking'). A red circle with a line through it tells you something you must not do.
- Fire-fighting signs – RED (for example, with white symbols or writing, such as a fire hose reel).
- Warning signs – YELLOW (for example, caution – hot surface). A black and yellow sign is used with a triangular symbol where there is a risk of danger.
- Mandatory signs – BLUE (for example, protective gloves must be worn). A solid blue circle with a white picture or writing gives a reminder of something you must do such as 'shut the door'.
- Hazard warning signs – YELLOW (for example, a corrosive substance).
- Emergency/escape and first-aid signs – GREEN, with a white picture or writing.

When using chemicals that could harm you, hazard symbols may be displayed on the container:

- corrosive – could burn your skin
- toxic – may cause serious harm if swallowed
- irritant – may cause itching or a rash if in contact with skin
- oxidising /flammable – easily flammable liquids that can catch fire and burn.

No smoking on site

Hair covering must be worn

Fire exit

▲ Safety signs seen in kitchens

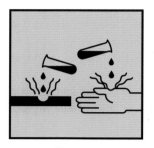

Corrosive   Flammable   Harmful   Toxic

▲ Hazard symbols used on the packaging of chemical products

Risk assessment

Assessing risk and producing a formal risk assessment is the key to effective health and safety in the workplace. This means nothing more than a careful examination of what in the workplace could cause harm or injury to people, so that an assessment can be made of the necessary safeguards and precautions to be put in place to prevent harm.

The five steps to assessing risk and producing a risk assessment are outlined below. There is more detail on the HSE website and also a template to assist in producing a risk assessment.

Five steps to assessing risk

1 **Identify the hazards (the things that could cause harm)**. Assess your work areas and decide what could cause harm. Suggestions from staff who work in the area could be very useful as they may have noticed things that are not immediately obvious to others. Check equipment manufacturers' instructions or data sheets for chemicals as they can be very helpful in identifying hazards and putting in place the necessary precautions. Checking previous accident and ill-health records can also help to identify hazards.

2 **Decide who might be harmed and how.** For each hazard, be clear about who could possibly be harmed. It is not necessary to name everyone but it is helpful to identify groups of people such as those spending time in cold storage rooms or those moving deliveries. Remember employees with different needs such as new and young workers, migrant workers and people with disabilities. Also consider visitors, contractors and maintenance workers who may not be in the workplace all the time, and customers/guests. Record the type of injury or ill health that might occur to people in the specific areas.

3 **Evaluate the risk and decide if existing precautions are adequate or if more is needed.** Having identified the hazards, decisions need to be taken about what actions are necessary. The law requires that everything 'reasonably practicable' is done to prevent harm. Consider the controls already in place and how the work is organised. Decide if this is the best practice (advice is available on HSE website). Consider if more could be done to achieve best practice. For example, could the hazard be eliminated altogether? If not, how could it be controlled so that harm is unlikely? Consider applying the following to each hazard:
- Try a less risky option.
- Prevent access to the hazard.
- Organise work to reduce exposure to the hazard.
- Issue personal protective equipment.
- Provide appropriate welfare facilities such as first aid and washing and changing facilities.

Remember to involve staff so that you can be sure that what is proposed will work in practice and won't introduce new hazards or difficulties.

4 **Record the findings and implement them**. Recording the results of the risk assessment and sharing them with other staff will formalise the decisions. Keep the recording and written account clear and simple so everyone understands it. When preparing the written risk assessment, make sure the following is done:

- make a thorough check
- fully consider who might be affected
- deal with all the obvious significant hazards, taking into account those who could be involved
- ensure the precautions are reasonable and workable and the remaining risk is low.

5 **Regularly review the risk assessment and revise it as necessary.** Areas, equipment and procedures are subject to change. A new menu, new procedures or a new piece of equipment could lead to new risks, so review the risk assessment at the time of the change. This can easily be forgotten so set dates through the year for regular risk assessment reviews when any changes can be considered and the risk assessment amended. Also consider any new or changed hazard/risk pointed out by staff.

Considering the risks

Hazards/risks will fall into one of the following categories:

- Minimal risk – safe conditions with safety measures in place.
- Some risk – acceptable amounts of risk, however attention must be given to ensure safety measures operate
- Significant risk – where safety measures are not fully in operation (also includes food most likely to cause food poisoning). Requires immediate action.
- Dangerous risk – processes and operations of equipment to stop immediately. The system or equipment to be completely checked and recommendations made for improvement.

When considering risks the following points should be considered:

- Assess the risks.
- Determine preventative measures.
- Decide who carries out safety inspections.
- Decide frequency of inspection.
- Determine methods of reporting back and to whom.
- Detail how to ensure inspections are effective.
- See that on-the-job training in safety is related to the actual work being done.

Risk assessment must be recorded and remain a 'live document'. This means that it should be regularly updated and changes made as appropriate (see above). The health and safety risk assessment will also be kept in the establishment's records as part of 'due diligence'.

> **Key term**
>
> **Due diligence** – when a person or organisation who may be subject to legal proceedings can establish a defence to show that they have taken 'all reasonable precautions and exercised due diligence' to avoid committing an offence.

The purpose of the exercise of assessing the possibility of risks and hazards is to prevent accidents. It is necessary to monitor the situation and to have regular and random checks to see that the standards set are being complied with.

Should an incident occur, it is essential that an investigation is undertaken to determine the cause or causes and any defects in the system remedied at once. Immediate action is required to prevent further accidents.

All personnel need to be trained to be actively aware of the possible hazards and risks and to take positive action to prevent accidents occurring.

> **Activity**
>
> Management have asked you to list the main pieces of large and small equipment in your kitchen and to categorise them into low, medium or high risk. Draw up the list in three columns, then select one high risk item and describe how you would minimise the risks when it is in use.

Emergencies in the workplace

Emergencies that might happen in the workplace include:

- serious accidents
- outbreak of fire
- bomb threat
- failure of a major system such as water or electricity.

An organisation will have systems in place to deal with emergencies. Key staff members are usually trained to tackle emergencies. There will be fire marshals and first aiders. These people will attend regular update meetings and specific training. Evacuation procedures will also be held so that employees can practise fire drill and fire alarms will be tested regularly. Ensure that you know the evacuation procedures in your establishment. If you have to leave the premises, ensure that the following procedures are adhered to:

- Turn off the power supplies: gas and electricity. Usually this means hitting the isolation/cut-off in the kitchen or turning off all appliances individually.
- Close all windows and doors.
- Leave the building by the nearest emergency exit. DO NOT USE THE LIFTS.
- Assemble in the designated area, away from the building.
- Check the roll-call of names to establish whether all personnel have left the building safely.
- Do not re-enter the building until you are told it is safe to do so.

First aid

When people at work suffer injuries or fall ill, it is important that they receive immediate attention and that, in serious cases, medical help is called. The arrangements for providing first aid in the workplace are set out in the Health and Safety (First Aid) Regulations 1981. First aiders and facilities should be available to give immediate assistance to casualties with injuries or illness.

As the term implies, first aid is the immediate treatment given on the spot to a person who has been injured or is ill. Since 1982 it has been a legal requirement that adequate first-aid equipment, facilities and personnel to give first aid are provided at work. If the injury is serious, the injured person should be treated by a medical professional as soon as possible.

First-aid equipment

A first-aid box, as a minimum, should contain:
- a card giving general first-aid guidance
- individually-wrapped, sterile, adhesive, waterproof dressings of various sizes
- 25g cotton wool packs
- safety pins of various sizes
- two triangular bandages
- two sterile eye pads, with attachment
- four medium-sized sterile un-medicated dressings
- two large sterile un-medicated dressings
- two extra-large sterile un-medicated dressings
- disposable rubber gloves
- scissors
- report forms to record all injuries, if these are not kept elsewhere.

First-aid boxes must be easily identifiable and accessible in the work area. They should be in the charge of a responsible person, checked regularly and refilled when necessary.

Large establishments may have medical staff such as a nurse and a first-aid room. The room should include a bed or couch, blankets, chairs, a table, sink with hot and cold water, towels, tissues and a first-aid box. Hooks for clothing and a mirror should be provided. Smaller establishments should have members of staff trained in first aid and in possession of a certificate. After a period of three years, trained first-aid staff must update their training to remain certificated. Courses are run by St John Ambulance Association, St Andrew's Ambulance Association or the British Red Cross Society and individual training providers.

Kitchen environment

Kitchens must have the most appropriate working environment for the work being completed. For further information on this, see Chapter 2, Supervising food safety.

Induction

Further information on induction training is provided on page 35. Inductions are a good opportunity to explain the relevant health and safety/food safety documentation and processes, as well as emergency evacuation procedures. The following information should be included on a health and safety induction course:
- fire safety emergency evacuation procedures
- leaflets on health and safety and where to find further information
- accident and hazard reporting procedures
- the organisation of health and safety policy
- first-aid facilities
- communication systems of health and safety
- who the enforcing authority is
- complying with legislation.

Recording and reporting procedures

Make sure you understand the health and safety management system and its procedures. Certain documentation must be kept, updated and be available for inspection. These documents may also be required as part of a due diligence defence. Documents contained within a health and safety management system would be:
- the organisation's health and safety policy
- risk assessments and their findings
- COSHH assessments
- company rules and procedures
- systems of safe working
- monitoring records
- accident records
- health and safety training records
- health surveillance records

- staff training and competency records
- first-aid records
- list of first aiders with dates of qualifications
- a fire risk assessment
- fire and emergency procedures
- fire inspection records
- examination and test certificates for work equipment
- maintenance and repair records for equipment
- Health and Safety law poster
- Health and Safety committee minutes
- details of visit by enforcement officers
- Health and Safety legislations 1974
- Health and Safety at Work Act
- approval codes of practice.

Having an up-to-date system of documentation is evidence of a strong commitment to health and safety. Always record key facts and ensure all record keeping is accurate.

The supervisor as a leader

The supervision role is a very important one in kitchen areas and hospitality in general. Supervisors tend to be placed where the actual day-to-day work is being completed and so have an excellent understanding of the working area and the procedures within it. However, in addition to this they often complete some management tasks and procedures, as well as communicate frequently with those at management levels. This puts them in a valuable and unique position to be able to liaise between the different employee levels easily. Those working on more practical and everyday tasks such as in a kitchen often feel more confident and comfortable communicating with the supervisor they see and work with every day than they would with someone in a more senior management position.

The purpose of supervision

On a daily basis the supervisor is the person who ensures everything runs as smoothly as possible and that all the necessary tasks are allocated logically among the available staff, so the job gets done. Chefs tend to work within very tight deadlines and time constraints, yet the demands of customer expectations and requirements must still be met. The supervisor is there to ensure this happens and that targets for individuals, the team and the wider organisation are met. All of this must be done with the welfare of the workforce in mind, avoiding too much stress or pressure on individuals. Supervisors play an important part in leading others, communicating effectively and contributing to the business being successful and running efficiently.

The supervisor plays a very important part in making the environment as safe and as pleasant to spend time and work in as possible. A good working environment tends to be valued by employees and this will be reflected in their attitudes to work and the quality of work they produce. Good working conditions and environments also contribute to staff loyalty and a lower staff turnover.

Hospitality workplaces are governed by significant amounts of legislation and the supervisor needs to have good knowledge of this and also know where to find out about the details they are not fully sure of. A good knowledge of relevant legal matters allows the supervisor to apply this to the working area, ensuring that areas, equipment and procedures meet legal requirements. They will also use their knowledge of legal requirements as part of staff training.

Characteristics of leadership

Certain leadership qualities are needed to enable the supervisor to carry out the role effectively. A good supervisor is someone who is:
- a good team member who can lead by example
- open to new ideas, can initiate new ideas and is fair
- well informed and knowledgeable
- well organised and able to organise others
- a good communicator and mediator
- motivational and able to inspire
- consistent and reliable
- an effective planner
- a good decision maker
- understanding and approachable
- supportive and respectful to the rest of the team
- able to establish policies and procedures
- aware of current legislation
- able to work under pressure
- able to cooperate well at different levels
- commands loyalty and respect.

The supervisor will use their skills to lead and motivate others so the required goals are met. It would be very difficult if not impossible to supervise others if you were not a good communicator. Effective communication is at the heart of good supervision and leadership and effort should be put into communicating well at all times.

The good supervisor will need to develop excellent 'people skills' – building trust and earning respect from those they are supervising. They need to be someone who is approachable, with good listening skills who can be spoken to with complete confidentiality, and can identify and deal with problems, conflict and unrest that may occur in their area. The effective supervisor is always proactive, sets the standards and leads the team by good example.

The supervisor should be able to obtain the best from the team they have responsibility for. They also need to completely satisfy the management of the establishment that an effective job is being done.

Supervisory tasks

The person completing the supervision role may well be known by their colleagues as the head chef, sous chef, chef de partie, kitchen supervisor or some other job title. The kitchen supervisor will be responsible to the catering manager, while in hotels and restaurants, a chef de partie will be responsible to the head chef. Some may be supervising a specific area of the kitchen or have specific supervisory tasks such as overseeing food safety. In a commercial catering role, they may supervise a section or sections of the food production system. Generally the head chef will have both managerial and supervisory skills and he/she will determine kitchen policies.

The exact details of the job will vary according to the different areas of the industry and the size of the various units, but generally the supervisory role is essentially that of an overseer. It involves three functions:

- **Technical function** – culinary skills and the ability to use kitchen equipment are essential for the kitchen supervisor. Most will have worked their way up through a kitchen team before gaining supervisory responsibility. The supervisor needs to be able to actually do the job themselves and do it well, and to impart these skills to others in the team.
- **Administrative function** – the supervisor will, in many kitchens, be involved with the menu planning, including the ordering of foodstuffs and accounting for and recording materials used. The administrative function includes the allocation of duties and may also include the writing of reports.
- **Social function** – the supervisor needs to motivate the staff in the team. This means keeping them interested and keen to achieve and progress. Supervisors often take responsibility for staff training and the recording of training. Training may be delivered by supervisors themselves or they may organise others to complete it where specialist knowledge is needed.

Activity

What are the qualities you may notice in a member of the kitchen team that makes you think they would make a good supervisor in the future? How could you develop this person towards a supervisory role?

The primary role of the supervisor is to ensure that a group of people work together to achieve the goals set by the business. Managing physical and human resources to achieve customer service goals requires planning, organising, staffing, directing and controlling.

Forecasting

Before making plans it is necessary to look ahead, to foresee possible and probable outcomes and to allow for them. For example, if a chef knows that their assistant will not be in work the next day, the chef will look ahead and plan accordingly. Forecasting is the good use of judgement acquired from previous knowledge and experience. An example could be that because many people are on holiday in August, fewer meals will be needed in the office restaurant; but a warm summer's day will increase sales at coastal outlets significantly. These events need appropriate forecasting; contingency plans may need to be in place.

Planning

From the forecasting comes the planning: how many meals to prepare; how much food and other items to have in stock, how many staff will be needed; which staff and at what times? Do the staff possess the required skills for the menus being offered? This is where training in a variety of skills is so important. Plan for additional training sessions when a new menu is introduced or new equipment installed.

Organising

Organisational skills are applied to food, to equipment and to staff. Organising means ensuring that what is required is where it is needed, when it is needed, in the right amount and at the right time. The supervisor may be involved in many tasks including duty rotas, training programmes, menu planning, food ordering and cleaning schedules.

Consider the supervisor's part in organising a wedding function where a reception is to be held. The guests require a hot meal and in the evening a dance will be held, during which a buffet will be provided. The supervisor would need to organise enough staff to be available, make sure that food and drinks were available for them and consider their transport home from the function. Calor gas stoves and portable refrigeration may be needed and the supervisor would need to arrange for these to be delivered and in good working order and for the equipment to be cleaned and returned after the function. The food would need to be ordered so that it arrived in time to be prepared. Consideration would need to be given about how much of the preparation and cooking would be done on site and if there is enough space and equipment to do this. Staff may need additional training to work in this different environment and maybe with types of food they have not handled before. The correct quantities of food, equipment and cleaning materials would also have to be at the right place when needed and if all the details of the situation were not organised properly, problems could occur.

Commanding

Food provision is a disciplined and time-constrained operation. The supervisor needs to give instructions to staff about:

- How specific tasks need to be completed.
- Exactly what needs to be done.
- When each process needs to be completed and individual items need to be ready.
- Where: many kitchens will have food items going out to different locations.

The successful supervisor is able to give instructions effectively, having made decisions and established the basic priorities. They need to communicate this to their teams.

Co-ordinating

Co-ordinating is a skill required to get all of the tasks completed to come together at the time required and for staff to co-operate and work together in harmony. To achieve this, the supervisor has to be interested in the staff, to deal with their queries, to listen to their problems and to be helpful. They also need to maintain good relationships with other areas and departments. Good service is dependent on effective co-ordination.

Controlling

This includes the controlling of people and products, managing security within the area, as well as improving performance. It is about:

- Checking that staff arrive on time, do not leave before their allocated time without permission and do not misuse time in between.

- Checking that the food product is of the required standard, the correct quantity and quality.
- Checking to prevent unnecessary waste and also to ensure that staff operate the portion control system correctly.

These aspects of the supervisor's function involve observing and inspecting, and require tact.

Delegation

By giving a certain amount of responsibility to others, the supervisor can be more effective. The supervisor needs to be able to judge the person capable of responsibility before any delegation can take place. But then, having recognised the abilities of an employee, the supervisor who wants to develop the potential of those under his or her supervision must allow the person entrusted with the job to get on with it.

Activity

1 List three advantages of a supervising chef delegating some of his duties to a more junior chef.
2 Suggest four tasks or duties a sous chef or kitchen supervisor could delegate to a chef de partie.

Motivation

Motivation means stimulating desire and energy in people to keep them interested and committed to a job role or subject, or to strive to attain a goal. Motivation can come from:

- The levels of desire or need.
- The incentives of the reward or achieving a goal.
- The expectations of the individuals, their managers or peers.
- Receiving encouragement or stimulation from others.

Although not everyone is capable of or wants significant amounts of responsibility, the supervisor still needs to motivate those who are less ambitious. Most people are prepared to work to improve their standard of living or status but getting satisfaction from the work they do is also a motivating factor. The supervisor must be aware of why people work and how different people achieve job satisfaction and then be able to act on this knowledge. Involving staff in decision making and planning, and making them aware that they are valued goes a long way in achieving this. Opportunities must be made to extend their abilities and learn new skills. Everyone should be kept fully occupied and the working environment must be conducive to producing their best work.

Professional tip

There are many symptoms of poor motivation. In general terms they reveal themselves as lack of interest in getting the job done correctly and within the required time. Although these may be indicators of poor motivation, the lack of efficiency and effectiveness could also be a result of the staff overworking, personal problems, poor work design, repetitive work, lack of discipline, interpersonal conflict, lack of training or failure of the organisation to value its staff. An employee may be highly motivated but may find the work physically impossible.

Welfare

People always work best in good working conditions and these include freedom from fear: fear of becoming unemployed, fear of failure at work, fear of discrimination. Job security and incentives such as opportunities for promotion, bonuses, profit sharing and time for further study all encourage a good attitude to work. People need to feel valued and that what they do is important. The supervisor is in an excellent position to ensure that this happens.

Personal worries affect individuals' performance and can have a very strong influence on how well or how badly they work.

The physical environment will naturally cause problems if, for example, the atmosphere is too hot and humid, the work area ill-lit, cluttered or too noisy, and there is constant rush and stress, staff are more liable to be irritable and aggressive and the supervisor needs to consider how these factors might be improved.

Problem solving

The supervisor needs to anticipate possible problems and build up a good team ethic to overcome problems. This means that work needs to be allocated according to each individual's ability.

Communication

The supervisor must be able to communicate effectively. In order to convey orders, instructions, information and manual skills, the supervisor should possess a positive attitude to those with whom they need to communicate. The ability to convey orders/instructions in a manner that is acceptable to the recipient is dependent not only on the words but on the emphasis given to the words, the tone of voice, the time selected to give them and on who is present when they are given. This is a skill that supervisors need to develop. Instructions and orders can be given with authority without being dictatorial.

Ethical issues

The supervisor needs to treat everyone in the team with equality and fairness and ensure there is no discrimination within the team, especially on issues such as gender, race, religion and beliefs, sexual orientation and ability.

A supervisor must be consistent when handling staff, avoiding favouritism and perceived inequity. Such inequity can arise from the amount of training or performance counselling given, from the promotion of certain employees and from the way in which shifts are allocated. Supervisors should engage in conversation with all staff, not just a selected few, and should not single out some staff for special attention. Ethical treatment of staff is fair treatment of staff. A good supervisor will gain respect if they are ethical.

Confidentiality

Confidentiality is an important issue for the supervisor. Employees or customers may wish to take the supervisor into their confidence and should be able to trust in the confidentiality being maintained. Confidentiality and security of information and data is now a legal requirement under the Data Protection Act 1988.

Leadership and supervisory styles

Staff must be encouraged and motivated to follow required procedures. This can be done in a positive way by offering incentives or rewards, or a more disciplinary approach may sometimes be necessary. Both methods can be effective and can be used by supervisors to achieve their required goals.

Leadership style is the way in which the functions of leadership are carried out – the way in which the supervisor typically behaves towards members of the team. There are many dimensions to leadership and many possible ways of describing leadership style. These are:

- **Dictatorial** – a supervisor who is dictatorial is autocratic and can often be oppressive and overbearing.
- **Bureaucratic** – a bureaucratic supervisor is one who follows official procedure and can be considered 'office-bound'. The bureaucrat sticks to the rules and operates best within a hierarchical system.
- **Benevolent** – a benevolent supervisor is kind, passionate, human, kind-hearted, good, unselfish, charitable.
- **Charismatic** – a supervisor who is charismatic has personality and special charm that inspires loyalty and enthusiasm from the team.
- **Consultative** – the consultative supervisor discusses issues with the team through team or individual meetings.
- **Participative** – a participative supervisor gets involved with the team, taking part in activities and issues and making an active contribution to the success or failure of the team.
- **Unitary** – the unitary supervisor unites the team, bringing them together as a whole unit.
- **Delegative** – the delegative supervisor entrusts others in the team to make decisions, assigning responsibility or authority to others.

- **Autocratic** – this is where the supervisor holds on to power and all the interactions within the team move towards the supervisor. The supervisor makes all the decisions and has all the authority.
- **Democratic** – the team has a say in decision-making. The supervisor shares the decision-making with the team. The supervisor is very much part of the team.
- **Laissez-faire** – this style is where the supervisor observes that members of the group are working well on their own and leaves them 'to get on with it'. The supervisor passes power to the team members, allows them freedom of action, does not interfere but is available for help if needed. The term 'laissez-faire' is sometimes wrongly used to describe the type of supervisor who does not care, keeps away from trouble and does not want to get involved.

Developing positive working relationships

The supervisor is responsible for developing good teamwork and acting as a catalyst in maintaining good relationships within the team. As individuals working within an organisation, we can achieve very little, but working within a group we are able to achieve a great deal more. Those managing/supervising the kitchen must develop effective teams in order to achieve the high standard of work that is required to satisfy both the organisation's management and the customers' demands.

Each member of a team must regard themselves as being part of that team. They must interact with one another and perceive themselves as part of the group. Each must share the purpose of the team; this will help build trust and support and will, in turn, result in an effective performance. Co-operation is therefore important in order for the work to be carried out.

People in groups will influence one another – within the group there may be a leader and/or hierarchical system. This is even more apparent in a formal kitchen where a structured system is in place. The pressures within the group and the styles of supervision and leadership used may have a major influence on the behaviour of the team.

The characteristics of a good team

- The objectives and the goals of the team should be clear and understood by everyone. Individuals need to fully understand what is expected of them and how to achieve their own personal goals.
- Everyone needs to understand what their responsibilities are, as well as the responsibilities of others and the team as a whole. There needs to be a culture where everyone is working together as a whole team towards consistent improvement and achieving the highest standards.

- There needs to be good communication from team leaders, supervisors and managers, as well as good communication between team members and with those from other departments or teams. Communication channels must remain open and discussion must be positive and supportive.
- There will be people working at different levels with a good mix of skills. These skills should be used in harmony with each other; there needs to be a culture of 'skills sharing' and teaching new skills to less experienced people.
- A good team will have members with mutual respect and trust for each other and an appreciation of each other's skills and what each person has to offer, contributing to the overall success of the team.
- There should be team spirit, with the mutual goal of striving towards success and moving the whole team forward to achieve excellence.

Benefits of team development

Selecting and shaping teams to work within the kitchen is very important and requires management skills. Matching each individual's talent to the task or job is an important consideration. A good, well developed team will be able to do the following:
- create useful ideas
- analyse problems effectively
- get things done
- communicate with each other
- respond to good leadership
- evaluate logically
- perform skilled operations with technical precision and ability
- understand and manage the control system.

Maintaining the success of the team and developing it further demands constant attention. The individual members of a group will never become a team unless effort is made to ensure that the differing personalities are able to relate to one another, communicate with each other and value the contribution each employee or team member makes.

The chef, as a team leader, has a strong influence on his/her team or brigade. The chef in this position is expected to set examples that have to be followed. They need to work with the brigade, often under pressure, and sometimes dealing with conflict, personality clashes, change and stress. The chef has to adopt a range of strategies and styles of working in order to build loyalty, drive, innovation, commitment and trust in team members.

The team leader or supervisor needs to identify the team's strengths and weaknesses, and develop ways to help to overcome any weaknesses there may be.

Motivating a team

A chef must motivate his/her team by striving to make their work interesting, challenging and demanding. People must know what is expected of them and what the standards are.

Rewards should be linked to effort and results. Any such factors must also work towards fulfilling:
● the needs of the organisation; and
● the expectations of team members.

Improved performance should be recognised by consideration of pay and performance.

The chef in charge should attempt to intercede on behalf of his/her staff; this in turn will help to increase staff members' motivation and their commitment to the team. For the chef to manage his/her staff effectively, it is important to get to know them well, understand their needs and aspirations and attempt to help them achieve their personal aims.

If a chef is able to manage the team by co-ordinating its members' aims with the corporate objectives – by reconciling their personal aspirations with the organisation's need to operate profitably – this chef will manage a successful team and in addition, will enhance their own reputation.

Effective communication

Because communication is so important to everything we do, even small improvements in the effectiveness of our communicating are likely to have significant benefits. In the kitchen, most jobs have some communication component. Successful communication is vital when working to build working relationships. Training and developing the team is about communicating.

Breakdowns in communication can be identified by looking at the 'intent' and the 'effect' separately. It is when the intent is not translated into the effect that communication can break down. Such breakdowns affect staff and team relationships and individuals' attitudes towards and views of each other. Awareness of the potential gap between intent and effect

can help clarify and prevent any misunderstanding within the group.

By bridging the gap between the intent and the effect, you can begin to change the culture of the working environment: the processes become self-reinforcing in a positive direction; the staff begin to respect each other in a positive framework; they listen more carefully to each other, with positive expectations, hearing the constructive intent and responding to it.

One of the main goals of a hospitality team is to satisfy customers' demands and expectations. Good communication within the organisation assists in the development of customer care. The kitchen must communicate effectively with the restaurant staff so that they, in turn, can communicate with the customer. Any customer complaints must be handled positively: treat customers who complain well, show them empathy. Use customer feedback – good or bad – positively. This may further develop the team and help solve any problems within the team.

As a chef/manager, the emphasis must be on achieving results through the team and communication is the key to this. A great deal of the chef's/supervisor's time will be taken up with communicating with large numbers of people.

Communication, therefore, plays a major part in the chef/supervisor's role. Communication at work needs to be orientated towards action – getting something done.

Diversity

The hospitality industry is becoming more diverse and, for this reason, the team must celebrate and welcome diversity and embrace equal opportunities. The free movement of labour within the European Union, for example, has meant that large numbers of people from different cultures now work together in the hospitality industry. Diversity in the kitchen can contribute positively to the development of the team, bringing to it a range of skills and ideas from different cultures.

Diversity recognises that people are different. It includes not only cultural and ethnic differences, but differences in gender, age, disability and sexual orientation, background, personality and work style.

Effective working relationships should harness such differences to improve creativity and innovation and be based on the belief that groups of people who bring different perspectives will find better solutions to problems than groups of people who are the same.

Minimising conflict

A conflict with the manager or with a colleague can easily get entangled with issues about work and status, both of which can make it difficult to approach the problem in a rational and professional way. One of the skills of the chef/manager is the need to identify conflict so that plans can be put in place to minimise it. The following are just some of the points that should be borne in mind:

- Conflict arises where there are already strained relationships and personality clashes between members of the team.
- Conflict often occurs in a professional kitchen when the brigade is understaffed and under pressure, especially over a long period. Pressure can also come from, say, restaurant reviews and guides, when a chef is seeking a Michelin star or other special accolade.
- Conflicts damage working relationships and upset the team and this will eventually show up in the finished product.

The chef and supervisor must also be aware of any conflict that may be going on around them in less obvious places. Covert conflicts (those that take place in secret) can be very harmful. Although this type of conflict is often difficult to detect, it will undermine the team's performance. Many conflicts start with misunderstandings or a small upset that grows and escalates out of all proportion.

It is important to reflect on and analyse the nature of the conflict and individual attitudes to it. Conflicts can be very damaging and upsetting but there can also be some positive outcomes. A conflict can be a learning curve that a chef has to enter into; it has to be handled properly and focused on in order to achieve the desired outcome.

Activity

A junior member of the kitchen team comes to you with concerns that there is on-going conflict between two more senior members of the team. This appears to be causing problems with team performance, lapses in communication and negative attitudes between team members. What measures would you take to:
- reassure the junior member of staff
- deal with the conflict issues?

Training

Training can be described as the process of preparing an individual to achieve the desired skill, knowledge and competence relevant to their job role. Supervisors are frequently responsible for training or parts of staff training and often advise managers what training is required. Suitable training must always be provided for new employees, when people are exposed to new and increased risks in the workplace and also when new equipment is purchased and installed. Any training undertaken by staff must be recorded and the records must be kept.

Training and personnel development can help to:
- Ensure a knowledgeable and skilled workforce which makes fewer mistakes, thereby reducing risks and the likelihood of injury.
- Reduce accidents and ill health in the workplace, resulting in fewer days off due to sickness and injury.
- Ensure that the employer complies with legislation on relevant training and guidance.
- Provide awareness of hazards and safe working procedures.
- Increase confidence, motivation, enthusiasm and commitment of staff, improving employee retention rates.
- Provide an increase in work productivity.
- Provide recognition, enhanced responsibility and the possibility of further career development.
- Give a feeling of personal satisfaction and achievement.

Planned and on-going training is essential and therefore is a key element of improved organisational performance. Training improves knowledge, skill, confidence and competence.

Induction

Effective induction processes are vital in ensuring that a new employee becomes effective in the shortest time. This is especially important in a kitchen area where the pace tends to be fast-moving, with potentially dangerous equipment and many procedures to introduce to a new employee.

Proper induction training is increasingly a legal requirement. Employers have a formal duty to provide new employees with all relevant information and training relating to their new role, and health and safety requirements are particularly important.

This is also a good opportunity to explain the relevant health and safety/food safety documentation and processes, as well as emergency evacuation procedures. New staff may benefit from being provided with an induction folder/pack which contains key employment and health and safety information. This might include a tick-box format that shows that the employee has covered the range of induction information and understood the importance placed on health and safety. A signed and dated copy of this should be placed in health and safety records.

After the initial induction, some more formal training may take place and with new employees who are not familiar with the kitchen, a mentor (or more than one) should be assigned to the new recruit to give them a reference point to ask questions and check on their own understanding and progress. The supervisor may coordinate the mentoring of the new recruit and will probably observe and monitor their progress. Regular feedback sessions with the new recruit, the mentor and the supervisor will always be beneficial and will reassure the new member of staff that the necessary support and information is available to them.

Methods of training

Different people will respond differently to various styles of training and people also have different learning styles. While some respond well to theoretical training, others prefer 'hands on', more practical training; generally most kitchen employees will fall into this second category. The supervisor needs to bear this in mind and adapt training to what the team responds to and learns from best. It will be beneficial to use varying styles and methods to keep the training interesting and address all learning styles.

Training could be delivered in any of the following ways or by a combination of methods:
- group training sessions
- individual training
- formal training with awarding body certification

- case studies
- shift handover topics
- role play – for example, acting out health and safety practice
- discussion groups
- interactive computer-based courses
- progress/learning log
- video clips
- visiting speakers such as Environmental Health Officers
- demonstrations
- group tasks.

Other methods of training to consider may include:
- Close observation of 'role models' (sometimes called 'shadowing').
- Being placed in challenging situations which require initiative and positive leadership. This, of course, needs to be carefully monitored.
- Provision of a good mentor. This can also help personal development. A good supervisor is also able to mentor and train members of the team to achieve their goals and objectives.
- Related CPD (continuing professional development) activities such as visits to farms and fisheries, attending seminars, reading relevant materials, being set a topic to research.
- Being seconded to other establishments on work placement and for networking and development opportunities.

Typical types of training needed by kitchen staff are:
- Mandatory training – such as health and safety or food safety. Training courses for these can be completed at different levels, depending on the individual's needs and their job role.
- Practical skills training – in-house or from an outside provider.
- Current and topical information such as food sustainability and managing food waste, allergies.
- Communication skills and teamwork skills.
- Workplace behaviour.
- Legislative requirements.
- Individual company policy and requirements.

Activity

Plan a one-day health and safety induction programme for a new apprentice starting their training in your kitchen. Remember that they may not have had any previous health and safety training.

Other training terminology:

- **Output training** – investing in a new employee or new machine to generate efficient output as quickly as possible.
- **Task training** – involves selected individuals being sent on short training or college-based courses such as hygiene courses, health and safety courses, financial training.
- **Performance training** – when the organisation has grown substantially and becomes well established. Training is viewed positively, with a person responsible for overseeing training. Plans and budgets are now some of the tools used to manage the training process.
- **Strategic training** – when the organisation recognises and practises training as an integral part of the management of people and the culture of the organisation.

Supervisors should encourage continuous training to improve knowledge, skills and encourage positive attitudes. This can lead to many benefits for both the organisation and the individual. Supervisors must also keep their own skills and training up to date and seek appropriate skills and supervisory training for their own development.

Training should be viewed as an investment in people. Training requires the co-operation of managers and supervisors, with a genuine commitment from all levels in the organisation.

Activity

Suggest two areas of practical research a junior chef could complete that could build his/her confidence and be beneficial to the establishment (for example, tracking diners coming into the restaurant for a week and establishing the busiest times and the most popular dishes).

Test yourself

1 What are the main requirements of the Health and Safety at Work Act 1974 for:
   (a) employers
   (b) employees?
2 What is meant by COSHH and what are the employer's responsibilities in relation to this? Which items used in a kitchen would be covered by COSHH?
3 What are the four standard colours of health and safety warning signs and what type of warning does each colour cover? For each colour group, state where one could be found in a kitchen.
4 What six costs could there be to employers if a serious accident occurred in their workplace?
5 What is meant by an improvement notice? What information must go onto an improvement notice?
6 What are the most frequent kinds of accidents that occur in kitchens and hospitality premises? Give four examples of how these could happen in a kitchen.
7 Why is induction important for new kitchen employees? Suggest five topics you would cover in an induction programme.
8 What are four qualities needed in a good supervisor? Give a practical kitchen example of how each of the qualities would be beneficial in the supervisory role.
9 When considering leadership, what is meant by:
   (a) dictatorial
   (b) autocratic
   (c) democratic
   (d) participative?
10 Suggest six advantages to the output of a kitchen in having good, effective teamwork.

2 Supervising food safety

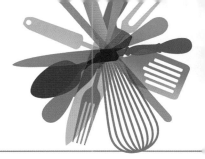

This chapter covers the following units:

NVQ:
→ Maintain food safety when preparing, storing and cooking food.
→ Ensure food safety practices are followed in the preparation and serving of food and drink.

VRQ:
→ The principles of food safety supervision for catering.

Introduction

This chapter will provide the information needed to ensure good food safety practice in food businesses, including explaining the principles of supervising others. It takes a practical approach to working safely with food and supervising others in achieving the required food safety standards. There is information about food safety law and the practical measures that must be adopted for a business to remain within the law.

Learning objectives

By the end of this chapter you should be able to:
→ Understand the importance of food safety management procedures.
→ Understand the role of the supervisor in ensuring compliance with food safety legislation.
→ Apply and monitor good hygiene practice, including temperature control, controlling cross-contamination, personal hygiene, cleaning, disinfection and waste disposal, food premises design requirements and pest control.
→ Implement food safety management procedures.
→ Understand the role of the supervisor in staff training.

Food safety legislation

What is meant by food safety?

Food safety means putting in place all of the measures needed to make sure that food and drinks are suitable, wholesome and safe to eat through all of the processes – from selecting suppliers and delivery of food, right through to serving the food to the customer. Everyone dealing with food must:

● Protect the food they are working with from contamination.
● Put measures in place to prevent any micro-organisms already in food multiplying to levels where they could cause illness or harm.
● Use cooking and other procedures to destroy micro-organisms in food that could possibly be harmful.

For many reasons discussed in this chapter, it remains of utmost importance for a chef/kitchen supervisor to have a sound understanding of the principles of food safety. It is also essential to have the knowledge to plan and apply these principles effectively and to train/supervise others working in food areas to achieve the same high standards. The chef or supervisor will be leading by example.

The importance of food safety management procedures

In any food business, food safety procedures must be planned, organised and monitored. It involves protecting food from the time it is delivered, through its storage, preparation, cooking, hot-holding, chilling and serving to avoid the risk of causing the consumer any illness or harm. It is also essential to enable the business to comply with legal obligations, avoid possible legal action against them and their employees, including civil action and the serving of notices from local authority enforcement officers.

It is a legal requirement for food safety management systems to be in place in all food businesses. The food safety management system should be based on **HACCP** (hazard analysis and critical control points – see page 58) and will identify, assess and monitor the **critical control points** in kitchen procedures, ensure corrective actions are in place and that systems are frequently verified, with accurate documentation available for inspection. These documents could be used as part of a 'due diligence' defence.

Key term

Due diligence – this is the main defence available under food safety legislation. It means that a business can prove that it took all reasonable care and precaution to comply with the 1990 and 2006 Food Safety legislation. To prove due diligence it is necessary to keep a range of documents as part of the HACCP based food safety system. (Also see page 62.)

Critical control points – stages in food production where food safety needs to be controlled.

For more information on food safety management procedures, including Safer Food, Better Business, see pages 62–3.

Employer responsibilities

Employers and owners of food businesses must register the business and all their premises and the vehicles used for transporting food with the local authority so that the appropriate licences can be issued. Employers must also:

- Ensure that, where a full HACCP system is established, at least one person trained in the principles of HACCP is involved in the planning, design and setting up of the system.
- Ensure proper implementation of food safety management procedures.
- Have full records of all suppliers used.
- Retain records of staff training commensurate with the different job roles and responsibilities of employees, dates of when the training took place and the topics covered.
- Put policies in place for planning and monitoring new and on-going staff training.
- Ensure appropriate levels of recruitment, staffing levels and supervision are in place.
- Provide an adequate supply of materials and equipment for staff use, including PPE (personal protective equipment).
- Provide sufficient ventilation, clean drinking water supplies and adequate drainage.
- Provide washing and cleaning facilities for premises, equipment and food, as well as separate hand-washing facilities and hygiene/welfare facilities for staff.
- Have systems in place for record keeping and accident/incident reporting.
- Remain compliant with requirements of environmental health officers/environmental health practitioners.

Employee responsibilities

Employees also have food safety responsibilities in order to comply with the law. Employees must:

- Work in such a way that will not endanger or contaminate food.
- Comply with and follow instructions put in place by employers for processes and procedures.
- Keep food safe.
- Attend and partake in planned training, instruction and supervision procedures.
- Maintain high standards of personal hygiene.
- Report any illnesses to supervisors/managers before starting work (see page 35).
- Report any shortfalls or omissions in the business's operating procedures that could affect food safety. These may be problems with deliveries and food storage, equipment, premises/infrastructure, staff hygiene or kitchen cleaning facilities.

Activity

Produce an information card or leaflet for new apprentice chefs explaining what their own legal responsibilities are with regard to food safety.

Procedures for compliance

A food safety management system based on the seven principles of HACCP (see page 60) is a legal requirement for food businesses and it must cover the full range of procedures within the business and their implementation. The system must be able to show written documentation relevant to the food safety of a business, as well as working practices, staff recruitment, training and supervision, and records of equipment used and its maintenance.

Enforcing food safety legislation

Legislation concerning food safety covers a wide range of topics including:

- controlling and reducing outbreaks of food poisoning
- registration of premises/vehicles
- content and labelling of food
- preventing manufacture and sale of harmful food
- food imports
- prevention of food contamination and equipment contamination
- training of food handlers
- provision of clean water, sanitary facilities, washing facilities.

Food Safety Act 1990 and Food Safety (England, Scotland, Wales, Northern Ireland) Regulations 2006

The main legislation of food safety in the UK is the Food Safety (England, Scotland, Wales, Northern Ireland) Regulations 2006. It is based on the earlier Food Safety Act 1990 and updated Regulations of 1995, and many of the Regulations are the same.

The Regulations provide a framework for EU legislation to be enforced in the UK. The legislation includes the requirement to have an approved food safety management procedure in place, with up-to-date permanent records available, including staff training and supervision records. All records must be reviewed and monitored regularly, especially if there is a change in procedures.

Take it further

For more information on food safety legislation, see the Food Standards Agency website www.food.gov.uk. The following information will be useful:
- Food hygiene: a guide for business
- Food law, inspections and your business
- Food hygiene legislation.

Food safety legislation under the 2006 Regulations and previous Acts is enforced by local authorities through inspection by environmental health officers (EHOs) or environmental health practitioners (EHPs), who are empowered to serve enforcement notices (including hygiene improvement notices, hygiene prohibition orders and hygiene prohibition notices) through criminal and civil courts. Other legislation will be in place relating to working practices, procedures and training and may well involve Trading Standards or the Health and Safety Executive.

Environmental Health Officers/ Environmental Health Practitioners

Enforcement officers may visit food premises as a matter of routine, as a follow-up when problems have been identified, or after a complaint. The frequency of visits depends on the type of business and food being handled, possible hazards within the business, the risk rating or any previous problems or convictions. Generally, businesses posing a higher risk will be visited more frequently than those considered low risk.

EHOs/EHPs can enter a food business at any reasonable time without previous notice or appointment, usually – but not always – when the business is open. The main purpose of these inspections is to identify any possible risks from the food business, to assess the effectiveness of the business's own hazard controls and also to identify any non-compliance of regulations so this can be monitored and corrected.

The role of the EHO/EHP is to:
- Offer professional food safety advice to a food business on routine visits and supply food safety information and training materials.
- Advise on and deliver food safety training.
- Advise on food safety legislation and compliance with legislation.
- Investigate complaints about food safety in the business.
- Ensure food offered for sale is safe and fit for human consumption.
- Monitor food operations within a business and identify possible sources of contamination.
- Observe the effectiveness of the food safety management systems and the keeping of essential records.
- Deal with food poisoning outbreaks or other food-related problems.
- Deal with non-compliance by formal action/serving notices.

The EHO/EHP also has the power to:
- Close a business and seize/remove food and records.
- Issue notices and orders for non-compliance and instigate prosecution.

A Hygiene Improvement Notice will be served if the EHO/EHP believes that a food business does not comply with regulations. The notice is served in writing and states the name and address of the business, what is wrong, why it is wrong, what needs to be done to put it right, and the time in which this must be completed (usually not less than 14 days). This does not apply to cleaning. It is an offence if the work is not carried out in the specified time without prior agreement.

A Hygiene Emergency Prohibition Notice is served if the EHO/EHP believes that there is an imminent risk to health from the business. This would include serious issues such as sewage contamination, lack of water supply or rodent infestation. Serving this notice would mean immediate closure of the business for three days, during which time the EHO/EHP must apply to magistrates for a Hygiene Emergency Prohibition Order to keep the premises closed. Notices/orders must be displayed in a visible place on the premises. The owner of the business must apply for certificate of satisfaction before they can re-open.

A Hygiene Prohibition Order prohibits a person such as the owner/manager from working in a food business.

Magistrates' courts can impose **fines and penalties for non-compliance**. This can include fines of up to £5,000, a six-month prison sentence, or both. For serious offences such as knowingly selling food dangerous to health, magistrates could impose fines of up to £20,000. In a Crown Court unlimited fines can be imposed and/or two years' imprisonment.

> **Take it further**
>
> The *Industry Guide to Good Hygiene Practice* gives advice (in plain, easy-to-understand English) to food businesses on how to comply with food safety law. The Guide has no legal force, but food authorities must give it due consideration when they enforce the Regulations. It is intended to help business owners and managers understand and use the information to meet legal obligations and to ensure food safety. Printed copies are available from HMSO Publications Offices or an online version at www.tsoshop.co.uk

The Food Standards Agency

The Food Standards Agency was established in 2000 with a role 'to protect public health from risks which may arise in connection with the consumption of food and otherwise to protect the interest of customers in relation to food'. The Agency is committed to put customers first, be open and accessible and be an independent voice on food-related matters. They provide information to public and government agencies on food safety and nutritional matters from 'farm to fork' and also protect consumers through enforcement and monitoring.

They are responsible for:
- research
- food safety, contaminants, nutrition, additives and labelling
- animal feeds
- the performance of local authority enforcements
- meat and butchery hygiene.

> **Take it further**
>
> A wide variety of information can be found on the Food Standards Agency website www.food.gov.uk. This includes:
> - various food safety topics
> - nutritional information
> - product and ingredient information (including food additives).

> **Activity**
>
> 1 Compile a list of all the advantages you can think of for a business to have good food safety standards in place.
> 2 Compile a list of disadvantages to a business of not having good standards of food safety.

Applying and monitoring good hygiene practice

The importance of temperature control

Effective use of temperatures (high or low) is one of the most useful means available to a food business to make food safe to eat and ensure that it remains safe. High temperatures can be effective in killing pathogens and other organisms, while low temperatures are effective in preventing (or slowing to an acceptable level) multiplication of pathogens in food and organisms causing spoilage of food. Uses of both high and low temperatures are of great importance in reducing the likelihood of a food poisoning outbreak. (For more information on pathogens and food poisoning, see pages 44–52.)

Legislation requires food premises to have sufficient temperature control in place for the levels of business to be conducted. Businesses and food handlers must keep food safe and use sufficient temperature control equipment for the work to reduce the possibility of food poisoning and to fulfil due diligence requirements. When dealing with the enforcement of food safety legislation, temperature control will be one of the main issues considered by the EHO/EHP. Recording of temperatures will form part of a food safety management system and could be part of a due diligence defence.

It is important that food is kept out of the 'danger zone temperatures' (5°C–63°C) as much as possible (see page 49). When cooking and cooling foods, they must be taken through these danger zone temperatures quickly. Failure to do this can allow multiplication of dangerous bacteria and possible formation of spores and the release of toxins.

Correct temperature control will also reduce wastage and retain the quality of food for longer.

Heat food quickly

- Cook in smaller quantities where appropriate to allow for faster heating to safe temperatures.
- Ensure that heat sources are adequate for the task and quantity.
- Make sure that heating and cooling equipment is in good repair and regularly serviced – record the servicing of equipment and keep in the food safety management records as part of temperature management control.
- Make use of thermostats to assist in achieving required equipment temperature and food temperatures.
- Use sanitised and calibrated temperature probes to check food temperatures, and record these temperatures.

Cool food quickly

- Cool cooked food to temperatures below 8°C, and preferably below 5°C, within 90 minutes and be sure to protect it from contamination during its cooling time. Use blast chillers where available and appropriate to cool food quickly and safely.
- Break food amounts down into smaller, thinner portions to allow for quicker and more efficient cooling. Use ice trays, ice packs, containers under cold running water, iced water containers or other appropriate methods to ensure that food is cooled quickly and safely.
- Cooled foods intended for freezing need to be placed in a freezer operating at or below –18°C and be kept frozen at this temperature.

> **Professional tip**
>
> To allow cooling in the required 90 minutes, joints of meat that are to be cooked and cooled should not exceed 2.25 kg in weight.

> **Professional tip**
>
> The law requires that chilled foods and refrigerators are kept below 8°C. However 'danger zone' temperature, that is, the temperature at which pathogenic bacteria may start to multiply, is 5°C. Therefore good practice is to keep refrigerators and chilled foods below 5°C.

Equipment for temperature control

There is a wide variety of temperature control equipment available to a food business (see Table 2.1).

Table 2.1 Temperature control equipment

Hot	Cold
Stoves and combi ovens, bain marie, hot cupboards, hot service counters, water-baths, steamers, hot cabinets, thermal lamps, temperature probes, microwave equipment, thermal food flasks, heated trolleys, heat retaining pellets and mats.	Refrigerators, freezers, blast chillers, chilled display equipment and cabinets, trolleys and food transportation vehicles. Temperature probes Thawing cabinets for defrosting frozen foods Computerised cold storage systems

▲ Heating food in a water bath

Recording and logging temperatures is of the greatest importance to comply with the requirements of HACCP and cover due diligence requirements. Temperatures from refrigerator or freezer displays can be recorded manually or by the use of a computerised temperature management system.

All food temperature control equipment must be serviced and maintained regularly to ensure it is working well and achieving the required temperatures. Service contracts with reputable companies will ensure this is achieved. Keep servicing and repair records with other food safety documentation.

Controlling contamination and cross-contamination

Contamination of food

Possible contamination of food is a major hazard in any food-related operation. A contaminant is anything that is present in food that should not be there. Food contaminants can range from inconvenient or slightly unpleasant, through to dangerous and even fatal. In any food business, it is essential to protect food from contamination and remove or destroy contaminants already present (such as possible pathogens in raw poultry). For more information on contamination, see page 44.

Cross-contamination

Cross-contamination occurs when pathogenic bacteria (or other contaminants) are transferred from one place to another. This is often from contaminated food, usually raw food, equipment, preparation areas or food handlers, and is transferred to ready-to-eat food. It is the cause of significant amounts of food poisoning and care must be taken to avoid it. Cross-contamination could be caused by:

- Foods touching – for example, raw meat touching cooked meat.
- Raw meat or poultry dripping onto high-risk foods.
- Soil from dirty vegetables coming into contact with high-risk foods.
- Contaminated cloths, staff uniforms or equipment.
- Contaminated cleaning equipment such as brooms and mops.
- Equipment used for raw then cooked food, e.g. chopping boards or knives.
- Hands – touching raw then cooked food, not washing hands between tasks.
- Pests spreading bacteria from their own bodies or droppings around the kitchen.
- Different people touching hand contact surfaces, for example, fridge or cupboard doors.

Ways of controlling contamination and cross-contamination

- Having separate working areas and storage areas for low-risk, raw and high-risk foods is highly recommended. If this is not possible, keep them well away from each other and make sure that working areas are thoroughly cleaned and disinfected between tasks.
- Protect food from any possible contamination while in storage, preparation, cooking or serving areas.
- Make sure there is good visibility in the kitchen so hazards will be seen and to allow effective cleaning to take place.
- Separate washing facilities are needed for food items, equipment and area cleaning, dishwashing and for hand-washing. Some establishments also have sterilising sinks with heating elements underneath but these are used less than they used to be since dishwashing machines became more popular.
- Ensure careful control of chemicals in food areas through COSSH (control of substances hazardous to health). COSSH regulations cover the control of chemicals and staff training in their correct use.
- Areas and equipment should be fit for purpose, smooth with no crevasses, impervious, non-corrosive, non-tainting and easy to clean.
- Large and small equipment and utensils need to be kept in good repair and well maintained. Equipment should be easy to take apart for cleaning and installed to allow for thorough cleaning of the surrounding areas. Proper storage for portable equipment and utensils must be provided to protect them from contamination.
- Swabbing procedures to monitor and record microbial presence on surfaces and equipment is good practice. Various kits and equipment are available designed specifically for kitchen use.
- Good personal hygiene practices by staff, especially frequent and thorough hand washing, are very important in controlling cross-contamination.
- Clear and strict policies on the wearing of correct, clean kitchen clothing are essential.
- Vegetables should be washed before preparation/peeling and again afterwards. Leafy vegetables may need to be washed in several changes of cold water to remove all of the soil.
- Enforce clear, strict policies for visitors to the kitchen. For example, restrict access and issue of white coats and hats for visitors.
- Ensure that pest control policies are in place with regular audits and reports.
- Kitchen waste can be a significant cause of cross-contamination and must be managed correctly. See page 59.
- Colour-coded equipment – colour-coded chopping boards are a good way to keep different types of food separate. Chopping boards will come into contact with the food being prepared, so need special attention when cleaning and disinfecting. Make sure that chopping boards are in good condition – cracks and splits trap bacteria and this could be transferred to food. As well as colour-coded chopping boards, some kitchens also provide colour-coded knives, cloths, cleaning equipment, storage trays, bowls and even staff uniforms to help prevent cross-contamination.

Much of the above can be achieved by setting high standards, effective supervision and planned, on-going staff training.

▲ Colour-coded chopping boards

> **Key term**
>
> **Swabbing** – taking a sample from a surface, hands or equipment to establish if bacteria are present.
>
> **Pesticides** – chemical products to kill specific pests.
>
> **Insecticides** – chemical products to kill insects.

Professional tip

A further colour is being added to the colour-coded equipment in some kitchens. The equipment is usually purple and is used for preparation of foods associated with allergies such as nuts.

▲ A purple chopping board for working with allergens

Activity

List six ways that cross-contamination could take place in a kitchen you are familiar with. For each one, suggest good practice procedures that would prevent cross-contamination taking place.

Personal hygiene

All humans can be a source of food poisoning bacteria and everyone working with food must be aware of the importance of personal hygiene in line with the tasks they complete. It is up to the supervisor to make staff aware of this and to lead by example.

Staphylococcus aureus organisms are frequently present on humans – in the nose, mouth, throat, on skin and hair – and could easily be transferred onto food, where it could then multiply and cause illness. It is therefore very important that hair, nose, ears and mouth are not touched while preparing food. Cuts, burns and boils are also likely to be a source of bacteria so must be covered.

The human intestine can be a source of bacteria such as salmonella and even e coli, so very thorough hand washing is essential after using the toilet to avoid transfer of bacteria by faecal/oral routes. This is when pathogens normally found in faeces are transferred to ready-to-eat foods, which can then cause food poisoning.

Anyone suffering from or who is a carrier of any disease that could be transmitted through food must not handle food. This would include diarrhoea and/or vomiting or any food poisoning symptom, infected cuts, burns or spots, bad cold or flu symptoms. Food handlers with any such illness, infection or wound must report this to their supervisor before starting work or going anywhere near food. This is a legal requirement. They should also report any illness that has occurred on a recent holiday and illness of other members of their family. Any food handler with a stomach-related illness such as diarrhoea and vomiting must not return to food-handling duties until 48 hours after the last symptom.

All supervisors must be aware that food handlers can be **carriers** of dangerous pathogens that could get into the food they prepare. All known carriers must not handle food.

- **Convalescent carriers** are recovering from an illness but still carry the bacteria and can pass these on to the food they handle.
- **Healthy carriers** show no signs of illness but can still contaminate the food as above. These people can remain carriers for long periods or all of their lives.

Those working with food can also be the cause of physical contamination, for example, jewellery, finger nails or hair getting into food. They can also taint or spoil food by the use of strong perfumes, aftershave or hair products.

Before starting a job involving food handling, supervisors/head chefs need to make all staff aware that they must practise high standards of personal hygiene. They must always arrive at work clean (daily bath or shower) and with clean hair. Not only is this a basic requirement for a food handler, it also shows pride in the work they do and consideration to others they work with.

Protective clothing should only be worn in the relevant food areas. Clothing needs to be suitable for the work to be carried out, completely cover the wearer's own clothing, be comfortable and washable at high temperatures. Jackets and coats should have press-stud closures rather than buttons and any pockets should be on the inside. Hair should be completely covered with a suitable hat or net. Protective clothing could actually be a cause of food contamination. If it becomes badly stained or dirty it must be changed for clean clothing. Staff must be provided with suitable areas/lockers to change and store their own clothing and belongings.

They must also:

- keep nails short and clean –no nail varnish or false nails
- remove jewellery and watches before handling food (a plain wedding band is permissible but could still trap bacteria)
- avoid wearing cosmetics and perfumes
- never smoke in food areas (ash, smoke and bacteria from touching the mouth area could get into food)
- never eat food, sweets or chew gum when working with food – this may also transfer bacteria to food
- always cover cuts, burns or grazes with a waterproof dressing, then wash hands thoroughly
- report any illness to the supervisor as soon as possible.

Food handlers will also need to be instructed on the importance of frequent hand washing and how to do this properly. They need to know that contamination of food, equipment and surfaces from hands can happen very easily and care must be taken with hand washing to avoid this.

Hand washing procedure

1 Use a basin provided just for hand washing.
2 Wet hands under warm running water.
3 Apply liquid soap to spread across the hands.
4 Rub hands together and rub between fingers and thumbs.
5 Remember, fingertips, nails and wrists as shown.
6 Rinse off under the running water.
7 Dry hands thoroughly on a paper towel and use the paper towel or your elbow to turn off the tap.

Hand-washing technique with soap and water

Wet hands with water

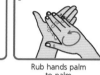
Apply enough soap to cover all hand surfaces

Rub hands palm to palm

Rub back of each hand with palm of other hand with fingers interlaced

Rub palm to palm with fingers interlaced

Rub with back of fingers to opposing palms with fingers interlocked

Rub each thumb clasped in opposite hand using a rotational movement

Rub tips of fingers in opposite palm in a circular motion

Rub each wrist with opposite hand

Rinse hands with water

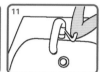
Use elbow to turn off tap

Dry thoroughly with a single-use towel

Hand washing should take 15–30 seconds

cleanyourhands campaign

NHS
National Patient Safety Agency

© Crown copyright 2007 283373 1p 1k Sep07

Adapted from World Health Organization *Guidelines on Hand Hygiene in Health Care*

Staff involved in food handling must know that thorough hand washing should take place:

- When they enter the kitchen, before starting work and handling any food.
- After a break (using the toilet, in contact with faeces).
- Between different tasks but especially between handling raw and cooked food.
- After touching hair, nose, mouth or using a tissue for a sneeze or cough.
- After application or change of a dressing on a cut or burn.
- After cleaning preparation areas, equipment or contaminated surfaces.
- After handling kitchen waste, external food packaging, money or flowers.

Activity

1 As kitchen supervisor/senior chef you wish to raise standards of personal hygiene of your kitchen staff. What are the ways you could do this? How could you involve the staff themselves and how could you monitor improvements taking place?
2 Suggest the food safety topics that should be included in an Induction Handbook for new kitchen assistants who will have some food handling duties.

Cleaning and disinfection

Clean food areas play an essential part in the production of safe food. Cleaning needs to be planned, organised and recorded so that areas, surfaces and equipment are properly cleaned – from the delivery areas, all the storage, preparation and cooking areas right through to where the food is served. Clean premises, work areas and equipment are essential to:

- control the organisms that cause food poisoning
- reduce the possibility of physical and chemical contamination
- make accidents less likely, for example, slips on a greasy floor
- create a positive image for customers, visitors and employees
- comply with the law
- avoid attracting pests to the kitchen
- create a pleasant and hygienic working environment.

Effective cleaning uses one or more of the following:

- **physical energy**/human effort of the cleaner carrying out the task
- **chemicals** such as detergents, disinfectants and degreasers
- **mechanical methods** – machines such as dishwashers and floor-cleaning machines
- **turbulence** – movement of liquids such as in a dishwasher
- **thermal energy** – hot water and steam
- **CIP** (clean in place) – e.g. for very large equipment.

Health and safety

⚠️

Always make sure equipment is disconnected from the electricity source before cleaning.

When done properly, cleaning is effective in removing dirt, grease, debris and food particles. However, cleaning alone is not effective in the removal of micro-organisms, this will need **disinfection**. Disinfection is often carried out after the cleaning process, but sometimes the two are done together by the use of sanitiser. It is difficult to be certain that disinfection will kill all micro-organisms but it will bring them to a safe level.

Disinfection may be completed with:

- chemicals – use only those recommended for kitchen use
- hot water – for example, 82°C+ for 30 seconds as occurs in a dishwasher
- steam – use of steam disinfection is good for equipment and surfaces that are difficult to dismantle or reach, e.g. clean in place (CIP).

All food contact surfaces, equipment and hand contact items (e.g. fridge handles) should be cleaned and disinfected regularly according to your cleaning schedule. Other areas and items may need thorough cleaning but not disinfection, e.g. floors and walls. It is important to clean and disinfect the actual cleaning materials after use.

Key term

Cleaning – the effective removal of dirt, grease, debris and food particles usually done by using hot water and detergent. Detergent breaks the surface tension and holds grease and dirt in suspension in the cleaning water. Cleaning can be completed by hand or by using mechanical equipment such as dishwashers and floor scrubbers.

Disinfection – destroys pathogens and other organisms and brings them down to a safe level. Disinfection can be achieved with chemical disinfectants, use of very hot water, such as in a dishwasher or by directed steam as in a steam gun.

Activity

Make two lists from the following, showing which items you would tell staff to **clean only** and which you would tell them to **clean and disinfect.**

- Windows
- Chopping boards
- Store room floor
- Mirror in staff changing room
- Work surfaces in pastry section
- Serving containers from salad bar
- Kitchen ceiling
- Kitchen hand wash basin
- Meat slicer
- Food probe
- Milk jugs
- Spoons and slices for buffet
- Computer screen
- Dish washing brush
- Dessert dishes
- White information board
- Balloon whisk
- Can opener
- Electrical flexes.

Cleaning products used in cleaning and disinfection

There are different cleaning products designed to complete different tasks.

- **Detergent** – This is designed to remove grease and dirt and hold them in suspension in water. It may be in the form of liquid, powder, gel or foam and usually needs to be added to water to use. Detergent will not kill pathogens (but the hot water it is mixed with may help to do this), however, it will clean and de-grease so disinfectant can work properly. Detergents work best in hot water.
- **Disinfectant** – This is intended to destroy bacteria when used properly. Disinfectants must be left on a cleaned, grease-free surface for the required amount of time to be effective and usually work best in cool water.
- **Sanitiser** – This cleans and disinfects and usually comes in spray form. Sanitiser is very useful for work surfaces and equipment, especially between tasks.
- **Steriliser** – This can be chemical or through the action of extreme heat and will kill all living micro-organisms.

Health and safety !

Before cleaning any areas, make sure that all loose items, small equipment and food are removed from the area. Never use spray cleaners near open food. This will avoid chemical contamination.

The supervisor must ensure that staff are aware of the possible dangers and hazards from cleaning chemicals. Recognised COSHH (Control of Substances Hazardous to Health) training for staff will increase their awareness of the risks of the possible dangers from the chemicals they use as part of their job role. Staff must be made aware through on-going training and supervision of the correct use of chemicals including:

- carefully following manufacturers' instructions for storage, use and disposal
- mixing and dilution of chemicals according to manufacturers' instructions
- use of PPE (personal protective equipment)
- dealing with chemical spillages safely
- proper chemical storage procedures – storage areas may have limited access
- keeping chemicals well away from food preparation areas in separate and lockable storage
- keeping chemicals in their original containers
- safe and proper disposal of chemicals in the correct manner according to company policy, Health and Safety Regulations and advised COSHH procedures.

Health and safety !

Data safety sheets will list all chemicals used, what they are used for and how they should be used. They will also outline how they should be stored, the procedures for accidents with the chemical, spillages and safe disposal.

Clean as you go

As a food supervisor, it is essential to train all staff in the importance of 'clean as you go' and not allow waste to accumulate in the area where food preparation and cooking are being carried out. Staff need to understand that it is very difficult to keep untidy areas clean and hygienic and it is more likely that cross-contamination will occur in untidy areas. Waste should never be allowed to remain in kitchen areas overnight.

As part of a food safety management system, cleaning of areas and equipment needs to be planned and recorded in a **cleaning schedule**. The cleaning schedule needs to include the following information:

- **What** is to be cleaned.
- **Who** should do it (name if possible).

- **How** it is to be done and how long it should take.
- **When**, such as the time of day, as well as the frequency.
- **Materials** to be used including chemicals, dilution, cleaning equipment, protective clothing to be worn.
- **Safety** precautions necessary.
- **Signatures** of the cleaner and the supervisor checking their work, also the date and time. The cleaning schedule should be completed and signed when a task is finished and the schedule kept as part of due diligence.

For effective cleaning to take place, staff must be properly trained, with emphasis given to use of the cleaning schedule, correct methods, use of chemicals and use of PPE (personal protective equipment).

Activity

Produce a kitchen cleaning schedule with all the necessary information in chart form showing the cleaning of: a large oven, a floor-standing mixer, chopping boards used for meat, the dry stores floor and a refrigerator used for salad items.

Dishwashing

The most efficient and hygienic method of cleaning dishes and crockery is the use of a dishwashing machine as this will clean and disinfect items that will then air-dry, removing the need for cloths. The dishwasher can also be used to clean/disinfect small equipment such as bowls and chopping boards. The stages in machine dishwashing are:

- remove waste food
- pre-rinse or spray
- load onto the appropriate racks
- the wash cycle, which will run at 50–60°C using a detergent, and the rinse cycle at 82–88°C

The very high rinse temperature will disinfect items and allow them to air dry so no drying cloths will be needed.

Dishwashing by hand

If items need to be washed by hand, the recommended way to do this is:

- Scrape/rinse/spray off any food residue.
- Wash items in a sink of hot water. The temperature should be 50–60°C, which means rubber gloves will need to be worn. Use a dishwashing brush rather than a cloth. (The brush will help to loosen food particles and is not such a good breeding ground for bacteria.)
- Rinse in very hot water – if rinsing can be done at 82°C for 30 seconds it will disinfect the dishes.
- Allow to air dry, do not use tea towels.

Double sink dishwashing

Before dishwashers were so widely available, a double sink system of dishwashing was often used. Dishes were rinsed/pre-cleaned, washed by hand in one sink using hot water and detergent then loaded onto racks and plunged into a second sink of very hot rinse water (up to 82°C). The water in this sink was often heated by heating elements or pipes under the sink. Although this system can still be seen in some establishments, it has fallen from favour because it is labour intensive, creates condensation problems and causes health and safety concerns (open sinks of very hot water).

Cloths

Great care must be taken with kitchen cloths; they provide an ideal growing area for bacteria and can be the cause of cross-contamination if not used properly. It is good practice to use kitchen paper or single-use cloths in place of the more traditional kitchen dish cloths. Where cloths are used, the colour-coded types used in different areas will help to avoid cross-contamination. Even just having two colours and using one for raw food and the other for high-risk food would help.

If tea towels are used, do so with great care. Remember that they can also spread bacteria so don't allow them to be used as an 'all-purpose cloth', and don't allow the placing of cloths on the shoulder while working (the cloth touches neck and hair and these can be sources of bacteria).

Do not soak cloths or mops overnight or for long periods. Wash them, rinse, squeeze out the excess water and allow to air dry.

Cleaning surfaces and equipment

Because kitchen surfaces and equipment are likely to come into direct contact with food, it is essential that they undergo planned and recorded cleaning and disinfection. One of the methods in Table 2.2 is recommended for cleaning.

Avoiding hazards

Cleaning is essential to prevent hazards but if not managed properly, it can become a hazard in itself. Do not store cleaning chemicals in food preparation and cooking areas and take care with their use (*chemical contamination*). Make sure that items such as cloths and paper towels or fibres from mops do not get into open food (*physical contamination*). *Bacterial contamination* could occur by using the same cleaning cloths and equipment in raw food areas then in high-risk food areas or by not cleaning/disinfecting cleaning equipment properly.

Table 2.2 Cleaning stages for surfaces and equipment

Six stage	Four stage
1 Remove debris and loose particles.	1 Remove debris and loose particles.
2 Main clean to remove soiling grease.	2 Main clean, use hot water and sanitiser.
3 Rinse using clean hot water and cloth to remove detergent.	3 Rinse using clean hot water and cloth if recommended on instructions.
4 Apply disinfectant, leave for contact time recommended on container.	4 Allow to air dry or use kitchen paper to dry.
5 Rinse off the disinfectant, if recommended.	
6 Allow to air dry or use kitchen paper.	

▲ Storing waste oil for collection

Waste disposal

Kitchen waste should be placed in suitable waste bins with lids (preferably with a foot-operated lid). Bins should be made of a strong material, be easy to clean, pest-proof, kept in good condition and lined with a suitable bin liner. They should be emptied regularly to avoid waste build up as this could cause problems with multiplication of bacteria and attract pests. An over-full, heavy bin is also much more difficult to handle than a regularly emptied bin. Waste must never be left in kitchen waste bins overnight (this needs to be part of kitchen closing checks).

Outside waste bins also need to be strong, pest-proof, in good condition and waterproof, with a closely fitting lid to avoid attracting pests. Waste bins should stand on hard surfaces that can be easily hosed down and kept clean. There needs to be a suitable water supply with a hose connection to be able to do this. If possible, avoid placing the waste bins where they will be in direct sunlight as this will speed up the decomposition of the contents. The area also needs to be easily accessible for the vehicles collecting the waste. Planned regular emptying of the bins and cleaning of the area should be recorded in the cleaning schedule as part of the food safety management system.

Separating waste

All waste areas inside and outside need to be kept organised, clean and tidy. Kitchen waste is now often separated into different types and may be collected at different times. Much of the waste can now be recycled. Waste may be divided into food, paper and cardboard, oil, grease trap residues, plastics, glass, metals and more, depending on local authority facilities and policies.

Food premises

Premises for food production must be planned with care to allow good food safety practices to take place. Suitable buildings with good surrounding infrastructure and access are essential. The premises must have well planned fittings, layout and equipment to enable supervisors to plan for required food safety standards and for effective cleaning and disinfection to take place.

Certain basics need to be available if a building is to be used for food production.

- There must be reliable electricity supplies to support the efficient running of all equipment, lighting, ventilation and refrigeration. An alternative supply such as an emergency generator in case of breakdown could be considered.
- Mains or bottled gas supplies provide efficient fuel supply for food premises but are not absolutely essential.
- Reliable supplies of potable (clean and drinkable) water are essential; UK tap water meets the required standards for food premises.
- Efficient drainage for waste water and effluent.
- Suitable road access for deliveries and refuse collection. Necessary arrangements must be in place for regular collection of all types of waste.
- Surrounding areas and buildings should not be sources of contamination by pests, chemicals, smoke, odours or dust.

Layout and workflow

When planning food premises, a linear workflow should be in place, for example, delivery → storage → preparation → cooking → hot holding → serving. This means there will be no cross-over of activities that could result in cross-contamination.

- There must be adequate storage areas; sufficient refrigerated storage is especially important.
- Appropriate staff hand washing/drying facilities suitable for the work being carried out must be provided.
- Clean and dirty (raw and cooked) processes should be well segregated. Cleaning and disinfection should be planned with separate storage for cleaning materials and chemicals.

- All areas should allow for efficient cleaning, disinfection and pest control.
- Personal hygiene facilities must be provided for staff such as toilets, hand basins and preferably showers, as well as changing facilities and storage for personal clothing and belongings. There must be sufficient hand-washing basins, conveniently sited in the kitchen to help avoid possible cross-contamination.

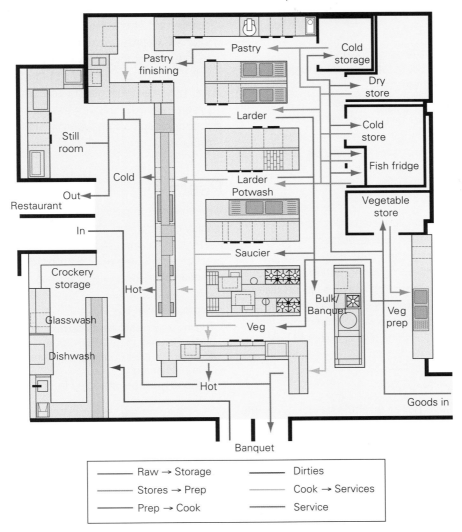

	Raw → Storage		Dirties
	Stores → Prep		Cook → Services
	Prep → Cook		Service

▲ A simple kitchen floor plan

Lighting and ventilation

Lighting (natural and artificial) must be sufficient for tasks being completed, to allow for safe working and so cleaning can be carried out efficiently. Good ventilation is essential in food premises to prevent excessive heat, condensation, circulation of air-borne contaminants, grease vapours and odours. If ventilation is poor, working conditions may be unpleasant and the likelihood of contamination greater. Ventilation systems should flow from clean areas to dirty areas.

Drainage

Drainage must be adequate for the work being completed without causing flooding. If channels, grease traps and gullies are used, they should allow for frequent and easy cleaning. The direction of drainage must be from clean to dirty.

Floors

These need to be durable and in good condition; they must be impervious, non-slip and easy to clean. The floor surface should be flat so no pools of water remain when cleaning and where necessary the floor must allow for efficient drainage into gullies. Suitable materials are non-slip quarry tiles, epoxy resins, industrial vinyl sheeting and granolithic flooring. Where the materials allow, edges between floor and walls should be coved to prevent debris collecting in corners.

Walls

Walls need to be impervious, smooth, easy to clean, non-toxic and preferably light in colour. Suitable wall coverings are plastic cladding, stainless steel sheeting, ceramic tiles, epoxy resin or rubberised painted plaster or brickwork or other smooth coatings. Lagging and ducting around pipes should be sealed and gaps sealed where pipes enter the building to prevent the entrance of pests.

Ceilings

Design often includes suspended ceilings so pipe work and wiring can be concealed. However, care must be taken to prevent pests from getting into these areas. Ceiling finishes must resist build-up of condensation which could encourage mould; they should be of a non-flaking material and be washable. Non-porous ceiling panels and tiles are frequently used and may incorporate lighting; non-flaking paints are also useful. Once again, edges should be coved where possible.

Windows and doors

These provide possible points of entry for pests into the building so should be fitted with suitable screening, strip-curtains, metal kick plates etc. Doors and windows should also fit well into their frames – again to stop pests gaining access.

Door handles and window handles need to be included in the cleaning schedule as they are hand contact surfaces and could be a cause of cross-contamination.

Fittings and surfaces

Surfaces and equipment in food areas should be smooth, impervious, corrosion resistant, non-tainting, non-toxic and easily allow effective and easy cleaning and disinfection to take place. All should be of suitable quality for the work to be done and be regularly maintained. Stainless steel is widely used and very popular for kitchen surfaces but sometimes ceramic and plastic surfaces are used. Surfaces and equipment are likely to come into direct contact with food so planned cleaning and disinfection recorded in the cleaning schedule is essential. Sinks for washing food must have hot and cold *potable* water.

Pests

When fines are imposed on food businesses or premises are forcibly closed down, an infestation of pests is often one of the reasons. Pest infestations and contamination of food by pests are one of the main reasons for prosecution for non-compliance on food safety grounds.

Professional tip

The law requires there to be pest control in premises under Regulation (EC) 852/2004:
- 'Effective pest control procedures must be implemented'
- 'Food must be protected from contamination'.

As pests can be a serious cause of contamination by pathogenic and spoilage bacteria, as well as carrying a number of diseases, they must be eliminated from food premises. Pests can carry food poisoning bacteria into food premises from their fur/feathers, feet/paws, saliva, urine and droppings. Other problems caused by pests include damage to food stock and packaging, damage to buildings, equipment and wiring, blockages in equipment and piping. Presence of pests in food premises can cause loss of business, poor reputation and develop low staff morale.

Pests can be attracted to food premises because there is food, warmth, shelter, water and possible nesting materials. All reasonable measures must be put in place to keep them out. Staff must be made aware of possible signs that pests may be present and that any sightings must be reported to the supervisor or manager immediately. Staff working with food must adopt 'clean as you go' methods of work and not allow waste to build up in the kitchen to avoid attracting pests. Premises should not have plants or shrubs close to the entrance as pests can hide in these places and enter the building when an opportunity arises. A clear path around the building helps with the detection of the presence of pests.

Pest management needs to be planned with a written pest control policy as part of the food safety management system. Regular visits from a recognised pest control company are recommended because they will offer advice, as well as deal with problems arising. Companies conducting a regular audit will provide a pest audit report which should be kept safe and could be used as part of due diligence documentation.

Table 2.3 Signs of pest presence and how to keep them out

Pest	Signs that they are present	Ways to keep them out
Rats and mice	Sightings of rodent, droppings, unpleasant smell, fur, gnawed wires etc. Greasy marks on lower walls, damaged food stock and packaging, paw prints and tail marks.	Block entry points, for example, no holes around pipe work, avoid gaps and cavities. Sealed drain covers, wire guards on top of pipes and soil stacks, metal kick plates on doors. Make sure any damage to building, fixtures and fittings is repaired quickly. Check deliveries/packaging for pests. Cut back outside vegetation. Baits and traps, inside and outside.
Flies and wasps	Sighting of flies and wasps, hearing them, dead insects (maggots).	Window/door screening/netting. Electronic fly killer.
Cockroaches	Sighting, dead or alive, also nymphs, eggs, larvae, pupae, egg cases. Live cockroaches often seen at night. Unpleasant smell.	Seal up holes around pipes, windows etc. Sealed containers, no uncovered food left out. Regular checks of ducting, pipe lagging etc.
Ants	Sightings and present in food and food stores. The tiny, pale-coloured Pharaoh's ant is difficult to spot but can still be the source of a variety of pathogens.	No build-up of waste in kitchen. Outside waste not kept too close to kitchen. Dry goods in sealed containers. No open food left in kitchen or store rooms.
Weevils	Sightings of weevils in stored products, e.g. flour/cornflour. Very difficult to see – tiny black insects moving in flour etc.	Effective stock control.
Birds	Sighting, droppings, in outside storage areas and around refuse.	Block entry to building, use of screens, netting and similar deterrent materials. Make sure outside refuse bins have close fitting lids.
Domestic pets	These must be kept out of food areas as the carry pathogens on fur, whiskers, saliva, urine etc.	Keep them out.

Pest control measures can also introduce food safety hazards. Bodies of dead insects or even rodents may remain in the kitchen (physical and bacterial contamination). Pesticides, insecticides and baits could cause chemical contamination if not managed properly. Pest control problems are best managed by professionals.

The role of a supervisor in pest management is mainly about good practice, good housekeeping and working with pest control contractors. These include:

- Reporting any damage to buildings and fittings and organising prompt repair.
- Keeping food areas clean (especially under/behind equipment and in corners) and not leaving out any traces of food or liquids overnight (closing checks).
- Making sure refuse areas are regularly checked and cleaned and that refuse containers have tight fitting lids and are emptied regularly.
- Effective stock control and regular cleaning of storage areas. Dry stores in sealed containers off the floor. Checking deliveries for any possible signs of pests.

Activity

Draw a sketch of a kitchen/restaurant that you are familiar with; include all doors and windows in the sketch. Mark up any possible ways that pests could enter (name the type of pests). Suggest the measures you could put in place to prevent pests from getting in.

Implementing food safety management procedures

Microbial, chemical, physical and allergenic hazards

In any food business there is going to be the possibility of food contamination which could include any of the following:

- **Microbial hazards (micro-organisms):** such as pathogens causing food poisoning or food-borne illness; also spores and toxins, moulds, viruses and parasites. These could lead to problems causing food poisoning or food-borne illness.
- **Physical hazards:** come from objects, such as machine parts or broken machinery, paperclips, fingernails, hair, insects, packaging materials, coins, buttons, blue plasters, parts of plants such as tomato stalks, insects getting into food. Often physical hazards are just cause for customer distaste and complaints, such as a hair found in food, but physical hazards can also cause a wide range of problems including choking, nausea, vomiting, broken teeth, cuts to the mouth and more.
- **Chemical hazards:** occur from various chemicals such as kitchen cleaning materials, disinfectants, insecticides, rodenticides, degreasers, agricultural chemicals, veterinary residues from treatment of animals, beer-line cleaners and others. Extreme care must be taken when using spray chemicals such as sanitiser or fly sprays that these are not used near uncovered food. Chemicals getting into food or residues on equipment and surfaces could cause abdominal pain, nausea, vomiting, skin, throat or eye irritation, breathing problems or burning sensations.
- **Allergenic hazards:** may come from nuts, wheat products, dairy products, shellfish, mushrooms, soft fruits and a range of other foods and products. Reactions may be serious such as anaphylactic shock, breathing difficulties, rashes, irritation, swelling of lips and throat, and symptoms similar to asthma causing difficulty with breathing.

Key term

Anaphylactic shock – This is the result of the immune system overreacting to a usually harmless substance, such as certain foods. A number of allergens can provoke this condition but the most common are nuts, especially peanuts. Symptoms often include faintness, skin irritation and swelling, especially in the nose and throat area which can lead to breathing difficulties and loss of consciousness. Anaphylactic shock should always be treated as an emergency.

All of these food safety hazards are a cause for concern and care must be taken not to allow contamination of food from them, but probably of the greatest concern is food poisoning from microbial sources. When implementing the Food Safety Management System, all types of hazards need to be considered and controls put in place to make them safe.

Food poisoning

This is an acute intestinal illness that is the result of eating foods contaminated with pathogenic bacteria and/or their toxins. Food poisoning may also be caused by eating poisonous fish or plants, chemicals or metals. Symptoms of food poisoning caused by different organisms are often similar and may include diarrhoea, vomiting, nausea, fever, headache, dehydration and abdominal pain. (For further information, see Table 2.4.)

Table 2.4 Common food poisoning bacteria

Major food poisoning pathogens	Sources of bacteria	Preferred temperature for growth	Illness onset time	Symptoms	Can it form spores?
Salmonella This used to be main cause of food poisoning in the UK (now campylobacter). There are many different types but most problems are caused by *Salmonella typhyimurium* and *Salmonella enteriditis*. These release toxins in intestines as they expire. Problems with food poisoning were reduced by measures taken to control salmonella in chickens and in eggs. Salmonella poisoning can also be passed on through human carriers (someone carrying salmonella but not showing any signs of illness).	Raw meat/poultry, raw egg, intestines and excreta of humans and animals. Sewage/untreated water. Pests (e.g. rodents, terrapins, insects and birds) as well as domestic pets. Food sources include raw meat and poultry, raw eggs, untreated milk and shellfish.	7–45°C	12–36 hours	Stomach pain, diarrhoea, vomiting, fever (1–7 days)	No
Clostridium perfringens Food poisoning incidents from this organism have occurred when large amounts of meat are brought up to cooking temperatures slowly then allowed to cool slowly for later use, or if meat does not get hot in the centre. Poultry where the cavity has been stuffed has also caused problems because the middle does not get hot enough to kill the bacteria. All these examples can lead to the formation of spores. Spores are very resistant to any further cooking and allow the survival of bacteria in conditions that would usually kill them.	Animal and human intestines and excreta. Soil and sewage. Insects. Raw meat and poultry. Unwashed vegetables and salads.	15–50°C	12–18 hours	Stomach pain and diarrhoea. (12–48 hours). Vomiting is rare	Yes
Staphylococcus aureus Produces a toxin in food; is heat-resistant and very difficult to destroy. To avoid food poisoning from this organism, food handlers need to maintain very high standards of personal hygiene.	Humans – mouth, nose, throat, hair, scalp, skin, boils, spots, cuts, burns etc.	7–45°C	1–7 hours	Stomach pain and vomiting, flu-like symptoms, maybe some diarrhoea, lowered body temperatures (6–24 hours)	No

Major food poisoning pathogens	Sources of bacteria	Preferred temperature for growth	Illness onset time	Symptoms	Can it form spores?
Clostridium botulinum A spore-forming organism also producing toxins but fortunately is fairly rare in the UK. Symptoms can be very serious, even fatal. It multiplies in conditions where there is no oxygen or very little oxygen (anaerobe) so is of concern to canning industries and where food is vacuum packed.	Soil, fish intestines, dirty vegetables and some animals.	7–48°C	2 hours – 8 days (usually 12–36 hours)	Difficulty with speech, breathing and swallowing. Double vision, nerve paralysis. Death.	Yes
Bacillus cereus Produces spores and two different types of toxin. One toxin is produced in food as the organisms multiply or expire and the other in the human intestine as the organisms expire. *Bacillus cereus* can survive whether oxygen is present or not (facultative anaerobe) and is more difficult to destroy when fats and oils are present. It is often associated with cooking rice in large quantities, cooling it too slowly and then reheating. The spores are not destroyed and further bacterial multiplication can then take place.	Cereal crops, especially rice, spices, dust and soil (unwashed vegetables).	3–40°C	1–5 hours for the toxin produced in food; 8–16 hours for the toxin produced in the human intestine	Vomiting, abdominal pain, maybe some diarrhoea (12–24 hours) Stomach pain, diarrhoea, some vomiting (1–2 days)	Yes

Food-borne illness

This is an illness caused by pathogenic bacteria and/or their toxins and also viruses, but in this case, pathogens do not multiply in the food, they just need to get into the intestine where they invade the cells and start to multiply. Only tiny amounts are needed and may be transmitted person to person, in water or airborne, as well as through food. Symptoms of food-borne illness are wide and varied and include severe abdominal pain, diarrhoea, vomiting, headaches, blurred vision, flu-like symptoms, septicaemia and miscarriage. (For further information see Table 2.5.)

Table 2.5 Food-borne illness and viruses

Bacteria or virus	Sources	Preferred temperature for growth	Illness onset time	Symptoms	Can it form spores?
Campylobacter jejuni This now causes more food-related illness than any other organism. One of the reasons thought to contribute to this is the increase in consumption of fresh chicken which is a significant source of this organism.	Raw poultry/ meat, untreated milk or water, sewage, pets and pests, birds and insects	28–46°C	2–5 days	Headache, fever, bloody diarrhoea, abdominal pain (mimics appendicitis)	No

Bacteria or virus	Sources	Preferred temperature for growth	Illness onset time	Symptoms	Can it form spores?
E. coli VTEC O157 There are many strains of E. coli but significant problems have been caused in recent years by *E. coli O157*. Symptoms can be very serious and can even be fatal.	Intestines and excreta of cattle and humans. Untreated water and sewage Untreated milk Raw meat, under-cooked mince Unwashed salad items and dirty vegetables	4–45°C	1–8 days (usually 3–4 days)	Stomach pain, fever, bloody diarrhoea, nausea. Has caused kidney failure and death.	No
Listeria monocytogenes This organism is of concern because it can multiply (very slowly) at fridge temperatures, i.e. below 5°C. It is also of concern because of the serious outcomes that poisoning from this organism can cause.	Pate, soft cheeses made from unpasteurised milk, raw vegetables and prepared salads Cook/chill meals	0–45°C	1 day – 3 months	Meningitis, septicaemia, flu-like symptoms, stillbirths	No
Norovirus An air-borne virus, widely present in the environment and highly contagious. Passed from person to person. Does not grow in food – viruses only grow in living cells.	Can survive on surfaces, equipment and cloths for several hours.	N/A	24–48 hours	Severe vomiting and diarrhoea	No
Typhoid/paratyphoid People who have suffered from this can become long-term carriers which means they could pass the organism on to others through food (food handlers need six negative faecal samples before returning to work after illness).	Sewage, untreated water Also dirty fruit/ vegetables	N/A	8–14 days	Fever, nausea, enlarged spleen, red spots on abdomen. Severe diarrhoea. Some fatalities	No
Shigella sonnei *Shigella flexneri* Causes dysentery	Infected carriers, sewage, water supplies, contaminated food, shellfish (faecal/oral routes).	N/A	12 hours – 7 days but usually 1–3 days	Bloody diarrhoea with mucus. Fever, abdominal pain, nausea, vomiting.	No
Hepatitis A This is a viral liver illness often contracted by contamination by a food handler with the virus.	Carriers, blood, urine, faeces, sewage, untreated water, shellfish, contaminated salads, dirty vegetables.	N/A	15–50 days	Nausea, vomiting, abdominal pain. Jaundice.	No

At-risk groups

Food poisoning and food-borne illness can be unpleasant for anyone but for some people, the illnesses can be very serious or even fatal. These high-risk groups include babies and the very young, elderly people, pregnant women, those with an impaired immune system and those who are already unwell or recovering from illness.

Pathogenic bacteria

The most common cause of food poisoning and food-borne illness is bacteria, though not all bacteria are harmful. Many are essential to maintain good health, for example, in the digestive system. Some are used in the manufacture of medicines and others in food production, for example, in making cheese, salami and yoghurt. The bacteria that can be harmful and cause food poisoning are referred to as **pathogenic bacteria**. These are dangerous once they get into food because they are not visible to the human eye and the appearance, smell and taste of the food may remain unchanged. Pathogenic bacteria are single-celled organisms using nutrients to thrive and enable multiplication. They secrete harmful toxins into food or the human body as they divide or as they die, resulting in illness.

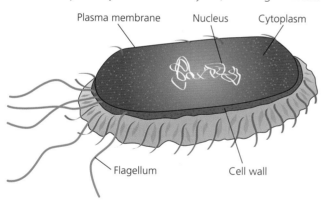

▲ The structure of a pathogenic bacterium

The nucleus in the centre of the cell is surrounded by a gel-like substance that protects it. This is surrounded by a cell wall and flagellae which are small hair-like protrusions allowing a very small amount of movement.

It is not possible to completely eliminate pathogenic bacteria from food premises as they are to be found widely in the environment, including in raw foods, dirty vegetables, animal pests and humans, to name just a few. They can very easily be transferred onto hands, surfaces, cooked foods, equipment and cloths (see cross-contamination on page 44). Bacteria may be carried from one place to another in the kitchen, for example, on hands, cloths or equipment. These are referred to as *vehicles* of contamination.

(see cross-contamination on page 44)

Key term

Food poisoning – an acute intestinal illness resulting from eating foods contaminated with pathogenic bacteria and/or the toxins produced by them. Food poisoning may also be caused by eating poisonous fish or plants, chemicals or metal deposits.

Food-borne illness – this term tends to describe an illness caused by pathogenic bacteria and/or their toxins and also viruses. However, these pathogens or viruses do not multiply in the food, they just need to get into the intestine where they invade the cells and start to multiply and cause illness. They can be transmitted person to person, in water or be airborne, as well as being transmitted through food.

Pathogenic bacteria – there are many thousands of strains of bacteria but pathogenic bacteria are those with the capacity to cause disease and reaction.

Multiplication of bacteria

Once pathogenic bacteria (pathogens) are present in food, given the required conditions (see below), they are able to multiply (vegetative reproduction) by dividing into two; this is called **binary fission**.

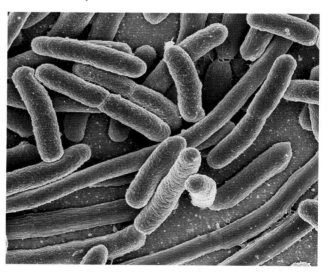

▲ Binary fission of E.coli bacteria

The conditions needed for bacteria to multiply in this way are:

- **Time** – Given the conditions they need, bacteria can divide by binary fission every 10–20 minutes.
- **Warmth** – The optimum temperature for bacteria to multiply is 37°C (body temperature). This is when binary fission occurs most quickly but some form of bacterial multiplication can occur between the temperatures of 5°C and 63°C; the temperature range referred to as the Danger Zone. Different bacteria prefer slightly different temperature ranges for growth (multiplication). See Table 2.4 for more details.

- **Moisture** – Bacteria need water to support their life cycle and processes. The amount of water in food is referred to as a_w (water activity). Other than dried foods, most food contains some moisture and a wide variety of foods contain enough water to support bacterial multiplication.
- **Food (nutrients)** – A wide selection of foods, especially protein-rich foods such as meat, poultry, fish and dairy foods, support the growth of bacteria. Foods with high concentrations of sugar, salt or acids such as vinegar lower the water activity in food so there is not enough moisture to support bacterial growth. This is why sugar, salt and vinegar are often used to preserve foods.
- **Oxygen requirement** – Some bacteria need oxygen in order to multiply; this group of bacteria are referred to as aerobes. Some can only multiply where there is no oxygen and these are called anaerobes. Others can multiply with or without oxygen and are called facultative anaerobes. Most bacteria fall into this group.
- **pH** – This is a measure of how acid or alkaline a food or liquid may be. The scale runs from 1.0 to 14.0. Acid foods are below 7.0 on the scale; 7.0 is neutral; and above 7.0 up to 14.0 is alkaline. Bacteria multiply best around a neutral pH and most bacteria cannot multiply at a pH below 4.0 or above 8.0 so foods such as citrus fruits would not support bacterial multiplication.

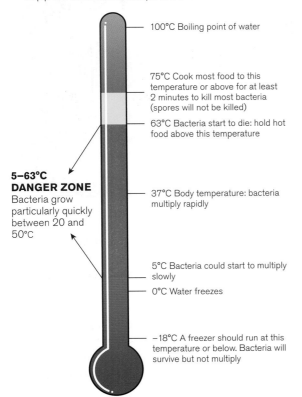

5–63°C DANGER ZONE
Bacteria grow particularly quickly between 20 and 50°C

100°C Boiling point of water

75°C Cook most food to this temperature or above for at least 2 minutes to kill most bacteria (spores will not be killed)

63°C Bacteria start to die: hold hot food above this temperature

37°C Body temperature: bacteria multiply rapidly

5°C Bacteria could start to multiply slowly

0°C Water freezes

−18°C A freezer should run at this temperature or below. Bacteria will survive but not multiply

▲ Important food safety temperatures

Activity

Anaerobic bacteria only multiply in conditions where there is no oxygen. Make a list of foods that would be in anaerobic conditions in a kitchen situation.

Vegetative bacterial growth curve

The phases bacteria go through in their multiplication are shown in the vegetative bacterial growth curve.

- **Lag phase** – When bacteria first enter suitable conditions, no immediate growth takes place: the bacteria are getting used to their surroundings.
- **Log phase** – In the presence of ideal conditions, bacteria will multiply rapidly.
- **Stationary phase** – Bacteria are multiplying but some are also dying off. There may also be competition between different bacteria for survival (food is running out) so numbers stay the same.
- **Decline phase** – Bacteria start to die off and numbers decrease.

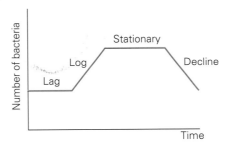

▲ Graph of the bacterial growth curve

Key term

Vegetative reproduction – the ability, in favourable conditions, of living bacteria to thrive and multiply by splitting in half. Each 'half' then becomes an independent cell with the ability to multiply by dividing itself.

Binary fission – when pathogenic bacteria (pathogens) are given the right conditions, they are able to multiply by dividing into two. This is referred to as binary fission.

Destroying pathogens

The majority of pathogenic bacteria will be destroyed by temperatures of 70°C+ (but spores and toxins will not be destroyed by these temperatures). To destroy pathogens in food effectively, this temperature must reach the centre (core) of the food and must be held for long enough (usually two minutes is recommended) before the temperature drops. Processes such as disinfection and sterilisation will also kill bacteria.

Controlling pathogens

It is a food handler's responsibility to keep food areas clean and hygienic at all times. Food handlers should work in a clean and tidy manner, 'clean as you go' and not allow waste to build up. Spills must be cleaned up straight away. All kitchen staff need to be aware of protecting food from bacteria and preventing bacterial growth by keeping food clean, cool and covered. The supervisor needs to check and regularly monitor that good practices are in place and always lead by example.

Toxins

Toxins (poisons) can be produced by some bacteria as they multiply in food; these are called **exotoxins**. Most toxins are very heat resistant and would not be killed by the normal cooking processes that kill bacteria (it would need prolonged cooking at boiling temperatures or above). So once formed, toxins often remain in the food and can cause illness. Some bacteria produce toxins as they die, usually in the intestines of the person eating the food. These are called **endotoxins**. Toxins are further identified as **neurotoxins**, that is, they affect the nervous systems, and **enterotoxins**, which affect the intestines.

Spores

Some bacteria are able to form spores when the conditions surrounding them become hostile, such as temperatures rising or in the presence of chemicals such as a disinfectant. A spore forms a protective 'shell' inside the bacteria, protecting the essential parts from the high temperatures of normal cooking, disinfection, dehydration etc. Once spores are formed the outer cell disintegrates and the organism cannot divide and multiply as before but simply survives until conditions improve such as when high temperatures drop to a level where cells can re-form and multiplication can start again.

Prolonged cooking times and/or very high temperatures are needed to kill spores. Food canning processes use 'botulinum cook' – a cooking time and temperature combination that will kill any spores, depending on the acidity and texture of the food, for example, 121°C for three minutes.

Time is very important in preventing the formation of spores. Large amounts of food such as meat for stewing, brought slowly to cooking temperature, allows time for spores to form; these are then very difficult to kill. An effective control is to bring food up to cooking temperature quickly, possibly by cooking in smaller quantities, and always cool food quickly.

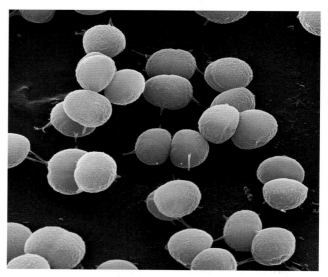

▲ Bacterial spores forming

Requirements for multiplication of different pathogens

- **Psychrophiles** (e.g. *listeria* and *Clostridium botulinum*) prefer temperatures below 20°C for multiplication.
- **Mesophiles** (e.g. *salmonella* and *Staphylococcus aureus*) prefer temperatures of 20–50°C for multiplication (optimum 37°C).
- **Thermophiles** (e.g. *Clostridium perfringens* and *Campylobacter*) multiply most rapidly above 45°C.

The oxygen requirements for the multiplication of these pathogens are:

- **obligate aerobe** (e.g. *Bacillus cereus*) multiplies only where there is oxygen present.
- **faculative anaerobe** can multiply with or without oxygen.
- **obligate anaerobe** (e.g. *salmonella* and *Staphylococcus aureus*) can multiply only in the absence of oxygen.

The rate at which bacteria multiply and survive may be partly dependent on other bacteria that are present as they will compete for the same conditions and nutrients. Some produce secretions as they multiply that prevent other types of organisms thriving in the same area.

Viruses

These are even smaller than bacteria and can only be seen with a very powerful microscope. They only multiply on living cells, not on food, though they may be transferred into the body on food or drinks and may live for a short time on hard surfaces such as kitchen equipment. Viruses can easily be passed from person to person and are sometimes associated with shellfish that come from contaminated water.

Moulds

Moulds are usually associated with food spoilage (page 62) rather than food poisoning but it is now known that some moulds can produce toxins called **mycotoxins**. Unlike bacteria, moulds will grow on acid, alkaline, sugary and salty foods. They grow best between 20°C and 30°C but will also grow at refrigerator temperatures.

Poisoning from metals

Occasionally metal elements may accidentally get into food. This may be from old water pipes, especially lead pipes or food placed in unsuitable metal containers – there can be a reaction between some metals and acid foods such as vinegar. Problems may also be caused when crops are sprayed with the wrong concentrations of chemicals containing metals. Metals that have been associated with causing discomfort, illness or adverse conditions when in contact with food or drinks are lead, tin, copper, zinc, cadmium and antimony.

Poisoning from fish and vegetable items

Scombrotoxic fish poisoning (SFP)

This is associated with the consumption of contaminated fish of the Scombroid family (including tuna, mackerel, herring, marlin and sardines). The poison builds up in the fish, especially in storage above 4°C. It is a chemical intoxication and symptoms occur within 10 minutes to 3 hours after eating the affected fish. Symptoms include rash on the face/neck/chest, flushing, sweating, nausea, vomiting, diarrhoea, abdominal cramps, headache, dizziness, palpitations and a sensation of burning in the mouth. Symptoms are usually gone within 12 hours. Some cases are severe and urgent medical attention is necessary. Antihistamine drugs may be used. These toxins are very heat resistant and are not affected by normal cooking. Scombrotoxins are thought to be responsible for up to 70 per cent of food poisoning from fish in the UK.

Paralytic fish poisoning

This is a very dangerous form of fish poisoning. It occurs when some bivalve shellfish, e.g. mussels, feed on poisonous plankton. The poison can survive normal cooking temperatures. Symptoms after eating affected shellfish may be nausea, vomiting, lowered temperature, diarrhoea and also a numbness of the mouth, neck and arms which can lead to paralysis and death within 2–12 hours.

Red kidney beans (black kidney beans)

It is possible to suffer poisoning from the toxin in red kidney beans (haemagglutinin). Poisoning occurs if the beans are not boiled rapidly at temperatures which kill the toxin for 10 minutes or more. After initial rapid cooking at high temperatures, the heat can then be reduced to finish the cooking process. The symptoms of this poisoning include nausea, vomiting, pain and diarrhoea. The symptoms usually disappear within a few hours. Problems with not getting kidney beans hot enough may also occur if cooking in very large quantities or if using a 'slow cooker'.

Poisoning from mycotoxins

Some moulds can produce toxins as they grow; these are called mycotoxins. Mycotoxins can cause serious illness and have been linked with cancers. They are invisibly present in mouldy food over a much wider area than the visible mould. Mycotoxin-producing moulds can tolerate a wide range of acid/alkaline conditions, low-water activity, temperature ranges from -6 to 35°C and very high temperatures. A wide range of foods may be affected by mycotoxin-producing moulds but the foods most frequently affected are cereal crops and products, nuts and fruit products such as juices, dried fruit and jams.

Food parasites

A variety of parasites can live on plants or animals from where they get their food. They present in different forms such as worms, eggs, lice or grubs. They can usually be destroyed by thorough cooking or very high temperature food preservation methods. In most cases, they will also be killed by freezing.

High-risk foods

Some foods support the rapid growth of bacteria more than others when given the required conditions; these are referred to as high-risk foods. They are usually foods that are ready to eat, have a high protein and moisture content and will not go through a cooking process that kills bacteria. They need to be stored in controlled temperatures and must be protected from contamination. They include:

- Cooked meat and poultry and products made from cooked meat/poultry – store at 1– 4°C.
- Eggs and egg-based products, especially where raw egg is used, e.g. fresh mayonnaise. Eggs may contain pathogens. Cook thoroughly and wash hands well after handling. Use pasteurised egg in items such as mayonnaise and mousses.
- Milk, cream and custards – keep custards hot (above 63°C) or cool quickly and refrigerate. Keep milk and cream under refrigeration as much as possible.
- Stock, sauces, gravy and soup – keep these hot (above 63°C) or cool quickly and refrigerate.
- Shellfish and seafood products – use a reputable supplier, cook or reheat thoroughly to kill bacteria, throw away shellfish with broken shells, and mussels or clams that do not open when cooked.
- Cooked rice – rice can contain the spores of *Bacillus cereus* which are not killed by normal cooking temperatures. If rice is cooled slowly in a warm kitchen, these spores germinate and multiply enough to cause food poisoning. Keep cooked rice above 63°C for service or cool quickly. Avoid cooking rice in very large quantities.

Low-risk foods

These are foods that are much less likely to be associated with food poisoning. For a variety of reasons, they do not support the growth of bacteria and could be:
- dry such as biscuits, dried pasta or dry mixes such as sponge mix
- preserved by canning or bottling
- acid foods such as pickled products or fruit
- foods with high sugar content
- salted, smoked or fermented foods (such as salami).

Food spoilage and food preservation

Most foods, once harvested, slaughtered, fished or manufactured, have a limited life and will eventually deteriorate. The deterioration may be caused by moulds, yeasts, enzymes or bacteria. Spoilage of food could also be due to exposure to oxygen moisture or chemicals, damage by pests or by poor handling and bad storage.

Unlike bacterial contamination which is impossible to detect in a normal kitchen situation, food spoilage can usually be detected by **organoleptic observation**. This means that the spoilage can be observed by sight, smell, taste or touch. Food spoilage includes mouldy, slimy, dried-up, over-wet fresh foods, blown cans and vac packs and food with freezer burn. Over many years, different methods have been developed to prolong the natural life of various foods and keep them fresh. These include:

- use of heat – cooking, canning, sterilisation, UHT, pasteurisation
- use of heat and sealing – canning, bottling, cooking, then vacuum packaging
- use of low temperatures – chilling, freezing
- exclusion of air – vacuum packaging
- changing gasses surrounding food – modified atmosphere packaging (MAP)
- removal of moisture – dehydrated foods
- use of acids, sugar or salt concentrations (see above)
- smoking – used on meat and fish (but this will only offer limited preservation)
- preservatives – e.g. nitrates and nitrites.

Food preservation will often combine methods to ensure effective preservation, e.g. smoked fish may be vacuum packed and will also be refrigerated; milk is pasteurised but also stored under refrigeration.

Careful use and control of temperature is essential in producing safe, high quality food for customers.

High temperatures

Normal cooking temperatures of 75°C+ will kill most pathogens (but not toxins and spores). Use of heat to make foods safe and to preserve it is one of the most useful procedures available in food production and manufacture. The following table shows some commonly used temperatures.

Methods and procedures for controlling food safety

Food delivered to food businesses will be taken through a number of stages before it reaches the customer. Careful control at each stage (critical control points) is essential to keep food safe and wholesome. For food to remain in best condition and be safe to eat, it is essential that the infrastructure of food areas, buildings, storage and equipment is fully compliant with required food safety standards.

Food deliveries

It is important that correct delivery and storage procedures are observed and fully understood by all kitchen staff. Full documentation systems need to be completed for all kitchen deliveries, in line with the food safety management system. This will ensure food is stored correctly and records will be available for inspections by the EHO and if necessary as part of 'due diligence'. Only approved suppliers should be used (traceability), who can assure that food is delivered

Table 2.6 Heat treatments used to make food safe

Cooking 75°C for 2 minutes	Cooked food does generally keep for longer than raw equivalents. However, treat cooked items with care as once cooked, foods may then be in the 'high risk' category. Normal cooking, for example, to 75°C for 2 minutes at the core of the food will kill most pathogens, but not spores and toxins.
Pasteurisation 72°C for 15 seconds	This involves heating food to a temperature similar to cooking temperatures but for a very short time, i.e. milk is heated to 72°C for 15 seconds then rapidly cooled. These temperatures will kill most pathogens, bringing them to a 'safe level'. Toxins and spores will not be killed. Because relatively low temperatures are used, milk that has been pasteurised will spoil more quickly than milk preserved by some of the methods described below and must be kept at refrigerator temperatures. Pasteurisation is also used for liquid egg, cream, ice cream, some fruit juices and wine. Because of the relatively low temperatures used in pasteurisation, the taste of the product remains mostly unchanged though the vitamin content is reduced by about 25%.
Sterilisation 100°C for 15–30 minutes	Sterilisation destroys all micro-organisms, heating food to 100°C for 15–30 minutes by applying steam and pressure. Because very high temperatures are used over a relatively long time, some caramelisation occurs; taste is altered and vitamin content is significantly reduced. Un-opened sterilised milk lasts much longer than pasteurised milk. Liquid or semi-liquid food such as soup, stock and sauces sold in pouches also go through a sterilisation process. These foods have a long shelf life and do not need refrigerated storage.
UHT 135°C for one second	UHT (ultra-heat treatment) gives milk or cream a long shelf life without the need for refrigeration (until the package is opened). The milk is heated under pressure to very high temperatures, typically 135°C for just one second, then cooled rapidly and sealed in sterile containers.
Canning 121°C for 3 minutes ('botulinum cook')	Canning efficiently preserves food by using very high temperatures and then sealing the food in a can (or pouch or bottle). A concern with canning is the survival of anaerobic bacteria, especially *Clostridium botulinum*, which is particularly dangerous. This is because any surviving anaerobic bacteria could thrive inside a can where there is no oxygen; also sources of *Clostridium botulinum* include some fish and vegetables, foods that are frequently canned. To overcome this, canned food is subjected to 'botulinum cook' (see page 50).

in the best condition, in suitable delivery vehicles fit for food purposes. The food must be properly packaged, date coded and at the correct temperature. The food delivery driver must also meet required standards of cleanliness and hygiene.

- All deliveries should be checked then moved to the appropriate storage area as soon as possible and chilled/frozen food within 15 minutes of delivery.
- Use a food probe to check the temperature of food deliveries: chilled food should be below 5°C and never above 8°C; frozen foods should be at or below -18°C. Many suppliers will now provide a printout of temperatures at which food was delivered. (Save these in kitchen records as they could be an important part of due diligence.)
- Dry goods should be in undamaged packaging, well within best before dates, be completely dry and in perfect condition on delivery.
- Remove food items from outer boxes before placing the products in refrigerator, freezer or dry store. Remove outer packaging carefully of items such as fruit and

vegetables, remaining fully aware of any possible pests that may have found their way into packaging.
- Segregate any unfit food from other food until it is thrown away or collected by the supplier. This is to avoid any possible contamination to other foods.

Storage

Refrigerated storage

Food safety legislation states that there must be sufficient refrigerated storage for the amounts of work being completed to keep food at safe temperatures. All refrigerators need to be sited correctly, appropriate to where food is being prepared, to avoid possible cross-contamination and in a convenient place for putting away deliveries. Refrigerators must be kept clean and this needs to be a part of the cleaning schedule. Inside the refrigerator should not be overloaded; to operate properly, cold air must be allowed to circulate between items. To keep the interior of the refrigerator at the required temperature, door seals must be in good condition and be regularly cleaned. Shelving must also be kept in good condition and not used if rusty or damaged and shelves must be kept clean at all times. All refrigerated equipment must be regularly maintained; this should be planned and recorded. Wherever possible, there should be different refrigerators for different types of food. This will help to prevent cross-contamination and the possibility of raw food dripping onto or touching other foods and contaminating them.

- **Raw meat and poultry** – Wherever possible store in a refrigerator just for meat and poultry storage, running at temperatures between 1°C and 4°C (butchers' refrigerators between -1°C and +1°C). If not already packaged, place on trays, cover well with cling film and label. If it is necessary to store meat/poultry in a multi-use refrigerator, make sure it is covered, labelled and placed at the bottom of the refrigerator running below 5°C and is well away from other items.
- **Fish** – A specific fish refrigerator is preferable (running at 1–4°C). Remove fresh fish from ice containers and place on trays, cover well with cling film and label. If it is necessary to store fish in a multi-use refrigerator, make sure it is well covered, labelled and placed at the bottom of the refrigerator well away from other items. Remember that odours from fish can permeate other items such as milk or eggs.
- **Dairy products/eggs** – Pasteurised milk and cream, eggs and cheese should be stored in their original containers in a refrigerator running at 1–4°C. Sterilised or UHT milk can be kept in the dry store following the storage instructions on the label. After delivery, eggs should be stored at a constant temperature and a fridge is the best place to store them. Prevent eggs from touching other items in the refrigerator and check that their shells are clean.

Multi-use refrigerators

If food cannot be stored in separate specific refrigerators and needs to be stored in multi-use refrigerators, it is absolutely essential that all staff know the correct procedures to store the food and this should become part of on-going staff training. Posters, pictures and charts near the refrigerator may help with this. Keep the refrigerator running at 1–4°C.

- Store raw foods such as meat, poultry and fish at the bottom of the refrigerator in suitable deep containers to catch any spillage. Cover with cling film and label with the name and the date. Do not allow any other foods to touch these raw foods.
- Store other items above raw foods. Again they should be covered and labelled. Keep high-risk foods well away from raw foods. Storing these foods above the raw foods removes the risk of the raw foods dripping onto other items and causing contamination.
- Wrap strong smelling foods very well as the smell (and taste) can transfer to other foods such as milk.
- As with other foods, check date labels and use strict stock rotation procedures.

Make sure that refrigerators are cleaned regularly – this needs to be part of the cleaning schedule. The procedure should be:

- Remove food to another refrigerator.
- Clean according to cleaning schedule using a recommended sanitiser (a solution of bicarbonate of soda and water is also good).
- Empty and clean any drip trays and clean door seals thoroughly.
- Rinse, then dry with kitchen paper.
- The refrigerator front and handle must be cleaned and disinfected to avoid cross-contamination.
- Make sure the refrigerator is down to temperature 1–4°C, before replacing the food in the proper positions (see above). Check dates and condition of all food before replacing.

Frozen foods

Store in a freezer running at -18°C or below. Make sure that food is wrapped or packaged. Separate raw foods from ready to eat foods in the freezer and never allow food to be re-frozen once it has thawed.

Fruit, vegetables and salad items

Storage conditions will vary according to type, e.g. sacks of potatoes, root vegetables and some fruit can be stored in a cool, well ventilated store room but salad items, green vegetables, soft fruit and tropical fruit would be better in refrigerated storage. If possible a specific refrigerator running at 8–10°C would be ideal to avoid any chill damage.

Dry (ambient) food stores

A dry or ambient food store is an area to store foods that generally have a longer shelf life than those needing refrigerated or frozen storage (though the dry stores area may also have areas for refrigerated and freezer storage). Items usually kept in the dry stores would include dry goods and cereal products, spices and dried herbs, packaged goods, e.g. biscuits, canned goods, bottled items, chocolate/cocoa, tea and coffee. Fruit and vegetables not requiring refrigeration may also be stored here. Refrigerated storage may be available for items needing temperature controlled storage. The room should be large enough to allow for correct storage of stock. It needs to be cool (8–15°C would be ideal), well ventilated, a light colour, well lit, protected to prevent entry of pests and easy to clean/disinfect efficiently.

When fitting out the room, surfaces for walls, ceilings and floors need to be smooth, impervious and easy to clean. Edges where walls join the floor or ceiling should be coved if possible to prevent build-up of debris in corners.

- Doors should be protected by metal kick plates and a plastic or chain curtain to prevent pests from gaining entry. (But also check packaging of deliveries for possible pests.)
- Windows should be well protected with wire gauze or netting and food items should not be stored in direct sunlight from windows.
- Shelves/racking must be made of a non-corrosive, easy to clean material, e.g. tubular stainless steel. Shelving must be deep enough to store the items required and the bottom shelf should be well raised from the floor to allow for ease of cleaning underneath. Never store items directly on the floor as this would prevent effective cleaning from taking place.

Dry goods such as rice, dried pasta, sugar, flour, grains etc. should be kept in clean, covered containers on wheels or in smaller sealed containers on shelves to stop pests getting into them. Retain packaging information as this may include essential allergy and storage advice.

Make sure the stock of dry goods is rotated effectively – **first in, first out**; this will ensure that existing stock is always used first. When new packages of dry stock are delivered, do not empty them into the container on top of what is already being stored. Use the older stock first, clean/sanitise the container, then re-fill. To avoid any possible cross-contamination, store high-risk (ready-to-eat) foods well away from any raw foods such as dirty vegetables and make sure all items remain covered.

Canned products

Cans are usually stored in the dry store area and once again rotation of stock is essential. Canned food will carry best before dates and it is not advisable to use them after this date. 'Blown' cans must never be used and do not use badly dented or rusty cans. Once opened, transfer any unused canned food to a clean bowl, cover and label it and store in the refrigerator for up to two days.

Cooked foods

These include a wide range of foods such as pies, paté, cream cakes, desserts and savoury flans. They will usually be 'high risk foods' so correct storage is essential. For specific storage instructions, see the labelling on the individual items, but generally keep items below 5°C. Store carefully, wrapped and labelled, in a refrigerator used only for high risk items where possible. If a multi-use refrigerator needs to be used, store well away from and above raw foods to avoid any cross-contamination.

Using cling film

Cling film is a very useful product for storing food hygienically, protecting from cross-contamination and preventing food from drying out. However, because cling film seals in moisture it can encourage growth of moulds on food. Do not leave cling film-wrapped or covered foods in direct light as it can increase the temperature of the food inside. This is referred to as 'the greenhouse effect' and it also applies to glass or Perspex display cabinets.

Because of concerns about migration of chemicals into food, it is recommended that foods are never cooked or reheated when wrapped in cling film unless a film specifically recommended for this is used.

Storage of chemicals – COSHH

All staff should receive COSHH training to enable them to use and store chemicals properly. When chemicals are delivered, make sure they are put away promptly in suitable lockable storage well away from food storage. Only the chemicals actually being used should be out in the kitchen area.

'First in, first out'

This term is used to describe stock rotation and is applied to all categories of food including those in refrigerated or frozen storage. It simply means that foods already in storage are used before new deliveries (providing stock is still within recommended dates and in sound condition). Food deliveries should be labelled with delivery date and preferably the date by which they should be used. Use this information along with food labelling codes. Written stock records should form a part of a food safety management system. Also apply 'First in, first out' to frozen foods, non-food deliveries and chemicals.

FRIDAY Viernes - Vendredi		**SATURDAY** Sabado - Samedi
Item:		Item:
Prep Date: ____ Time: ____ ☐AM ☐PM		Prep Date: ____ Time: ____ ☐AM ☐PM
Shelf Life: ____ ☐Shifts ☐Fresh Daily		Shelf Life: ____ ☐Shifts ☐Fresh Daily
Use By: ____ ☐4 PM ☐Close Emp: ____		Use By: ____ ☐4 PM ☐Close Emp: ____

▲ Examples of date labels

- **Use by** dates appear on perishable foods with a short life. Legally, the food must be used by this date and not stored or used after it.
- **Best before** dates apply to foods that are expected to have a longer life, for example, dry products or canned food. A best before date advises that food is at its best before this date and to use it after the date is still legal but not advised.

Professional tip

When referring to date codes, it is essential to observe storage requirement instructions. For example, if a food comes with a use by date, the date only applies if the food has been stored correctly such as at 1–4°C. If stored above this temperature, the food will deteriorate more quickly.

Activity

The following foods have just been delivered. Create a chart which includes instructions for a new member of staff about the correct way to store these items:
- frozen chicken breasts (need defrosting for tomorrow)
- sirloin steaks
- lamb's liver
- fresh whole sole
- smoked haddock
- live crabs
- butter
- strawberries
- bread rolls
- onions
- salad leaves
- rice
- canned tomatoes
- flour
- ice cream.

Food preparation

Food should not be prepared too far in advance of cooking. If food is prepared a significant time before it is to be cooked, control measures must be in place to ensure that this is safe. Preparation areas for raw and cooked food should be well separated, as should dirty and clean processes. Dirty vegetables and salad items need to be washed in several changes of deep, cold water. Temperatures of food need to be monitored before, during and after preparation ensuring that food being prepared is only out of temperature control such as refrigeration or cooking processes for the shortest time possible.

Avoid handling food unnecessarily – use disposable gloves, tongs, slices, spoons etc. where possible and wash hands frequently, especially between tasks. High standards of personal hygiene are essential for those handling food, especially when handling high-risk foods (see pages 35–6 for further details on personal hygiene standards). Clean as you go, but be careful not to allow any chemical contamination from items such as sanitiser sprays and disinfectants. There is also a need to be aware of allergenic contamination – some foods may need to be prepared in a separate area or eliminated altogether.

Use of colour-coded equipment is very useful in food preparation areas to avoid cross-contamination (see pages 34–5 for more information).

Defrosting

If you need to defrost frozen food, place it in a deep tray, cover with film and label it, describing what the item is and the date defrosting started. Place at the bottom of the refrigerator where thawing liquid can't drip onto anything else. Defrost food completely (no ice crystals on any part), then cook thoroughly within 12 hours. Make sure that you allow enough time for the defrosting process – it may take longer than you think (for example, a 2 kg chicken will take about 24 hours to defrost at 3°C).

Cooking

Cooking is one of the best measures available to destroy and control bacteria in food. The usual recommendation is to cook to a core temperature of greater than 75°C for at least two minutes but slightly lower temperatures for a longer time can be as effective. Dish specifications or personal preference may require lower temperatures than this (for example, in rare beef and some fish dishes). However, avoid undercooked dishes when dealing with groups vulnerable to the effects of food poisoning. The most usual way to check if required core temperatures have been achieved is with a calibrated and disinfected temperature probe but also use visual/physical checks such as amounts of steam and checking that no parts of a cooked chicken are pink; juices running off should be clear.

It is also essential to protect food from cross-contamination while it is cooking. For example, raw chickens being placed in the oven to cook may drip liquid onto food below that has finished cooking.

Staff should be aware of the danger zone temperatures and the need to heat food through these temperatures quickly to avoid formation of spores. It is also good practice to use lids on cooking pans to prevent heat escaping and to stir food frequently to keep temperatures even.

> **Key term**
>
> **Core temperature** – the temperature right at the centre of the food, in the thickest part. For most foods to be safe, it is necessary for foods to remain at the core temperature for two minutes or more. Core temperature is usually checked with a disinfected food probe.

Chilling

For food that is to be cooled or chilled, it is recommended that food is cooled to 8°C or below within 90 minutes; this is best done in a blast-chiller. If this is not available, food in containers can be plunged into ice water baths or placed on trays and surrounded with ice or use ice packs. Place cooling food in the coolest place available; fans can also be useful to speed up cooling time. To allow food to cool more quickly, it is recommended that:

- Items such as soups, stews and sauces are placed in small containers; also use shallow containers. Increased surface area allows the food to cool more quickly.
- Clear soups and stock can be strained through a chinois (conical sieve) lined with ice bags.
- Cook smaller joints of meat because these will cool more quickly. A maximum weight of 2.25 kg is recommended.
- Always protect food from cross-contamination while it is cooling and chilling.

As bacteria can only multiply very slowly at low temperatures (see danger zone diagram), good practice is to keep refrigerators running between 1°C and 4°C. The legal requirement is at or below 8°C. Refrigerator and freezer temperatures should be checked and recorded at least once daily; keep the recorded temperatures as part of food safety management records.

When food is cooled either to be served cold or to be re-heated, it must be done carefully and quickly, to avoid the formation/germination of spores and to reduce risk of cross-contamination as food is cooling.

Cold food can be kept out of temperature control for up to four hours on any single occasion such as at a buffet. This time includes preparation time and food must be disposed of at the end of the four hours.

> **Activity**
>
> The following foods have been cooked today to use on a buffet tomorrow. Write a set of instructions for staff explaining how these items should be cooled/chilled/reheated, including all timings and temperatures involved.
> - Roast beef, whole poached salmon, honey roast ham, roast turkey breast – these will be served cold.
> - Twenty-portion lasagne, asparagus flan, chicken and leek pie – these will be cooled then reheated tomorrow.

Holding temperatures

Hot, cooked food being held for service must not fall below 63°C; this is a legal requirement. Make sure that there is adequate equipment to keep food above this temperature, check the temperature frequently and record it. Hot food out of this temperature for more than two hours must be thrown away (this includes the time in the kitchen). Make sure that equipment for keeping food hot such as bain maries or hot cabinets are pre-heated and clean/disinfected; do not overfill food containers and allow all of the food to be used from them before topping up. Use lids on open containers to keep the heat in and stir items such as soups and stews to even the temperature.

There are similar rules for cold food being held for service or displayed. The food being held for service should not be above 5°C (legal requirement is 8°C). Make sure that there is adequate refrigeration or refrigerated display cabinets at the correct temperature; check the temperature and keep a record. Cold food above 8°C for four hours on one occasion must be thrown away (this includes the preparation time).

Reheating

Reheating should be done thoroughly and quickly to a core temperature of at least 75°C (Scottish law requires 82°C). Reheated food should then be served quickly and never reheated more than once. Check the core temperature of reheated items with a disinfected probe.

Serving food

Food must also be protected when it is being served. Do not keep food unprotected and out of temperature control in service areas longer than necessary. Make sure that all service equipment and surfaces are suitable for food service and are clean. Staff training in food safety is essential for food service staff as well as for those preparing the food.

Measuring temperature

Temperature probes

Electronic temperature probes are extremely useful to measure the temperature in the centre of both hot and cold food. They are also very useful for recording the temperature of deliveries and checking uniformity of food temperatures in fridges. Make sure the probe is clean and disinfected before use (disposable disinfectant wipes or sanitiser sprayed onto kitchen paper are useful for this). Place the probe into the centre of the food, making sure it is not touching bone or the cooking container. Allow the temperature to 'settle' before reading. Check regularly that probes are working correctly. This can be done electronically but a simple and low cost check is to place the probe in icy water – the reading should be within the range -1°C to +1°C. To check accuracy at high levels, place the probe in boiling water and the temperature reading should be in the range of 99°C to 100°C. If probes read outside of these temperatures they need to be repaired or replaced.

Record when temperature probes have been checked for accuracy and store this with food safety records.

▲ Using a temperature probe

Infra-red thermometers

These are also very useful in the kitchen area and give instant readings as they work by measuring radiant energy. They are very hygienic to use as they do not actually touch the food so there is no risk of cross-contamination or damaging the food.

Data loggers

These will record information about the temperatures of refrigeration over a set period. They record highs and lows of temperature in individual or different fridges/freezers and can provide a graph of trends. Systems are available that record all refrigerator, chill units and freezer temperatures in a business and send the information to a central computer.

Activity

Produce an information leaflet for new members of staff instructing them on the use of a food probe. Explain how it is calibrated, when it should be used, how it is disinfected, and instructions of how to use it for testing the temperature of a roast leg of lamb, a roast chicken, individual portions of cannelloni and a reheated 20-portion vegetable moussaka.

HACCP and food safety management systems

For some time it has been considered good practice for all food businesses to have a food safety management system in place. From January 2006 this became law. Article 5 of the Food Hygiene (England) Regulations 2005 gave effect to the EU Regulations and states that:

Food business operators shall put into place, implement and maintain a permanent procedure based on the principles of hazard analysis critical control point (HACCP). Food handlers must receive adequate instruction and/or training in food hygiene to enable them to handle food safety. Those responsible for HACCP based procedures in the business must have enough relevant knowledge and understanding to ensure the procedures are operated effectively.

This gave strength to the Food Standards Agency's commitment to significantly reduce food poisoning cases by 2020.

All systems must be based on HACCP. This is an internationally recognised food safety management system that looks at identifying the critical points or stages in any process and identifying hazards that could occur, i.e. what could go wrong, when, where and how.

Controls are then put in place at the stages where it is essential for intervention to be taken to deal with the risks. These are the **critical control points** – making possible risks safe.

Critical limits are the maximum limits within a process that have been set by management in the HACCP analysis. They must be absolute and measurable. For example, all cooked chickens are tested with a probe at the thickest point of the thigh; it must read no less than 75°C.

Corrective actions – if something goes wrong or is not working properly, for example, not meeting critical limits, do staff know what to do about it? What checks are in place and what should be done to put things right?

Procedures are kept up to date – confirm that they are still working.

Documents and records are kept that show the system is working and is regularly reviewed – a wide range of documents are used as part of the HACCP system.

Advantages of implementing HACCP

- Needed to comply with EU Legislation.
- Risks are reduced – less risk of civil action.
- System is internationally recognised.
- Demonstrates due diligence.
- A proactive system where action can be taken before problems occur.
- Generates a food safety culture, all staff are involved in procedures.
- Less food will be wasted.
- Assists with local authority (enforcement officer) inspections.
- Ensures correct records are kept and documentation is in place.

Prerequisites

Before setting up a new HACCP system, certain prerequisites need to be considered, i.e., what needs to be in place:

- **Suppliers** – these should be approved suppliers and wherever possible they should provide written specifications.
- **Traceability** – systems in place along with suppliers to trace the source of all foods; this will include supplier audits.

- **Premises, structure and equipment** – records that premises are properly maintained. A flow diagram needs to be produced showing the process from delivery to service, avoiding any crossover of procedures that could result in cross-contamination.
- **Storage and stock control** – raw ingredients to finished product. Effective stock control, stock rotation and temperature controlled storage must be in place.
- **Staff hygiene** – protective clothing, hand-washing facilities, toilets, changing facilities need to be provided. A policy for personal hygiene needs to be established with appropriate training on the standards to be achieved.
- **Pest control** – a written pest control policy ideally as part of a pest management system and involving a recognised contractor.
- **Cleaning/disinfection/waste** – a documented system in place that includes cleaning schedules and how waste removal will be managed.
- **Staff records** – records to include all staff training and refresher training with dates completed.

Setting up HACCP

A team of people suitably trained in HACCP procedures will be established to set up the system. If the business is small, just one person may be responsible. There are also a number of specialist HACCP consultancy companies who can complete and monitor the procedure.

The hazards identified and the controls put in place will be essential to food safety and safe production methods, e.g. core cooking temperatures, possible multiplication and survival of bacteria, time food spends in danger zone, cooling food etc.

Food handlers must receive food safety training and effective supervision commensurate with the tasks being completed. Staff training records must be kept.

There must be awareness that physical and chemical hazards could occur at any stage in the process and controls must include these.

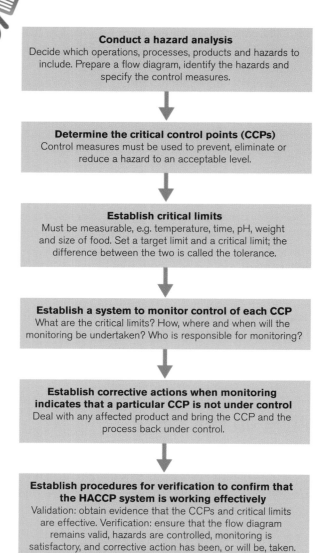

Conduct a hazard analysis
Decide which operations, processes, products and hazards to include. Prepare a flow diagram, identify the hazards and specify the control measures.

Determine the critical control points (CCPs)
Control measures must be used to prevent, eliminate or reduce a hazard to an acceptable level.

Establish critical limits
Must be measurable, e.g. temperature, time, pH, weight and size of food. Set a target limit and a critical limit; the difference between the two is called the tolerance.

Establish a system to monitor control of each CCP
What are the critical limits? How, where and when will the monitoring be undertaken? Who is responsible for monitoring?

Establish corrective actions when monitoring indicates that a particular CCP is not under control
Deal with any affected product and bring the CCP and the process back under control.

Establish procedures for verification to confirm that the HACCP system is working effectively
Validation: obtain evidence that the CCPs and critical limits are effective. Verification: ensure that the flow diagram remains valid, hazards are controlled, monitoring is satisfactory, and corrective action has been, or will be, taken.

Establish documentation and records of all procedures relevant to the HACCP principles and their application
This will be proportionate to the size and type of business. Documentation is necessary to show that food safety is being managed. Managers need records when auditing; enforcement officers and external auditors will also need to see them.

▲ Seven principles of HACCP

The system needs to provide a documented record of the stages all food will go through right up to the time it is eaten and may include purchase and delivery, receipt of food, storage, preparation, cooking, cooling, hot holding, reheating, chilled storage and serving. Once the hazards have been identified, corrective measures are put in place to control the hazards and keep the food safe.

The system must be updated regularly, especially when new items are introduced to the menu or systems change (e.g. a new piece of cooking equipment). Specific, new controls must be put in place to include them.

A flow diagram will need to be produced for each dish or procedure showing each of the stages (critical control points) that need to be considered for possible hazards.

Example: cooking a fresh chicken

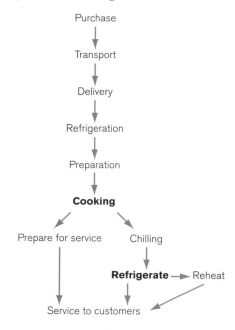

▲ Example of a hazard analysis flow diagram: cooking a fresh chicken

When dealing with the fresh chicken, it is necessary to recognise the possible hazards at all of the identified stages:

Hazard – pathogenic bacteria are likely to be present in raw chicken.

Control – the chicken needs to be cooked thoroughly to 75°C+ to ensure pathogens are killed.

Monitor – check the temperature where the thigh joins the body with a calibrated temperature probe (75°C+), make sure no parts of the flesh are pink and the juices are running clear, not red or pink.

Hot holding – before service the chicken must be kept above 63°C; this can be checked with a temperature probe.

Or chill and refrigerate – chill to below 10°C within 90 minutes. Cover, label and refrigerate below 5°C.

Documentation – temperatures measured and recorded. Hot holding equipment checked and temperature recorded. Record any corrective measures necessary.

For the cold storage of the chicken after cooking/chilling

Hazard – cross-contamination, multiplication of micro-organisms in cooked chicken.

Control – protect from cross-contamination, store below 5°C.

Monitor – check temperature of the chicken is below 5°C.

Corrective action – if the chicken has been above 8°C for four hours or more, it should be thrown away. Investigate why the temperature control has not worked so it can be put right if there is a fault.

Documentation – record temperatures at least once a day and record any corrective action.

Table 2.7 Example of an HACCP control chart

Process steps	Hazards	Controls	Critical limit	Monitoring	Corrective action
Purchase	Contamination, pathogens, mould or foreign bodies present	Approved supplier			Change supplier
Transport and delivery	Multiplication of harmful bacteria	Refrigerated vehicles		Check delivery vehicles, date marks, temperatures	Reject if > 8°C or out of date
Refrigerate	Bacterial growth Further contamination – bacteria, chemicals, etc.	Store below 5°C Separate raw and cooked foods Stock rotation	Food below 5°C	Check and record temperature twice a day Check date marks	Discard if signs of spoilage or past date mark
Prepare	Bacterial growth Further contamination	No more than 30 minutes in 'danger zone' Good personal hygiene Clean equipment, hygienic premises		Supervisor to verify at regular intervals Visual checks Cleaning schedules	Discard if > 8°C for 6 hours
Cook	Survival of harmful bacteria	Thorough cooking	75°C	Check and record temperature/time	Continue cooking to 75°C
Prepare for service	Multiplication of bacteria Contamination	No more than 20 minutes in 'danger zone'	2 hours	Supervisor to verify at regular intervals	Discard if > 8°C for 2 hours
Chill	Multiplication of bacteria Contamination	Blast chiller	90 minutes to below 10°C	Supervisor to verify at regular intervals	Discard if > 20°C for 2 hours
Refrigerate	Multiplication of bacteria Contamination	Store below 5°C Separate raw and ready-to-eat foods	8°C for 4 hours	Check and record temperature twice a day	Discard if > 8°C for 4 hours
Reheat	Survival of bacteria	Reheat to 75°C in centre	75°C (82°C in Scotland)	Check and record temperature of each batch	Continue reheating to 75°C

Due diligence

Due diligence is the main defence available under food safety legislation. It means that a business took all reasonable care and precautions and did everything reasonably practicable to prevent the offence or food poisoning outbreak; that is, 'exercised all due diligence'. To prove due diligence, accurate written documents must be kept and are essential. They may include:

- staff training records with dates and levels of training
- sickness and reported illness, previous employment and if possible post-employment records
- temperature records (delivery, cooking, reheating, holding and cold storage)
- records of food sampling policy and procedure
- details of suppliers and contractors (traceability)
- pest control policy and audits
- cleaning schedules, deep clean reports and monitoring of cleaning
- maintenance schedules and equipment repairs
- visitor and contractor records
- critical control points (CCPs) monitoring activities (only CCPs to avoid excessive paperwork)
- deviations, corrective actions and recalls
- modifications to the HACCP system
- customer complaints/investigation results
- calibration of instruments
- waste management policy including recycling, disposal of glass, oils, chemicals and other relevant items.

Training

Food businesses must ensure that all staff who handle food are supervised and instructed and/or trained in food hygiene commensurate (appropriate) to the work they do. The person responsible for the food safety management of a business must also be responsible for staff training. Appropriate training can take place in-house or with a training provider. All records of staff training must be kept for possible inspection.

Checking that the HACCP system is effective

It is of great importance that the HACCP system must be regularly reviewed and monitored. This will include:

- internal and external audits of the HACCP systems
- regular observation and spot checks by supervisors
- swabbing procedures and recording and acting upon the findings

- advisory information and inspection visits from Environmental Health Personnel
- meetings, briefings, information sessions and feedback with food handling staff
- supervisor and team handover and records (handover book, Safer Food, Better Business, electronic handover systems).

Reviewing and monitoring the food safety management systems and records on a regular basis will ensure the on-going safety of food production from the time suppliers are selected right through production procedures to service of the food. It will also ensure that company policies and legal requirements are being met. Reviews and any corrective actions taken must, like everything else in the process, be recorded and available for inspection.

The supervisor or others should check, and record where appropriate, that controls are being adhered to and are working properly. Many monitoring procedures will be part of the food safety management system, such as monitoring fridge temperatures and adherence to the cleaning schedule. Other methods of monitoring include:

- visual inspections
- checks of food stocks
- checklists
- walk-through checking procedures
- checking that required tasks such as temperature recording are completed.

In some establishments, bacterial monitoring is completed by swabbing surfaces, equipment or processes to establish whether specific pathogens are present; this is referred to as total viable counts (TVCs).

Professional tip

Swabbing kits are available from food safety management/equipment companies for checking and recording the levels of bacteria on a surface or piece of equipment. This can be done at regular intervals and the findings used to improve cleaning and disinfection of areas. The findings also provide a good tool for staff training and meetings.

Auditing is a more formal procedure, taking the form of an inspection. This could be done by an internal auditor (part of the organisation) or an external auditor or consultant, but it should be someone who does not carry out or monitor the day-to-day systems. Auditing is often used to verify that the HACCP or similar system is working properly and recorded recommendations may be made for improvement.

Safer Food, Better Business (SFBB)

The HACCP system described above may seem complicated and difficult to set up for a small or fairly limited business. With this in mind, the Food Standards Agency launched their Safer Food, Better Business system for England and Wales. This is based on the principles of HACCP but in an easy-to-understand format, with pre-printed pages and charts to enter the relevant information such as temperatures of individual dishes. It is divided into two parts. The first part is about safe methods referred to as the 4Cs: cross-contamination, cleaning, chilling and cooking. The second part covers opening and closing checks, proving methods are safe, recording safe methods, training records, supervision, stock control and the selection of suppliers and contractors.

Once the basic information has been recorded, for example, suppliers and staff training, the actual diary pages are very easy to complete and just need confirmation that opening and closing checks have been completed and have been dated and signed. The only other entries in the diary are the recording of problems that have occurred and what will be done about them. If no problems occur that day, nothing needs to be entered. This is called 'management by exception' or 'exception reporting'.

A copy of Safer Food, Better Business is available from www.food.gov.uk.

A similar system called CookSafe has been developed by the Food Standards Agency (Scotland) and Safe Catering in Northern Ireland.

Scores on Doors

This is another strategy that has been introduced by the Food Standards Agency to raise food safety standards and help reduce the incidence of food poisoning. At inspection, a star rating for food safety (ranging from 0 to 5 stars) is awarded based on the following three criteria, taken from the Food Standards Agency's rating system:

- level of compliance of food hygiene practices and procedures
- level of compliance relating to structure and cleanliness of premises
- confidence in management of the business and food safety controls.

It is expected that the **Scores on Doors** scheme will have a lasting positive impact on food safety standards. No matter how good the food in a particular establishment is, few people will want to eat there if the food safety score is low.

The role of the supervisor

Effective planning of food safety is essential to ensure that high standards are maintained, ensure compliance with the 2006 Food Safety Regulations, and to avoid the possibility of food safety-related problems. The supervisor plays a key role in implementing and managing food safety procedures and is an important link between management and the actual food operation. Although supervisors may not be the actual policy-makers, it is likely that they will be involved in devising, setting and managing the day-to-day food safety procedures. This will involve implementing the food safety management system (HACCP or similar) including:

- Overseeing formal and informal staff training including relevant training for HACCP.
- Managing the various temperature controls and recording.
- Putting measures in place to avoid contamination of food and cross-contamination.
- Setting required standards for personal hygiene and requirements for protective clothing.
- Monitoring standards for premises and equipment and safe disposal of waste.
- Monitoring and managing the correct storage of food and rotation of stock.
- Managing cleaning and disinfection of premises and equipment and the proactive control of pests.

Communication

Effective communication of food safety matters to staff is of absolute importance and can be achieved through induction procedures, staff training, supervision, mentoring, information posters, leaflets, film clips, information/training from EHO/EHP, and by making food safety issues part of staff meetings, briefings and handovers. Communication must be planned, on-going and consistent. Where shortfalls or discrepancies in required food safety standards are observed and identified, corrective action must be taken and retraining may be necessary.

Because the supervisor is part of the day-to-day food operation, he/she is the obvious person to communicate food safety matters to business owners, managers, other departments, suppliers, contractors and enforcement officers. Keeping up-to-date and accurate records will help with effective communication.

The importance of effective communication of food safety procedures

- Effective and recorded training forms part of due diligence (this includes HACCP training).
- Staff will be made aware of food safety legislation and their own legal obligations.
- There will be less likelihood of food poisoning/food-borne illness outbreaks.
- Staff will be made aware of the company's procedure and policies.
- Food-handling staff can be made aware of how to eliminate risks or bring them to a safe level.
- Staff ability and food safety awareness will be raised.
- Good working relationships can be established through training.

Food safety training

As well as being a legal requirement, food safety training for staff 'commensurate' to their job roles is essential in any food business. Until suitable and recorded food safety training has been completed, any member of staff handling food must be closely supervised and observed. Specific training sessions, both formal and informal, may be delivered or organised by the supervisor to meet the need of the actual business and specific staff, as well as satisfying legal requirements.

The training methods chosen will depend on the type of business, the activities carried out and the previous training that staff have completed. Training could take place in one or a selection of the following ways:

- Food safety management companies who undertake part or all of the food safety requirements for a business, including training at different levels.
- Partaking in food safety qualifications accredited by awarding organisations such as CIEH, RSPH, City & Guilds, Highfield and others. Courses are available at colleges and universities, through independent training providers, local authorities and adult education centres. Training and testing for these can also take place in the workplace. Qualifications are available at a range of levels, from Level 1 for those new to food handling, through to Level 4 for senior supervisors requiring a high level food safety qualification.
- On-line or computer-package training, often with end tests and certification.
- Food safety training packs, including books, workbooks, DVDs, activities and tests.
- Use of a variety of materials now widely available such as posters, leaflets, films, interactive games and puzzles.
- Training programmes delivered by the EHO/EHP.

Induction

Induction plays a very important role in quickly integrating new members of staff into their role, ensuring that food safety standards are fully met and that new employees completely understand the requirements of their position. Fully recorded, planned and organised induction of staff will form part of the due diligence requirements of the employer and is increasingly becoming a legal requirement. Many organisations such as large hotel groups run centralised induction, presenting the entire company, its policies and procedures. However, even where this takes place, it is still essential to have thorough induction into the actual requirements and the specific procedures and policies involved. Information and topics covered at induction should be revisited regularly to remind staff of company policies and their own responsibilities.

On-going training

Observation and spot checks by the supervisor must be completed regularly to ensure that the required standards for food safety are being met and that the procedures carried out meet legal requirements. Any identified shortfalls in the required standards need to be recorded and the proper corrective action taken.

Planned retraining/refresher sessions are essential at all levels for a variety of reasons:

- Staff will be reminded of the required food safety standards and how these can be achieved.
- Specific company procedures can be reinforced.
- New procedures and processes can be introduced safely.
- Problems can be identified, recorded and addressed.

- Staff can be reminded of their legal obligations and the legal responsibilities of the employer.
- Most accredited food safety qualifications have an expiry date; for example, most Level 2 qualifications need to be updated within three years.

Specific on-going and refresher training is also essential for those involved in setting up the HACCP system and all food-handling staff should be trained in the requirements of HACCP systems and be made aware of its importance.

Test yourself

1. An Environmental Health Officer/Practitioner has issued a Hygiene Emergency Prohibition Notice.
   (a) Why might this notice have been issued?
   (b) What needs to happen immediately after the notice has been issued?

2. UK food safety legislation was amended in 2006. Give three examples of how these changes affect supervision of food safety in a large kitchen.

3. When setting up new food production premises, what water supply/washing facilities will be needed throughout the premises?

4. What are the ways you would advise staff to store and display chilled foods? What are the records you would need to keep relating to these foods?

5. In a large kitchen using a wide range of different foods, what are the procedures that could be implemented and the equipment put in place to reduce possible risks of cross-contamination?

6. In a large staff restaurant/kitchen area the waste containers (bins) need to be replaced. What are the waste bins that will be needed for the whole area, inside and outside? What qualities should you look for?

7. Explain what is meant by the following and what their significance is in a food production area:
   (a) HACCP
   (b) FIFO.

8. You would like all kitchen staff to be aware of the signs that may indicate that there may be a pest problem.
   (a) What are the signs you would tell them to look for?
   (b) What advice would you give them on preventing pests from getting into the building?

9. The following foods have been cooked for a buffet on the following day: roast beef, whole poached salmon, honey-baked ham. These will be served cold. There are also, 20 portions of lasagne, asparagus flan and chicken and mushroom pie. These will be cooled, chilled, then reheated. Write an explanation of how to complete the cooling, storage and reheating safely, including all temperatures and times.

10. A small business is using Safer Food, Better Business.
    (a) What are the advantages to a small business of using this?
    (b) What needs to be recorded in the diary pages?

11. Your kitchen is about to have an external audit and you have been asked for all of the records relevant to due diligence.
    (a) What is meant by due diligence?
    (b) What records should you have available for the audit?

3   Exploring gastronomy

This chapter covers the following units:

NVQ:
→ Contribute to the control of resources.

VRQ:
→ Exploring gastronomy.

Introduction

The aim of this chapter is to introduce you to the subject of gastronomy, which covers an in-depth range of disciplines. It will give you an understanding of the meal and dining experience which is so important in the study of hospitality. It will assist you in menu planning and design and give you an understanding of the customer experience and expectation.

Learning objectives

By the end of this chapter you should be able to:
→ Define gastronomy.
→ Identify the factors that make a good dining experience.
→ Explain the different meal types and dining experiences.
→ Identify the types of beverages that complement different foods.
→ Explain the influence that different cultures/religions, science and technology, changes in lifestyle and the media have had on eating and drinking.
→ Describe the contributions of individuals who have made significant impact on professional cookery.
→ Explain the considerations to take into account when choosing suppliers.
→ Describe the effect geography has on local produce.
→ Explain the impact that the development of transport/transportation has had on food.
→ Contribute to the control of resources.

What is gastronomy?

In very simple terms, gastronomy is the study of how food influences habits, and the relationship between culture and food. It includes the study of how sociology, history, economics, geography, anthropology, marketing, science and technology impact on eating and drinking.

Choosing what to eat is a complex development process, which we learn from childhood and the way we are socialised into food habits through family and relationships. This is how our taste for certain foods is developed.

What is practical gastronomy?

This is when the theories of gastronomy are applied in a practical situation. This includes planning and writing menus and creating a practical dining experience for customers. It involves understanding different types of customers, their background, age, culture, social group and their dietary needs.

Taste

Why do we eat what we eat? Why do we select one dish from the menu in preference to another, choose one particular kind of restaurant or use a takeaway? Why are these dishes on the menu in the first place? Is it because the chef likes them, the customer or consumer wants them, or is this the only food available? What dictates what we eat?

Hospitality reflects the eating habits, history, customs and taboos of society, but it also develops and creates them. You have only to compare the variety of eating facilities available on any major high street today with those that were available just a few years ago. Taste affects food choice and is based on biological, social, economic and cultural perspectives.

Taste results from the stimulation of the cells that make up the taste buds. The taste buds are mostly on the upper surface of the tongue. Different parts of the tongue are particularly sensitive to different primary tastes; an exact map has yet to be created as the facts are inconclusive. The overall taste of food is made up of one or more primary tastes, of which there are five. These are:
- sweet
- sour
- salt
- bitter
- umami.

In the West, salt, sweet, sour and bitter were the known basic tastes. In Japan, they talk of the fifth taste called 'umami' or 'xian', as it is known in China. Asian cuisine is based on umami-rich ingredients. The Chinese have been referring to umami for more than 1,200 years. Umami is found naturally in many foods. It is a combination of proteins, amino acids and nucleotides, which include not only glutamates, but also inosinates and guanylates. When the proteins break down through cooking, fermenting, ageing or ripening, the umami flavours intensify.

In his book *The Physiology of Taste*, published in 1825, Brillat-Savarin makes reference to osmosone, generally considered a forerunner to the concept of umami.

Taste is a complicated issue; scientists continue to research it and refine our understanding of how it works. In recent years, there has been controversy over the concept of umami: some scientists are not convinced of its existence.

Take it further

To find out more about umami, visit the Umami Information Centre website at www.umamiinfo.com

It is important for chefs to assess the quality of dishes by tasting. In this way they learn about flavour and become skilled in blending and mixing different flavour components. Using our senses to evaluate food is called **organoleptic assessment**.

Key term

Organoleptic assessment – using the senses to evaluate food.

The colour of food

The colour of food is important to our enjoyment of it. People are sensitive to the colour of the food they eat and will reject food that does not have the accepted colour. Colouring is sometimes added to food to enhance its attractiveness. There is a strong link between the colour and the flavour of food – our ability to detect the flavour of food is very much connected with its colour. If the colour is unusual, our sense of taste is confused. For example, if a fruit jelly is red it is likely that the flavour detected will be that of a red-coloured fruit, such as raspberry or strawberry, even if the flavour is lemon or banana.

The depth of colour in food also affects our sense of taste. We associate strong colours with strong flavours. For example, if a series of jellies all contain the same amount of a given flavour, but are of different shades of the same colour, then those having a stronger colour will seem also to have a stronger flavour.

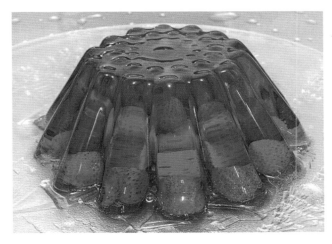

▲ The link between colour and flavour: strawberry jelly

Smell

Because the nose shares an airway (the pharynx) with the mouth, we smell and taste food simultaneously, and what we call the flavour or the 'taste' of the food is really a combination of these two sensations. To quote Brillat-Savarin:

Smell and taste form a single sense, of which the mouth is the laboratory, the nose the chimney or to speak more exactly, of which one serves for the tasting of actual bodies and the other for the savouring of their gases.

With taste and smell, then, we first decide whether a particular food is edible and then go on to sample its chemistry simply to enjoy it.

Our sensitivity to the flavour of food in our mouth is greatest when we breathe out with the mouth closed; air from the lungs passes along the back of the mouth on its way to the nose and brings some food vapours with it.

Temperature

The temperature of food also affects our sensitivity to its taste. Low temperatures decrease the rate of detection; we are most sensitive to the taste of food between 22 and 44°C. Sweet and sour are enhanced at the upper end of that range; salt and bitter at the lower end. At any temperature, however, we are much more sensitive to bitter substances than we are to sweet, sour or salty ones, by a factor of about 10,000. Synthetic sweeteners are effective at concentrations nearer to bitter substances than to table sugar.

It is important when assessing food to remember that taste, smell and colour are closely linked and contribute to the overall assessment of the dish. Training and knowledge are essential to develop a discriminating palate and the ability to identify individual flavours.

Activity: the tasty bit test

1 Take a piece of food (sweet, crisp, biscuit etc.) and bite while holding your breath. (Open your mouth and place the bite on your tongue while pinching your nose, so that you can breathe only through your mouth. If you have good control, you can prevent air from entering your nose by lifting the palate at the back of your mouth while you continue to breathe through your mouth. The requirement by any method is to prevent any air you are breathing out from going through your nose.)

2 With the food on your tongue, you will detect it with some of your senses. If it is sweet, you will be able to detect the sweet taste. Your sense of touch will indicate that the food is resting on your tongue, and whether it is hard or soft, hot or cold.

3 Now unpinch your nose – immediately you will experience the flavour of the food. The sudden whiff of flavour can be surprising. Almost without realising it, you will have let the air shoot from the back of your mouth. Receptors in the nasal cavity stimulate your sense of smell. The faintest whiff of outward breathing will produce this result. You have proved that smell is the major component of flavour.

Factors that make a good dining experience

The factors that affect the meal experience include:

- **Food and drink on offer** – the range of foods, choice, availability, flexibility for special orders and the quality of the food and drink.
- **Level of service** – the level of service should be appropriate to the needs people have at the time. For example, a romantic night out may call for a quiet table in a top-end restaurant, whereas a group of young friends might be seeking more informal service. This factor also takes into account services such as booking and account facilities, acceptance of credit cards and the reliability of the product quality.
- **Level of cleanliness and hygiene** – of the premises, equipment and staff. Recent media focus on food production and the risks involved in buying food have heightened awareness of health and hygiene aspects.
- **Perceived value for money and price** – customers have perceptions of the amount they are prepared

to spend and relate these to differing types of establishments and operations.

- **Atmosphere of the establishment** – composed of a number of factors such as design, décor, lighting, heating, furnishings, acoustics and noise levels, the other customers and the attitude of the staff.

Identifying the meal experience factors is important because it considers the product from the point of view of the customers. Food service operators can get caught up in the provision of food and drink, spend vast amounts of money on design, décor, equipment etc., but ignore the actual experience the customer will have.

Factors affecting how and what we eat

There are many factors influencing our choice of what we eat. These include our individual preferences, our relationships and emotional needs. Other factors such as what is acceptable to us as food, images of food, as well as the needs and preferences of people we are eating with also affect our choice.

The individual

Everyone has individual needs and wishes. Tastes and habits in eating are influenced by three main factors:

- upbringing
- peer group behaviour
- social background.

For example, children's tastes are developed at home according to the eating patterns of their family, as is their expectation of when to eat meals. Teenagers may frequent hamburger or other fast-food outlets and adults may eat out once a week at an ethnic or high-class restaurant, steakhouse or gastro-pub.

How hungry an individual feels will affect their choice of what, when and how much to eat – although some people in the Western world overeat and food shortages cause under-nourishment in poorer countries. Everyone ought to eat enough to enable body and mind to function efficiently; if you are hungry or thirsty, it is difficult to work or study effectively.

Health considerations may influence an individual's choice of food, either because they need a special diet for medical reasons or because of a desire for a nutritionally balanced diet. For example, many people feel it is healthier to avoid meat or dairy products; others are vegetarian or vegan for moral or religious reasons.

What people choose to eat says something about them as a person; it creates an image. Why do we choose to eat what we do when there is choice? One person will perhaps avoid trying snails because of ignorance, because they do not know how to eat them or because the idea is repulsive, while another will select them deliberately to show off to other diners. One person will select a dish because it is a new experience; another person will choose it because they have previously enjoyed eating it.

Relationships

Eating is a necessity, but it is also a means of developing social relationships. Often the purpose of eating, either in the home or outside it, is to be sociable and to meet people, to renew acquaintances, or provide the opportunity for people to meet each other. There may be an occasion (such as birthday, anniversary or wedding) which requires a special party or banquet menu, or it may just be that a few friends choose to have a meal at a restaurant.

Business is often conducted over a meal, usually at lunchtime but also at breakfast and dinner. Eating and drinking help to make work more enjoyable and effective – it is often in canteens, dining rooms and restaurants that relationships develop.

Providing suitable dishes for pupils at school mealtimes in an appropriate environment can be a means of developing good eating habits and fostering social relationships.

Activity

Consider how you use food and have learnt how to use food to build relationships.

Emotional needs

Sometimes we eat, not because we need food, but to meet an emotional requirement:

- For sadness or depression – in eating a meal we may comfort ourselves or give comfort to someone else. For example, after a funeral people eat together to comfort one another.
- For a reward or treat, or to give encouragement to oneself or to someone else – an invitation to a meal is a good way of showing appreciation.

Consumer behaviour

A range of other factors affect why consumers buy certain foods and services, from economics (the consumer's income and the price of the product) to physiology (for example, food intolerance or allergy).

Depending on individual products, dishes, types of restaurants and the circumstances of the consumer, all of these factors will have some bearing on what is selected, the quality, when and how often.

Examining consumer behaviour helps us to understand the relationship between how individuals make decisions about how to spend their available resources (their money, time and effort) on food, goods and services and how producers such as the hospitality industry act to meet or create their needs and wants. For example:

- Who buys what.
- When they buy.
- Why they buy.
- How they buy.
- Where they buy.
- How often they buy.

An example of four basic stages in choosing food is shown in Table 3.1.

Table 3.1 Examples of four basic stages in choice

1	Noticing	I'm hungry
		That looks tasty
2	Choosing	I feel like a snack
		I like that brand/flavour
3	Acting	I'll buy that to eat now
		Just a small portion will do
4	Assessing	I prefer the item I usually buy
		That was good value for money

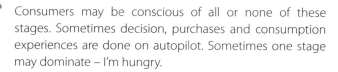

Consumers may be conscious of all or none of these stages. Sometimes decision, purchases and consumption experiences are done on autopilot. Sometimes one stage may dominate – I'm hungry.

Beverages that complement different foods

Matching wine with food

The matching of food and wine is a matter of personal taste – there are no set rules, just some basic guidelines. It is easier to think of wine as a sauce and match the strength of the flavours and weight of the dish with the wine. To achieve the best match, it is necessary to analyse the basic components in both the wine and the food. The idea is to try to balance them so that neither the food nor the wine overpowers the other. Many wine styles evolved to complement the cuisine of the region. Some food and wine combinations work very well together – here are some examples.

Table 3.2 Classic food and wine combinations

Food	Wine
Asparagus	Sauvignon blanc
Consommé	Fino sherry or manzanilla sherry
Foie gras	Sauternes and Barsac
Goat's cheese	Sancerre or Pouilly Fumé
Parma ham and melon	Italian pinot grigio
Roast lamb	Red Bordeaux
Roquefort	Sauternes
Stilton	Port and Madeira
Sushi	Riesling

Some foods are difficult to match with wine. Examples include artichokes, capers, chilli, eggs, fennel, horseradish, olives, spinach, truffles and yoghurt.

Beers and food

Certain specialist beers are suitable to serve with food. It may involve alcoholic strength, malt, character, hop, bitterness, sweetness, richness or 'roastiness'. Combinations often work when they share some common flavour or aroma elements. Foods that have a lot of sweetness or fatty richness (or both) can be matched by various elements in beer. Delicate dishes work best with delicate beers. Carbonation is also effective in cutting richness.

Cultural influences on our choice of food

People's ideas about food and meals and about what is and what is not acceptable vary according to where and how they were raised, the area in which they live and its social customs.

The great variety of cultures in the world each have their own ways of cooking. Different societies and cultures have conflicting ideas about what constitutes good cooking, what being a good chef means and the sort of food a good chef should provide. For example, the French tradition of producing fine food and their chefs being highly regarded continues to this day; whereas other countries traditionally may have less interest in the art of cooking and less respect for chefs.

Individuals' ideas of what constitutes a snack, a proper meal or a celebration will depend on their backgrounds, as will their interpretation of terms such as lunch or dinner. One person's idea of a snack may be another person's idea of a main meal; a celebration for some will be a visit to a hamburger bar; for others, a meal at a fashionable restaurant.

Eating etiquette

The idea of what is 'the right thing to do' when eating varies with age, social class and religion. To certain people, it is right to eat with the fingers, while others use only a fork. Some will have cheese before the sweet course; others will have cheese after it. Some religions do not eat certain foods. Knowledge of this is essential for those working in the catering industry because:

- Tourism has created a demand for a broader culinary experience.
- Many people from overseas have opened restaurants using their own foods and styles of cooking.
- The development of air cargo means perishable foods from distant places are readily available.
- The media, particularly television, have stimulated an interest in world cooking.

Food and celebrations

Food is used in celebrations to convey a range of emotions (for example, love, happiness, joy, satisfaction) in all parts of the world, regardless of culture or religion. Food can strengthen community bonds and help to maintain a common identity amongst groups of people. Different countries use food in different ways to help celebrate special occasions. Tradition may dictate the food that is served, or how it is served – for example, in the United Kingdom, a birthday cake is often decorated with candles.

▲ Roast turkey is often served to celebrate Christmas Day in the UK and USA

Religion and food and drink

Many religions have a particular cuisine or tradition of cookery associated with their culture. In many religions food has symbolism and meaning; for example, bread and wine in Christianity represents the body and blood of Christ; baklava is associated with the fasting month of Ramadan and Eid al-Fitr in the Balkans and Ottoman Empire.

Certain religions place restrictions on eating and drinking habits. For example:

- Hindus do not eat beef or drink alcohol and mainly eat vegetables.
- Muslims do not eat pork or shellfish and do not drink alcohol. They only eat halal meat (requires a Muslim to be present at the slaughter).
- Sikhs do not eat beef or drink alcohol; they only eat meat slaughtered with one blow to the head.
- Jews do not eat pork and only eat fish with fins and scales; meat must be kosher and meat and dairy produce are not eaten together.
- Rastafarians do not consume any animal products except milk; they do not eat canned or processed foods; no salt should be added to their food and foods should be organic.

Activity

Explain why a chef should understand the implication of different cultures when planning menus.

Lifestyle and its influence on eating and drinking

Factors affecting lifestyles

Lifestyles are patterns of behaviour; lifestyle groups are constructed by measuring consumers' activities, interests and options, which affect their food choice. Demographics (for example, age or marital status) are also factors.

The late 20th century and the beginning of the 21st century have seen a change in consumer behaviour. Supermarkets now stock many different types of food, including pre-packed ready meals. As increasing numbers of women have entered and remained in the workplace and have little time to prepare food from scratch at home, pre-prepared food has become more and more popular. Takeaway restaurants make many foods readily available, and eating out as a social activity has increased. (See Table 3.3.)

The growth in the number people going abroad for holidays has led to people tasting new foods and dishes, and when returning home wanting them to be available on the supermarket shelf.

With more types of food readily available, including high quality fine foods, there has been a significant change in eating habits and buying behaviour.

Fashions/fads

Fashions and fads affect our choice of food and it is not always clear if catering creates or copies these trends. Once foods become plentiful and varied, fashion takes over and we are lured towards novelty foods. Television programmes and celebrity chefs popularise certain foods. Restaurants also follow fashion – some types of restaurants become fashionable, especially if they are patronised by celebrities as the place to be seen and the place to go.

Vegetarian diets, nouvelle cuisine, high-fibre diets and cuisine minceur have all become popular recently. All masquerade as healthy: in fact they are nutritionally suspect, but fashionable. Sushi is another food that has become very fashionable.

Table 3.3 Cultural values and their effects on food production

Core cultural values	Food production
A more casual lifestyle with less formality	De-structuring of meal occasions and more individual autonomy over what is eaten, multiple product choices consumed at the same meal time by different people and less formal meals and meal times
Pleasure seeking and novelty – a desire for products and services which make life more fun	Constant innovation and product differentiation in all aspects – taste, texture, portion size, packaging, advertising, branding, product concepts etc. Food as entertainment
Consumerism – increased concern over value for money with rising expectations about quality and performance	Rise in 'grocerant' products – restaurant-style food available to take home, for 'eating out, staying in'. Increase in functional foods, e.g. energy drinks, vitamin-enriched products
Instant gratification – living for today and intolerance of non-immediate availability	Rise in treats, indulgence in luxury items, super premium lines. More convenient access and availability through wider distribution of food
Simplification – a removal of time and energy spent on 'unnecessary' things or tasks	More pre-prepared, pre-packaged, processed and added-value lines for consumption at once or after microwaving to cut down effort in product selection, preparation, cooking and clearing away
Time conservation – time has to be used effectively	Pre-/part-prepared, complete or partially ready meals
Concern with appearance and health, youth, keeping fit and looking good	Expansion/creation of calorie-light product meals. Increase in low fat, low calorie, low salt, high fibre products and substitutes. More product innovation in the area of functional foods and nutraceuticals. Meat reduction and meat substitutes (mycoprotein, soya, tofu). Eat yourself healthy campaign – 'five fruits and vegetables a day', Mediterranean 'superfoods' (e.g. garlic, olive oil, red wine, red peppers, sun-dried tomatoes, pasta, rice, fish and shellfish)

Source: R.K. Proudlove, 2012 *The Science and Technology of Foods* **(Forbes)**

▲ Sushi has become fashionable

Money, time and facilities

How much money an individual has available or decides to spend on food is crucial to their choice of what to eat. Some people will not be able to afford to eat out; others will be able to eat out only occasionally; while for others, eating out will be a frequent event. The money that individuals allocate for food will determine whether they:

- cook and eat at home
- use a takeaway
- go to a pub or restaurant.

The amount of time people have to eat at work will affect whether they use any facilities provided, go out for a snack or meal during their lunch break, or take in their own food to the workplace. People working longer hours has led to an increase in the consumption of ready meals and takeaway food, as people have less time available to prepare and cook food.

The ease of obtaining food, the use of convenience and frozen food, and the facility for storing foods have all led to the availability of a wide range of foods in both the home and catering establishments. It is possible to freeze foods that are in season and use them throughout the year, so eliminating spoilage in the event of a glut of items.

Organic food

Customer demand for 'organic' food has increased. All foods are organic, but the term has become restricted to mean those grown without the use of pesticides or processed without the use of additives.

Vegetarianism

A vegetarian does not eat any meat, poultry, game, fish, shellfish or crustacean, or slaughter by-products such as gelatine or animal fats. Vegetarians live on a diet of grains, pulses, nuts, seeds, vegetables and fruits. They may or may not use dairy products and eggs.

Strict vegetarians will not wish their food to be prepared or served with utensils contaminated with any animal products. Chefs have to take special care in the preparation of vegetarian foods in a kitchen that also caters for meat and fish eaters.

Healthy eating initiatives

Everyone should follow a diet that is rich in fresh fruit and vegetables, fish, wholegrain breads, pulses, rice and pasta, and that is low in fats of all types, especially animal fats. Healthy eating initiatives may affect the types of food people eat. Nutritionists inform us about foods that are good and necessary in the diet, what the effect of particular foods will be on the body and how much of each food we require. This helps to produce an 'image' of food. This image changes according to research, availability of food and what is considered to constitute healthy eating. For more information on healthier dishes, see Chapter 18.

The influence of the media

The media influences what we eat: television, radio, newspapers, magazines and literature of all kinds have an effect on our eating habits.

Nutrition, hygiene and outbreaks of food poisoning are publicised in all forms of media; experts in all aspects of health, including those extolling exercise, diet and environmental health, state what should and should not be eaten.

Information on food packets and advertising influence our choices. Food outlet marketing is designed to target specific groups of people. The media contributes to our knowledge about eating and foods, as do the family, teachers, school, college, and the experience of eating in different areas and different countries.

> **Activity**
>
> Explain how the media affects people's food choices.

The influence of science and technology

Science and technology have had a major impact on food availability, sources and how we store food. With technology it is possible to speed up the ripening process. Fertilisers and pesticides have helped to increase farming yields for crops and prevent the destruction of crops by pests. Intensive farming methods have also increased yields and scientific methods of rearing animals for slaughter have produced a reduction in the fat content of livestock.

Food science has developed food products which are safe to eat with a longer shelf life. The food industry relies on a range of additives, emulsifiers and stabilisers to develop products. Today some chefs have adopted a scientific approach and use these ingredients to enhance their creativity and innovation. Recipe and product development is a continuous process. Customers demand choice and are looking for different styles of food. Food technology is having an impact on recipe development, covering ingredient performance, stability, safety, nutrition, texture, flavour, convenience, ease of preparation and storage. Chilling, freezing and other preservation techniques also continue to be developed.

Food preservation methods

The development of preservation methods has increased the shelf life for many foods. Examples include:

- **Drying** – removing a food's water, which inhibits the growth of micro-organisms (for example, sundried fruits).
- **Sublimation** – the process in which a solid changes directly to a vapour without passing through a liquid phase.
- **Curing** – salting, crystallisation. Preserving through the use of salt and drying; sugar, spices or nitrates may also be added.

- **Fermentation** – culturing or fermenting a food involves the chemical process of breaking down a complicated substance into simpler parts, usually with the help of bacteria, yeasts or fungi. Fermented food is considered a 'live' food and the culturing process continues during storage to enhance the food's nutrient content. During fermentation the bacteria also produce B vitamins and enzymes that are beneficial for digestion. Throughout Asia, vegetables are commonly fermented. In North America cucumbers, olives and cabbage are often preserved by fermentation.
- **Pickling** – using vinegar to preserve foods.
- **Edible coating** – adding a thin layer of edible material such as natural wax, oil or petroleum-based wax, which acts as a barrier to gas and moisture.
- **Canning** – foods are processed at high temperatures sealed in the can. This provides a long shelf life from one to five years; the deterioration of the product after that is normally because of chemical reactions within the can, not bacterial contamination.
- **Bottling** – normally preserves food by acidity and pasteurisation; used for pickles, chutneys and fruits.
- **Sterilisation** – the elimination of all micro-organisms through extended heating at high temperatures.
- **Refrigeration** – slows down the biological, chemical and physical reactions leading to deterioration.
- **Freezing** – makes water unavailable to micro-organisms. The chemical and physical reactions leading to deterioration are slowed by freezing.
- **Pasteurisation** – kills off pathogenic bacteria, yeasts and moulds. Liquids are heated to 63°C for 30 minutes or 72°C for 15 seconds.
- **Ohmic heating** – an electric current is passed through food, generating enough heat to destroy micro-organisms.
- **Irradiation** – foods are treated with low doses of gamma rays, x-rays or electrons. The energy absorbed by the food causes the formation of short-lived molecules known as free radicals, which kill bacteria that cause food poisoning. They can also delay fruit ripening and help stop vegetables, such as potatoes and onions from sprouting. The Food Irradiation (England) Regulations 2009 set out the requirements for producing, importing and selling of irradiated food in the UK. High pressure processing of foods inactivates food-borne micro-organisms at low temperatures without the use of chemical preservatives.
- **Pascalisation** – utilises ultra-high pressures to inhibit the chemical processes of food deterioration.
- **Ozonation** – ozone is an oxidising agent. It is an effective disinfectant and sanitiser for many food products.

- **Aseptic and modified atmosphere packaging (MAP)** – hermetically sealed foods that have been surrounded by gases that slow down the deterioration of food, then packaged airtight by a commercial sealing process.

▲ The RADURA symbol indicates that food has been irradiated

Use of additives

There has been an increase in the use of food additives. Additives are chemicals (both synthetic and natural); they are used to give various functional properties to foods. Some additives are very widespread in nature, for example, pectin and ascorbic acid (vitamin C); others come from specific sources, for example, gums from certain seaweeds.

Additives in a food can alter its:

1 physical characteristics
2 sensory characteristics, such as flavour, texture, colour
3 storage life (preservatives include sorbic acid, sulphur dioxide, potassium nitrate, benzoic acid)
4 nutritional status.

Some additives can fulfil more than one function, e.g. thickeners such as starch provide a food with both its physical and sensory characteristics. Vitamin C is an antioxidant, but is also a vitamin so gives nutritional value.

Emulsifiers and stabilisers

These are important in many recipes and food products as without them foods would become unstable and separate out into watery and fatty layers. Examples of emulsifiers include lecithin, pectin and vegetable gums. Examples of stabilisers (starch) include gums and gelatine.

Antioxidants

Over time, fats and oils in foods may be affected by oxygen, turning the food rancid. Antioxidants are added to food to slow down the process of rancidity (off flavours).

Common antioxidants include ascorbic acid (vitamin C), propyl gallate and butylated hydroxyanisole.

Genetically modified (GM) foods

Scientists can create plants that nature itself has never created – plants that are resistant to chemicals that kill weeds (herbicides), plants that produce chemicals to kill insects (pesticides) and plants that last longer after harvesting.

The methods used to produce these new crops involve changing or modifying the crop's genes. Genes are contained in the cells of all living things; they guide how living things are made and how they function. They act as codes for different traits such as the size or colour of fruit. The desired gene is chosen and isolated – this might be, for example, a gene that makes strawberries sweeter by producing more sugar. Cells of the crop plant are modified and then grown into plants with the gene present.

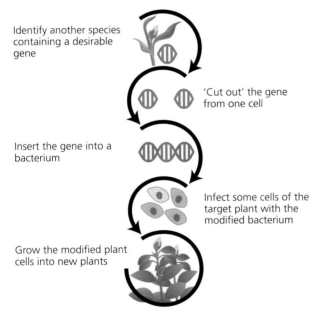

Identify another species containing a desirable gene

'Cut out' the gene from one cell

Insert the gene into a bacterium

Infect some cells of the target plant with the modified bacterium

Grow the modified plant cells into new plants

▲ One of the most common processes of genetic modification of food crops

The most common GM crops grown at the moment are those that are resistant to herbicides. The second most common are the crops that can kill pests. A bacterium called *Bacillus thuringiensis* (Bt) produces a toxin that kills insects but is harmless to people. Putting the Bt toxin gene into maize plants allows them to make their own poison, which kills a crop pest called the corn borer.

In future, GM crops that grow in poor, dry or salty soils may be developed. This would make huge areas of worthless land productive. Potatoes modified to contain less starch could make healthier chips because they would not absorb so much fat in cooking. GM vegetables produced with added nutrients may help to fight off heart disease and cancer.

There is some opposition from various groups and individuals to modifying food crops in this way.

Molecular gastronomy

The development of molecular gastronomy in modern cuisine is discussed in Chapter 17.

A short history of gastronomy

The Ancient Greeks

Among the Ancient Greeks there were many famous people, but because Greece never had any prime beef, rich butter or fresh cream (because of its climate and soil being unsuitable for grazing), they recorded very little about food and wine. The main meal of the day for Ancient Greeks was the evening meal and it could last for hours. It was an occasion for reunion and recreation.

The Ancient Romans

In Ancient Rome, meals were at first simple but once they conquered the known world, they adopted the culinary methods of the countries they occupied and introduced many delicacies into Italy. When the Roman Senate was divided on whether or not to embark upon the Third Punic War, Cato the Elder silenced the protests against war partly by producing some fine specimens of African figs, which were evidently considered well worth fighting for. Besides figs, Africa gave the coarse ribbed melon to the Romans, who cultivated it at Cantaloupe, hence the name by which it is still known today.

The Romans preferred boiling and stewing their meat rather than smoky, fatty roasting.

They had a taste for rich and well cooked food, sweet-and-sour sauces made of honey, fruits, vinegar, garum (nuoc mam) seasoned with herbs (such as coriander, mint, oregano), spices and flour. Common spices included pepper, ginger, asafoetida, and cardamom.

They only used olive oil in cooking: recipes listed it just as oil. The Romans designated places of origin for olive oil, as the French now do for wine. For Apicius, the first Roman cook to record recipes, the best olive oil came from Liburnia (modern day Croatia). He even gives a recipe for transforming simple olive oil from Spain to mimic that from Liburnia.

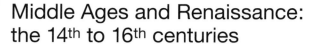

Middle Ages and Renaissance: the 14th to 16th centuries

Medieval cuisine started to diverge from Roman principles. Vegetables were eaten less, bacon and lard replaced olive oil. Spices became more varied.

Meat became the central part of the meal (replaced by fish on the 150 days of fasting per year when meat was prohibited by the Church). Meat and fish were roasted or boiled, served with a light acidulated sauce. It was also common to cook them in a tart or sweet-and-sour sauce with verjuice or vinegar, sugar or fruits.

Spices, a sign of luxury and good nutrition, were preferred to herbs. Common spices included cinnamon, ginger, galangal, cloves, nutmeg, cardamom and saffron.

The cooks of these times were Taillevent, Muestro, Martino and Maître Chiquart. Over a hundred books on cuisine were written in Europe between the 14th and 16th century.

The 17th and 18th centuries

Medieval and Renaissance tastes remained popular in Europe but French cuisine distinguished itself by rejecting these tastes. The cooks of the *Grand siècle* promoted natural flavours, undercooking to respect the product, and developed some of the standard culinary preparations: stocks, meat juices and coulis. Roux and emulsified sauces (beurre blanc, hollandaise) arrived. Béchamel sauce was invented in 1735.

- In 1651, François Pierre dit la Varenne wrote *Le Cuisinier François* – he also invented puff pastry.
- In 1656 Pierre de Lune wrote *Le Cuisinier* – he was credited with inventing the bouquet garni and roux sauce.
- In the 18th century more cookbooks were published that promoted a new classical cuisine.
- Vincent La Chapelle published, in English, *Le Cuisinier Moderne* (1733).
- Menon popularised la nouvelle cuisine by publishing three tomes: *Nouveau Traité de la Cuisine* (1739), *La Cuisinière Bourgeoise* (1746) and *Les Soupers de la Cour* (1755).

During this period the dining room began to become a permanent feature of the house.

Cuisine classique: the 19th and 20th centuries

Cuisine classique is a mix of aristocratic and French bourgeois cuisine. This type of culinary standard developed in Europe during the 19th century up to the beginning of the 20th century. Cuisine classique is the traditional cuisine with which we are familiar. It is a mix of advanced recipes and local dishes.

- During the 18th and 19th century, luxury hotels were developed and with them grand restaurants.
- The first food writers emerged; these included the most famous – Brillat-Savarin and Grimod de la Reynière

The 19th century is known as the Golden Age of French gastronomy. The first celebrity chefs spread the influence of French cuisine internationally. Antonin Carême codified French cuisine in his work which culminated in his masterpiece *L'Art de la Cuisine Française au XIXe Siècle* (1833).

Almost a century later, Auguste Escoffier modernised Carême's work in his *Guide Culinaire* (1903) and reorganised the workforce in the kitchens of the luxury restaurants and hotels all over Europe. Escoffier created the division of kitchen labour, known as the Partie System. Escoffier worked at The Savoy and The Carlton in London, The Ritz in Monte Carlo and Lucerne. Service techniques changed cooking methods and methods for preservation: cork stoppers were introduced in 1700, corkscrews in 1750, the freezer in 1845, pasteurisation in 1860 and the refrigerator in 1933.

The invention of the motor car led to the concept of food tourism and popularised regional food.

During the latter part of the 20th century, restaurants thrived, cookbooks flourished (over 500 cookbooks were published during the last 20 years of the 20th century).

Nouvelle cuisine

The term 'nouvelle cuisine' was first introduced in 1973 in an article entitled *Nouveau Guide Gault et Millau – Vive la nouvelle cuisine français*. Ten Commandments for nouvelle cuisine were declared, the most important of which were a rejection of long cooking times, heavy sauces, spices and marinades that mask the natural flavours of the food stuffs. New techniques were embraced, food pairing conventions challenged (red wine/red meat, white wine/fish, lamb/beans) and new products, cooking techniques and presentations were welcomed. In 1976, over 100 chefs were listed in the guide under the school of nouvelle cuisine.

Nouvelle cuisine from 1970 to 1980 did not break completely from the cuisine classique. While Bocuse and his contemporaries promoted la cuisine du marché, respect for the product and lightly cooked sauces, their recipes were still similar to those found in cuisine classique.

The discovery of Japanese cuisine by these chefs helped them further transform cuisine into something new: they reduced cooking time, paid more attention to the dietary needs of diners, emphasised artistic dish presentations and described dishes in their menus with extensive details on the provenance of ingredients.

A breakthrough in gastronomy was led by European chefs born in the 1950s and 1960s: Marc Veyrat developed emulsions, replaced gravies with infusions and introduced wild herbs from the Alps to the world of gastronomy.

In the 21st century, working conditions in professional kitchens have become more chef-friendly and ergonomic and there has been a rise of women to the rank of great chefs (for example, Hélène Darroze, Anne-Sophie Pic and Reine Sammut).

For much of the history information in this chapter, the authors are indebted to *A History of Gastronomy* by Andre Simon and *A History of European Cuisine and Gastronomy* by Bruce Lee.

The supply and use of commodities

The effect of geography on local produce

Indigenous plants and animals are often the main sources of local food. What grows locally depends on the climate, the soil and the terrain. Climate change will affect what can be grown locally and will also affect yields. Climate and rainfall are incredibly important to agriculture: too much rain or too little rain can seriously affect the overall yield.

Rivers can be a rich source of fish which may be farmed locally in rivers and lakes. A country surrounded by sea is often rich in readily available fresh fish.

Considerations when choosing suppliers

Most companies find that 80 per cent (by value) of their purchases of goods and services come from 20 per cent of their suppliers. Suppliers of food include:

- Producers – for example the farmers who sell their produce direct through farm shops or to supermarkets or large catering companies.
- Wholesalers – suppliers who buy from the producers, food manufacturers, and so on and sell to the caterer.

Cash and carry wholesalers are an example of this type of operation, as are produce markets such as Billingsgate.
- Retailers – large retail supermarkets stock a wide range of ingredients. This type of buying may be suitable for a small restaurant operation.

Activity

List the local suppliers in your area for fresh fruit and vegetables.

The selection of suppliers is an important part of the purchasing process. It is important to consider how a supplier will be able to meet the needs of your operation. Consider:
- price
- delivery
- quality/standards.

You may obtain information on suppliers from other purchasers. Buyers should visit suppliers' establishments. When interviewing prospective suppliers, you need to question how reliable a supplier will be under competition and how stable under varying market conditions.

The buyer

This is the key person who not only makes decisions regarding quality, amounts required, price and what will satisfy the customers, but also what will make a profit. The wisdom of the buyer's decisions will be reflected in the success or failure of the operation. The buyer must be knowledgeable about the products and must have the skills to deal with sales people, suppliers and other market agents. The buyer must be prepared for hard and often aggressive negotiations.

The responsibility for buying varies from company to company according to its size and management policy. Buying may be the responsibility of the chef, manager, storekeeper, buyer or buying department.

A buyer must understand the company's internal organisation, especially the operational needs, and be able to obtain the products needed at a competitive price. Buyers must also acquaint themselves with the production procedures and how these items are going to be used in the production operations, in order that the right item is purchased. For example, the item required may not always have to be of prime quality, for instance, tomatoes for use in soups and sauces.

A buyer must also be able to make good use of market conditions. For example, if there is a glut of fresh salmon at low cost, has the organisation the facility to make use of extra salmon purchases? Is there freezer space? Can the chef make use of salmon by creating a demand for the product on the menu?

Knowing the market

Since markets vary considerably, a buyer must know the characteristics of each market.

A market is a place in which ownership of a commodity changes from one person to another. This exchange of ownership could occur while using the telephone, on a street corner, in a retail or wholesale establishment, or at an auction.

It is important that the buyer has knowledge of the items to be purchased, such as:

- where they are grown
- seasons of production
- approximate costs
- conditions of supply and demand
- laws and regulations governing the market and the products
- marketing agents and their services
- processing
- storage requirements
- commodity and product, class and grade.

Buying and negotiation

Buying involves selecting suppliers and negotiating terms. The buyer will need to understand the importance of quality, as well as securing value for money. Other important aspects include research and seeking out local, regional and national suppliers.

 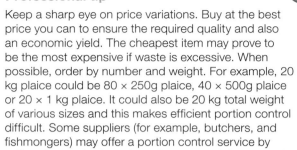
Follow up and expediting

The chef or purchaser needs to develop a good working relationship with their suppliers. They must have close contact with quality management, sales planning and sales support staff and will be involved with delivery and quality targets, as well as detailed specifications (traceability) – see also Chapter 2.

Purchase research

Purchase research is undertaken by an individual or group that looks beyond the day-to-day requirements and considers alternative ingredients, sources of supplier, long-term price trends or at improving supplier appraisal. Those involved must have up-to-date knowledge of future marketing plans and menu design.

Centralised or decentralised purchasing

Whether you decide to centralise or decentralise, the purchasing procedure will depend on the size of the business. Well managed centralised buying reduces costs by

- consolidating quantities to achieve lower prices from your suppliers
- reducing overall stock levels through using central facilities.

Centralisation should avoid duplication of administrative effort at each site, thereby reducing errors and bringing further savings. Equally important is that centralisation will often allow a buyer to specialise.

A purchasing consortium operates a similar principle as centralised buying. Purchasing consortiums are organisations set up to negotiate prices for goods and commodities for hospitality companies to enable them to obtain the best possible price for the desired quality.

Decentralised buying is about local control where:
- Sites/operations are some distance apart.
- Purchasing is bound up with other functions that are part of a specialist unit requiring different types of ingredients.
- Corporate culture needs people at the sharp end of the operation to have maximum control, where the chef responsible for the ordering reports directly to the client he or she is serving.

In a small company where there is a need to purchase only a limited number of commodities and ingredients, the purchase department may well be the chef or one person and an assistant. In a large organisation with several hundred employees, however, it would not be unusual to have ten or more people in the purchasing department. The overall size of the firm in turnover terms can be a major factor, as can the total value of the purchases. The main determinants tend to be the type of business, the complexity of the ingredients and the commodities.

Activity

1 List the advantages of centralised buying to a large organisation.
2 Why do you think many chefs prefer local, decentralised buying?

Professional tip

Compare purchasing by retail, wholesale and contract procedures to ensure the best method is selected for your own particular organisation.

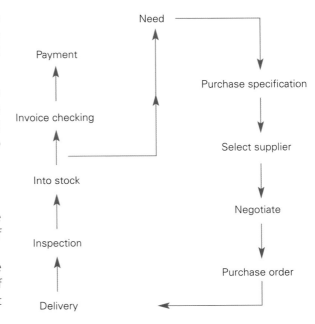

▲ The purchasing cycle

Buying methods

Buying methods depend on the type of hospitality operation and the market they serve. Purchasing procedures are usually formal or informal. See Table 3.5.

Table 3.5 Informal and formal methods of buying

Informal buying	This usually involves oral negotiations, talking directly to sales people, face to face or using the telephone. Informal methods vary according to market conditions.
Formal buying	Known as competitive buying, formal buying involves giving suppliers written specifications and quantity needs. Negotiations are normally written.

Both methods of buying have advantages and disadvantages. Informal methods are suitable for casual buying, where the amount involved is not large, and speed and simplicity are desirable. Formal contracts are best for the purchase of commodities over a long period of time; prices do not vary much during a year, once the basic price has been established. Prices and supply tend to fluctuate more with informal methods.

Professional tip

Organise an efficient system of ordering, ensuring that you keep copies of all orders for cross-checking, whether orders are given in writing, verbally or by telephone.

Principles of purchasing

A menu dictates an operation's needs for commodities; based on the menu, the buyer searches for a market that can supply the company. After locating the right market, the buyer investigates the various products that are available and may meet the company's needs. The right product of the right quality must be obtained to meet those needs. Other factors that might affect food production include:

- type and image of the establishment
- style of operation and system of service
- occasion for which the item is needed
- amount of storage available (dry, refrigerated or frozen)
- finance available and supply policies of the organisation
- availability, seasonality, price trends and supply

In making your choice as a buyer, you should also take into account the skill of the employees and the chefs who work at the establishment. You should consider the condition of the product and the processing method for which it is to be used, the suitability of the product to produce the item or dish required, and the storage life of the product.

Methods of choosing suppliers

Visit local producers, farms, dairies, factories. Speak to other chefs in the area; visit other restaurants, taste their dishes and assess what they have to offer. Use the internet to research ingredients and suppliers. You may also find information about ingredients and suppliers from:

- Directories – trade organisations produce directories of suppliers.
- Company representatives – invite company representatives to see you and ask them to bring samples of their produce.
- Trade shows – attend local and national trade shows, where there are always suppliers present. Country and regional shows are also good venues to meet producers and growers.
- The media – always read local newspapers. Listen to local radio and farming news. Refer to trade press, where there is always information on new products. Often these products are tested and evaluated by specialist panels of consumers.
- Recommendations – from other chefs, food and beverage managers. Word of mouth is one of the best ways to get reliable information about suppliers.
- The trade press – *Restaurant Magazine, Caterer and Hotelkeeper, Cost Sector Catering, The Grocer, Food Manufacturing*. These are examples of trade magazines where you are able to get information on suppliers and new products.

Quality assurance

Quality assurance of suppliers involves being confident that you will receive the goods you want, performing as specified at the time you want them. When evaluating suppliers, consider for example:

- Are they financially stable?
- Do they have the ability to deliver?
- Do they understand what you want?
- Are they easy to deal with?

How good are your existing suppliers? Formal measurement is called 'vendor rating', which gives a good guide in helping to review the purchasing base.

Setting up a vendor rating system involves:

- writing down the (preferably measureable) qualities you expect from a good supplier
- recording performance
- looking at results from time to time
- acting on them.

Table 3.6 Example of a vendor rating system

	Maximum	Achieved
Quality achievement	20	20
Price	20	5
Delivery on time	20	20
Packaging temperature	20	12
Efficient paperwork	10	8
Good communication	10	6
Total	100	71

Source: Defra, *Putting it into Practice*, 2008. Reproduced under the terms of the click-use licence.

The above table shows that problems can relate to the supplier's price, packaging, paperwork, and how the purchaser should measure these objectively.

Quality can be measured at goods inwards. Price may be measured from competitive tenders, delivery performance from orders and delivery notes and so on.

These assessments should be done on a regular basis – at least every six months. Inform suppliers how they are being measured. If improvements are required, set them a time limit in which to improve.

Professional tip
- Buy perishable goods when they are in full season, as this gives the best value at the cheapest price.
- To help with purchasing the correct quantities, it is useful to compile a purchasing chart for 100 covers from which items can be divided or multiplied according to requirement. An indication of quality standards can also be incorporated in this chart.

How many suppliers can you afford?

It takes time to research a supplier, to establish relationships and to set up the purchasing system. In order to get the best out of your main suppliers, you will need to develop and educate suppliers about your needs, listen to any suggestions they have, track their performance and correct any problems.

One approach is to move towards single sourcing. This involves bringing together the requirements from all parts of the business in order to make larger, more regular orders from fewer suppliers. Where you purchase the same ingredients from two or more suppliers, consider choosing one main supplier. In other areas, where purchases are small in value, think about buying from a distributor who handles a range of goods. With fewer suppliers you can afford more time for each one. They will respond well to more personal contact and the possibility of larger orders. The danger of this approach is that if you have fewer suppliers, you are more dependent on their performance. Some hospitality companies do not adopt this approach, as it is considered to be too high a risk.

Professional tip
Keep the number of suppliers to a minimum but have at least two suppliers for every group of commodities, when possible. The principle of having competition for the caterer's business is sound.

Building relationships with suppliers

Areas to investigate with a supplier are:
- Where do they source goods from at the moment? For example, country of origin, farms and so on.
- What do they buy in season? Most fruit and vegetable wholesalers will be able to show you their 'buying plan', which will show you where and when they buy in the UK and from abroad. When fresh produce is in season, it is at its most abundant and therefore at its cheapest.

- What are the production strengths of your region? Some regions are better at growing things than others. For example, Sussex has a good reputation for its South Downs lamb, salad production and fresh sweet corn, whereas Kent is known for apples, soft fruit and brassicas. As with seasonality, local abundance makes these items more affordable.
- Can they supply more fresh items? Using frozen goods increases the likelihood that the goods in question have travelled a considerable distance within the UK or have come from abroad. Blast freezing also uses a considerable amount of energy and storing frozen goods is less energy-efficient than storing chilled fresh goods.
- What are the delivery arrangements? Can you, the purchaser, manage with less frequent deliveries?
- Are they willing to arrange producer visits and visit your team at work? Supplies are an important part of the food culture in your organisation and seeing how food is produced first hand is something that most teams find inspiring.

Assessing commodities and ingredients

When deciding to purchase a commodity or ingredient, it is important to assess whether it is suitable for your needs and whether it will satisfy the demands of your menu. Ensure that you check:
- Quality and flavour – is it affordable? Is it cost effective? Will it give the number of portions required? Will it give the yield required?
- Terms of supply – what are the terms of supply, for example, delivery times, payment requirements and so on?
- Supply meets demand – is the supplier able to regularly supply to your requirements? Can they supply the quantity required on a regular basis?
- Hygiene, hazard analysis critical control points – visit the supplier before committing to purchase. Inspect the supplier's premises for hygiene. Do they have an HACCP policy? How effective is the policy? Do they have appropriate records?
- Supplier's reputation – who are they already supplying to? What do these establishments think of the supplier and their produce? Are they reliable? What type of packaging do they use? How efficient are they?

Ethical considerations when choosing suppliers

When choosing suppliers, it is important to consider what kind of production methods are used to produce the commodities; for example, free range farming or factory farming. What state are the animals kept in and how well are they looked after? How energy efficient are the production methods? How green are the transport methods used? How are the supplier's employees treated? Do they have an equal opportunity policy, a health and safety policy and a welfare policy for staff? What are their employees' working conditions like? Do they receive a fair wage?

▲ Are your meat and dairy products produced to high animal welfare standards?

Sustainability

This is ensuring that food is purchased, consumed and prepared with as little impact on the environment as possible, for a fair price, and which makes a positive contribution to the local economy.

Simple guidelines to follow when purchasing food:
- Use local, seasonal and available ingredients as standard to minimise transport, storage and energy use.
- Specify produce from farming systems that minimise harm to the environment such as certified organic.
- Limit foods of animal origin such as meat, dairy products, eggs, as livestock farming is one of the most significant contributors to climate change. Promote meals rich in fruit, vegetables, pulses and nuts.
- Ensure that meat, dairy and egg products are produced to high environmental and animal welfare standards.
- Exclude fish species identified as most 'at risk' by the Marine Conservation Society and specify fish only from sustainable sources.
- Buy Fair Trade certified products and drinks imported from poorer countries to ensure a fair deal for disadvantaged producers.
- Avoid bottled water. Serve plain, filtered tap water. This will minimise transport and packaging waste.

Source: Defra, *Putting it into Practice*, 2008, reproduced under the terms of the click-use licence

The charity LEAF (Linking Environment and Farming) promotes sustainable farming and aims to bring together farmers, the food industry and consumers. See www.leafuk.org

Activity

Design a poster or leaflet to explain a sustainable food policy.

Free trade

This means governments have to treat local and foreign producers in the same way, for example, by not creating barriers to importing goods, services or people from other countries, or giving national businesses and farmers an advantage over foreign firms by offering them financial support. In practice, truly free trade has never existed and the reduction of trade barriers is always subject to intense political negotiation between countries of unequal power.

Ethical trade

This involves companies finding ways to buy their products from suppliers who provide good working conditions and respect the environment and human rights.

Fair trade

This encourages small-scale producers to play a stronger role in managing their relationship with buyers, guaranteeing them a fair, financial return for their work. Some corporate buyers help to set up schools and health centres on the farms in countries where the food is produced.

Organic

Organic farming requires farms to restrict the use of pesticides, avoiding the use of chemical fertilisers. It creates living soil, full of life with microorganisms, fungi, worms and termites. The soil is rich in macro and micro elements, trace elements and organic matter. It also relies on crop rotations, crop residues, animal manures and the biological system of nutrient mobilisation and plant protection.

Organic farmers must follow high standards of animal welfare. The Soil Association certifies organic inspection. All certifying bodies must comply with European Organic Regulations.

Red Tractor Assurance

Suppliers and farmers sign up to this whole-chain assurance scheme, which provides responsible production standards covering the environment, food safety, animal welfare and traceability. UK farmers and suppliers who sign up to the scheme are able to use the Red Tractor logo to promote their produce, which indicates that food has met these standards and provides an independent sign of origin.

▲ The Red Tractor logo

Free range

Refers to food produced from animals that have access to outdoor spaces. Usually free range also means that animals have free access to graze or forage for food. Free range, unlike organic, is not a certification. Organic food is free range, meaning animals must have access to pasture. Free range food doesn't have to meet any legal requirements.

Slow Food

Slow Food is a movement that links the pleasure of food with a commitment to community and the environment. It is a not-for-profit organisation which promotes a better way to eat, reconnecting people with where their food comes from and how it is produced so they can understand the implications of the choices they make about the food they put on their plates. It means encouraging people to choose nutritious food from sustainable local sources. See the Slow Food website www.slowfood.org.uk

Sustainable fishing

Consumers are becoming increasingly concerned about sustainability and sources of supply. Good restaurants should stop purchasing fish from over-fished stocks or badly managed fisheries.

Always obtain information from your fish supplier about:

● the source of the fish
● how it was caught.

Some fishing methods are environmentally damaging and some fishing practices devastate the marine environment and include bottom trawling, by catch. For more information see Chapter 10.

> **Take it further**
>
> You can download a list of fish to avoid from the Marine Conservation Society at www.fishonline.org/fish-advice/avoid

Activity

1 Ask all the team in the kitchen to review the menu to see what changes can be made to favour more seasonal produce.
2 Organise a visit to a farm and producer in your local area whose produce you use so that you can understand the products better. Take photographs to promote the food within your organisation and local press.
3 Organise promotions that reflect seasonality or celebrate your success in obtaining local produce. Celebrate seasonal produce with special menu items, new recipes to celebrate a new ingredient or new source of supply.
4 Get the chefs to sit down together and eat seasonal food, to learn about food and ingredients.
5 Invite the producer or farmer to explain how the food is produced.

The impact of transportation on food

Food miles

Food miles are the distance that food travels from producer to consumer. In the UK, for example, food travels an amazing 30 billion kilometres each year. This includes imports by boat and air, and transport by lorries and cars. You can, for example, buy asparagus and strawberries all year round thanks to refrigeration, heated greenhouses and of course global food transportation.

The problems with products that have a long supply chain include their contribution to climate change, compromised animal welfare standards and a deeper industrialisation of food and food culture. By importing food, we generate large amounts of CO_2, contributing to global warming. The direct social, environmental and economic costs of food transport are estimated at over £9 billion each year.

▲ Air transportation of food and other cargo

But shortening food miles can neglect the social and economic benefits associated with trade in food, especially for developing countries. Exporting countries like Africa use the money from the crops they sell to build schools and communities. This has transformed communities so they wish to continue growing and exporting food to the West. And while there is an emphasis on rediscovering local seasonal produce, people still want diversity and choice.

The relationship between food and sustainable development is complex and food miles are just one variable. However, food miles are important as they capture a wide range of concerns about the food system.

Measuring the success of sustainability

Key success indicators for sustainability are:
- percentage of food sourced locally, nationally and abroad
- decrease in food wastage
- reduction in food miles
- financial contribution to the local economy
- increase in recycling
- increase in food sales.

Controlling resources

Monitor and control receipt of goods

Staff responsible for receiving goods should be trained to recognise the items being delivered and to know if the quality, quantity and specific sizes, and so on, are those ordered. This skill is acquired by experience and by guidance from the departmental head, for example, the head chef, who will use the items.

Purchasing specifications detail the expected standards of the goods to be delivered. However, the chef, supervisor, storekeeper or whoever is responsible for controlling receipt of goods, needs to check that the specification is adhered to. If the establishment does not have purchasing specifications, a check must still be made when goods are delivered that the expected standards have been achieved. In the event of goods being unsatisfactory, they should not be accepted.

Receipt of goods

Receipt of deliveries must be monitored to ensure that goods delivered correspond with the delivery note and there are no discrepancies. It is essential that items are of the stipulated size or weight, since this could affect portion control and costing. For example, 100g fillets of plaice will need to be that weight and melons to be used for four portions should be of the appropriate size.

It is necessary to make effective control possible. This means that:
- Delivery access and adequate checking and storage space are available.
- These areas are clean, tidy and free from obstruction.
- Staff are available to receive goods.

It is important that the standard of cleanliness and temperature of the delivery vehicles are also satisfactory. If these are not up to the required standard, the supplier must be told at once.

Temperature

Vehicles weighing over 7.5 tonnes must have an internal temperature of 5°C or below when being used to deliver food outside their locality. For local deliveries of food, the temperature of vehicles under 7.5 tonnes should be 8°C or under. On receipt, goods should be transferred as soon as possible to the correct storage area (see Chapter 2).

> **Professional tip**
>
> Deliveries must all be checked against the orders given for quantity, quality and price. If any goods delivered are below an acceptable standard, they must be returned, either for replacement or credit.

Control storage of goods

Frozen items should be stored at the optimum temperatures, as shown in the table below.

Table 3.7 Optimum storage temperatures for frozen items

Commodity	Temperature (°C)
Meat	-20
Vegetables	-15
Ice cream	-18 to -20

Refrigerator temperatures should be 3–4°C and larders provided for cooling of food should have a temperature of no higher than 8°C.

Every place of work will have a security procedure to ensure that goods are stored safely. The business needs a system for reporting non-compliance with the procedure. It is important that staff are aware of the system and who they should report any deviation to.

If control of stock is to be effective, there should be ample storage space with adequate shelving bins, and so on, to enable the correct storage of goods. The premises must be clear and easy to keep clean, well lit and well ventilated, dry, secure and safe. Space should be available for easy access to all items which should not be stored at too high a level. Heavy items should be stored at a low level.

Stock rotation is essential so as to reduce waste; last items in must be the last items to be issued. Any deterioration of stock should be identified, action taken and reported. In order to keep a check on stock, there is a need for a system of documentation which states:

- the amount in the stores
- the amounts issued, to whom and when
- the amounts below which stock should not fall.

Shelf life and 'use by' and 'best before' date information should be complied with. As a guide to storage, consider the points in Table 3.8.

Persons responsible for controlling the storage of stock, in addition to monitoring staff as they use the stores, must also check the correct storage temperatures of store rooms, refrigerators, deep freezes and so on. The establishment's policy and food safety management system may require records of temperature checks.

It is now recommended that chilled foods are stored at refrigeration temperatures between 1 and 4°C (even though food safety legislation only requires storage below 8°C). Refrigerators intended just for meat or fish storage are often run at the lower end of this temperature range, that is 1–2°C.

Checking stock

An essential aspect of the supervisory role is the full stock audit and spot check of goods in the stores, to assess deterioration and losses from other causes. Spot checks by their very nature are random, stock audits will occur at specified times during the year. Some establishments have a system of daily records of stock in hand.

Control the issue of stock and goods

A system of control is needed so that a record is kept of stock issued – how much, to whom and when. This enables control over the amount of items issued and records how much of each item is used over a period of time. This should help to avoid over-ordering and thus having too much stock on the premises. It should also diminish the risk of pilfering.

To be effective, the requisition document should include the date, the amount of the item or items required, and the department, section or person to whom they are to be issued. Usually a signature of the superior, for example, chef, chef de partie or supervisor, is required. It may be desirable to draw a line under the last listed item so that unauthorised items are not then added.

Cost control and spending limits

It is important to work within the resource budget, food, labour, fixed costs (for example, rent) and variable costs. Failure to control costs results in a business being unviable and therefore will be forced to close.

Table 3.8 Guide to storage conditions and time (also refer to 'use by' dates on packaging)

Item	Storage conditions and storage time
Canned goods	Store up to 9 months. Discard damaged, rusted, blown tins
Bottles and jars	Store at room temperature. Store in refrigerator once opened
Dry foods	Dry room temperature. Humid atmosphere causes deterioration
Milk and cream	Refrigerate and use within three days
Butter	Up to one month refrigerated
Cheese	According to the manufacturer's instruction. Soft cheese should be used as soon as possible
Salads	Keeps longer if refrigerated or in a dark, well ventilated area
Meat and poultry	Up to one week in refrigerator
Meat products	For example, sausages and pies, can be refrigerated for up to three days
Fish	Use on day of purchase ideally or up to 12 hours if refrigerated
Ice cream	Deep freeze for a week
Frozen foods	Six months; meat -18°C; fruit and vegetables -12°C

Menu engineering

One technique for sales analysis is 'menu engineering'. This uses two key factors of performance in the sales of individual menu items: the popularity and the gross profit (GP) cash contribution of each item. The analysis fits each menu item into one of four categories: Stars, Plough Horses, Puzzles and Dogs. This provides a simple way of graphically indicating the relative cash contribution of individual items on a matrix.

Cash gross profit (GP) contribution

	Low	High
High Popularity	**Plough horses** Items of high popularity but with low cash GP contribution	**Stars** Items of high popularity and also high cash GP contribution
Low	**Dogs** Items of low popularity and with low cash GP contribution	**Puzzles** Items of low popularity but with high cash GP contribution

▲ The menu engineering matrix

Source: Adapted from Kasavana and Smith, 1990

There are computer-based packages that will automatically generate the categorisation, or some development of it, usually using data directly from the electronic point-of-sale control systems.

In order to place an item on the matrix, two things need to be calculated. These are:
● the cash GP category
● the sales percentage category.

The cash GP category for any menu item is calculated by reference to the weighted average cash GP. Menu items with a cash GP which is the same as or higher than the average are classified as high. Those with lower than the average are classified as low cash GP items. The average also provides the axis separating Plough Horses and Dogs from Stars and Puzzles.

The sales percentage category for an item is determined in relation to the menu average, taking into account an additional factor. With a menu consisting of 10 items one might expect, other things being equal, that each item would account for 10 per cent of sales. Any item which reached at least 10% of the total menu items sold would therefore be classified as highly popular; any item which did not achieve the rightful share of 10 per cent would be categorised as having a low popularity. With this approach, half of the menu items would be shown as below average in terms of their popularity. This could lead to frequent changes to the menu. For this reason, Kasavana and Smith (1982) recommended a 70 per cent formula. Under this approach, all items which reach at least 70 per cent of their rightful share of the menu mix, are categorised as highly popular. For example, where a menu consists of 20 items, any item that reached 3.5 per cent or more of the menu mix (70 per cent of 5 per cent) would be regarded as enjoying high popularity. While there is no theory that supports the 70 per cent figure rather than some other percentage, experience has shown some merit in this approach.

Interpreting the categories

Items in each of the four categories of the matrix might be viewed like this:
● **Stars** – These are the most popular items, which may be able to yield even higher GP contributions by careful price increases or cost reduction. High visibility is maintained on the menu and standards for the dishes should be strictly controlled.
● **Plough Horses** – These are solid sellers, which may also be able to yield greater cash profit contributions through marginal cost reduction. Lower menu visibility than Stars is usually recommended.

● **Puzzles** – These are exactly what they 'say on the tin'. Items such as flambé dishes or a particular speciality can add an attraction in terms of drawing customers, even though the sales of these items may be low. Depending on the particular item, possible strategies range from accepting the current position because of the added attraction provided, to increasing the price further.

● **Dogs** – These are the worst items on the menu and the first reaction is to remove them. An alternative, however, is to add them to another item as part of a special deal. For instance, adding them in a meal package to a Star may have the effect of lifting the sales of the Dog item and may be a relatively low cost way of adding special promotions to the menu.

PLU No.	Menu item	Selling price	Selling ex VAT	Cost price	Cost total	Selling total ex. VAT	Qty sold	Cost %	GP%	Category identified	Under av. cost %
	Starter										
123	Soup	£5.00	£4.17	£1.00	£11.00	£45.83	11	24.00%	76.00%		yes
124	Melon	£4.50	£3.75	£1.20	£30.00	£93.75	25	32.00%	68.00%	STAR	yes
125	Chilli prawns	£6.50	£5.42	£1.80	£21.60	£65.00	12	33.23%	66.77%		yes
126	Paté	£3.80	£3.17	£0.90	£1.80	£6.33	2	28.42%	71.58%		yes
127	Mushrooms	£8.00	£6.67	£3.50	£63.00	£120.00	18	52.50%	47.50%		no
128	Bisque	£5.20	£4.33	£2.20	£121.00	£238.33	55	50.77%	49.23%		no
				Total	£248.40	£569.25	Average	36.82%	63.18%		
	Main course										
223	Salmon	£10.25	£8.54	£2.20	£33.00	£128.13	15	25.76%	74.24%		yes
224	Pork fillet	£15.00	£12.50	£3.20	£89.60	£350.00	28	25.60%	74.40%		yes
225	Braised beef	£13.50	£11.25	£3.80	£45.60	£135.00	12	33.78%	66.22%		no
226	Chicken	£9.25	£7.71	£1.90	£104.50	£423.96	55	24.65%	75.35%	STAR	yes
227	Turkey escalopes	£11.50	£9.58	£2.25	£4.50	£19.17	2	23.48%	76.52%		yes
228	Aubergine	£7.95	£6.63	£1.50	£48.00	£212.00	32	22.64%	77.36%		yes
				Total	£325.20	£1,268.25	Average	25.98%	74.02%		
	Dessert										
330	Trifle	£4.95	£4.13	£1.25	£35.00	£115.50	28	30.30%	69.70%	STAR?	yes
331	Fruit salad	£5.80	£4.83	£1.90	£114.00	£290.00	60	39.31%	60.69%		yes
332	Brandy snaps	£6.00	£5.00	£2.20	£44.00	£100.00	20	44.00%	56.00%		no
333	Choc. mousse	£3.60	£3.00	£1.90	£3.80	£6.00	2	63.33%	36.67%		no
334	Bavarois	£4.50	£3.75	£1.70	£20.40	£45.00	12	45.33%	54.67%		no
335	Gateau	£4.90	£4.08	£1.20	£13.20	£44.92	11	29.39%	70.61%		yes
				Total	£230.40	£601.42	Average	41.94%	58.06%		
		Overall cost % average		35%				Overall GP% average		65%	
		Grand total of cost totals		£804.00				Grand total of selling totals		£2,438.92	

(Please note the following increment boxes will only change the selling price ex. VAT)	Increase ALL starters by	(inc. VAT)	Ex. VAT:
	Increase ALL mains dishes by	(inc. VAT)	Ex. VAT:
	Increase ALL desserts by	(inc. VAT)	Ex. VAT:

▲ Using a spreadsheet to analyse the performance of menu items

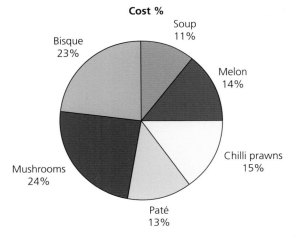

▲ Pie chart showing the cost of each starter as a percentage of the total cost; the melon is a star because it was 14% of the food cost but 25 were sold with a GP of 68%

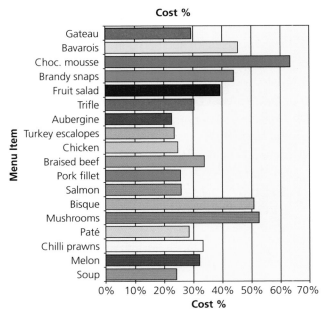

▲ Bar chart showing the cost percentage of each menu item

The physical stock take

The purpose of a physical stock-taking procedure is to check that the documentation of existing stock tallies with the actual stock held on the premises. The reason for this exercise is to prevent capital being tied up by having too much stock in hand. It also tests the accuracy of the system and thus indicates where modification could be made.

Discrepancies may become apparent that would then be investigated. Items such as returned empties, damaged stock and credit claims will be reconciled so that an accurate record is made. This may mean that both the storekeeper and the manager responsible will take action on the stock-take details. Records must be accurate, legible and carefully maintained in order to achieve the aim of the exercise.

To be effective, every item should be recorded, indicating the appropriate detail such as weight, size and so on and the number of items in stock.

Activity

Explain why it is important to keep accurate documentation.

Design a documentation system for a simple storekeeping process for a small organisation which does not have a sophisticated computer system.

Waste reduction and recycling

England generates 228 million tonnes of waste every year. This is a poor use of resources and costs businesses and households money. It also causes environmental damage (for example, waste sent to landfill produces methane, a powerful greenhouse gas). Businesses must take more responsibility for ensuring that a proportion of the goods they produce are recycled to reduce waste. The UK has laws that require some businesses to make sure that a proportion of what they sell is recovered and recycled. These producer responsibility regulations are based on EU legal requirements and cover producers of packaging, batteries, electrical and electronic equipment and vehicles.

Test yourself

1 What is meant by practical gastronomy? How can gastronomy be explored to develop menus and enhance the customer experience?
2 Name four factors which affect customer decision making when purchasing food in a restaurant.
3 What do you understand by the meal or dining experience?
4 Describe a food additive.
5 Give two uses for xanthan gum.
6 Describe the difference between formal and informal buying.
7 State the difference between fair trade and free trade.
8 What is meant by organic farming?
9 What do you understand by the Slow Food movement and its links with sustainability?
10 In menu engineering, what is a:
   (a) Star
   (b) Dog
   (c) Plough horse
   (d) Puzzle

4 Complex cold products, canapés and cocktail products

This chapter covers the following **NVQ** units:
→ Prepare, cook and present complex cold products.
→ Prepare, cook and present canapés and cocktail products.

Cold products, canapés and cocktail products in the **VRQ** Diploma are incorporated and assessed in a range of units. Content in this chapter covers a range of advanced skills and techniques related to cold products, canapés and cocktail products, which will cover the following units:
→ Advanced skills and techniques in producing vegetable and vegetarian dishes.
→ Advanced skills and techniques in producing meat dishes.
→ Advanced skills and techniques in producing poultry and game dishes.
→ Advanced skills and techniques in producing fish and shellfish dishes.

Learning objectives

By the end of the chapter you should be able to:
→ Identify a range of ingredients and equipment needed to prepare a range of cold products, canapés and cocktail products.
→ List the important quality points needed to select the ingredients required to produce cold products, canapés and cocktail products.
→ Prepare, cook and present a range of cold products, canapés and cocktail products.
→ Identify common faults and describe what to do if there is a problem.
→ Demonstrate the correct storage requirements and holding temperatures for cold products, canapés and cocktail products.
→ Systemically organise the equipment and ingredients to work professionally when preparing cold products, canapés and cocktail products.

Recipes included in this chapter

The larder, kitchen or garde manger

Traditionally the larder chef was responsible for the receiving and storage of a range of ingredients used in the professional kitchen, including hors d'oeuvres, starters and, in some establishments, the butchery and fishmongery. Many establishments would also have a separate cheese room. In the modern professional kitchen, with all the cost constraints and a very competitive environment, the larder or cold sections have been scaled down. Most butchery and fishmongery is now brought in already prepared. The modern larder chef will therefore generally be responsible for food storage, cold buffet items, starters and hors d'oeuvres. They will often keep stock records and may be responsible for ordering food.

Starters, appetisers or hors d'oeuvres

Hors d'oeuvres

Hors d'oeuvre means outside of the main meal. It is intended to introduce the meal, create interest and stimulate the taste buds. Hors d'oeuvres may be served at the dinner table as part of the meal (referred to as 'table hors d'oeuvres'), or before seating by waiters and food service staff, or passed around ('butler-style' hors d'oeuvres).

Appetisers

These are small and often bite-sized portions of food served at the table and designed to stimulate the appetite, create interest in the food and to excite the customer. They are usually part of the main meal but can be served at different times – for example, they may be served before the main meal or as a sweet appetiser before the dessert.

Amuse-bouche

Also known as amuse-gueule, these are single, bite-sized hors d'oeuvres. They are different from appetisers in that they are not ordered from a menu by the customer, but when served are done so for free and according to the chef's selection alone. The term is French – literally translated as 'mouth amuser'.

International terms for starters as hors d'oeuvres

Different countries and cultures have different ways of introducing the meal and different names to describe starters, appetisers or hors d'oeuvres. These are often served at the beginning of the meal and they are also intended to stimulate the appetite, to excite and introduce the customer to a range of flavours and customs.

- **Antipasto (Italy) or antepasto (Portugal)** – a selection of cooked cured meats, fish and vegetables.
- **Banchan (Korea)** – small disks of vegetables, cereals and meats served either before or alongside the meal.
- **Leng Pan (China)** – 'cold plate'. A Mandarin first course.
- **Meze** – Middle Eastern, Mediterranean and Balkan cuisine.
- **Zakuski (Russia)** – a range of small dishes presented buffet style, often consisting of cured meats and fishes, various pickled vegetables and breads.
- **Zensai (Japan)** – meaning 'before dish'.
- **Chat (India)** – small dishes of food eaten at all times of the day.
- **Dim Sum (China)** – refers to a style of Cantonese food prepared as a small bite-sized or individual portion, traditionally served in small steamer baskets or on small plates. Eating dim sum at a restaurant is usually known as going to 'drink tea' or 'yum cha', as tea is typically served with dim sum.
- **Tapas (Spain)** – a wide variety of appetisers or snacks. They may be cold (for example, olives, cheeses or pickled vegetables), or warm (for example, chopitos or fried baby squid). In central American countries they are known as bocas. The serving of tapas is designed to encourage conversation, because people are not focused on eating an entire meal. The word 'tapas' is derived from the Spanish verb *tapar* – 'to cover'. Tapas have evolved through Spanish history by incorporating ingredients and influences from many countries and cultures.

> **Professional tip**
>
> Appetisers should be kept simple using fresh, high quality ingredients served at the correct temperature.

Types of hors d'oeuvres

There is a wide variety of foods, combinations of foods and recipes available for the preparation and service of hors d'oeuvres. Hors d'oeuvres can be divided into three categories:

- **Single cold items** – for example, smoked salmon, avocado, pear, caviar, pâté, shellfish cocktails and melon. These should be well presented and garnished to make the dish attractive and appetising.
- **Seasoned hot dishes** – these can include vegetables, meat, fish, egg, pasta, dairy (cheese), soufflés, tartlets, puff pastry and choux pastry items.
- **Hors d'oeuvres variés** – a variety of salads, pickles, meats and fish, which are carefully selected and combined, cut neatly and delicately combined and

well seasoned with fresh herbs, dressings, vinaigrettes, mayonnaise, cream, yoghurt or fromage frais.

These may be served for lunch, dinner or supper and the wide choice, colour, appeal and versatility of the dishes make many items and combinations of items suitable as snacks and salads at any time of the day.

When a chef prepares hors d'oeuvres, it is very important to take account of a range of customer needs, lifestyles, special dietary requirements (for example, vegan, vegetarian or religious requirements).

Cold products

Cured meats

Curing is a term usually applied to meat which involves the development of colour, flavour and enhanced keeping qualities. It is one of the oldest methods of preservation. Curing brine is added to meat, which comprises salt (sodium chloride), sodium nitrate and sugar. The nitrate is converted to nitrite which combines with the pigments in the meat to produce the characteristic pink/red colour. The salt penetrates the meat, and this removes water – because the outside environment is saltier than the inside of an organism, the water diffuses out to dilute the salt. This process is known as **osmosis**. If there are micro-organisms in the food, water will pass from their cells (a dilute solution) into the more concentrated solution surrounding them and this dehydration will destroy them.

> **Key term**
>
> **Osmosis** – the movement of water from a diluted solution to a concentrated one, through a semi-permeable membrane.

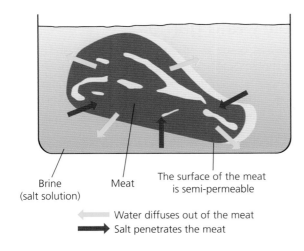

Brine (salt solution) Meat The surface of the meat is semi-permeable

⬅ Water diffuses out of the meat
➡ Salt penetrates the meat

▲ The role of osmosis in curing food

There are different methods of curing meats. **Dry curing** involves the curing ingredients being rubbed into the meat. Dry curing draws out moisture; it reduces ham weight by at least 18%, usually 20–25%. This results in a more concentrated ham flavour. This process of curing is slower than other methods. Examples of dry cured meats are Serrano, Parma, San Daniele and Bayonne.

Other types of dry curing use other additives in small amounts – sugar with sodium nitrate; sodium nitrate. Sugar can cause undesirable acidity during prolonged storage due to active *lactobacillus* bacteria. The addition of phosphates, especially in combination with salt, increases the water-binding capacity of raw meat and can contribute to improved texture in the final product. The curing mix is applied directly to the meat and is used for a wide variety of products: meat, poultry, game and some fish; while curing, the product is kept at 5°C.

Smoked meats

Smoking adds colour and flavour and can be used for a range of commodities. There are two types of smoking:

1 **Cold smoking** – smoke enters the smoking chamber at approximately 30–35°C for a period of around 6 hours (depending on the product; it can take days). The prolonged smoking process yields a complex flavour development, but also requires careful work to do it correctly, as the perfect conditions for cold smoking are also ideal for the promotion of bacterial growth. Cold smoking is widely used in Scandinavian countries. Since cold smoking does not technically cook or cure the meat, it is usually brined in salt first.

2 **Hot smoking** – this is smoking at a higher temperature; the products are smoked in the same chamber as the burning wood, whereas cold smoked products are held in an unheated chamber through which smoke is pumped. The temperature of the smoke can vary from 70–104°C, but is usually around 93°C. This produces a stronger smoked flavour. The process takes approximately 30 minutes to 6 hours, depending on the products being smoked.

Alder, apple, cherry and hickory are the most traditional woods used for smoking, but others such as maple, oak and pecan can also be used. Other flavours are also added to the smoke chamber, for example, sage, rosemary, juniper and spices such as cinnamon and mixed spices.

> **Professional tip**
>
> Cold smoke ingredients to impact a smoky flavour to foods that don't need to be cooked (for example, cheese), or that will be cooked later on the grill or in the oven. Hot smoke adds flavour and cooks the product.

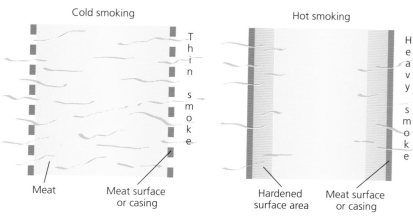

Cold smoking

Hot smoking

Thin smoke

Heavy smoke

Meat

Meat surface or casing

Hardened surface area

Meat surface or casing

▲ Hot and cold smoking

Table 4.1 The two methods of smoking

Cold smoking	Hot smoking
Cold smoking allows total smoke penetration inside of the meat. Very little hardening of the outside surface of the meat or casing occurs and smoke penetrates the meat easily	Hot smoking dries out the surface of the meat, creating a barrier for smoke penetration

Cold cooked meats

Roast meats may be served on a buffet by allowing them to cool quickly directly from the oven, either at ambient temperature or by blast chilling to ensure the meat is cool and succulent for eating. The meats are often garnished with salad, vegetables and fresh herbs.

Cold cooked meat should be sliced as near to serving time as possible and arranged neatly on a dish. Finely diced or chopped aspic may be placed around the edge. It may be decorated with a bunch of picked watercress, or presented in the piece with three or four slices out.

Whole joints, particularly ribs of beef, are often placed on a buffet table. They should either be boned or have any bones that may hinder carving removed before being cooked. They should be trimmed if necessary, strings removed and, after glazing with aspic jelly or brushing with oil, dressed on a dish garnished with watercress and lettuce leaves. Fancy cut pieces of tomato may also be used to garnish the dish. Fillet of beef Wellington and roast suckling pig are popular cold buffet dishes.

Fatty meats, such as pork and bacon, act as a good lubricant and give a good texture when used in pâtés, terrines and stuffings. Veal has a very delicate flavour and when used

in pies and terrines requires the addition of a fatty meat such as ham to produce the lubricant which will add to the eating quality.

Quality points

● Meats must be fresh, with no unpleasant odours and no slime. They should be at the correct temperature and delivered with a bright colour.
● Look for marbling in certain meat, especially beef – this is not a sign of tenderness. Fat is a lubricant and when it is dispersed in the muscle, it improves the eating quality of the meat.

Health and safety

Raw and cooked meat must be stored separately and prepared separately to avoid cross-contamination of bacteria. Meat is high in protein and is therefore at considerable risk of bacterial contamination

Poultry and game

Poultry and game are used as buffet items and in cold preparations. Quality is very important – the poultry and game must be fresh, free from unpleasant smells and stored correctly. Cold meats are cooked fresh and allowed to cool, but not left in the refrigerator for long periods as this affects the eating quality.

Game has a very distinctive flavour. Game animals – because of their tougher muscle structure – need a much longer period of conditioning and are therefore hung for longer periods than other animals. The ageing process depends on the type of game, usually from 7 to 14 days. (For more information, see Chapter 9.)

Serving cold roast chicken or duck

- When serving individual portions it is usual to serve either a whole wing of chicken neatly trimmed, or a half chicken. If a half is served, the leg is removed, the wing is trimmed, and the surplus bone removed from the leg, which is then placed in the wing (1½ kg chickens). Larger chickens may be cut into four portions, the wings in two lengthwise and the legs in two joints. Sometimes the chicken may be requested sliced. It is usual to slice the breast only and then reform it on the dish in its original shape.
- If the chicken is to be displayed whole at a buffet, it may be brushed with aspic or oil. It is then dressed on a suitably sized oval dish with watercress and a little diced aspic jelly.
- Serve duck with sage and onion dressing and apple sauce.
- To keep duck moist and succulent, roast 2 hours before service and cool within 90 minutes.

Serving game

Larger birds, such as pheasant, may be sliced or served in halves or quarters; small birds are served whole or in halves. The birds are served with watercress and game chips. Traditionally, the smaller birds are served on a fried bread croute, spread with a little of the corresponding pâté, farce au gratin, or pâté maison.

Serving cold turkey or goose

For display, cold turkey or goose may be brushed with jelly or oil, but otherwise it is normally served with the dark meat under the white, chopped jelly and watercress. Serve turkey with a dressing and cranberry sauce; serve goose as for duck.

Foie gras

Foie gras comes from the Périgord district of France and is classified as a delicacy. It is produced from the fattened liver of the duck or goose. These birds are force fed, which causes the liver to fatten. **The force feeding of animals is illegal in the UK, but is permitted in France.**

Foie gras can be purchased in the following ways:
- Raw – sliced and served grilled or lightly sautéd.
- Fresh – freshly prepared and cooked.
- Mi-cuit – semi-cooked. The foie gras is cooked at a temperature of 70–80°C; it is then bottled and canned and must be used within a short period of time (usually up to three months).
- Cooked – cooked at a higher temperature of 120°C in a can. Canning will preserve the foie gras for several years.
- Pâté de foie gras – a pâté produced from foie gras; purchased in cans or made from fresh foie gras. For this process, the liver must be well cleaned and all veins removed.

> ## Quality points
> - Firm to the touch but retaining the pressure mark.
> - The fewer blemishes, the higher the grade.
> - A light yellow to amber colour (as a result of the fattened liver).

Ballotines

Ballotines are boned-out, stuffed legs of poultry (usually chicken, duck, or turkey), which can be prepared and stuffed with a variety of forcemeat stuffings, cooked, cooled and prepared for service cold. They can be served:
- simply, garnished with a suitable salad
- decorated and coated with aspic or brushed with oil.

When preparing ballotines, keep the skin long as this will help to form a good shape, which can be long, round or like a small ham.

Eggs

A range of different egg varieties can be used in cold and buffet work, including hen's eggs, quail's eggs and duck eggs. They may be served boiled or poached with salads – for example, poached eggs with asparagus salad is a popular starter.

> **Health and safety**　⚠
>
> When selecting and purchasing eggs, check the source of supply and the farm. Salmonella-free eggs cannot be guaranteed. Producers should take precautions to keep the risk of infection down by good husbandry.

Farinaceous products and cereals

Couscous, bulgur wheat, quinoa, barley and pasta all make good, interesting salads. (For more information on pasta, see Chapter 5.)

Fish

Fish is frequently used as a buffet item. Examples of fish used include smoked salmon, smoked halibut, smoked eel and smoked trout. Whole, cooked fish such as a dressed salmon is sometimes also used. Round fish should have the back bone removed, leaving the head and tail intact on the skin. The cavity can then be filled with a suitable stuffing, for example:
- trout with a salmon or lobster forcemeat with diced mushrooms
- red mullet with a white fish forcemeat with chopped fennel

- sea bass with crayfish forcemeat and crayfish forcemeat with diced crayfish.

The fish is then reshaped (held in shape with greased greaseproof paper if necessary) and gently poached in a little stock or stock and wine in the oven. When cooled, the fish may be served:

- plain with a suitable accompanying cold sauce
- the skin removed, the fish cleaned, garnished with salad
- decorated and coated with a fish aspic or fish wine aspic, or brushed with oil.

Shellfish such as crab, lobster, shrimp, prawns, mussels, cockles and squid may also be presented.

Health and safety

Fish must be purchased fresh and stored separately; fresh fish should be kept in a separate refrigerator from cooked fish.

Dairy products

Many varieties of dairy products are used in cold and buffet work, including milk, cheeses, yoghurt, cream, fromage frais and crème fraiche.

A selection of different cheeses is offered in many restaurants. When selecting cheese, think about local cheeses and balance the variety: hard and soft; mild and strong; blue veined; from cows' milk, goats' milk and ewes' milk. Also include cheeses with different rinds: washed (rind soaked or rubbed with alcohol or brine) and downy (cheeses with a white dappled rind that are left to ripen).

Cheeses must be stored in the appropriate condition to prevent them drying out: wrapped in wax paper or cheese cloth, depending on the type of cheese and its manufacturing process. This will allow the cheese to breathe and continue to ripen. Cheese must be stored in the refrigerator and served at room (ambient) temperature to allow the customer to fully experience the characteristics of the cheese: its flavour, texture and smell. It is these characteristics which give each individual cheese its unique identity.

Vegetables and fruit

Vegetables play an important role in buffet and cold preparation work. A wide range of vegetables and different cooking methods are used to create a variety of flavours and textures. Salads are prepared from a wide variety of leaves. Herbs can be used to add colour and flavour to cold dishes.

Fruits are also used to add colour, texture and flavour to salads and cold starters.

Fungi

Different varieties of mushrooms have distinctive characteristics and provide a variety of flavours. Varieties include oyster, shiitake, chanterelle, porcini, portobello, morel, enoki, crimini and agaricus. Mushrooms can make a delicious food but many are poisonous. Poisoning by wild mushrooms is common and may produce mild gastrointestinal disturbance or a slight allergic reaction. It can also be fatal and therefore it is important that every mushroom intended for eating be accurately identified.

Truffles are a type of black fungi or white fungi and are highly prized as a food; they are often used as a garnish. The French Black or Périgord truffle (tuber melanosporum) is prized for its aromatic and fruity qualities. When fresh, it has a brown–black exterior with white veins on the inside; it ranges from the size of a pin to that of an orange and can weigh up to one kilo. The rare Italian White or Piedmont truffle (tuber magnatum) has the strongest smell of truffle. It has a smooth, dirty beige surface that ages to brown. Its size ranges from that of a walnut to that of an apple and it weighs up to half a kilo.

Rice

Rice is now a widely used in cold preparations, especially sushi (see p. 120). Other types of rice such as basmati, paella, patna and Carolina are all used in cold dishes and salads. They provide an important source of carbohydrate to accompany meat and fish protein.

Health and safety

Uncooked rice can be stored on the shelf in a tightly sealed container. Refrigerator storage is recommended for longer shelf life. Once cooked, it is important to keep hot (above 65°C for no longer than two hours) or cool quickly (within 90 minutes) and keep cool, below 5°C. If this is not done, *Bacillus cereus* may multiply in the cooked rice.

Pulses

Pulses provide a valuable source of protein for vegetarians in cold preparation and buffet work and are used in a wide variety of salads.

Salads

Salads add variety to the diet and menu. They provide a full range of nutrients and enhance buffets and cold tables, providing texture, flavour, colour and variety of ingredients. Salads are classified into two main types:

- **Simple salads** – these focus on one ingredient and have a simple dressing. They can be made up from a variety of ingredients. Classical simple salads include tomato salad, cucumber salad and green salad.
- **Compound salads** – these use a variety of ingredients that provide a range of different interesting textures and flavours and enhance cold preparations and the culinary experiences. The most common compound salads are Russian salad, Waldorf salad, meat salad, fish salad and vegetables à la Greque (Greek style).

Recently, salads have been served as a range of different types of ingredients on one plate; these ingredients are not blended together but are substantial and often served as a main meal. Such salads are called **combination** salads.

A more recent innovation has been to serve salads warm (in French, these are known as a **salade tiède**). An example would be a selection of salad leaves, lightly seasoned with a vinaigrette dressing, topped with freshly sautéed chicken livers; or fresh asparagus salad topped with a warm poached egg and served with balsamic dressing.

> **Professional tip**
>
> In a busy kitchen it is important to prepare the cold items of the salad in advance and top the salad with the hot ingredients just as it is about to be served.

Many of our gastronomic influences come from the USA and salads are no exception. Side dishes are popular there; salads such as Waldorf often accompany main courses.

Examples of **American salads** are:
- spinach walnut salad (baby spinach, griddled pear slices, crumbled blue cheese, red onion, sour dough croutons, walnuts, tossed in fig vinaigrette)
- 1905 salad (beef tomatoes, lettuce, ham, Swiss cheese, olives and garlic dressing)
- South American salads (e.g. quinoa salad with chicken and black beans)
- grapefruit and avocado salad (with ginger cassis dressing)
- quinoa salad (with fresh hearts of palm in a vinaigrette and parsley dressing).

> **Healthy eating tip**
>
> Highly nutritious salads are becoming more popular, especially as people are increasingly weight conscious and aware of the link between diet, health and nutrition. Many people now require salads which are healthy and low in calories.

Quality points

- Assess whether the salad is appropriate to serve to different customers, e.g. children, the elderly. Consider cultural aspects – how is quality assessed by different customers?
- Good, nutritious salads must be flavoursome and use a variety of seasonings.
- Use a combination of fresh ingredients to create an interesting quality salad. Textures can be crisp, soft, hard, moist, dry etc.
- Overall balance of ingredients, texture and flavour is important. Remember that portion size contributes to the balance of the meal.
- Add the dressing just before serving to avoid the salad items (leaves, herbs) becoming limp and losing their crispness.

> **Professional tip**
>
> When creating salads, remember the cost of the various ingredients. Some salads can be costly to produce if using a variety of ingredients. However, the energy costs of producing salads can be quite low.

Cold dressings and sauces

Vinaigrette and mayonnaise are used extensively for salads, but sour cream, tofu and yoghurt may also be used. For more information on sauces and dressings used in cold dishes, please see Chapter 6.

> **Healthy eating tip**
>
> Salad dressings can be high in calories, especially mayonnaise and vinaigrette. Use low-fat mayonnaise, combined with fromage frais or natural yoghurt as an alternative.

Pâtés and terrines

Pâtés and terrines are standard features in any French charcuterie and provide the basis of many lunches for those travelling in that country.

It is difficult to make a firm distinction between a pâté and a terrine: a fine-textured liver pâté is almost always called a pâté; yet a coarse, meaty product might be called either a *pâté de champagne* or a *terrine maison*. The difference is the actual cooking method and the receptacle in which they are cooked. Originally, a pâté was always enclosed in pastry (dough) and was made with almost any sort of meat or fish. Later it came to be baked in an earthenware dish (a terrine), which was lined with thinly sliced pork fat to keep the mixture moist.

The filling consists of a forcemeat prepared from the required meat, poultry or game, which is well seasoned with herbs, spices and any other garnish that may be relevant to the particular pâté or terrine. Most pâtés and terrines contain a good proportion of pork, especially pork belly fat.

Fat is essential to the texture of the pâté or terrine and enables it to be sliced without crumbling. It also means that a pâté or terrine is served without butter and just with crusty French bread or good wholewheat bread, plus some gherkins or olives, or a crisp salad. It is sometimes served as a first course and in that case, it is usually accompanied by thin, crisp Melba toast, or warm, freshly made white or brown toast, or a selection of savoury crackers.

When making a pâté or terrine, if possible choose a particular cut of meat and either mince (grind) it yourself or have it minced (ground) by the butcher, rather than buying ready-minced (ground) meat.

Preparing a pâté en croûte

1 Line a raised pie mould with pie pastry, then with thin slices of larding bacon.

2 Add the forcemeat with the garnish in between layers until the mould is full, the last layer being forcemeat.

3 Cover the top with larding bacon and pie pastry, neatly decorated.

4 Make one or more 1cm holes in the top and insert short, oiled, metal or stiff-paper funnels to enable steam to escape during cooking.

5 Eggwash the top 2–3 times and bake for 1¼–1½ hours at 190°C.

6 When cold, fill the pie through the holes in the top with a well-flavoured aspic jelly or a flavour to suit the pie.

Terrines

These are cooked in ovenproof dishes, fitted with a lid. Once the mixture is in the dish, the lid may be replaced and sealed with a plain flour and water paste to prevent the steam escaping. Terrines can be made from chicken, duck, veal, hare, rabbit, turkey and chicken livers. A binding agent is needed if making a pressed terrine (i.e. cooking all components separately, adding them together warm and pressing overnight).

Quality points

Use of ingredients is important: for example, the use of fresh herbs and half-cooked onions will send the terrine sour very quickly; under-cooked foie gras or chicken livers will make the terrine bitter.

Professional tip

When making a terrine, have a picture of how the terrine should look when sliced and be mindful that you will cut laterally through your layers, revealing a mosaic finish. Layering is essential to the visual presentation.

Preparing and cooking a terrine

The forcemeat is made in the same way as for a pâté, as is the filling, which is marinated in the liquor and spices.

1 Line the bottom and sides of the terrine with thin slices of larding bacon.
2 Add half the forcemeat (if this is too dry, moisten with a little good stock).
3 Neatly lay in the marinated garnish.
4 Add the remainder of the forcemeat and spread it evenly.
5 Cover the top with larding bacon and add a bay leaf and a sprig of thyme.
6 Put a thin layer of flour and water paste around the rim of the terrine and press the lid down firmly to seal it.
7 Place the terrine in a bain marie and cook it in a moderate oven, 190°C for approximately 1¼ hours. If the fat that rises to the top of the terrine is perfectly clear (shows no signs of blood) when lightly pressed with the fingers, this indicates that it is cooked; using a food probe, it should have reached at least 70°C.

8 When cooked, remove from the bain marie. Remove the lid and add a piece of clean wood that will fit inside the dish; place a weight of 1–1½ kg to press the meat down evenly. Allow to cool.
9 When cool, remove the weight and board, and all the fat from the surface.
10 To remove the terrine from the dish, place it in boiling water for a few seconds.
11 Turn the terrine out, trim and clean it and cut it into slices as required.

Note: If the terrine is to be served in its cooking dish, then wash and thoroughly dry the dish before returning the terrine to it. If the terrine is then to be covered with a layer of aspic jelly, all fat must be removed from the top beforehand.

Health and safety

Always use strict levels of hygiene. Wear hygienic latex gloves if putting together a pressed terrine as the terrine will not be heated again, so any bacterial transfer from hands to food will only multiply and not be destroyed.

Professional tip

When seasoning a pressed terrine, over season slightly. Your taste buds are more active for savoury notes when the temperature range is between 38°C and 47°C and because the terrine is served cold, the savoury notes are less susceptible to being picked. Slightly over-seasoned ingredients when warm will decant into a flavoursome and moderately seasoned product.

For storage, the terrine is wrapped tightly with cling film to prevent oxidisation. When slicing, slice through the cling film to prevent crumbling and cross-contamination from hand to terrine. If you are right-handed, the left part of the terrine will be touched up to 25 times before being served – this will increase the bacterial count for the last remaining terrines.

Cold mousses

A mousse is basically a purée of the bulk ingredient from which it takes its name, with the addition of a suitable non-dairy or cream sauce, cream and aspic jelly. The result should be a light creamy mixture, just sufficiently set to stand when removed from a mould. Mousses are set using a setting agent such as agar agar or gelatine.

Professional tip

Convenience aspic jelly granules may be used for all mousse recipes.

Various mousses are used as part of other dishes, as well as dishes on their own. A mould of a particular substance may be filled with a mousse of the same basic ingredient.

Whole decorated chickens may have the breast reformed with a mousse such as ham, tomato or foie gras. Mousse may be piped to fill cornets of ham, borders for chicken suprêmes or cold egg dishes.

Although most recipes specify a lined mould for the mousse to be placed in when being served as an individual dish, mousses may often be poured into a glass bowl (or even smaller dishes for individual portions) to be decorated on top when set, then glazed.

Professional tip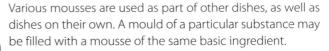

Care must be taken when mixing not to curdle the mixture as this will produce a 'bitty' appearance with small white grains of cream showing. The cream should only be half whipped, as a rubbery texture will otherwise be obtained. If fresh cream is over-whipped, the mixture will curdle.

Sushi

Sushi has been developed and refined in Japan over the centuries using local ingredients. In recent years, sushi has become one of the most popular grab-and-go dishes, snacks and cold buffet items.

Sushi ingredients

- Rice – sushi should only be made using short grain Japonica rice from Japan, Carolina or Italy. See Recipe 5 for how to make sushi rice.
- Nori – seaweed resembling sheets of black paper.
- Soy sauce – shoya or soy sauce has anti-bacterial and antioxidant properties, vitamins and anti-carcinogens.
- Wasabi – has anti-bacterial preservatives. Wasabi is a member of the Brassicaceae family. The root is used as a condiment.
- Gari – is a pickled ginger. It is also used as a palate cleanser between mouthfuls of sushi. It contains an anti-bacterial agent that can help prevent food poisoning and can improve circulation and metabolism.
- Fish – maguro or tuna is perhaps the quintessential sushi fish.
- Shellfish – such as hotategal (scallop), mirugai (gaper), torigai (cockle) and akagai (ark shell) are generally served raw. The popular sushi shellfish ebi (prawn) is usually lightly boiled on a skewer, ama-ebi or sweet prawn is often served raw. All are served as nigirizushi and some as chirashizushi.
- Fish roe – most fish roe for sushi is served as gunkan maki, wrapped in nori which ensures that the eggs do not fall off.
- Green tea – is a traditional accompaniment to sushi; beer and saki are also drunk when eating sushi.
- Vegetarian options – a number of non-fish ingredients appear in traditional authentic sushi, these include tamago or egg served as nigirizushi, futomaki and kyuri (cucumber) and kampyo (gourd) served as hosomaki (cucumber hosomaki is known as kappamaki). More recently, western innovations include avocado and red pepper which appears in futomaki and California roll.

Types of sushi

▲ Mixed sushi, including gunkan maki (front centre) and nigirizushi (front left and right)

Nigirizushi – the most common type of sushi. This is sushi topped with raw or cooked fish or shellfish, placed on a finger of vinegared sushi rice smeared with wasabi paste.

Makizushi – rolled sushi where a filling of fish or vegetables is enclosed by vinegared sushi rice and wrapped in toasted nori seaweed. Main varieties include the thin hosomaki, thicker futomaki and the California roll, where the nori is on the inside and rice on the outside.

Gunkan maki – this translates as 'battleship roll'. This is where a finger of vinegared sushi rice is surrounded by a strip of nori so that toppings, generally fish roe, can be placed on top without falling off. The end result is said to resemble a battleship.

Temakizushi – hand-rolled sushi. A small amount of sushi rice is spread in a square of nori, filling placed on top and the whole thing rolled into a cone shape.

▲ Temakizushi

Chirashizushi – literally means 'scattered sushi'. This differs from other kinds of sushi in that it is served in a bowl with the vinegared sushi rice being topped with a variety of raw fish.

▲ Chirashizushi

Inarizushi – vinegared sushi rice is served in a pocket of abura-age or sweetened deep fried tofu. Inari is the fox god of the Japanese indigenous Shinto religion and because foxes are traditionally believed to like abura-age, he lends his name to the sushi.

Sushi equipment

- uchiwa – paper fan
- handai – a piece of equipment for placing cooked rice in preparation for adding vinegar and cooling
- makisu – a simple bamboo rolling mat, for making makizushi.

Sushi rolling

1 With moistened hands, place the rice on a nori sheet which is positioned shiny side down on a bamboo mat.

2 Spread the rice evenly, leaving a 1cm gap at the edge on the far side.

3 Line up the fillings neatly below the line of wasabi at the centre. The fillings here are pickled radish, cucumber, smoked salmon and cooked mushrooms.

4 Roll up the ingredients in the bamboo mat, making a firmly packed cylinder.

5 Press the roll into a neat cylinder.

6 Cut the roll into 4–6 slices, using a sharp knife, moistening the blade after each slice.

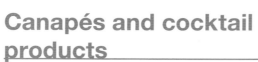

Canapés and cocktail products

These are small items of food, hot or cold, that are served at cocktail parties, buffet receptions and may be offered as an accompaniment to drinks before any meal (luncheon, dinner or supper). An interesting variety should be offered and in order to give an appetising presentation, they should be prepared as near to service time as possible (in which case they need not be refrigerated). Items to be baked should be served warm but not too hot, as guests may have to eat them with their fingers. Because, at most receptions, the guests remain standing, items should not be larger than a comfortable mouthful.

Although an attractive display should always be provided, in the interests of hygiene, cold snacks, canapés etc., should not be allowed to go into the reception until the last possible moment.

Hot snacks are normally offered after the guests have arrived, during the reception, and are sent out from the kitchen in small quantities.

The following are typical items for cocktail parties and light buffets:
- hot savoury pastry patties (of lobster, chicken, crab, salmon, mushroom, ham etc.), small pizzas, quiches
- hot sausages (chipolatas); these may contain various fillings or be wrapped in bacon, skewered and cooked under the salamander
- fried goujons of fish
- game chips, gaufrette potatoes, fried fish balls, celery sticks spread with cheese
- sandwiches, bridge rolls (open or closed but always small)
- sweets such as trifles, charlottes, jellies, bavarois, fruit salad, gateaux, strawberries and raspberries with fresh cream, ice creams, pastries
- beverages – coffee, tea, fruit cup, punch bowl, iced coffee.

Canapés

Canapés may be served on neat pieces of buttered toast, puff or short pastry, blinis or rye bread. A variety of foods may be used – slices of hard-boiled egg, thin slices of cooked meats, smoked salmon, fish anchovies, caviar, prawns etc. The size of a canapé should be suitable for a mouthful.

Bouchées

Bouchée fillings are numerous as bouchées are served both hot and cold. They may be served as cocktail savouries or as a first course, a fish course or as a savoury. All fillings should be bound with a suitable sauce, for example, mushroom, shrimp, prawn, lobster, chicken, ham or vegetable.

Brochettes

This refers to food cooked in and sometimes served on skewers or brochettes. The term derives from French cuisine; other terms used include shish kebab, satay or souvlaki. In Portugal they are known as *espetada*.

Filled items

A variety of savouries may be served either as hot appetisers (at a cocktail reception) or as the last course of an evening meal.

Examples of filled savouries:
- A **tartlet** or **barquette** may be made from thinly rolled short pastry and cooked blind. The cooked tartlets or barquettes should be warmed through before service, the filling prepared separately and placed neatly in them. They may then be garnished with a sprig of parsley. Examples of fillings include shrimps in curry sauce, chicken livers in a light juice or devilled sauce and mushrooms in béchamel, supreme or aurora sauce.
- **Bouchées** and small **vol-au-vents**, filled with a variety of fillings, are served both hot (for example, ratatouille, mushroom, chicken and lobster) and cold (for example, prawn, lobster and tomato).
- Small **samosas** are made with filo or spring roll pastry, filled with minted spiced lamb.
- **Dartois** – these savoury puff pastry slices are made and cooked in a long strip, then cut after cooking. Fillings include smoked haddock, chicken and mushroom, chicken, sardines, tuna, anchovies, ratatouille and mushroom.
- Small **puff pastry turnovers** can be filled with cheese and apple chutney.
- Spinach and feta cheese mini calzones.
- Small, bite-sized brioches may be made for appetisers. The centre is scooped out and various fillings may be used, including fish, meat and vegetables.

▲ Bite-sized brioches filled with creamed smoked salmon (back) and mushrooms in cream sauce (front)

Presentation

Cold buffets

Attractive, inviting cold buffets should give pleasure to customers and help stimulate their appetites. When preparing and decorating dishes, bear in mind that they are to be presented and served in front of customers; therefore, ease of service is important and should be considered when choosing the method of decoration. A cold buffet should not look a wreck after a handful of customers or one or two portions have been served.

As the chef has prepared all these dishes, he/she should be involved in their display and the following points borne in mind:

- Display food under refrigeration if possible. If not, then keep in cool/cold storage until the last possible moment, bearing in mind that cold buffet food is a favourite target for bacteria. (Refer to Guidelines for the Catering Industry on the Food Hygiene Regulations Amendments 1990, 1991 and 1995, HMSO.)
- Select the most outstanding dish as the centrepiece.
- Consider carefully how the food is to be served when placing all dishes in position. If the customers are to help themselves, make sure that all dishes are within reach.
- Ensure that the various complementary dressings and salads are by the appropriate dishes, otherwise customers will move backwards and forwards unnecessarily and cause hold-ups.
- On self-service buffets, dishes quickly become untidy. Have staff on hand to remove and replace or tidy up dishes as required

Hot buffets

Many establishments provide a range of heated foods for buffet style self-service, particularly for tourists. This food is frequently presented in containers with a lift-up cover and is provided with a small amount of heat under the container. The advantage of this is that food is available over a period of time and there is a minimum need for service staff. However, to ensure a satisfactory standard of food safety it is essential that all the food is hot, not just that closest to the heat source. The amount of food in the container should not exceed that which can be kept hot, so the buffet should be replenished with very hot food a small amount at a time. In hot climates this is particularly important, even where there is air conditioning.

Aspic jelly

Aspic is a savoury jelly that may be used on cold egg, fish, meat, poultry, game and vegetable dishes that are prepared for cold buffets so as to give them an attractive appearance. It prevents food from losing its moisture and can also be used as a garnish for certain dishes when chopped or cut into shapes.

For meat dishes, a beef or veal stock is made; for fowl, chicken stock; and for fish, fish stock. Aspic jelly is produced with the addition of gelatine. Vegetarian aspic jelly is produced from vegetable stock with the addition of agar agar as a setting agent.

Health and safety ⚠

Great care must be taken when using aspic jelly as it is an ideal medium for the growth of micro-organisms. Therefore the following procedures should be observed:

- Always use fresh aspic and bring to a simmer; temperature must be 82°C for 15 seconds. Cool quickly and use sparingly.
- Avoid using warm aspic, especially over long periods.
- Do not store aspic for long periods at room temperature.
- If required for further use, chill rapidly and store in the refrigerator.
- If stored in refrigerator, simmer for 10 minutes before further use. Discard after 24 hours storage.
- Once a dish has been glazed with aspic or any jelly, it may be kept refrigerated for up to 8 hours.
- Once removed from the refrigerator, it must be consumed within 2 hours and then, if uneaten, discarded.
- It is advisable, if possible, to make only the quantity required for use and so avoid storage. Where possible, display in refrigerated units.

1 Smoked salmon, avocado and walnut salad

Ingredient	4 portions	10 portions
Smoked salmon	100g	250g
Avocado pears	2	5
Walnuts	50g	125g
Fennel or parsley	1 tsp	1 tbsp
Vinaigrette	1 tbsp	2 tbsp
Radicchio		
Curly endive		

Energy	Cal	Fat	Sat fat	Carb	Sugar	Protein	Fibre
1413 kJ	341 kcal	32g	4.8g	2.7g	1.4g	10.7g	4.2g

1 Cut the smoked salmon in strips and neatly slice the peeled avocado.

2 Carefully mix together with the walnuts, chopped fennel or parsley and vinaigrette.

3 Neatly pile on a base of radicchio and curly endive leaves.

Note: flaked cooked fresh salmon or flaked smoked mackerel can be used to create a variation of this salad.

2 Beetroot and orange salad

Ingredient	4 portions	10 portions
Large beetroot, raw	200g	500g
Valencia oranges	2	5
Vinaigrette	30ml	75ml

1 Wash the beetroot and steam for approximately 1 hour. Check that they are cooked by using the tip of a small knife to effortlessly pierce their flesh.

2 Allow to cool, then peel.

3 Cut the beetroot into 1cm dice.

4 Peel and segment the oranges.

5 Just before serving, mix all the ingredients together and season well.

Energy	Cal	Fat	Sat fat	Carb	Sugar	Protein	Fibre
115 kJ	27 kcal	1.4g	0.2g	3.4g	3.3g	0.5g	1.0g

3 Salmon and asparagus salad with a rose petal dressing and shiso

1 Peel and clean the asparagus and place into boiled, salted water with thyme to add some flavour. Leave until the tips are cooked.

2 Remove and place into a bowl of iced water to refresh.

3 Cut the smoked salmon into slices around a small plate, to form a circle.

4 Wash and mix the salads together and cut the rose petals into strips (ready to add to the dressing).

5 Place the salmon neatly on the plate with asparagus arranged on top.

6 Add the rose petal dressing (see below) to the salad, drain and place salad in the middle of the salmon and sprinkle with shiso.

Note: Shiso is a small-leafed purple-coloured herb similar to mustard and cress with a basil-like flavour. The dish may also be finished with fine slices of radish and herb mayonnaise

Ingredient	4 portions	10 portions
Asparagus spears	20	50
Thyme	20g	50g
Scotch smoked salmon	200g	500g
Shiso	100g	250g
Curly endive (spider lettuce)	100g	250g
Oakleaf	40g	100g
Lollo rosso	2	5
Rose head	4	10

Energy	Cal	Fat	Sat fat	Carb	Sugar	Protein	Fibre
679 kJ	161 kcal	4.5g	0.9g	12.2g	10.1g	18.6g	6.7g

Rose petal dressing

Ingredient	4 portions	10 portions
Rose petal vinegar	1 tsp	2½ tsp
Virgin oil	1 tsp	2½ tsp
Shallot finely cut	30g	75g

1 Mix the oil and vinegar together, then add the shallots and finely cut rose petals.

This recipe is based on one by Anthony Marshall.

4   Tian of green and white asparagus

Ingredient	4 portions	10 portions
Garlic	10g	25g
Thyme	½ tsp	1 tsp
Black pepper	Pinch	10g
Gelatine	1 sheet	3
Green asparagus (medium)	400g	1 kg
White asparagus (medium)	400g	1 kg
Salt	Pinch	½ tsp
Shallots	100g	250g
Extra virgin olive oil	20 ml	50ml
Sundried tomatoes in oil	30g	75g
Chives, blanched	8	20

Energy	Cal	Fat	Sat fat	Carb	Sugar	Protein	Fibre
651 kJ	157 kcal	10.2g	1.4g	5.7g	4.9g	11.1g	5.2g

To make the jelly

1 Bring 500ml of water to the boil in a pan. Add a touch of garlic, thyme and black pepper.

2 Melt the gelatine in warm water and add to the bouillon.

3 Pass through a fine sieve and leave to cool.

To make the tian

1 Trim and peel the asparagus.

2 Cook the colours separately in salted water infused with thyme for approximately 4–5 minutes. Cool down immediately in iced water.

3 Trim each length of asparagus to exactly 8cm long (keep the trimmings).

4 Slice the trimmings of asparagus and sauté with the shallots and remaining garlic, then leave to cool.

5 Using a metal ring (approximately 7cm in diameter), place the spears of asparagus around the inside of the ring in alternate colours. Use the sautéed mixture to fill the centre of the ring and pack tightly.

6 Pour the jelly over the top of the mixture inside the metal ring and place in the refrigerator to set.

To make the dressing

1 Cut the tomatoes into strips and add to the olive oil. Season well.

To plate

1 Heat the metal ring of asparagus very quickly with a flame torch to loosen the edges. Place in the centre of a plate and tie with blanched chives.

2 Carefully spoon the dressing in a circle around the edge. Garnish with micro-herbs.

This recipe was contributed by Anthony Marshall.

5 Sushi rice

Ingredient	
Uncooked, matured Japanese or Californian short grain rice	500g
Cold water	625ml
Sushi vinegar	120ml

Use one part rice to one and a quarter parts water.

1 Place the rice in a bowl, pour in some water and swill the bowl to remove any impurities, then drain the water off.

2 Pour in more water and wash the rice by stirring, drain again. Repeat 4 to 5 times until the water remains clear.

3 When draining for the final time, use a sieve and let the rice drain for 30 minutes to absorb the moisture on the surface. Another method is to leave the rice in a bowl of water for 30 minutes to absorb water.

4 Place the rice and 625ml of water in a saucepan with a tight-fitting lid. Bring to the boil, simmer for approximately 10 minutes. Remove from heat and leave for a further 10 minutes. Leave the lid on.

5 Remove the rice from the pan and place in a shallow container, ideally in a traditional Japanese cypress wood rice tub known as a handai. Spread the rice evenly across the pan using a spatula.

6 Add the sushi vinegar, pouring it in as evenly as possible over the rice. Using a spatula, make gentle cutting and folding movements to mix the vinegar thoroughly into the rice, being sure not to crush the kernels by mashing or stirring.

7 As you cut and fold the vinegar through the rice, fan the rice with a Japanese paper fan, called an *uchiwa*, to cool the rice. Make sure the rice is evenly coated with vinegar.

8 Use within an hour of preparation.

Sushi vinegar

Ingredient	Quantity	
Rice vinegar	200ml	
Granulated sugar	120g	
Salt	2g	

Boil all the ingredients together and allow to cool.

6 Aspic jelly

Ingredient	4 portions	10 portions
Whites of egg	2–3	5
Strong, fat-free seasoned stock (as required: poultry, meat, game or fish)	1 litre	2 litre
Vinegar	1 tbsp	2½ tbsp
Sprigs tarragon	2	5
Leaf gelatine	75g	190g

1 Whisk the egg whites in a thick-bottomed pan with half a litre of cold stock and the vinegar and tarragon.

2 Heat the rest of the stock, add the gelatine (previously soaked for 10 minutes in cold water) and whisk until dissolved.

3 Add the stock and dissolved gelatine into the thick-bottomed pan. Whisk well.

4 Place on the stove and allow to come gently to the boil until clarified.

5 Strain through muslin.

6 If the jelly is not clear, repeat the whole procedure using egg whites only.

7 Celeriac remoulade

Ingredient	25 portions
Large celeriac, cut into fine julienne	2 (400g)
Horseradish relish and mayonnaise, in equal quantities, to bind	200ml
Lemon, juice of	1
Seasoning	

Energy	Cal	Fat	Sat fat	Carb	Sugar	Protein	Fibre
235 kJ	57 kcal	5.4g	2.6g	0.3g	0.3g	1.8g	0.1g

1 Mix all ingredients together and correct the seasoning.

8 Tomato mousse

Ingredients	4 portions	10 portions
Onion, finely chopped	50g	125g
Butter or oil	50g	125g
White stock or consommé	500ml	2½ litres
Tomatoes, skinned, deseeded and diced	250g	600g
Tomato purée	25g	60g
Salt and pepper		
Pinch of paprika		
Velouté	125ml	600ml
Aspic jelly	125ml	600ml
Whipping cream, half whipped	125ml	600ml

1 Sweat the onion in the butter or oil without colour.

2 Moisten with the stock or consommé, and reduce by half.

3 Add the tomatoes and tomato purée. Simmer for approximately 20 minutes, season and add paprika.

4 Add the velouté and aspic, simmer for 2 minutes, then liquidise.

5 Place in a basin on ice, stir until setting point, and fold in the half-whipped cream.

6 Use as required: either place in individual moulds, allow to set, turn out onto individual plates, decorate and serve as a first course, or use as part of a cold dish.

Energy	Cal	Fat	Sat fat	Carb	Sugar	Protein	Fibre
1691 kJ	410 kcal	39.4g	24.3g	9.7g	5.3g	4.9g	1.5g

9 Smoked fish platter

Ingredient	4 portions	10 portions
Fillets of smoked mackerel	2	5
Fillets of smoked trout	1	2–3
Smoked eel or halibut	200g	500g
Smoked salmon	100g	250g
Lemon	1	2
Mayonnaise with horseradish	60ml	150ml

Energy	Cal	Fat	Sat fat	Carb	Sugar	Protein	Fibre
2440 kJ	589 kcal	48.9g	9.6g	0.5g	0.4g	36.7g	0.1g

1 Carefully remove the skin from the mackerel, trout and eel or halibut fillets, and divide into four pieces.

2 Arrange with a cornet of salmon on each plate.

3 Garnish with a quarter of lemon.

4 Serve mayonnaise sauce separately, containing finely grated horseradish.

Variation: Shellfish platter

A selection of shellfish (e.g. lobster, crab, prawns and shrimp) neatly arranged and served with quarters of lemon and mayonnaise sauce separately.

▲ A shellfish platter

10 Red mullet with tomatoes, garlic and saffron

Ingredient	4 portions	10 portions
Red mullet, small	4	10
Salt and pepper		
Dry white wine	125ml	300ml
Vegetable oil	60ml	150ml
Tomatoes, skinned, deseeded and diced	150g	375g
Clove of garlic (crushed)	1	2–3
Sprig of thyme, bay leaf		
Pinch of saffron		
Peeled lemon	4 slices	10 slices

Energy	Cal	Fat	Sat fat	Carb	Sugar	Protein	Fibre
1228 kJ	298 kcal	18.7g	2.0g	1.4g	1.4g	26.8g	0.4g

1 Clean, prepare and dry the fish.

2 Place in a suitable oiled dish. Season with salt and pepper.

3 Add the white wine, oil, tomatoes, garlic, herbs and saffron.

4 Cover with aluminium foil and bake in the oven at 220°C for approximately 7 minutes. Allow to cool in dish.

5 Serve on individual plates with a little of the cooking liquor, garnished with a slice of peeled lemon.

11 Hot water pastry

This paste may be used for pâté en croûte, game pie (Recipe 12) or pork pie (Recipe 13).

Ingredient	4 portions	10 portions
Strong plain flour	250g	625g
Salt		
Lard or margarine (alternatively use 4 parts lard and 1 part butter or margarine)	125g	300g
Water	125ml	312ml

1 Sift the flour and salt

2 Make a well in the flour with your fingers. Boil the fat in water, then pour immediately into the well

3 Mix the ingredients together using a spoon until the mixture cools down

4 Mix until a smooth dough is formed

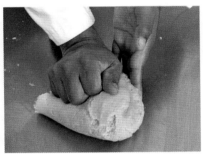

5 Knead the dough before rolling it out for use. Use while still warm

12 Game pie

Ingredients	4 portions	10 portions
Forcemeat		
Game flesh	200g	500g
Fat bacon	200g	500g
Beaten egg	1	2-3
Salt	10g	25g
Spices		
Filling		
Game fillets	300g	750g
Larding bacon	200g	500g
Brandy or Madeira	60ml	150ml
Salt	15g	35g
Spices		

Game pies can be made from hare, rabbit or any of the game birds. The filling should be marinated in the liquor, salt and spices for 1–2 hours.

Energy	Cal	Fat	Sat fat	Carb	Sugar	Protein	Fibre
6714 kJ	1612 kcal	109.9g	46.8g	97.3g	2.0g	56.7g	5.2g

Prepare, cook and finish as for pâté en croûte (see page 96).

13 Raised pork pie

Certain pies are not cooked in moulds but are hand raised using a hot water pastry. It is advisable to use a mould for presentation.

Ingredient	4 portions	10 portions
Hot water paste (Recipe 11)		
Filling		
Shoulder of pork (without bone)	300g	750g
Bacon	100g	250g
Allspice, or mixed spice and chopped sage	½ tsp	1¼ tsp
Salt and pepper		
Bread soaked in milk	50g	125g
Stock or water	2 tbsp	5 tbsp

Energy	Cal	Fat	Sat fat	Carb	Sugar	Protein	Fibre
2867 kJ	683 kcal	41.8g	17.1g	54.1g	1.5g	26.1g	3.2g

Video: pork pies
http://bit.ly/1h9jKb3

1 Cut the pork and bacon into small, even pieces and combine with the rest of the ingredients.
2 Keep one-quarter of the paste warm and covered.
3 Roll out the remaining three-quarters and carefully line a well-greased raised pie mould.
4 Add the filling and press down firmly.
5 Roll out the remaining pastry for the lid.
6 Eggwash the edges of the pie.
7 Add the lid, seal firmly, neaten the edges, cut off any surplus paste.
8 Decorate if desired.
9 Make a hole 1cm in diameter in the centre of the pie.
10 Brush all over with eggwash.
11 Bake in a hot oven, 230–250°C for approximately 20 minutes.
12 Reduce the heat to moderate 150–200°C and cook for 1½–2 hours in all.
13 If the pie colours too quickly, cover with greaseproof paper or aluminium foil. Remove from the oven and carefully remove tin. Eggwash the pie all over and return to the oven for a few minutes.
14 Remove from the oven and fill with approximately 125ml of good hot stock in which 5g of gelatine has been dissolved.
15 Serve when cold, garnished with picked watercress and offer a suitable salad.

14 Terrine of bacon, spinach and mushrooms

Ingredient	12 portions
Collar of bacon	1 kg
Carrot	1
Onion clouté	1
Bouquet garni	1
Celery	2 sticks
Peppercorns	8
Fresh spinach	500g
Butter	50g
Mushrooms (preferably morels)	200g

Energy	Cal	Fat	Sat fat	Carb	Sugar	Protein	Fibre
1436 kJ	347 kcal	30.4g	7.2g	1.2g	1.0g	17.0g	1.4g

Healthy eating tips
- Soaking the bacon overnight will remove some of the salt.
- Use a minimum amount of salt to season the vinaigrette.

1 If necessary soak the bacon overnight. Drain.

2 Place the bacon in cold water, bring to the boil, add the carrot, onion, bouquet garni, celery and peppercorns.

3 Poach until tender.

4 Pick some large leaves of spinach to line the terrine, blanch the leaves, refresh and drain.

5 Lightly cook the rest of the spinach gently, refresh, drain and shred.

6 Alternatively, shred the spinach raw, quickly cook in butter, drain and then blast chill.

7 Cook the mushrooms in a little butter, chill well and season.

8 When cool, remove the bacon from the cooking liquor, chop into small pieces.

9 Line the terrine with cling film, then the spinach leaves. Layer with bacon, mushrooms and spinach. Cover with spinach leaves and cling film. Place weights evenly on top, to press. Wait for 12 hours or overnight.

10 When ready, turn out, slice and serve on plates with leek and mushroom vinaigrette.

15 Cured pork

Ingredient	15 portions (depending on size of cut)
Salt	150g
Sugar	50g
Spanish smoked paprika	20g
Pork tenderloin, trimmed	1 (1.2–1.5 kg)
Confit oil (equal parts olive oil and vegetable oil)	2 litres
Thyme	1 sprig
Bay leaves	3
Garlic	3 cloves

1 Mix the salt, sugar and paprika together, and rub into the pork belly, meat side.

2 Wrap in cling film and allow to cure for 1½ hours. Rinse quickly under running water, but don't try to wash out all the salt, just pat dry.

3 Roll the belly into a cylindrical shape and tie in even sections.

4 Pre-heat oven to 87°C. Meanwhile, heat the confit oil on the stove, adding the thyme, bay leaves and garlic. In a thick-bottomed pan, seal the pork all over until golden brown then immerse in the confit oil. Place in the oven for 3 hours.

5 Test if cooked by squeezing with your forefinger and thumb. The meat should just give. Remove carefully and allow to cool at room temperature. Once at room temperature, wrap tightly in cling film and refrigerate.

6 This can be served cold with an acidic chutney or pan fried and served with a sauerkraut.

Energy	Cal	Fat	Sat fat	Carb	Sugar	Protein	Fibre
6079 kJ	1475 kcal	153.6g	23.3g	4.2g	3.5g	19.4g	0.1g

16 Chicken galantine

Ingredients	8 portions
Chicken, whole (1½ kg)	1
Lean veal	100g
Belly of pork	100g
Bread soaked in 125ml milk	75g
Egg	1
Salt, pepper and nutmeg to season	
Double cream	250ml
Thin slices of fat bacon or lardons	
Ham	25g
Tongue cut into ½cm batons	25g
Bacon	25g
Blanched and skinned pistachio nuts	12g
Chicken suprême	1
Chicken stock	

1 Clean and carefully skin the whole chicken, and place the skin in cold water to remove blood spots.

2 Bone the chicken. Make sure the chicken meat is free from all bone.

3 Pass the veal, pork and squeezed, soaked breadcrumbs through a fine mincer.

4 Remove into a basin, mix in the egg and seasoning and pass through a sieve.

5 If using a food processor, add the egg while the mixture is in the processor and continue to chop until very fine.

6 Place into a basin over a bowl of ice, add the cream slowly, mixing well between each addition.

7 Place a clean damp cloth on the table; arrange the chicken skin on the cloth. Cover with slices of fat bacon, to about 5cm from the edge. Lay the boned-out chicken over this.

8 Spread on one-third of the mixture.

9 Garnish with alternate strips of ham, tongue, bacon, pistachio nuts and the chicken suprême, also cut into ½cm batons.

10 Place another layer of mixture on top and repeat the process.

11 Finish with a one-third layer of the mixture.

12 Roll the galantine up carefully. Tie both ends tightly.

13 Poach in chicken stock for approximately 1½ hours.

14 When thoroughly cold, remove cloth.

15 Cut into slices, serve garnished with salad.

Energy	Cal	Fat	Sat fat	Carb	Sugar	Protein	Fibre
2794 kJ	673 kcal	49.7g	19.8g	6.5g	1.6g	50.1g	0.3g

Galantine may be served as a starter with an appropriate garnish such as a small tossed salad, or suitable chutney.

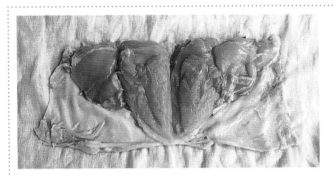

Bone out the chicken through the back and lay it on the cleaned chicken skin (on a clean, damp cloth)

Start to lay the minced meat mixture on top of the chicken

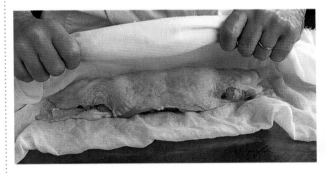

Carefully roll up the galantine

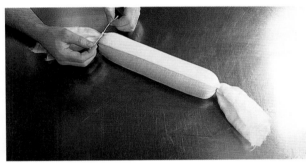

Tie the cloth at each end

Variations

To enhance the flavour of the mixture, some fresh chopped herbs, such as tarragon and chervil, may be added.

The galantine may be made without the whole boned-out chicken, just using the skin and the chopped and minced fillings. However, the chicken meat gives the galantine more bulk and improves the texture and eating quality.

17 Spring rolls

Ingredient	4 rolls	10 rolls
Spring roll paste	4 sheets	10 sheets
Sesame oil	50g	125g
Finely chopped onion	25g	60g
Sliced bamboo shoots	50g	125g
Cooked, diced pork	50g	125g
Finely diced celery	25g	60g
Groundnut oil	1 tbsp	2½ tbsp
Ve-tsin (optional)	½ tbsp	1¼ tbsp
Peeled shrimps	100g	250g

1 Prepare the filling: heat the sesame oil in a suitable pan, add the finely chopped onion and heat without colour for 2 minutes.

2 Add all other ingredients except the shrimps, mix well. Simmer for 3 minutes.

3 Turn out into a clean basin and leave to cool. Add the shrimps.

4 Put a small spoonful of filling onto a sheet of spring roll paste, roll up, tucking in and sealing the edges with eggwash. Place on a tray and chill well for approximately 20 minutes.

5 Deep fry in hot oil approximately 190°C, until golden brown, drain well and serve immediately on a dish paper.

Ve-tsin is a flavour enhancer, similar to monosodium glutamate. For the nutritional analysis, soy sauce was used instead.

Energy	Cal	Fat	Sat fat	Carb	Sugar	Protein	Fibre
1170 kJ	281 kcal	19.9g	3.8g	12.0g	2.3g	14.4g	1.0g

Carefully separate one sheet of pastry from the rest

Place the filling on one half of the sheet, then eggwash the edges of the pastry

Fold the tip of the pastry over the filling, then eggwash again

Fold over one side of the pastry, eggwash again, then fold over the other side

Eggwash, then gently roll up the spring roll

18 Vegetable samosas

Ingredient	18 portions	
Filling		
Oil	1 tbsp	
Carrots, peeled and julienned	2	
Mooli, peeled and julienned	50g	
Savoy cabbage, julienned	50g	
Spring onions, julienned	4	
Beansprouts	50g	
Sultanas	10g	
Ginger root, finely diced	3g	
Toasted sesame seeds	3g	
Dark soy sauce	50ml	
Pastry		
Filo or spring roll pastry	6 sheets	

Energy	Cal	Fat	Sat fat	Carb	Sugar	Protein	Fibre
237 kJ	57 kcal	3.6g	1.3g	5.5g	1.5g	0.9g	0.8g

For the filling

1 Heat the oil in a large saucepan.

2 Sweat the carrots and mooli for 2 minutes.

3 Add the cabbage, spring onions, beansprouts and sultanas.

4 Sweat on a high heat until dry (approximately 4 minutes).

5 Add the ginger, sesame seeds and soy sauce and season.

6 Reduce liquid by half and chill.

To make the samosas

1 Cut the paste into strips approx. 13 × 7cm. Fold over the end of the pastry

2 Eggwash the upper side of the pastry, then fold over the top part to form a pocket

3 Flip over the pocket, ease it open and fill it

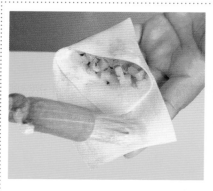

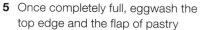

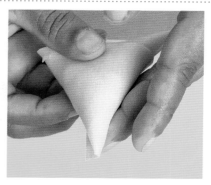

5 Once completely full, eggwash the top edge and the flap of pastry

6 Fold over the pastry and wrap it round to seal the samosa

7 Deep fry until golden brown and crisp

19 Goat's cheese crostini

Ingredient	15 portions
Thin baguette bread, thinly sliced, brushed with olive oil	12 slices
Soft goat's cheese	200g
Red pepper, roasted, skinned, cut into neat diamonds	100g

Energy	Cal	Fat	Sat fat	Carb	Sugar	Protein	Fibre
531 kJ	126 kcal	4.1g	2.4g	17.5g	1.5g	5.8g	0.8g

1 Toast or bake the baguette slices until crisp.

2 Spoon or pipe the cheese onto the baguette slices.

3 Top with diamond-shaped slices of red pepper and a fresh herb leaf.

20 Soba noodle norimaki

Ingredient	50 portions
Soba noodles	225g
Coriander, chopped	4 tbsp
Spring onions, chopped	1 tbsp
Soy sauce	2 tbsp
Fresh ginger, chopped	1 tbsp
Red wine vinegar	2 tbsp
Pickled ginger, chopped	2 tbsp
Nori sheets	
Cucumber, peeled, seeded and cut into julienne	1
Red pepper, seeded and cut into julienne	1

Energy	Cal	Fat	Sat fat	Carb	Sugar	Protein	Fibre
80 kJ	19 kcal	0.3g	0.0g	3.7g	0.3g	0.6g	0.2g

1 Cook the soba noodles in boiling water and then cool.

2 Mix the coriander, spring onions, soy sauce, ginger, vinegar and pickled ginger together.

3 Add the drained noodles then mix well.

4 Lay out a sheet of nori and lay the noodles onto it, approximately 0.5cm thick and running lengthways.

5 Lay some cucumber and red peppers on top, then roll neatly and tightly as for sushi rolls.

6 Chill well in refrigerator, and when ready to serve, cut into 1cm thick rounds on the slant.

7 It may be easier to lay the nori sheets on cling film first, and then roll with the cling film. This will help achieve a tighter roll and assist in shaping. Remove the cling film after cutting.

21 Mirin-glazed tuna and pickled ginger

Ingredient	40 portions		
Tuna	1½ kg		
Teriyaki marinade			
Tamari soy sauce	225ml		
Lime juice and zest	2		
Brown sugar	3 tbsp		
Garlic cloves, chopped	4		
Fresh ginger, chopped	1 tbsp		

Energy	Cal	Fat	Sat fat	Carb	Sugar	Protein	Fibre
256 kJ	61 kcal	1.7g	0.5g	2.2g	2.0g	9.2g	0.1g

1 Mix all ingredients together in a pan, bring to the boil, reduce to a syrupy consistency, allow to cool.

2 Cut the tuna into bite-size pieces, 1–2cm square, place in a tray, marinate in the teriyaki marinade for at least 1 hour.

3 Remove from the marinade, drain quickly, seal and char the tuna on a grill pan, leaving the centre raw.

4 Allow to cool, place into open chopsticks, decorate with a little wasabi and a slice of pickled ginger.

22 Sweet potato kofta

Ingredient	100 portions
Cumin seeds	4 tsp
Coriander seeds	4 tsp
Sweet potato	1 kg
Potatoes	400g
Chickpeas, cooked	300g
Chickpea flour	400g
Ginger, finely chopped	80g
Red chilli brunoise	8
Coriander	2 bunches
Seasoning	

Energy	Cal	Fat	Sat fat	Carb	Sugar	Protein	Fibre
120 kJ	28 kcal	0.4g	0.0g	5.4g	0.8g	1.3g	1.2g

1 Roast the cumin and coriander seeds then crush.
2 Steam the potatoes until just cooked, then cool and peel.
3 Grate the potatoes into a bowl.
4 Purée the chickpeas in a food processor and then place into a bowl.
5 Add the chickpea flour to the chickpeas and mix well.

6 Add the remaining ingredients and mix well.
7 Take teaspoons of the mixture and shape into even-sized balls.
8 Place onto a tray with chickpea flour.
9 When needed for service, deep fry in hot oil at 180°C until golden brown, drain, serve as required with cucumber raita.

23 Parma ham and tarragon tart

Ingredient	20 tarts
Thinly sliced Parma ham	100g
Chopped tarragon	1 tsp
Small short crust pastry tartlets	20
Seasoning	
Egg custard	
Milk	100ml
Single cream	40ml
Whole egg	1
Yolk	1
Seasoning	

Energy	Cal	Fat	Sat fat	Carb	Sugar	Protein	Fibre
765 kJ	183 kcal	11.6g	3.8g	16.7g	0.6g	4.1g	0.9g

1 Combine all the ingredients for the egg custard and pass through a fine sieve (care should be taken not to over-season with salt as the ham will be salty).

2 Cut the ham into julienne, mix with the tarragon and place in the bottom of each of the tartlets.

3 Fill the tartlets up with the custard mixture and place in the oven on a 180°C setting for 6–8 minutes until the custard just sets.

4 Remove and serve warm.

24 Parmesan tuiles

Ingredient	Makes 30 medium-sized tuiles
Parmesan Reggiano	250g

1 Finely grate the Parmesan and sprinkle into cutters on baking paper to form rounds 4cm wide by 0.5cm thick.

2 Bake at 180°C until golden brown. Remove immediately and shape into tile-like pieces, then allow to cool and serve.

Note: Reggiano must be used, as Padano, another cheese type, will not set hard due to its fat content.

25 Salmon tartare

Ingredient	20 portions
Organic salmon, fresh and free from bone, skin and blood line	300g
Chives, chipped	1 tsp
Crème fraiche	50g
Seasoning	
Thin croutes	30
Chervil	
Caviar (optional)	

Energy	Cal	Fat	Sat fat	Carb	Sugar	Protein	Fibre
554 kJ	132 kcal	5.3g	1.2g	16.0g	0.6g	6.0g	0.7g

1 Chop the salmon into 3mm cubes.

2 Mix in the chives and crème fraiche.

3 Check seasoning.

4 Spoon a small mound of salmon mix onto each croute and garnish with chervil.

5 If using caviar, place a small amount on each croute and serve immediately.

26 Lamb, peach and cashew bitoks

Ingredient	4 portions	10 portions
Dried peaches	100g	250g
Minced lamb	450 g	1¼ kg
Garlic, crushed and chopped	1 clove	2–3 cloves
Salt and pepper		
Egg	1	2–3
Breadcrumbs	50g	125g
Cashew nuts	50g	125g

Energy	Cal	Fat	Sat fat	Carb	Sugar	Protein	Fibre
1504 kJ	359 kcal	19.0g	6.0g	29.0g	14.9g	20.0g	3.8g

Variations

Use pine kernels in place of cashew nuts.

Use beef or chicken instead of lamb.

1 Reconstitute the peaches in boiling water for 5 minutes. Drain well. Chop finely and add to the lamb.

2 Add the garlic and season. Bind with the egg and breadcrumbs.

3 Form into small cocktail pieces, insert a cashew nut into each and mould into balls. Flatten slightly.

4 Fry the bitoks in a shallow pan in vegetable oil; drain.

5 Place a cocktail stick in each or place on croutes. Serve a yoghurt and cucumber dressing separately.

27 Wild mushroom risotto balls

Ingredient	40 portions
Mixed wild mushrooms (washed and trimmed)	300g
Oil	50ml
Shallots, diced	2
Garlic clove, chopped	1
Cooked and cooled risotto	300g
Pané anglaise (flour, beaten egg and fresh white breadcrumbs)	
Seasoning	

Energy	Cal	Fat	Sat fat	Carb	Sugar	Protein	Fibre
125 kJ	30 kcal	2.0g	0.3g	2.7g	0.1g	0.4g	0.2g

1 Cut the wild mushrooms into ½cm dice.

2 Heat the oil and sweat off the shallots and garlic.

3 Add the mushrooms and cook slowly until soft. Once cooked, allow to cool.

4 Mix the risotto and mushrooms together.

5 Check seasoning and mould into 1½cm spheres. Allow to set in the fridge for 30 minutes.

6 Pass twice through the pané.

7 Chill, deep fry, drain well and serve. Garnish with cheese and flat parsley.

28 Cherry tomatoes with goat's cheese

Ingredient	24 portions
Cherry tomatoes	24
Shallots	2
Garlic clove, chopped	1
Olive oil	
Goat's cheese	250g
Double cream	50ml
Salt and pepper	

Energy	Cal	Fat	Sat fat	Carb	Sugar	Protein	Fibre
191 kJ	46 kcal	3.9g	2.6g	0.5g	0.5g	2.4g	0.2g

1 Cut the tomatoes in half and scoop out the seeds with a spoon.

2 Sweat the shallots and garlic in olive oil for 2 minutes and cool.

3 Put the goat's cheese and double cream in a bowl and mix together until smooth.

4 Place some of the shallot mix in the bottom of each tomato.

5 On top, pipe a little of the goat's cheese mix.

6 Garnish with chervil or flat parsley and serve.

Test yourself

1 Give an example of an amuse-bouche.

2 Describe the following:
 - Dim sum
 - Tapas
 - Antipasto.

3 What is meant by curing?

4 What is added to brine to give meat its pink colour?

5 What is foie gras?

6 At what temperature should cheese be stored?

7 What type of rice should be used for the preparation of sushi?

8 State the difference between a simple salad and a compound salad.

9 List four quality points to look for when assessing salads.

10 List four types of fish suitable for a smoked fish platter.

11 List six types of cold canapés.

12 State the food safety precautions that need to be taken into account when preparing and serving cold buffet items.

5 Advanced pasta dishes

This chapter covers the following **NVQ** units:
→ Prepare, cook and finish fresh pasta dishes.

Pasta dishes in the **VRQ** Diploma are incorporated and assessed in a range of units. Content in this chapter covers a range of advanced skills and techniques related to pasta dishes, which will cover the following units:
→ Advanced skills and techniques in producing vegetable and vegetarian dishes.
→ Advanced skills and techniques in producing meat dishes.
→ Advanced skills and techniques in producing poultry and game dishes.
→ Advanced skills and techniques in producing fish and shellfish dishes.

Learning objectives

By the end of this chapter you should be able to:
→ Select the correct type, quality and quantity of pasta needed for the dish.
→ Describe what quality points to look for in pasta and other ingredients.
→ Select the appropriate tools and equipment and use correctly.
→ Be able to prepare, cook and present fresh pasta dishes.
→ Make sure the dish has the correct flavour, colour, consistency and quantity.
→ State healthy eating options when preparing, cooking and finishing complex pasta dishes.
→ Explain how to minimise and correct common faults with fresh pasta dishes.
→ State the correct temperature for holding and serving complex pasta dishes.
→ Describe how to store complex pasta dishes.

Recipes included in this chapter

Pasta composition

Dried pasta is made from durum wheat, which has a 15 per cent protein content. This makes it a good alternative to rice and potatoes for vegetarians. Pasta also contains carbohydrate in the form of starch, which gives the body energy. Eating more pasta is in line with the recommendation to 'eat more starchy carbohydrates'.

When most wheat is milled, the endosperm (wheat kernel) breaks down into flour, but the endosperm of durum wheat holds together, and the result is semolina. Durum wheat kernels are amber coloured and larger than those of other wheat varieties.

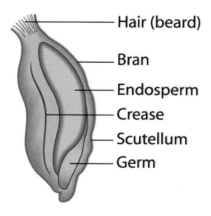

- Hair (beard)
- Bran
- Endosperm
- Crease
- Scutellum
- Germ

▲ The structure of a grain of wheat

The secret of good pasta dough is to use the strongest flour, which has the highest proportion of gluten proteins (gliadin and glutein when hydrated with water produce gluten), which gives the dough its elasticity.

00 flour

This is an Italian grading of flour that is used to make pasta. Italian flour is graded by colour, technically called the **extraction rate** (that is the extent to which the bran and the germ are extracted from the flour). It is marked 00 to 04; grade 00 is really white; 04 is closer to wholemeal. English flour is graded by colour – white, brown and wholemeal – and by the gluten content.

The amount of water the pasta dough can absorb depends on the quality of the protein and the starch. Semolina is used in the process of making pasta dough; the size of the semolina is important – semolina with too small a grain will result in a sticky dough; too large and it will be dry, giving white spots to the finished product.

Types of pasta

There are basically four types of pasta:
- Dried durum wheat pasta, made from the durum variety of wheat
- Fresh egg pasta
- Semolina pasta, made from the finest durum wheat
- Wholewheat pasta, made from flour that uses the entire grain seed; contains slightly more protein than pasta made from refined wheat.

Each of these may be plain, or flavoured with spinach or tomato.

Almost 90 per cent of the pasta eaten in Italy is dried; the remainder is homemade. There is a wide variety of types of dried pasta (pasta asciutta), especially if you include all the regional variations.

Professional tip

A rule of thumb for dried pasta is to allow 90–100g per portion as a starter course. If larger portions are required, increase accordingly. Traditionally, pasta was eaten as a starter, but now it is used much more as a main course or stand-alone dish.

Filled/stuffed pasta

Filled/stuffed pasta is filled with a combination of ingredients that complement the pasta, adding flavour and texture. Examples include:
- **Agnolini** – small half-moon shapes usually filled with ham and cheese or minced meat.
- **Cannelloni** – squares of pasta, poached, refreshed, dried and stuffed with a variety of fillings (for example, ricotta cheese and spinach), rolled and finished with an appropriate sauce.
- **Cappelletti** – shaped like little hats, are usually filled as agnolini, and are available dried.
- **Ravioli** – usually square with serrated edges. A wide variety of fillings can be used (fish, meat, vegetarian, cheese etc.).
- **Tortellini** – a slightly larger version of cappelletti, are also available in dried form.
- **Tortelloni** – is a double sized version of tortellini.

Pasta that is to be stuffed must be rolled as thinly as possible. The stuffing should be pleasant in taste and plentiful in quantity. The edges of the pasta must be thoroughly sealed, otherwise the stuffing will seep out during poaching.

The list of possible stuffings is almost endless, as every district in Italy has its own variations and with thought and experimentation, many more can be produced.

All stuffed pasta should be served in or coated with a suitable sauce. Depending on the type of recipe, it may be finished 'au gratin' – by sprinkling with freshly grated Parmesan and browning lightly under the salamander.

Making tortellini

1 Place the filling in the centre of a piece of pasta. Lightly brush the pasta with beaten egg.

2 Fold the pasta over the filling, creating a semi-circle, and press down around the filling to seal.

3 Trim the excess pasta with a pastry cutter.

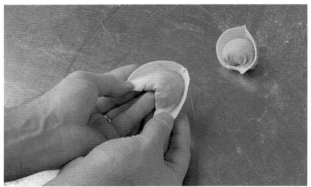

4 Bring the two ends of the long edge together, eggwash the seam and firmly press the ends together to seal.

Making ravioli

1 Large ravioli are made individually. Pipe the filling onto a piece of pasta.

2 Place another piece of pasta, the same size, over the top.

3 Seal the edges, then trim the excess pasta with a pastry cutter. Pinch the edges up.

Small ravioli may be made in sheets and then cut up.

Noodles

Noodles are probably the world's oldest fast food; they are versatile and quick to cook. They may be steamed, boiled, pan fried, stir fried or deep fried. A staple food in the Far East, they are also popular in the West.

Noodles are high in starch (carbohydrate). They provide some protein, especially those made from hard wheat and beans. The addition of egg also increases noodles' protein content. Examples of the nutritional content of some noodles are given in Table 5.1.

Table 5.1 Nutritional content of different types of noodles

100g dry weight	Calories	Carbohydrate	Protein	Fat	Sodium/salt
Rice sticks	380	88%	6%	0	
Rice vermicelli	363	85.5%	7.3%	0	
Egg noodles	341	70%	11%	1.9%	0.8%
Wheat noodles	308	60% +fibre 10%	12.5%	2%	0.8%
Bean thread	320	65%	20%	0	
Pasta	350	75%	11.5%	0.3%	

Sauces and accompaniments for pasta

Sauces to go with pasta include:
- tomato sauce
- cream, butter or béchamel based
- rich meat sauce
- olive oil and garlic
- soft white or blue cheese
- pesto.

Cheeses used in pasta cooking include the following:
- **Parmesan** – the most popular hard cheese for use with pasta, ideal for grating. The flavour is best when it is freshly grated. If bought ready-grated, or if it is grated and stored, the flavour deteriorates.
- **Pecorino** – a strong ewe's milk cheese, sometimes studded with peppercorns. Used for strongly flavoured dishes, it can be grated or thinly sliced.
- **Ricotta** – creamy white in colour, made from the discarded whey of other cheeses. It is widely used in fillings for pasta such as cannelloni and ravioli and for sauces.
- **Mozzarella** – traditionally made from the milk of the water buffalo. Mozzarella is pure white and creamy, with a mild but distinctive flavour and usually round or pear shaped. It will keep for only a few days in a container half filled with milk and water.
- **Gorgonzola and Dolcelatte** – distinctive blue cheeses that can be used in sauces.

Preparing and cooking dried pasta

1 Bring plenty of water (at least 4 litres for every 600g of dry pasta) to a rapid boil. Add 1 tablespoon of salt per 4 litres of water, if desired.
2 Add the pasta in small quantities to maintain the rapid boil. Stir frequently to prevent sticking. Do not cover the pan.
3 Follow package directions for cooking time. Do not overcook. Pasta should be *'al dente'* (meaning literally 'to the tooth' – tender, yet firm). It should be slightly resistant to the bite, but cooked through.
4 Drain pasta to stop the cooking action. Do not rinse unless the recipe specifically says to do so. For salads, drain and rinse pasta with cold water.

Storage

Dry pasta can be stored almost indefinitely in a tightly sealed package or a covered container in a cool, dry place. If fresh egg pasta is to be stored, it should be allowed to dry, and then kept in a clean, dry container in a refrigerator.

If cooked pasta is not to be used immediately, drain and rinse thoroughly with cold water. If the pasta is left to sit in water, it will continue to absorb water and become mushy. When the pasta is cool, drain and toss lightly with salad oil to prevent it from sticking and drying out. Cover tightly and refrigerate or freeze. Refrigerate the pasta and sauce separately or the pasta will become soggy.

To reheat, put pasta in a colander and immerse in rapidly boiling water just long enough to heat through. Do not allow the pasta to continue to cook. Pasta may also be reheated in a microwave.

1 Fresh egg pasta dough

Ingredient	Makes 500g
Strong flour	400g
Medium eggs, beaten	4
Olive oil as required	approx. 1 tbsp
Salt	

Note: If using a pasta rolling machine, divide the dough into three or four pieces. Pass each section by hand through the machine, turning the rollers with the other hand. Repeat this five or six times, adjusting the rollers each time to make the pasta thinner.

1 Sift the flour and salt.

2 Make a well in the flour. Pour the beaten eggs into the well.

3 Gradually incorporate the flour.

4 Mix until a dough is formed. Only add oil to adjust to required consistency. The amount of oil will vary according to the type of flour and the size of the eggs.

5 Pull and knead the dough until it is of a smooth elastic consistency.

6 Cover the dough and allow to rest in a cool place for 30 minutes.

7 After resting, knead the dough again.

8 Roll out the dough on a well floured surface to a thickness of ½mm.

9 Trim the sides and cut as required using a large knife.

Professional tip

Fresh egg pasta requires less cooking time than dried pasta. When cooking fresh pasta, the addition of a few drops of olive oil in the water will help prevent the pasta from sticking together.

Variations

- Add 75–100g finely puréed, dry, cooked spinach to the dough.
- Add 2 tablespoons of tomato purée to the dough.
- Other flavours used include beetroot, saffron and black squid ink.
- For wholewheat pasta, use half wholewheat and half white flour.

2 Ravioli stuffed with spinach and cumin, with lemon sauce

Ingredient	4 portions	10 portions
Pasta dough	200g	500g
Filling		
Butter	25g	75g
Garlic cloves, crushed	1	3
Cumin seeds	1 tsp	2½ tsp
Salt, pepper, nutmeg		
Spinach, chopped	250g	625g
Ricotta	100g	250g
Parmesan, grated	1 tbsp	2½ tbsp
Olive oil	1 tbsp	2½ tbsp
Lemon sauce		
Unsalted butter	175g	500g
Garlic, crushed	1	2
Vegetable stock	100ml	250ml
Lemon zest and juice	1	2
Parsley, chopped	1 tbsp	2 tbsp

Energy	Cal	Fat	Sat fat	Carb	Sugar	Protein	Fibre
3086 kJ	744 kcal	62.7g	34.6g	33.7g	3.4g	13.7g	3.6g

1 For the filling, heat the butter in a suitable pan with the garlic and cumin seeds.

2 Add the spices to taste, and heat for 30 seconds.

3 Add the spinach and cook over a low heat for 3–4 minutes.

4 Place the filling in a bowl and allow to cool.

5 Add the ricotta and Parmesan. Mix well. Season with salt, pepper and nutmeg.

6 Roll out the pasta dough on a lightly dusted surface of semolina. Cut into rounds, approximately 8cm across. Brush half of the rounds of dough with water, then place on pieces of the filling. Cover each one with a second round of dough. Press down the edges well to seal the dough. Place the ravioli on a tray dusted with semolina.

7 Cook the ravioli in boiling, salted water for 2–3 minutes. Remove and drain well on a cloth.

8 Place ravioli onto plates, drizzle with olive oil, and keep warm.

9 For the sauce, melt the butter with the garlic in a pan, add the stock, and bring to the boil. Simmer for 5 minutes.

10 Add the lemon juice and zest. Season.

11 Pour the sauce over the ravioli, and serve garnished with julienne of zest of lemon and flat parsley. Alternatively, garnish with basil leaves and add a little chopped basil to the lemon sauce.

Video: ravioli
http://bit.ly/1cMC0dD

3 Tagliatelle with pesto and squid

Ingredient	4 portions	10 portions
Pesto		
Sea salt	1 tsp	2½ tsp
Pine nuts, plus extra for garnish	25g	60g
Garlic cloves, chopped	2	5
Basil plant (large), leaves only	1	2
Freshly grated Parmesan cheese	75g	180g
Freshly grated pecorino cheese	25g	60g
Extra virgin olive oil	100–200ml	250–500ml
Tagliatelle		
Fresh tagliatelle	200g	500g
Squid		
Olive oil	2 tbsp	5 tbsp
Squid, cleaned and scored	2	5
Lemon juice	½	1

Energy	Cal	Fat	Sat fat	Carb	Sugar	Protein	Fibre
3523 kJ	845 kcal	58.5g	11.9g	41.9g	2.2g	40.3g	1.7g

1 Using a mortar and pestle or a blender, prepare the pesto by pounding together the salt, pine nuts and garlic. Work in the basil leaves, pounding until you have a smooth paste. Add the cheeses, and then beat in the olive oil until you have a thick, dense sauce. Adjust the amount of oil depending on the texture of pesto you prefer. Set aside.

2 Bring a large pan of salted water to the boil. Drop in the tagliatelle and boil for 4–5 minutes, or as per the packet instructions.

3 For the squid, heat the oil in a medium frying pan. Add the squid and fry for 2–3 minutes. Squeeze in half the lemon juice.

4 Drain the tagliatelle and tip into the pan with the squid. Mix in the pesto and divide the pasta between four (or ten) bowls. Scatter over the pine nuts to finish.

5 Store the remaining pesto in a jar in the refrigerator. Parmesan was used in place of pecorino for the nutritional analysis.

4   Tripoline with peppers and aubergine

Ingredient	4 portions	10 portions
Aubergines, small	700g	1.75 kg
Olive oil	80ml	200ml
Tripoline	400g	1 kg
Fresh basil, shredded		
Red pepper sauce		
Red peppers	600g	1.5 kg
Clove of garlic, crushed and chopped	1	3
Olive oil	1 tbsp	2 tbsp
White vinegar	2 tbsp	5 tbsp
Balsamic vinegar	2 tsps	5 tsps
Caster sugar	1 tsp	3 tsp
Chicken stock	125ml	300ml
Single cream	80ml	200ml

Energy	Cal	Fat	Sat fat	Carb	Sugar	Protein	Fibre
2882 kJ	693 kcal	29.7g	6.2g	93.6g	14.4g	12.6g	7.9g

1 Cut the aubergines into 1cm slices, place in a colander, sprinkle with salt and stand for 30 minutes.

2 Rinse slices under cold water and drain well.

3 Shallow fry aubergines in hot oil until brown on both sides.

4 Cook the tripoline in boiling water until al dente and drain well.

5 Quarter the peppers and remove the seeds and membranes.

6 Place the peppers under the grill, skin side up, until the skin blisters.

7 Peel the skin.

8 Purée the peppers in a processor with the garlic, oil, vinegars and sugar until smooth.

9 Place in a pan with the stock and cream and stir until heated through.

10 Serve the tripoline by heating in a little oil. Season, and add the pepper sauce.

11 Serve arranged on a plate with the slices of aubergine. Garnish with shredded fresh basil.

5 Penne with pancetta and mushrooms

Ingredient	4 portions	10 portions
Dried porcini mushrooms	30g	75g
Boiling water	375ml	1 litre
Penne pasta	500g	1250g
Olive oil	2 tbsp	5 tbsp
Onions, shredded	300g	750g
Button mushrooms, sliced	100g	250g
Single cream	300ml	750ml
Pancetta, sliced	100g	250g
Parmesan, grated	50g	125g
Basil leaves, shredded		

Energy	Cal	Fat	Sat fat	Carb	Sugar	Protein	Fibre
3077 kJ	734 kcal	35.6g	14.5g	81.7g	7.9g	27.4g	1.8g

1 Place porcini mushrooms in a suitable bowl, cover with boiling water and stand for 30 minutes.

2 Drain the mushrooms and chop.

3 Cook the pasta in boiling water until al dente and drain.

4 Heat the oil in a suitable pan and add the onions. Sweat but do not colour. Stir in the porcini and button mushrooms and cream, stir well.

5 Stir in the pasta and the pancetta until heated through.

6 Serve sprinkled with cheese and garnished with shredded basil.

6 Tagliatelle with smoked salmon and asparagus

Ingredient	4 portions	10 portions
Fresh asparagus	250g	625g
Smoked salmon, sliced	200g	500g
Butter	50g	125g
Leek, shredded	200g	500g
Clove of garlic, chopped	1	2
Brandy	2 tbsp	3 tbsp
Single cream	500ml	1.25 litres
Tomato purée	1 tsp	3 tsp
Tabasco sauce	½ tsp	1½ tsp
Fresh basil, chopped	1 tbsp	2 tbsp
Salmon roe	20g	50g
Capers	1 tsp	2 tsp
Tagliatelle pasta	500g	1250g

Energy	Cal	Fat	Sat fat	Carb	Sugar	Protein	Fibre
3696 kJ	880 kcal	39.4g	22.5g	99.1g	9.9g	36.0g	8.0g

1 Cut the asparagus into 3cm lengths.

2 Cook the asparagus in boiling water or steam it.

3 Cut the salmon into 2cm strips.

4 Heat the butter in a suitable pan. Add the leek and garlic, and sweat until cooked without colour. Add the brandy, and cook for a further minute.

5 Add the cream, tomato purée, and Tabasco. Simmer for 5 minutes.

6 Add the asparagus and salmon.

7 Cook the pasta in boiling water until al dente and drain.

8 Place the pasta on suitable plates, cover with the sauce, and garnish with salmon roe and capers.

In the nutritional analysis, peas were used instead of capers.

7 Mussels with pasta, turmeric and chervil

Ingredient	4 portions	10 portions
Olive oil	50ml	125ml
Chilli, finely chopped	½	1
Garlic clove, finely chopped	1	3
Spring onions, chopped	3	8
Double cream	3 tbsp	8 tbsp
Turmeric	1 tsp	2½ tsp
Mussels, cooked	100g	250g
Pasta, cooked	100g	250g
Salt and freshly ground black pepper		
Fresh chervil, chopped	1 tbsp	2 tbsp

Energy	Cal	Fat	Sat fat	Carb	Sugar	Protein	Fibre
1196 kJ	289 kcal	26.3g	10.0g	7.7g	1.0g	6.0g	0.7g

1 Heat the oil in a wok and sauté the chilli, garlic and spring onion until soft.

2 Add the double cream, turmeric and mussels and cook until the mussels are heated through and the cream has reduced slightly.

3 Add the pasta, season well with salt and freshly ground black pepper and stir together to combine.

4 To serve, transfer to a serving bowl and sprinkle over the chervil.

8 Smoked fish carbonara

Ingredient	4 portions	10 portions
Fresh tagliatelle	400g	1 kg
Smoked pancetta	200g	500g
Mushrooms, sliced	100g	250g
Spring onions, chopped	3	7
Cherry tomatoes, quartered	4	10
Smoked cod or haddock fillet, fresh or defrosted, skinned and cubed	400g	1 kg
Eggs	4	10
Egg yolks	4	10
Double cream	210ml	525ml
Olive oil	1 tbsp	2½ tbsp
Pecorino cheese, grated	150g	375g
Black pepper		
Fresh parsley, chopped	2 tbsp	5 tbsp

Energy	Cal	Fat	Sat fat	Carb	Sugar	Protein	Fibre
5943 kJ	1417 kcal	70.1g	32.7g	131.9g	7.9g	73.0g	8.0g

1 Place a large pot of water on the stove and bring to a rolling boil to cook the tagliatelle.

2 While the pasta is cooking, heat a frying pan and fry the pancetta without any extra oil until it is crisp and golden (about 5 minutes). Add the onions and mushrooms and cook for a further 5 minutes. Add the tomatoes and diced fish and stir around the pan for a minute or two (at this point it is only partly cooked).

3 Whisk the eggs, yolks and cream in a bowl and season generously with black pepper, then whisk in the cheese.

4 When the pasta is cooked, drain it quickly in a colander, leaving a little of the moisture still clinging. Quickly return it to the saucepan and add the pancetta, mushroom, onion and fish mix and any oil in the pan, along with the egg and cream mixture and the chopped parsley.

5 Stir very thoroughly, so that everything gets a good coating – what happens is that the liquid egg cooks briefly as it comes into contact with the hot pasta, and the fish will cook through at this point.

6 Serve the pasta on really hot, deep plates with some extra grated pecorino or Parmesan cheese.

Spaghetti or other types of pasta may be used.

9 Herb pasta with Parma ham

Ingredient	4 portions	10 portions
Chlorophyll		
Watercress, picked	40g	90g
Parsley, picked	2g	6g
Baby spinach	10g	25g
Rocket	3g	6g
Water	5ml	12ml
Pasta		
00 pasta flour	240g	550g
Egg yolks	5	12
Chlorophyll (see instructions below)	12g	35g
Parma ham, sliced	50g	125g

Energy	Cal	Fat	Sat fat	Carb	Sugar	Protein	Fibre
1229 kJ	292 kcal	9.0g	2.6g	42.8g	0.9g	12.4g	2.6g

Chlorophyll

1 Pick the vegetables. Blanch for 3 minutes and refresh.

2 Blitz and pass through a chinois.

3 Add water to the required consistency.

To complete the dish

1 Place the pasta flour into a food processor (e.g. Robot Coupe) and turn on.

2 Slowly pour the whisked egg yolks and chlorophyll into the flour until it forms a paste.

3 Turn out of the food processor and knead on a bench for 5 minutes to work the gluten and make the pasta more pliable for use.

4 Roll in a pasta machine and turn into linguini.

5 Cook for 2 minutes in boiling salted water, then toss in a little oil.

6 Finish with some sliced Parma ham.

10 Farfalle alla papalina

Ingredient	4 portions	10 portions
Fresh farfalle	400g	1 kg
Onion, finely chopped	50g	125g
Prosciutto crudo (Parma, San Daniele or similar)	125g	300g
Butter	50g	125g
Eggs	3	8
Double cream	3 tbsp	8 tbsp
Parmesan cheese, grated	60g	150g
Salt, pepper and nutmeg		

Energy	Cal	Fat	Sat fat	Carb	Sugar	Protein	Fibre
2358 kJ	565 kcal	37.4g	20.2g	33.4g	1.9g	25.8g	2.8g

1 Shallow fry the onion and the prosciutto in half the butter.

2 Beat the eggs, cream and half the cheese together. Pour into a large frying pan.

3 Cook the pasta in boiling salted water until 'al dente'.

4 Place the frying pan with the eggs and cream over a low heat – do not allow it to scramble.

5 Drain the pasta and add to the pan. Mix thoroughly, then add the ham mixture and the remaining butter.

6 When it is mixed thoroughly, sprinkle with remaining cheese over the top and serve.

Note: It is advisable to use pasteurised egg for this recipe.

11 Pappardelle with anchovy sauce

Ingredient	4 portions	10 portions
Pappardelle		
Semolina flour	400g	1 kg
Egg	1	3
Water	100ml	250ml
Anchovy sauce		
Olive oil	3 tbsp	200ml
Onion, finely sliced	1	3
Garlic cloves	1	2
Anchovy fillets	100g	250g
Fresh parsley, chopped	1 tsp	2 tsp
Pepper		

Energy	Cal	Fat	Sat fat	Carb	Sugar	Protein	Fibre
2599 kJ	619 kcal	26.1g	3.7g	81.4g	2.6g	19.6g	3.8g

1 To make the pasta dough, combine the flour, egg and water. Mix and knead well, adding extra water if needed, until the dough is smooth and elastic. Turn out onto a lightly floured surface and knead again. Divide the pasta dough into four pieces and then knead each piece well.

2 Roll each piece out as thinly as possible, then cut into thin strips, using more flour if necessary to prevent sticking. Set aside while making the anchovy sauce.

3 Heat the olive oil in a pan. Add the onions and sweat gently before adding the garlic. Cook over a low heat for 10 minutes, stirring occasionally.

4 Add the anchovies to the pan, stir and cook for a further 10 minutes, stirring occasionally until the anchovies have puréed among the onions and created a rich sauce.

5 Meanwhile, bring a large pan of salted water to the boil and cook the pasta for 3–4 minutes or until 'al dente' – cook in batches if necessary.

6 Remove the garlic clove from the pan and add the chopped parsley to finish.

7 Drain the pasta and add it straight into the pan with the sauce, add a couple of tablespoons of the pasta water to loosen. Mix through and season with black pepper. Serve immediately.

12 Spaghetti with piquant sauce

Ingredient	4 portions	10 portions
Spaghetti	400g	1 kg
Smoked bacon cut into lardons	125g	300g
Onion, finely shredded	100g	250g
Fresh sage, finely chopped	1 tsp	3 tsp
Olive oil	60mls	150mls
Canned plum tomatoes, chopped	300g	750g
Chilli powder	¼ tsp	½ tsp
Wine vinegar	1 tbsp	3 tbsp
Capers	2 tbsp	5 tbsp
Dried marjoram	¼ tsp	½ tsp
Butter	25g	60g
Pecorino cheese, grated	50g	125g

Energy	Cal	Fat	Sat fat	Carb	Sugar	Protein	Fibre
2792 kJ	665 kcal	30.7g	9.8g	79.3g	7.1g	23.0g	5.4g

1 Shallow fry the bacon, onion and sage in the oil until the onion is cooked but not coloured.

2 Add the tomatoes and simmer for 10 minutes.

3 Season with salt and chilli powder.

4 Add the vinegar, capers and marjoram and cook for a further 2 minutes.

5 Cook the pasta in boiling salted water and drain.

6 Place pasta in a suitable pan, add the butter, sauce and cheese and mix thoroughly and serve.

In the nutritional analysis, peas were used instead of capers.

13 Cappelletti romagnoli (in the style of Romagna)

Ingredient	4 portions	10 portions
Filling		
Chicken or turkey breast meat, finely diced	250g	625g
Minced pork	65g	170g
Butter	50g	125g
Fresh sage	5 leaves	12 leaves
Fresh rosemary	1 sprig	2 sprigs
Ricotta cheese	50g	125g
Fresh watercress	50g	125g
Parmesan cheese, grated	50g	125g
Lemon rind, grated	1 tsp	2½ tsp
Salt and pepper, grated nutmeg		
Eggs	2	5
For the pasta		
Plain flour (strong)	400g	1 kg
Eggs	4	10
Salt		
To serve		
Fresh tomato sauce (see page 180)	500ml	1.25 litres
Parmesan cheese, grated	75g	180g

Energy	Cal	Fat	Sat fat	Carb	Sugar	Protein	Fibre
3844 kJ	915 kcal	39.8g	20.5g	87.8g	10.7g	69.5g	8.0g

1 Add the butter to a pan, then add the chicken or turkey, the pork, sage and rosemary.

2 When cooked, mince with the ricotta cheese, watercress, Parmesan and lemon rind.

3 Season with salt, pepper and nutmeg, and bind with beaten egg.

4 Make the pasta by sieving the flour, adding the beaten eggs and salt.

5 Roll half the pastry out onto semolina, into a large square approximately 30cm × 30cm and divide into squares approximately 8cm × 8cm. Place a small amount of filling on each square. Eggwash around the filling. Filling should be approximately 25mm apart.

6 Roll out the remaining pastry and cover the first sheet. Cut into neat squares.

7 Take each square and fold into a triangle. Wrap each triangle around your finger and press the corners together. A little water or eggwash will help them stick together.

8 Cook the cappelletti in plenty of boiling, salted water, drain and transfer onto a warm platter. Mask with a fresh tomato sauce or other suitable sauce, e.g. Bolognaise or cream sauce. Sprinkle with Parmesan and serve.

14 Fried ravioli with asparagus spears

Ingredient	4 portions	10 portions
For the filling		
Large aubergine	1	3
Onion, finely chopped	50g	125g
Butter	50g	125g
Plain, strong flour	50g	125g
Milk	120ml	300ml
Walnuts, chopped	4	10
Egg yolks	2	5
For the pasta		
Plain, strong flour	400g	1 kg
Eggs	4	10
Salt		
Oil for deep frying		
To serve		
Asparagus spears	300g	750g
Plain flour	15g	40g
Butter	25g	60g
Tomato sauce	500ml	1.25 litres
Parmesan cheese	50g	125g

Energy	Cal	Fat	Sat fat	Carb	Sugar	Protein	Fibre
3594 kJ	857 kcal	38.5g	16.6g	99.3g	9.2g	34.5g	11.3g

1 Slice the aubergines, sprinkle with salt and leave for 2 hours.

2 Rinse under cold water, dry on absorbent paper, and cut into dice.

3 Shallow fry the onion in butter. Add the diced aubergine and fry until golden brown. Remove the aubergine, and add the flour. Stir until smooth; add boiling milk, gently stirring between each addition. This should be a smooth, paste-like consistency and may require more milk.

4 Add the walnuts and egg yolks, mix well, then add the aubergine. Mix the filling together well, and season with salt and pepper.

5 Make the pasta dough with the flour, eggs and salt. Roll out half the dough on semolina into a large square 30cm × 30cm and divide into squares approximately 5cm × 5cm. Add the filling, eggwash and cover with the remaining dough. Allow to rest for 10 minutes.

6 Deep fry the ravioli in hot oil at 180°C until crisp.

7 Serve the ravioli on a suitable plate.

8 Toss the asparagus in flour, fry quickly in butter until tender and arrange round the ravioli.

9 Serve the tomato sauce over the top or separately.

10 Sprinkle the ravioli with cheese or serve separately.

Test yourself

1 Which part of the wheat grain is the endosperm?

2 What is 00 flour and why is it used to make pasta dough?

3 Name three common Italian cheeses used in pasta dishes.

4 What does the term 'al dente' mean?

5 How should cooked pasta be stored when not for immediate use?

6 Name two types of stuffed pasta.

7 Explain why semolina is used to roll out pasta dough.

8 Why is it important to select the current quality and quantity of ingredients in order to meet the dish requirements and specification?

9 How should dry pasta be stored?

10 When making complex pasta dishes, why is it important to consider the balance of colour, texture and flavour of the dish?

11 List the health and safety points that must be considered when preparing, cooking and serving pasta dishes.

12 Name two different types of sauces that may be served with pasta dishes.

6 Advanced soups, sauces and dressings

This chapter covers the following **NVQ** units:
→ Prepare, cook and finish complex hot sauces.
→ Prepare, cook and finish complex soups.
→ Prepare, cook and finish dressings and cold sauces.

Soups, sauces and dressings in the **VRQ** Diploma are incorporated and assessed in a range of units. Content in this chapter covers a range of advanced skills and techniques related to soups, sauces and dressing which will cover the following units:
→ Advanced skills and techniques in producing meat dishes.
→ Advanced skills and techniques in producing poultry and game dishes.
→ Advanced skills and techniques in producing fish and shellfish dishes.

Introduction
This chapter will help you develop a range of advanced skills and knowledge needed for the preparation, cooking and serving of hot and cold sauces, soups and dressings.

You will cover ingredients, equipment, processes, systems, hygiene and health and safety.

Learning objectives
By the end of the chapter you should be able to:
→ Select the correct type, quality and quantity of ingredients to be used in the preparation, cooking and serving of soups, hot and cold sauces and dressings.
→ Describe the correct tools and equipment to use when preparing soups, sauces and dressings.
→ Describe what quality points to look for in relation to advanced soups, hot and cold sauces and dressings.
→ Prepare, cook and finish soups, hot and cold sauces and dressings to meet requirements.
→ State healthy eating options when preparing, cooking and finishing soups, hot and cold sauces and dressings.
→ Describe how to store soups, sauces and dressings.
→ Explain how to minimise common faults with soups, sauces and dressings.

Recipes included in this chapter

Soups

History of soup

Originally soup provided basic sustenance and, in most houses, it would be a one-pot meal that would consist of scrag ends of meat, vegetables, pulses and roots. It would be cooked for a while to extract lots of flavour. Soup was originally seen as a meal in itself – served with a generous chunk of bread and enjoyed beside a warm hearth.

There are many classes of soup: the classic velouté thickened with a liaison; purée of lentils with ham hock stock; broth with a clear liquid; or the crystal clarity of a consommé. All of these methods still grace modern restaurant tables. However, while 15–20 years ago most soups were thickened by either purée or by roux (the latter method the most common), today's dining takes a lighter and more sophisticated approach. Many restaurant chefs are now making soups without using flour and fat to thicken. A careful approach to thickening will yield a modern soup.

Soups often open the menu and therefore must stimulate the appetite. They should be interesting with light, delicate flavours.

Types of soup

- Bisque: a very rich soup with a creamy consistency; usually made of lobster or shellfish (crab, shrimp, etc.).
- Bouillabaisse: a Mediterranean fish soup/stew, made of multiple types of seafood, olive oil, water, and seasonings like garlic, onions, tomato and parsley.
- Broth: an unpassed soup containing vegetables and sometimes meat or fish (e.g. Scotch broth).
- Chowder: a hearty North American soup, usually with a seafood base.
- Consommé: a clear, unthickened soup, with an intense flavour derived from meat or fish bones and a good stock, clarified by a process of careful straining.
- Cream: based on a velouté or a purée, and finished with cream.
- Dashi: the Japanese equivalent of consommé; made of giant seaweed, or *konbu*, dried bonito and water.
- Gazpacho: a Spanish tomato soup served ice cold.
- Minestrone: an Italian vegetable-based soup.
- Potage: a French term referring to a thick soup.
- Puréed soup: a soup of vegetable base that has been puréed in a blender or liquidiser; typically altered after blending with the addition of stock, cream, butter, sour cream or coconut milk.
- Velouté: a velvety French sauce made with stock; can be adapted into a soup and thickened with a liaison.
- Vichyssoise: a simple, flavourful, puréed potato and leek soup, thickened with the potato itself. If served hot, this is leek and potato soup, and a little cream or crème fraiche may be added to give a richness to the soup; however, served cold, it is classified as vichyssoise and fat is not added, because if it were, the dish would leave a fatty residue on the palate and offer a less than clear mouth feel.

Combination soups

Combination soups can be created in several different ways:
- By combining finished soups, e.g. adding a watercress soup to a tomato soup.
- Combining the main ingredients together in the initial preparation, e.g. tomatoes and watercress.
- Carefully using different herbs to introduce a subtle flavour, e.g. basil, rosemary, chervil.
- Using a garnish that is varied, e.g. blanched watercress leaves, chopped chives and tomato concassé, and finishing with non-dairy cream or yoghurt.

Examples of combination soups include watercress and lettuce, leek and broccoli, potato and endive, tomato and cucumber, chicken and mushroom.

Table 3.1 Guide to soups, their preparation and presentation

Soup type	Base	Passed or unpassed?	Finish	Example
Clear	Stock	Strained	Usually garnished	Tomato and garlic consommé (Recipe 2)
Broth	Stock Cut vegetables	Unpassed	Chopped herbs	Scotch broth Minestrone
Purée	Stock Fresh vegetables Pulses	Passed	Croutons	Lentil soup Potato soup
Velouté	Blond roux Vegetables Stock	Passed	Liaison of yolk and cream	Chicken velouté with mushrooms and tongue (Recipe 8)
Cream	Stock and vegetables Vegetable purée	Passed	Cream, milk or yoghurt	Cream of asparagus (Recipe 15)
Bisque	Shellfish Fish stock	Passed	Cream	Lobster bisque (Recipe 18)
Miscellaneous				Cockle chowder (Recipe 19) Gazpacho (Recipe 11)

Quality points: soups

- Soups should never be too thick and should easily flow from the spoon.
- Only the best quality stock of the appropriate flavour should be used to enhance the soup. Care should be taken to preserve the flavour of the main ingredients – they should not be overpowered by an over-strong stock.
- Always use fresh, carefully selected, quality ingredients.
- Follow the recipe carefully to achieve the right balance of flavours and consistency. Use the correct quantities and ingredients.
- Portion size is usually approximately 250ml, but can be less if followed by a number of courses.
- Always check the flavour and consistency before service.

Dietary requirements

Many customers have special dietary requirements and consideration must be given to their needs:

- **Elderly people** need soups that are easily digestible, healthy, nutritious and light, but with carbohydrates for energy – for example, a good, light vegetable broth.
- **Diabetes**: when making soups for customers with diabetes include complex carbohydrates and use foods that are high in the glycaemic index, such as beans and wholemeal flour.

- **Coeliac disease**: for customers with coeliac disease, use potatoes and vegetables, or rice flour, to thicken soups.
- **Dairy intolerance**: in place of milk, use soya milk or rice milk.
- **Vegetarians and vegans**: only use vegetable stock and vegetable-based oils.

Health and safety

When preparing soups, sauces and dressings for vegetarians and vegans, use the colour-coding system to prevent any cross-contamination with meat.

Healthy eating tips

- Cream and velouté soups may be made with skimmed milk and finished with non-dairy cream, natural yoghurt or fromage frais.
- Purée soups may be thickened using vegetables and require no flour.
- Wholemeal flour can be used to thicken soups.
- Instead of butter, use unsaturated oils, including flavoured oils such as basil oil, chive oil, truffle oil.
- Vegetable broths and vegetable soups provide customers with a healthy option and can form part of your '5 a day'.

Garnishes and accompaniments

Most soups are served with a garnish and/or an accompaniment. Bread is usually offered but other garnishes and accompaniments include:

- **Croutons:** small cubes of bread – usually white, but wholemeal can be used. The crusts are taken off the bread before it is cubed. The croutons are then fried: traditionally in hot clarified butter, but vegetable oils, including flavoured oils (e.g. garlic oil) are increasingly being used. Croutons help to soak up the soup, add texture and enhance the eating quality.
- **Toasted flutes (croutes de flute):** these are very traditional. They are slices of bread taken from a thin baguette. They are usually toasted, but can be fried in clarified butter or oil, or drizzled with oil and baked in the oven.
- **Sippets:** again, these are very traditional. They are cut in triangles from the corners of bread. They are toasted or baked in the oven, drizzled with a little flavoured oil. They may also be sprinkled with chopped, mixed fresh herbs.
- **Diablotins:** rounds or squares of toasted bread, sprinkled with grated cheese and gratinated. They may also be spread with a meat or vegetable purée and coated with a light cream sauce.
- **Melba toast:** made from very thin slices of bread and toasted. The modern method is to slice the bread in a gravity feed slicing machine and toast under the salamander.

- **Rouille:** usually consists of olive oil, breadcrumbs, garlic, saffron and chilli peppers. Served as a garnish with fish soup.
- **Cheese straws:** made from strips of puff pastry flavoured with cheese and baked in the oven. They can be twisted or flat like a straw.
- **Julienne vegetable garnish:** a cut of fine strips, sized to fit in a soup spoon 35mm in length. Vegetables used include carrot, leek, celery and turnip. Carefully blanch in a little stock or water.
- **Brunoise vegetable garnish:** small dice approximately 2mm; used for consommés. For vegetable broths, the dice are made larger, approximately 4mm. Carrot, turnip, leek or celery can be used and blanched in a little stock or water.
- **Paysanne vegetable garnish:** squares and triangles of vegetables. Often celery is cut into thin slices to give a triangle effect. Other vegetables such as carrot, leek and turnip are cut into approximately 12mm squares.
- **Lettuce or spinach:** shredded finely and sweetened in a little oil or butter.
- **Profiteroles:** very small choux buns, filled with chicken mousse and vegetable purées.
- **Fritters:** small cheese or vegetable fritters to finish the soup as it is served.
- **Gnocchi:** small poached pieces of choux pastry, can be flavoured with cheese, mixed herbs, or tomato.

> **Professional tip**
>
> It is important to constantly toss croutons while they are frying to give an even colour. Drain well on kitchen paper.

Preparation methods

Passing and straining soups

This soup is a tomato and garlic consommé (see Recipe 2).

1 The tomatoes are chopped roughly.

2 The chopped tomatoes are blended, with the garlic, and then placed into a clean muslin.

3 Tie the muslin securely, forming a bag.

4 Passing through the muslin yields an almost clear, but intensely flavoured, consommé.

Holding and serving soups

Soups must be served at the correct temperature – hot but not boiling. In accordance with the Food Hygiene Regulations 2006, soup must be served at above 63°C, but 72°C is a better temperature to ensure that pathogenic micro-organisms are killed. Chilled soups should be served at 5°C.

All food on display for service must be discarded after 2 hours.

Storage of soups

Ideally, soup should be served immediately after cooking. Soups prepared in advance should be chilled immediately in a blast chiller, in a shallow container, reaching 3°C within 1½ hours of cooking. Store in a refrigerator. When required, reheat to at least 72°C for at least 5 minutes.

Regional recipe – Devon mussel broth

Contributed by David Beazley at City College Plymouth

Ingredient	4 portions	10 portions
Dry Devon cider	125ml	325ml
Mussels, cleaned	1 kg	2.25 kg
Devon double cream	125ml	325ml
Purslane or rocket, prepared	50g	125g
Sea beet, shredded	50g	125g
Rock samphire, blanched	25g	75g

Energy	Cal	Fat	Sat fat	Carb	Sugar	Protein	Fibre
592 kJ	142 kcal	8.5g	4.6g	3.2g	0.7g	12.4g	0.1g

Note: purslane has oval leaves with a sweet and sour flavour. It can be cooked or used in salads.

1 In a large saucepan, add the cider and reduce by half.

2 Add the cleaned mussels and cover with the lid, turn up the heat and leave for three minutes or until all of the mussels are open. Discard any that remain closed.

3 Strain the liquid from the mussels into a bowl. Remove the mussels from their shells.

4 Return the liquid to the saucepan, add the cream and bring to the boil, remove from the heat.

5 Add the leaves and the mussels and the soup is ready.

1   Consommé of duck and beetroot

This recipe is based on borscht, a traditional, unclarified broth of Eastern European origin. Borscht is served with sour cream, beetroot juice and small duck patties.

Ingredient	4 portions	10 portions
Duck leg, chopped and minced with no fat	200g	500g
Water	125ml	300ml
Small onion, chopped finely	1	3
Small carrot, chopped finely	1	3
Celery, small stick, chopped finely	1	3
Bouquet garni	1	3
Salt	¼ tsp	½ tsp
White peppercorns	4	10
Sherry vinegar	2 tbsp	5 tbsp
Raw beetroot, grated	200g	500g
Egg whites	2	5
Cold game stock	1.25 litres	3 litres
To finish		
Beetroot, cooked and diced	150g	375g
Duck breast, roasted	2	5

Energy	Cal	Fat	Sat fat	Carb	Sugar	Protein	Fibre
682 kJ	163 kcal	7.8g	2.5g	3.6g	3.2g	19.6g	0.7g

1 Mix all the ingredients together, excluding the stock.

2 Place into a saucepan. Add the stock and mix well.

3 Heat slowly to simmering point, whisking occasionally until the froth rises to the surface.

4 Remove the whisk, cover and simmer very gently for 45–60 minutes. Do not allow to boil or the froth will break and cloud the consommé.

5 Strain slowly through a muslin and if necessary strain again.

6 Re-heat and re-season if required; however, adding salt will make the consommé slightly cloudy.

7 Serve garnished with the diced, cooked beetroot and julienne of roast duck breast.

2 Tomato and garlic consommé

Ingredient	4 portions	10 portions
Plum tomatoes	3 kg	7 kg
Large onion, chopped finely	1	3
Celery stick, chopped finely	1	3
Garlic cloves, crushed	5	12
Salt		
To finish		
Tomato concassé	4 tbsp	10 tbsp
Basil, julienned		

Energy	Cal	Fat	Sat fat	Carb	Sugar	Protein	Fibre
44 kJ	10 kcal	0.1g	0.0g	2.0g	1.9g	0.3g	0.4g

1 Chop the tomatoes roughly.

2 Place in a food processor and using the pulse button, pulse the tomatoes for 5–10 seconds.

3 Mix the tomatoes with the rest of the ingredients. Sprinkle with a little salt and place in a clean, damp tea towel or muslin and hang overnight.

4 This will yield clear, intensely flavoured tomato water that can be served chilled or heated slightly.

5 Garnish with the tomato concassé and julienne of basil.

3 Mussel soup

Ingredient	4 portions	10 portions
Mussels	400g	1.25 kg
Fish stock	1 litre	2.5 litres
Shallots or onions	50g	125g
Celery	50g	125g
Leek	50g	125g
Parsley	10g	25g
Salt and pepper		
White wine	60ml	150ml
Cream	125ml	300ml
Egg yolk	1	2–3

Energy	Cal	Fat	Sat fat	Carb	Sugar	Protein	Fibre
834 kJ	202 kcal	18.8g	10.9g	3.2g	2.1g	5.0g	0.6g

1 Scrape and thoroughly clean the mussels.

2 Place in a pan with the stock, chopped vegetables and herbs; season.

3 Cover with a lid and simmer for 5 minutes.

4 Extract the mussels from the shells and remove the beards. Discard any closed shells.

5 Strain the liquid through a double muslin and bring to the boil; add the wine.

6 Correct the seasoning, finish with a liaison of egg yolk and cream and garnish with the mussels and more chopped parsley.

Variations

- Try using scallops in place of mussels, and fennel and dill in place of parsley.
- Prepare a potato soup using fish stock, garnish with mussels and finish with cream.

4 Watercress and beetroot soup

For the beetroot

1 Place the beetroot in a pan and cover with the water. Bring to the boil.
2 Turn down to a simmer until cooked (about 1½ hours).
3 While the beetroot is cooking, place the vinegar, water and sugar in a separate pan and bring to the boil.
4 Boil this for 5 minutes, then remove from the heat.
5 Once the beetroot is cooked, drain the liquid and peel the beetroot while it is still warm, then cut into dice.
6 Add the bay leaf to the vinegar/water/sugar mixture, and pour over the diced beetroot. Reserve for a least 2 hours before using.

For the watercress soup

1 Sweat the onion and leek without colour in the butter. Cook until very tender.
2 Add the potatoes and bring quickly to the boil with the water. Season with salt and pepper and allow to cool (blast chill).
3 Separately sweat the spinach and watercress in the butter until wilted. Transfer to a suitable container and add ice (to help preserve the colour).
4 When both are cool, liquidise each mix separately and pass through a fine strainer.
5 Add the potato purée to the watercress purée until the correct consistency and flavour has been achieved.
6 Correct seasoning.
7 Meanwhile place a small amount of the drained beetroot in each serving bowl and top with the soup, which should be heated to at least 75°C.

> **Professional tip**
> The beetroot for this dish is best left overnight to develop the flavour

Ingredient	4 portions	10 portions
Beetroot		
Large beetroots	2	5
Water for cooking		
White wine vinegar	150ml	375ml
Water	100ml	250ml
Sugar	150ml	375ml
Bay leaf	1	2
Watercress soup		
Onions	160g	400g
Leeks	175g	430g
Butter (for onions and leeks)	125g	300g
Potatoes, diced small	750g	1.8 kg
Water	1.8 litres	4.5 litres
Salt	15g	30g
Pepper to taste		
Spinach	125g	300g
Watercress	600g	1.5 kg
Butter (for spinach and watercress)	125g	300g

Energy	Cal	Fat	Sat fat	Carb	Sugar	Protein	Fibre
2822 kJ	681 kcal	53.4g	32.8g	42.3g	9.8g	10.3g	8.0g

5 Lentil and mushroom soup

Ingredient	4 portions	10 portions
Butter	50g	125g
Large onions, chopped	2	5
Carrots, peeled and finely chopped	3	7
Garlic cloves, finely chopped	2	5
Piece of smoked bacon or pancetta	100g	250g
Soaked puy lentils	200g	500g
White wine	150g	375g
Mushroom nage	300ml	750ml
Vegetable nage	300ml	750ml
Salt and pepper to taste		
To finish		
Cooked sliced mushrooms (sweated in a little oil)	250g	625g

Energy	Cal	Fat	Sat fat	Carb	Sugar	Protein	Fibre
1258 kJ	301 kcal	13.0g	7.3g	34.4g	11.4g	13.8g	5.2g

1 Add the butter to the pan and sauté the vegetables, bacon and lentils for 5 minutes.

2 Add the wine and nages, and cook for approximately 50 minutes on a low heat.

3 Remove the piece of bacon and retain for garnish.

4 Purée in a food processor, adding more stock if necessary, if the soup becomes too thick.

5 To finish, add the cooked mushrooms and cooked bacon, diced.

6 Serve immediately.

6 Red lentil and bacon soup

Ingredient	4 portions	10 portions
Baby shallots	3	8
Leeks	100g	250g
Celery sticks	50g	125g
Oil	100ml	250ml
Pancetta bacon, chopped into small pieces	200g	500g
Red lentils	400g	1 kg
Chicken stock	1.2 litres	3 litres

Energy	Cal	Fat	Sat fat	Carb	Sugar	Protein	Fibre
1620 kJ	390 kcal	28.8g	4.2g	17.9g	3.4g	15.9g	2.0g

1 Slice the shallots, leek and celery into 1cm dice.

2 Heat about 100ml of oil in a pan. Add the vegetables and bacon, and cook until they are slightly coloured.

3 Add the lentils and cover them with the chicken stock. Bring to the boil and turn the heat down to a very slow simmer.

4 Cook this until all the lentils have broken down.

5 Allow to cool for 10 minutes and then purée until smooth.

6 Correct consistency. Garnish with crispy fried bacon and serve immediately.

7 Roast butternut squash soup

1 Cut the squash into thick pieces, place on a lightly oiled baking sheet and roast for 20–25 minutes in a hot oven until the flesh is soft and golden brown.

2 Sweat onions and garlic without colouring, approximately 5 minutes.

3 Add bacon and lightly brown.

4 Add roasted squash, pour in stock, bring to boil, simmer 20 minutes.

5 Allow to cool, liquidise or blend until smooth.

6 Season lightly, add yoghurt or cream, reheat gently and serve.

Ingredient	4 portions	10 portions
Butternut squash, peeled and de-seeded	600g	2 kg
Onion	100g	250g
Olive oil	2 tbsp	5 tbsp
Garlic (optional), finely chopped		
Bacon, back rashers, in small pieces	4	10
Chicken or vegetable stock	625ml	1.5 litres
Salt and pepper		
Cream or thick natural yoghurt	90ml	225ml

Energy	Cal	Fat	Sat fat	Carb	Sugar	Protein	Fibre
923 kJ	220 kcal	12.0g	3.5g	19.6g	13.1g	9.9g	2.8g

Variation
Add 3–4 saffron strands soaked in 1 tablespoon hot water at step 4.

Healthy eating tip
Use unsaturated oil (sunflower or olive), lightly oil the pan to sweat the garlic and onions. Drain off any excess fat after cooking the bacon. Use low-fat yoghurt to reduce the fat. Add a little cornflour to stabilise the yoghurt before adding to the soup.

8 Chicken velouté with mushrooms and tongue

Ingredient	4 portions	10 portions
Chicken velouté	1 litre	2.5 litres
Mushroom trimmings	200g	500g
Yolk of egg – liaison	2	5
Cream	125ml	312ml
Salt and pepper		
Mushrooms	25g	60g
Chicken – julienne garnish	25g	60g
Tongue	25g	60g

Energy	Cal	Fat	Sat fat	Carb	Sugar	Protein	Fibre
1948 kJ	469 kcal	40g	10.7g	20.7g	1.6g	7.8g	1.4g

1 Prepare a chicken velouté (see right), adding the chopped mushroom trimmings at step 2.

2 Liquidise and add the liaison by adding some soup to the liaison of yolks and cream and returning all to the pan.

3 Bring almost to the boil, stirring continuously and being careful not to boil, then strain into a clean pan.

4 Correct the seasoning and consistency, add the garnish and serve.

To make chicken velouté

Ingredient	1 litre	2.5 litres
Butter	60g	150g
Flour	60g	150g
White stock	1 litre	2.5 litres
Seasoning		

1 Melt the butter in a suitable pan, add the flour and cook to a blond roux.

2 Gradually add the boiling stock, stirring continuously with a spoon, bring to the boil between each addition of stock.

3 When all the stock is added, skim and gently simmer for approximately 30 minutes.

4 Season.

5 Pass through a fine strainer into a clean pan and reboil.

Professional tip

100g raw minced chicken may be cooked in the velouté then liquidised with the soup to give a stronger chicken flavour (for 10 portions, increase the proportion 2½ times).

9 Roasted plum tomato and olive soup

Ingredient	4 portions	10 portions
Plum tomatoes	400g	1.5 kg
Small onions	1	2
Cloves of garlic	1	2
Sprigs of basil	2	4
Tomato purée	2 tbsp	4 tbsp
Olive oil	25g	50g
Black and green olives	50g	100g
Balsamic vinegar	1 tsp	3 tsp
Water	500ml	1.5 litres
Salt and pepper		
Sugar	10g	25g
Croutes	2	3
Sundried tomato paste	1 tbsp	3 tbsp
Parmesan	25g	75g
Black olives	2	5
Chopped parsley	25g	50g

Energy	Cal	Fat	Sat fat	Carb	Sugar	Protein	Fibre
873 kJ	208 kcal	10.7g	2.6g	22.9g	9.4g	6.6g	2.9g

1 Roughly chop the plum tomatoes, onion, garlic and basil.

2 Place into a roasting tray with tomato purée and a few drops of olive oil. Roast at 204°C for 10 minutes.

3 Remove from the oven and put into a saucepan. Add water.

4 Simmer for 20 minutes, stirring occasionally.

5 Liquidise for 2–3 minutes with olives. Pass through a conical sieve.

6 Check the consistency of the soup, add seasoning, a pinch of sugar and a few drops of balsamic vinegar.

7 Prepare the croutes.

8 Slice dinner rolls into rounds and lightly toast both sides.

9 Spread with sundried tomato paste and Parmesan.

10 When the soup is required, bring to the boil. Put into soup cups and top with croutes and a slice of black olive and chopped parsley.

10 Vichyssoise

Ingredient	4 portions	10 portions
Onions, finely sliced	160g	400g
Leeks, finely sliced	175g	430g
Butter	125g	300g
Potatoes, diced small	750g	1.8 kg
Water/vegetable stock	1.8 litres	4.5 litres
Salt	15g	30g
Pepper to taste		
Garnish		
Whipped cream		
Chives, chopped		

Energy	Cal	Fat	Sat fat	Carb	Sugar	Protein	Fibre
1653 kJ	397 kcal	26.4g	16.3g	36.9g	4.5g	5.3g	4.0g

1 Sweat the sliced onion and leek without colour in the butter. Cook until very tender.

2 Add the potatoes and bring quickly to the boil with the water or vegetable stock.

3 Liquidise in food processor and allow to cool.

4 Check seasoning. Note that seasoning needs to reflect the serving temperature.

5 Serve cold with whipped cream and chopped chives.

11 Chilled tomato and cucumber soup (Gazpacho)

This soup has many regional variations. It is served chilled and has a predominant flavour of cucumber, tomato and garlic.

Ingredient	4 portions	10 portions
Plum tomatoes, ripe	2.5 kg	6.25 kg
White onion, roughly chopped	1	2
Cucumber, peeled, roughly chopped	1	2
Garlic clove, crushed	½	1
Red peppers, peeled, deseeded	550g	1.3 kg
Salt	40g	80g
Cayenne pepper	2g	5g
Chardonnay vinegar or white wine vinegar	6g	15g
Sugar (to taste, depending on season)	30g	75g

Energy	Cal	Fat	Sat fat	Carb	Sugar	Protein	Fibre	Sodium
861 kJ	203 kcal	2.7g	0.8g	40.5g	38.7g	6.7g	12.5g	3.2g

1 Mix all the ingredients together and leave to marinate overnight in the fridge.

2 Blitz the ingredients in a food processor and strain through a chinois.

3 Discard the remaining pulp into a colander lined with muslin, to catch the extra juices that come from the pulp.

4 Use the juices from the pulp to thin out the gazpacho until it reaches the correct consistency.

5 Check seasoning. Store in the refrigerator. Serve well chilled.

Variations
- The soup can be served unpassed for a different texture.
- The soup may also be finished with chopped herbs.
- A tray of garnishes may accompany the soup, e.g. chopped red and green pepper, chopped onion, tomato, cucumber and croutons.

12 Petite marmite

Ingredient	4 portions	10 portions
Chicken winglets	4	10
Lean beef (cut in 1cm dice)	50g	125g
Good strength beef consommé	1 litre	2.5 litres
Carrots	100g	250g
Celery	50g	125g
Leeks	100g	250g
Cabbage	25g	60g
Turnips	100g	250g
Slices of beef bone marrow	8	20
Toasted slices of flute	50g	125g
Parmesan cheese (grated)	25g	60g

Energy	Cal	Fat	Sat fat	Carb	Sugar	Protein	Fibre
1098 kJ	261 kcal	16.2g	7.1g	13.0g	4.4g	16.5g	3.2g

This is a double strength consommé garnished with neat pieces of chicken winglet, cubes of beef, turned carrots and turnips and squares of celery, leek and cabbage. The traditional method of preparation is for the marmites to be cooked in special earthenware or porcelain pots ranging in size from 1–6 portions. Petite marmite should be accompanied by thin toasted slices of flute, grated Parmesan cheese and a slice or two of poached beef marrow.

1 Trim the chicken winglets and cut in halves.

2 Blanch and refresh the chicken winglets and the squares of beef.

3 Place the consommé into the marmite or marmites.

4 Add the squares of beef. Allow to simmer for 1 hour.

5 Add the winglet pieces, turned carrots and squares of celery.

6 Allow to simmer for 15 minutes.

7 Add the leek, cabbage and turned turnips; simmer gently until all the ingredients are tender. Correct seasoning.

8 Degrease thoroughly using both sides of 8cm square pieces of kitchen paper.

9 Add the slices of beef bone marrow just before serving.

10 Serve the marmite on a dish paper or a round flat dish accompanied by the toasted flutes and grated cheese.

13 Country-style French vegetable soup

Ingredient	4 portions	10 portions
Dried haricot beans	50g	125g
Olive oil	2 tbsp	5 tbsp
White of leek, finely shredded	100g	250g
Carrot	100g	250g
Turnip	50g	125g
Vegetable stock	1 litre	2.5 litres
French beans, cut into 2.5cm lengths	50g	125g
Courgettes	100g	250g
Broad beans	100g	250g
Tomatoes, skinned, de-seeded and chopped	4	10
Tomato purée	50g	125g
Macaroni	50g	125g
Seasoning – Pistou		
Basil	25g	60g
Clove of garlic	1	2½
Gruyere cheese	50g	125g
Olive oil	6 tbsp	15 tbsp

Energy	Cal	Fat	Sat fat	Carb	Sugar	Protein	Fibre
2341 kJ	564 kcal	46.1g	8.6g	26.9g	9.1g	12.2g	9.6g

1 Soak the haricot beans for approximately 8 hours.

2 Place the beans in a pan of cold water, bring to the boil and gently simmer for approximately 20 minutes. Refresh and drain.

3 Heat 2 tablespoons of oil in a suitable pan, add the white of leek and sweat for 5 minutes.

4 Add the carrot and turnip (both cut into brunoise) and cook for a further 2 minutes.

5 Add the haricot beans. Cover with the vegetable stock and add the remaining ingredients. Cook until everything is tender.

6 Make the pistou by puréeing the basil, garlic and cheese in a food processor. Gradually add the oil until well mixed and emulsified.

7 Correct the seasoning and consistency of the soup and serve with the pistou separately.

Variation: Cock-a-Leekie
Use half a litre good chicken stock and half a litre veal stock. Garnish with a julienne of prunes, white of chicken and leek.

14 Cream of green pea soup with rice, sorrel and lettuce

Ingredient	4 portions	10 portions
Shelled peas	400g	1.25 kg
Fresh mint	Sprig	Sprig
Onion	25g	60g
Bouquet garni		
Thin béchamel	500ml	1.25 litres
Cream	60ml	150ml
Lettuce	Quarter	Three-quarters
Sorrel	25g	60ml
Rice, cooked	50g	125g

Energy	Cal	Fat	Sat fat	Carb	Sugar	Protein	Fibre
1569 kJ	376 kcal	21.7g	13.7g	33.7g	11.1g	14g	4.9g

1 Cook the peas in water with a little salt, mint, onion and bouquet garni.

2 Remove the mint, onion and bouquet garni.

3 Purée the peas in a food processor or liquidiser and add to the béchamel.

4 Bring to the boil and simmer for 5 minutes.

5 Correct the seasoning; pass through a medium strainer.

6 Correct the consistency and add the cream.

7 Garnish with shredded lettuce, sorrel, cooked in butter, and cooked rice, and serve.

The nutritional analysis assumes that the béchamel is made with skimmed milk.

15 Cream of asparagus soup

Ingredient	4 portions	10 portions
Onions or white of leek	50g	125g
Celery	50g	125g
Butter	50g	125g
Flour	50g	125g
Chicken stock	1 litre	2.5 litres
Asparagus trimmings	400g	1.25 kg
Bouquet garni		
Salt and pepper		
Milk	250g	600ml
OR Cream	125ml	300ml

Energy	Cal	Fat	Sat fat	Carb	Sugar	Protein	Fibre
361 kJ	151 kcal	25.3g	11.9g	27.1g	8.1g	7.7g	2.5g

1 Sweat the sliced onion and celery in the butter without colour.

2 Remove from heat, add flour, return to heat and cook for a few minutes without colour.

3 Cool, gradually add the hot stock, stir to the boil.

4 Add the washed asparagus trimmings, the bouquet garni and season.

5 Simmer for 30 minutes. Remove the bouquet garni.

6 Purée in a food processor or liquidise and pass through a conical strainer.

7 Return to a clean pan, reboil, add the milk or cream.

8 Correct the seasoning and consistency and serve.

Variations

- Add 25g sprigs of mint with the asparagus. Garnish with blanched mint leaves, using 2 or 3 per portion.
- Add the flesh of 1½ avocado pears added with the asparagus trimmings. The remaining half of avocado pear is neatly diced and used as garnish. This may be served cold.
- Use 200g asparagus trimmings and 200g mushrooms. Slice 150g mushrooms and sweat them with the onion and celery. Garnish with the remaining 50g sliced mushrooms cooked in a little butter.

16 Brown onion soup

Ingredient	4 portions	10 portions
Onions	600g	1.5 kg
Butter or margarine		
Clove of garlic, chopped (optional)	1	2–3
Flour, white or wholemeal	10g	25g
Brown stock	1 litre	2.5 litres
Salt and pepper		
Flute	Quarter	Three-quarters
Cheese, grated	50g	125g

Energy	Cal	Fat	Sat fat	Carb	Sugar	Protein	Fibre
197 kJ	827 kcal	9.7g	5.8g	20.4g	8.1g	8.3g	3.1g

1 Peel the onions, halve and slice finely.

2 Melt the butter in a thick bottomed pan, add the onions and garlic and cook steadily over a good heat until cooked and well browned.

3 Mix in the flour and cook over a gentle heat, browning slightly.

4 Gradually mix in the stock, bring to the boil, skim and season.

5 Simmer for approximately 10 minutes until the onion is soft. Correct the seasoning.

6 Pour into an earthenware tureen or casserole or individual dishes.

7 Cut the flute (French loaf, 2cm diameter) into slices and toast on both sides.

8 Sprinkle the toasted slices of bread liberally on the soup.

9 Sprinkle with grated cheese and brown under the salamander.

10 Place the toasted bread on the soup and serve.

During cooking, the finely sliced onions first become translucent

As the onions continue to cook, they begin to brown without becoming crisp

When the onions are ready, they are well browned

Cook in the flour, and finally the stock

17 Langoustine and mussel soup

Ingredient	4 portions	10 portions
Raw langoustine tails (large), bodies and claws retained for the stock	20	50
Mussels, cleaned	400g	1 kg
Fish stock	300ml	750ml
Butter	80g	200g
Fresh bay leaves	2	5
Dry white wine	50ml	125ml
Celery sticks, cut into small dice	1	3
Rindless dry-cured, unsmoked bacon, cut across into short, fat strips	50g	125g
Potatoes, peeled and cut into small dice	225g	560g
Plain flour	20g	50g
Full-cream milk	300ml	750ml
Whipping cream	120ml	300ml
Salt and freshly ground black pepper		

Energy	Cal	Fat	Sat fat	Carb	Sugar	Protein	Fibre
2355 kJ	566 kcal	36.3g	20.8g	21.6g	5.7g	39.3g	1.2g

1 If using raw langoustines, put them into the freezer for 30 minutes to kill them painlessly. Then put the langoustines and mussels into a pan and add the stock.

2 Cover, bring to the boil and steam for 2 minutes.

3 Remove from the heat and tip the contents into a colander set over a clean bowl to retain the cooking liquid.

4 Check that all the mussels have opened and discard any that remain closed.

5 Melt about one-third of the butter in a large pan, add the langoustine shells and the bay leaves, cook hard for 1 minute.

6 Add the wine and the reserved cooking liquor and while it is bubbling away, crush the shells to release all their flavour into the cooking liquid.

7 Cook for 10–12 minutes.

8 Meanwhile, heat the rest of the butter in a pan, add the shallots, celery and bacon, cook gently until the shallots are soft but not coloured.

9 Add the diced potatoes and cook for 1–2 minutes, stir in the flour, then add the milk and cream.

10 Pass the cooking liquor into a clean pan and add to the roux base, stirring continuously to prevent lumps.

11 When all the stock has been added, cook out until the potatoes are soft.

12 Stir in the langoustines and mussels, and adjust the seasoning if necessary.

13 Ladle into warmed soup plates and serve with traditional sour bread or crusty bread.

Professional tip

Proceed with caution when using both mussels and langoustines as they overcook quickly and this will spoil the eating quality.

18 Lobster bisque

Ingredient	4 portions	10 portions
Live lobster	400g	1 kg
Butter	100g	250g
Onion	50g	125g
Carrot	50g	125g
Brandy	60ml	150ml
Flour	75g	190g
Tomato purée	50g	125g
White stock (beef or veal or chicken or a combination of any 2 or 3)	1.25 litres	3 litres
White wine (dry)	120ml	300ml
Bouquet garni		
Salt and cayenne pepper		
Cream	120ml	300ml

Bisque is the term applied to any thickened shellfish soups, e.g. crab, crayfish, prawn and shrimp.

Energy	Cal	Fat	Sat fat	Carb	Sugar	Protein	Fibre
1841 kJ	438 kcal	28.7g	17.2g	18.8g	4.2g	14.4g	1.2g

1 Wash the live lobster.

2 Cut it in half lengthwise tail first, then the carapace.

3 Discard the sac from the carapace, clean the trail from the tail and wash all the pieces.

4 Crack the claws and the four claw joints.

5 Melt the butter in a thick-bottomed pan.

6 Add the lobster and the roughly cut onion and carrot.

7 Allow to cook steadily without colouring the butter for a few minutes, stirring with a wooden spoon.

8 Add the brandy and allow it to ignite.

9 Remove from heat and mix in the flour and tomato purée.

10 Return to gently heat and cook out the roux.

11 Cool slightly and gradually add the white stock and white wine.

12 Stir until smooth and until the bisque comes to the boil.

13 Add the bouquet garni and season lightly with salt.

14 Simmer for 10–15 minutes. Remove lobster pieces.

15 Remove lobster meat, crush the lobster shells, return them to the bisque and allow to continue simmering for a further 15–20 minutes.

16 Cut lobster meat into large brunoise.

17 Remove bouquet garni and as much bulk from the bisque as possible.

18 Pass through a coarse and then fine strainer or liquidise, and strain through a fine strainer again. Return the soup to a clean pan.

19 Reboil, correct seasoning with a little cayenne and add the cream.

20 Add the brunoise of lobster meat and serve. At this stage, 25g butter may be stirred into the bisque as a final enriching finish.

Cut the live lobster in half

Remove the sac

Crack each claw

Fry the lobster with the other ingredients

Add the brandy and ignite it

Mix in the flour and tomato

Gradually add the liquid and then bring to the boil and simmer

Remove the lobster pieces, extract and set aside the meat, and crush the shells. The crushed shells will be returned to the bisque; the meat is used for a garnish

Pass the bisque through a coarse strainer, then a fine one

Variations

To produce a less expensive soup, cooked lobster shell (not shell from the claws) may be crushed and used in place of a live lobster.

An alternative method of thickening is to omit the flour and thicken by stirring in 75g rice flour diluted in a little cold water 10 minutes before the final cooked stage is reached.

Video: lobster bisque
http://bit.ly/1kJn5mE

19 Cockle chowder

Ingredient	4 portions	10 portions
Cockles		
Medium shallots, finely diced	2	5
Butter	50g	125g
Cockles, shells tightly closed	2 kg	5 kg
White wine or vermouth	200ml	500ml
Chowder		
Vegetable oil	50ml	125ml
Smoked bacon, cut into 1cm dice	50g	125g
Medium onion, cut into 1cm dice	1	3
Medium carrot, cut into 1cm dice	1	3
Garlic cloves, finely chopped	2	5
Celery sticks, cut into 1cm dice	1	3
Medium potato, peeled, cut into 1cm dice	1	3
Medium yellow pepper, cut into 1cm dice	1	3
Chicken stock	1 litre	2 litres
Whipping cream	100ml	250ml
Butter	50g	125g
Salt and pepper		

Energy	Cal	Fat	Sat fat	Carb	Sugar	Protein	Fibre
2223 kJ	536 kcal	45.0g	21.4g	13.9g	7.3g	19.9g	2.4g

For the cockles

1 Take a large saucepan with a tight-fitting lid and place over a medium heat, add the shallots and butter and cook for 1 minute without letting the shallots colour.

2 Add the washed cockles, shake the pan, then add the wine and place the lid on the pan immediately. Leave the cockles to steam for 1–2 minutes so that they open and exude an intense liquor.

3 Remove the lid and make sure all the cockles are open. Remove the pan from the heat and discard any with closed shells.

4 Place a colander over a large bowl, and pour the contents of the pan into the colander, reserving the liquor for the chowder.

5 Allow the cockles to cool. Pick out the meat, check carefully for sand and discard the shells. Store the cockle meat in an airtight container in the fridge until you are ready to serve the chowder.

For the chowder

1 In a large saucepan, heat the oil. When hot, add the bacon and cook for about 5 minutes until crisp and brown.

2 Using a perforated spoon, transfer the bacon onto kitchen paper to drain. Add the onion, carrot, garlic, celery and potato to the saucepan, reduce the heat to medium-low and cook the vegetables for 3–4 minutes without colouring.

3 Add the peppers and cook for 5 minutes. Pour in the reserved liquor from the cockles and the chicken stock.

4 Bring to the boil and simmer for 10 minutes or until the volume of liquid has reduced by about half.

5 Add the cooked bacon and cream, then bring to the boil and reduce for a further 2 minutes until the soup thickens slightly.

6 Just before serving, whisk in the butter.

To finish

1 While the chowder is cooking, carefully remove the meat from the shell, checking for sand, and place in a clean pan.

2 Combine the chowder base and the cockle meat together, reheat carefully and serve. May be garnished with a grilled rasher of bacon or pancetta.

Test yourself

Soups

1 List three quality points to look for in a good soup.
2 List the two main ingredients required to clarify a stock for consommé.
3 At what temperature should chilled soup be served?
4 List three safety points to be considered when puréeing a soup.
5 Why is it advisable to measure and weigh the ingredients when preparing a soup?
6 List four garnishes that are commonly used for soups.
7 List three different types of pulses which may be used in a purée soup.
8 When making a cream soup, list two alternatives to finishing the soup with cream.
9 Describe the soup petite marmite.
10 What type of soup is a bisque?

Sauces and dressings

A sauce is a flavoured liquid which may be thickened slightly to improve texture and flow. Not only do sauces enhance flavour and texture, they add variety and attractiveness to dishes.

The basis of any sauce is usually a good meat, fish or vegetable stock. Sauces can also be made from the cooking juices of meat, deglazed with wine and can be slightly thickened with a starch base such as arrowroot. Other sauces, such as hollandaise, are produced from clarified butter, flavoured with reduced vinegar and peppercorns, and thickened with egg yolks to form an emulsion. Beurre blanc is a sauce produced by melting and whisking the butter into an emulsion; white wine may be added for flavour. Cream sauces can be produced by reducing white or red wine and sometimes a spirit such as whisky.

The current trend is for lighter sauces: many sauces are produced from reduced stocks and butter is replaced by unsaturated oils such as olive oil, or a combination of 50 per cent butter, 50 per cent olive oil. Sauces produced from non-traditional cuisines tend to be thicker – salsas, for example, tend to be quite thick.

A dressing can be described as a combination of ingredients which enhance a dish. It can include a stuffing served to accompany a dish (for example, sage and onion stuffing with roast duck). It can also be a sauce to accompany a dish (for example, a vinaigrette or mayonnaise in a salad). Dressings also include chutneys and condiments such as Worcester sauce or tomato ketchup.

Preparation and cooking methods

Thickening sauces

- **Roux:** this consists of equal quantities of butter, flour or vegetable oil in flour. There are two basic types of roux: white roux (used for béchamel); and blond roux (used for velouté). Brown roux, though not usually used today, uses dripping and flour, but vegetable oil can also be used.
- **Beurre manié** (also known as raw roux): this means mixing equal quantities of butter and flour. Hard vegetable oil can also be used. It is whisked into the boiling liquid to gelatinise the starch and form a smooth sauce.
- **Sauce flour:** this is made from specific wheat so that it does not form glutinous lumps when heated with liquids. It is very good for fat-free cooking.
- **Vegetable starches:** examples of vegetable starches include cornflour, arrowroot, fecule and rice flour. These starches are mixed with cold water or cold stock to form a slurry. The slurry is then whisked into the hot sauce. The starch gelatinises and thickens the sauce, which should be smooth and free from lumps. Arrowroot is used in a number of sauces; it is a unique product which on thickening becomes transparent and shiny.
- **Vegetable gums:** examples include alginin, guar gum, locust bean gum and xanthan gum. One of the most remarkable properties of xanthan gum is that a small amount of gum, approximately 1 per cent, will greatly increase the viscosity of a liquid. It is often used in salad

dressings. For more information on xanthan gum, see Chapter 3.

- **Butter:** Sauces may be thickened by melted butter. The butter is added in small pieces to the hot liquid. This gives the sauce a glossy finish, but it must not be reheated otherwise the emulsion will split.
- **Egg yolks and cream:** This is a thickening agent known as a liaison. Egg yolks and cream are whisked together, and a small quantity of hot sauce is added to the egg yolks and cream, whisked well and added back to the hot liquid. The sauce at this stage must not be allowed to boil otherwise the emulsion will curdle. This is a classical finish to velouté sauces. The slight thickening comes from the viscosity of the cream and the emulsifying power of the egg yolks.

Skimming sauces

Scum or impurities often appear as foam or froth on the surface of the cooking liquid. Remove these by skimming with a ladle during cooking.

Straining sauces

Straining or passing removes any solids or small particles from the sauce, including ingredients from the cooking. This technique may also be used after puréeing. Sauces are usually strained through a fine conical strainer known as a chinois. A very delicate sauce is strained through a muslin cloth.

Types of sauces

Emulsified sauces

An emulsion is a mixture of oil and water. For example, milk is an emulsion of fat in water; butter is an emulsion of water in fat. An emulsified sauce is produced by dispersing fats or oils in a water base. An **emulsifying agent** will help the ingredients to emulsify together with the oil and water.

Emulsifying agents are substances whose molecules contain both a hydrophilic ('water loving') group and a lipophilic ('fat loving') group which is hydrophobic or 'water hating'.

Lecithin (a phospholipid) is an important natural emulsifier which favours oil/water emulsions and is present in egg yolk and many crude oils, particularly vegetable oils.

Sometimes stabilisers are added to products in addition to emulsifiers. These maintain an emulsion once it has been formed and improve the stability of the emulsion by increasing its viscosity. Stabilisers can be gelatine or complex carbohydrates such as pectins, starches, alginates and gums. Stabilisers will hold emulsions together for long periods.

It takes knowledge to blend the ingredients of an emulsified sauce successfully. Temperature control is crucial.

Hollandaise, béarnaise and vinaigrette are examples of emulsified sauces.

Compound butter sauces

Compound butter sauces can be warm, hot or cold.

Warm butter sauces include:

- **Hollandaise** – an emulsion of egg yolks and butter or oil, flavoured with vinegar and pepper. It is often served with vegetables, for example, asparagus, globe artichokes, and with grilled and poached fish. Derivatives include maltase and mousseline.
- **Béarnaise** – the same as hollandaise but thicker in consistency and flavoured with finely chopped shallots, pepper and tarragon. Usually served with grilled meat and fish.
- **Beurre blanc** – made from butter, vinegar and water, with chopped shallots (see Recipe 28); served with vegetables, fish and white meats, veal and chicken.

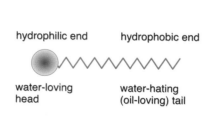

hydrophilic end   hydrophobic end

water-loving head   water-hating (oil-loving) tail

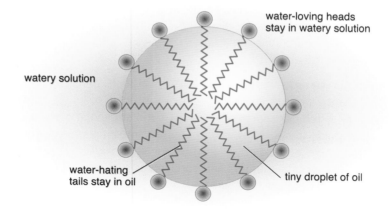

water-loving heads stay in watery solution

watery solution

water-hating tails stay in oil

tiny droplet of oil

▲ Why oil in water forms an emulsion

Reducing, passing and whisking sauces

This example shows the preparation of a hollandaise sauce.

1 Place peppercorns and herbs in vinegar, and then reduce it by one third.

2 Add the vinegar to the egg yolks, passing it through a fine strainer to remove the aromats.

3 Whisk the vinegar and egg yolks together.

4 Over a gentle heat, cook the mixture, whisking it continuously, to form a sabayon.

5 The sabayon is ready when it has reached the ribbon stage.

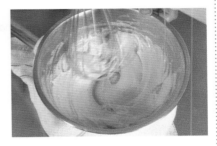

6 Over a bain marie, gradually whisk in melted butter until thoroughly combined.

7 If necessary, whisk in a little warm water to stabilise the sauce.

Hot butter sauces include:

- **Beurre fondu** – melted butter, which may be added to a little white wine. This is an emulsion between fat and liquid. Melted butter emulsified with any nage will give you a slightly thicker sauce that can be used to coat vegetables or fish.
- **Beurre noisette** – nut brown butter.
- **Beurre noir** – black butter, not exactly burnt, but cooked to noisette stage to which is added malt vinegar. Served with skate.

Cold butter sauces include:

- **Parsley butter** (also known as beurre maître d'hôtel) – creamed butter with lemon juice, chopped parsley and cayenne pepper. Spread onto greaseproof or aluminium foil and rolled; allowed to set in the refrigerator. The foil or paper is removed and the butter is sliced. Used for grilled meat.
- **Anchovy butter** – creamed butter with anchovy essence, prepared in the same way as parsley butter. Used for grilled meats such as entrecote steak and rump steak.

- **Velouté sauce** – made from equal quantities of butter and flour, cooked to a blond roux; good stock is gradually added. Served with white chicken, veal, fish or vegetable stock. Variations on velouté include suprême sauce (chicken velouté finished with cream); mushroom sauce (velouté sauce with mushroom essence garnished with sliced mushrooms finished with cream); and aurore (velouté sauce with purée of fresh tomatoes).

Brown sauces

Traditionally, a brown sauce is produced from a brown roux and brown stock, which is boiled out with additional stock to produce a **demi-glace**. However, today a demi-glace is usually produced from reducing brown stock and lightly thickening the stock with arrowroot.

This is similar to **jus-lié**, which is traditionally produced from brown veal stock, flavoured with tomatoes and mushrooms, browned chicken wings and other meat trimmings. Some chefs will add bacon trimmings to enhance the flavour. The jus-lié is thickened with arrowroot and strained.

There are many variations of demi-glace or brown sauce:

- **Chasseur** – flavoured with white wine shallots, fresh tarragon, mushrooms and tomato concassé.
- **Bordelaise** – a reduction of red wine, brown sauce and classically diced bone marrow added to the sauce.
- **Robert** – brown sauce with a reduction of white wine, flavoured with shallots and mustard. Serve with grilled pork and sautéed kidneys.
- **Charcuterie** – this is a sauce Robert with a garnish of gherkins.

Glazes

A glaze is a stock, fond or nage that has been reduced, removing a high percentage of the water through boiling and thus concentrating the solids and increasing the flavour. This yields an intense sauce that should be used sparingly to finish a dish. Alternatively, it can be refrigerated or frozen and added to a weak stock to give it a vibrant boost.

Oil-based sauces

Oil-based sauces fall into two categories:

1 Cold sauces such as mayonnaise and derivatives.
2 Hot sauces including vinaigrette and variations (for example, vinaigrettes used with some fish and vegetable dishes).

The availability of a wide range of oils allows chefs to combine some very unique and different flavours. Oils such as walnut, almond, olive, coconut and sunflower can be blended with a variety of herbs and spices to enhance their flavour. Flavoured oils add variety to dishes, provide texture and compliment the mouth feel. The oils can be flavoured in two ways:

- Place a bunch of blanched herb(s) into a bottle of oil, cork tightly and keep on a cool shelf.
- Warm the oil with the herb(s) for 15–20 minutes.

To extract the maximum flavour from spices, they are usually heated in a pan and added to the oil. Oils may also be coloured with natural colours from chopped herbs (for example, the colour chlorophyll and lycopene in tomatoes).

There is scope here to experiment with various herbs, spices and vegetables so that a mise-en-place of several flavoured oils can be produced:

- Red or white vinegars can be flavoured with herbs such as thyme, tarragon, dill, mint and rosemary.
- Fruit vinegars include raspberry, blackberry, peach, apple and cherry.
- Floral vinegars include elderflower and rose.
- Sharp vinegars include chilli, garlic and horseradish.

Vinaigrettes

Vinaigrettes are an emulsion of one part vinegar and two parts oil, flavoured with mustard (a natural emulsifying agent). They may also be flavoured with chopped herbs and spices. Some vinaigrettes reduce the amount of vinegar or replace some of the vinegar with sherry or white wine. Examples of vinegars used include balsamic vinegar, raspberry vinegar, sherry vinegar, white wine and red wine vinegar. Vinaigrettes may be served at room temperature or slightly warm to enhance flavour.

When making vinaigrettes for use in hot dishes, the skill lies in blending the ingredients and flavours of both oils and vinegars to complement the dish and ensuring that they do not dominate. For example, skate steamed, grilled or fried and lightly masked with a hot vinaigrette is a basic recipe to which variations can be applied, such as adding a small brunoise of vegetables or finishing with chopped fennel or dill.

Many vegetables simply boiled or steamed can be given additional flavours by finishing with a light dribble of suitably flavoured hot vinaigrette – for example, sliced or diced beetroot, carrots, turnips and swedes; a mixture of cooked vegetables or crisply cooked shredded cabbage.

Mayonnaise and derivatives

Mayonnaise is an emulsion of pasteurised egg yolks, oil, mustard vinegar and/or lemon juice. The egg yolks contain lecithin which emulsifies the sauce. Mayonnaise is very popular and versatile sauce.

> ### ⚠ Health and safety
>
> Mayonnaise must be made with pasteurised eggs to reduce any risk of salmonella food poisoning.

Some derivatives of mayonnaise are:

- Tartare – mayonnaise with chopped parsley, gherkins and capers. Served with fried, breaded fish.
- Aioli – with chopped garlic.
- Marie rose – mayonnaise with Worcester sauce, tomato ketchup and lemon juice. Used in fish cocktails.

Butter-thickened sauces (monter au beurre)

Many sauces can be given a final finish to enrich flavour and given sheen by the thorough mixing in of small pieces of butter at the last moment before serving. The traditional way of making à la carte classic fish sauces, for example, is to strain off the liquid in which the fish has been poached, reduce it to a glaze and then, away from the heat, gradually and thoroughly mix in small pieces of butter. If the sauce is to be glazed, then some lightly whipped double cream may be added.

Egg-based sauces

Once made, egg-based sauces such as béarnaise should be kept warm for no more than 2 hours. They should then be discarded.

Purée-based sauces

Chefs are now producing a range of vegetable sauces from a purée of vegetables, as a way to make healthier dishes. Examples include pepper sauces produced from red, green and yellow peppers flavoured with chopped herbs and garlic; and broccoli, courgette and celeriac sauces, which add flavour texture and colour to dishes. **Soubise** is the traditional name given to a purée or sauce made from onions. In contemporary use this has been broadened to include other examples.

Fruit purées fall into two categories:

1 Fruits that are cooked with the minimum of sugar to retain a degree of sharpness; they are used for serving with fish, meat or poultry dishes. For example, apple, cranberry and gooseberry.

2 Fruits that are puréed and used for sweet dishes, for example apricot, blackcurrant, blackberry, damson, plum and rhubarb. These are lightly cooked in a little water with sufficient sugar to sweeten, then liquidised and strained.

Coulis (or cullis) are similar to vegetable sauces. They can be a purée of vegetable, flavoured with oil, mixed herbs and spices, or can be made with fruit. A coulis may also be flavoured with white or red wine, a spirit such as brandy, or a fortified wine such as sherry. A coulis must be well mixed and thoroughly emulsified. Like vegetable sauces, they can be served hot or cold.

Foams (espumas)

The word espuma translates from Spanish to 'foam' or 'bubbles'. An espuma is created using a classic cream whipper – a stainless steel vessel with a screw top and a non-return valve which you charge with nitrogen dioxide. Once charged, it will whip cream like a whisk. The nitrogen dioxide forces the liquid out of the canister under pressure through two nozzles, agitating the fat and increasing the volume of the cream. Nitrogen dioxide has minimum water solubility and so it will not affect the product that is being charged. Preparation varies according to whether using a cold fat-based, warm fat-based or gelatine-based cream. For more information on espumas, see Chapters 3 and 14.

Cream-based sauces

Certain vegetables, for example, carrots and broad beans, when almost cooked can have double cream added and cooking completed to form a cream sauce. Cream sauces may also be produced from a béchamel or velouté base, for example, using half the amount of cream and an equal amount of chicken velouté.

Cold cream sauces are produced from whipped cream, yoghurt or fromage frais, flavoured with flavoured oils, herbs, spices, Worcester sauce, tomato ketchup or horseradish sauce. Cream-based sauces are used to accompany salads and hors d'oeuvres.

Natural yoghurt or fromage frais can be used in place of cream in any sauce. This considerably reduces the fat content of the sauce. Care must be taken when adding natural yoghurt – excessive heat will give it a curdled appearance.

Pesto and salsa

Pesto is widely used today in culinary work. Italian in origin, it is made from high quality virgin olive oil, garlic, pine nuts, chopped basil and Parmesan cheese. There are various versions which can include almonds, black and green olives, sun-dried tomatoes and sun-dried peppers.

'Salsa' means 'sauce' in Spanish. These sauces tend to be thicker and made from a variety of ingredients; they are often used as a dressing.

Reduced vinegars

In the modern kitchen, chefs are using specialist vinegars reduced to syrup. They are slightly viscous in appearance and should be used sparingly to add flavour and to enhance the quality of the dish. Examples are balsamic, sherry, red and white wine, cider and raspberry.

Assessing the quality of sauces

- Ensure the sauce is suitable for the dish it is to be served with or to accompany.
- The sauce should have a delicate flavour and should not overpower the dish.
- The sauce should add texture and improve the overall quality of the dish.

- The sauce should add moisture and flavour to compliment the other ingredients of the dish.
- The sauce should have a good flow and, if required, easily mask the main item or ingredient.
- Seasoning a sauce does not always mean adding salt and pepper – it can include the correct use of herbs, spices and natural flavourings, e.g. reduced stock.

Storing cold sauces and dressings

- Cold sauces are best stored chilled.
- They must be covered and accurately labelled.
- It is advisable to remove the sauces and dressings from the refrigerator approximately 1 hour before service to allow them to come to room temperature, as this will enhance the eating quality. Only take out the amount that will be used, and do not return left-over sauce to the refrigerator for future use.
- Although vacuum packing is used in some kitchens to store sauces, it is not advisable unless strict microbiological conditions are in place.

20 Fish stock (modern method)

Ingredient	Makes 2 litres
Fish bones, no heads, gills or roe (turbot, sole or brill bones are best)	5 kg
Olive oil	100ml
Onions, finely chopped	3
Leeks, finely chopped	3
Celery sticks, finely chopped	3
Fennel bulb, finely chopped	1
Dry white wine	350ml
Parsley stalks	10
Thyme	3 sprigs
White peppercorns	15
Lemons, finely sliced	2

1 Wash the bones thoroughly in cold water.

2 Heat the olive oil in a pan that will hold all the ingredients and still have a 1cm gap at the top for skimming. Add all the vegetables and sweat without colour for 3 minutes.

3 Add the fish bones and sweat for a further 3 minutes.

4 Add the white wine and enough water to cover. Bring to a simmer, skim off the impurities and add the herbs, peppercorns and lemon. Turn off the heat.

5 Infuse for 25 minutes, then pass into another pan and reduce by half. The stock is now ready for use.

21 Shellfish stock

Ingredient	Makes 2 litres
Olive oil	50ml
Unshelled prawns	500g
Crab bodies (gills removed), roughly smashed	1 kg
Carrots, chopped	2
Onion, chopped	1
Celery stick, chopped	1
Fennel, chopped	Half a bulb
Garlic cloves	2
Tomato paste	3 tbsp
White wine	200ml
Cold water	2 litres
Fish stock	500ml
Parsley, thyme and bay leaf	
Star anise	2

1 Place the stock pot on a medium heat and add the oil, whole prawns and crab bodies.

2 Sweat for 8 minutes until slightly golden. Add the chopped vegetables, garlic and tomato paste, and continue sweating for a further 8 minutes.

3 Pour in the wine, bring to the boil and reduce by half.

4 Add the water, stock, herbs and star anise, cook for 30 minutes, skimming frequently.

5 Strain the liquid through a fine sieve and leave to cool.

Remove the legs, then break open the body of the crab

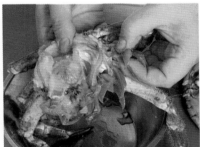

Remove the gills, then roughly cut up the crab

Crack open the claws

Sweat the shellfish pieces until slightly golden, then add the vegetables

Add the wine, then reduce by half

Add the stock and remaining ingredients, and continue to cook

22 Lamb jus (meat base)

Ingredient	Makes 1.5 litres
Thyme	Bunch
Bay leaves, fresh	4
Garlic	2 bulbs
Red wine	1 litre
Lamb bones	10 kg
Veal bones	5 kg
White onions, peeled	6
Large carrots, peeled	8
Celery sticks	7
Leeks	4
Tomato purée	6 tbsp

1 Pre-heat the oven to 175°C. Place the herbs, garlic and wine in a large, deep container. Place all the bones onto a roasting rack on top of the container of herbs and wine, and roast in the oven for 50–60 minutes. When the bones are completely roasted and have taken on a dark golden-brown appearance, remove from the oven.

2 Place all the ingredients in a large pot and cover with cold water. Put the pot onto the heat and bring to the simmer; immediately skim all fat that rises to the surface.

3 Turn the heat off and allow the bones and vegetables to sink. Once this has happened, turn the heat back on and bring to just under a simmer making as little movement as possible to create more of an infusion than a stock.

4 Skim continuously. Leave to simmer (infuse) for up to 12 hours then pass through a fine sieve into a clean pan and reduce down rapidly, until you have about 1.5 litres remaining.

23 Veal stock for sauce

No flour is used in the thickening process and consequently a lighter textured sauce is produced. Care needs to be taken when reducing this type of sauce so that the end product is not too strong or bitter.

Ingredient	Makes 2 litres
Veal bones, chopped	4 kg
Water	4 litres
Calves' feet, split lengthways (optional)	2
Carrots, roughly chopped	400g
Onions, roughly chopped	200g
Celery, roughly chopped	100g
Tomatoes, blanched, skinned, quartered	1 kg
Mushrooms, chopped	200g
Large bouquet garni	1
Unpeeled cloves of garlic (optional)	4

1 Brown the bones and calves' feet in a roasting tray in the oven.

2 Place the browned bones in a stock pot, cover with cold water and bring to simmering point.

3 Using the same roasting tray and the fat from the bones, brown off the carrots, onions and celery.

4 Drain off the fat, add the vegetables to the stock and deglaze the tray.

5 Add the remainder of the ingredients; simmer gently for 4–5 hours. Skim frequently.

6 Strain the stock into a clean pan and reduce until a light consistency is achieved.

24 Mushroom nage

Mushroom nage has a number of uses, including soups, sauces, risotto and consommé, and is an interesting variation or alternative to vegetable stock, especially with vegetarian dishes.

The flavour is very earthy, nutty and fragrant.

1 Sauté off the mushrooms, onion, garlic and thyme in a little oil and butter until nut brown.

2 Add the water and simmer for 20–25 minutes.

3 Remove from the heat and add the dried ceps. Cover with cling film until cold.

4 Refrigerate overnight and then pass through a fine chinois.

This may be stored in a refrigerator for up to one week or frozen for up to one month.

Ingredient	Makes 2 litres
Field mushrooms	2 kg
Onion sliced	350g
Garlic cloves	2
Thyme	2 sprigs
Butter	200g
Water	3 litres
Dried ceps	150g

25 Vegetable nage

Ingredient	Makes 2 litres
Onions, chopped	3
Celery sticks, chopped	2
Leeks, chopped	2
Carrots, chopped	4
Dry white wine	200ml
Garlic bulb, cut across middle	1
White peppercorns	10
Pink peppercorns	10
Star anise	3
Sprigs thyme	2
Sprigs parsley	2
Sprigs chervil	2
Sprigs tarragon	2
Cold water	2 litres

Vegetable nage is very useful for soups and sauces. The flavour you capture in this recipe comes primarily from the 12-hour infusion that takes place after cooking the stock. Once left to infuse, the nage will have a clean and distinct vegetable flavour with undertones of herbs and spices.

1 Place all the vegetables in a pan. Cover with water and bring to the boil.

2 When boiling, add white wine and return to the boil.

3 Turn down to a simmer for 10 minutes.

4 Remove from the heat; add all the aromats and cover with cling film. Cool thoroughly.

5 Place in refrigerator and leave for at least 12 hours before using, or overnight.

Store in a refrigerator for up to one week and freeze for up to one month.

26 Shellfish nage

Ingredient	Makes 2 litres
Olive oil	50ml
Unshelled prawns	500g
Cooked crab body, no claws, gills removed (roughly smashed)	1 kg
Carrots, chopped	2
Onion, chopped	1
Celery stick, chopped	1
Fennel, chopped	Half a bulb
Garlic cloves	2
Tomato purée	3 tbsp
Dry white wine	200ml
Cold water	2 litres
Fish stock	500ml
Star anise	2
Bay leaf	1
Sprigs parsley	5
Sprigs thyme	5

Shellfish nage can be used in soups, sauces and risottos, and can be reduced to a glaze. With a small amount of milk added, it can be 'cappuccinoed' using a hand blender. Light and fragrant, with a hint of aromatic spices and herbs, it gives an excellent flavour to all types of fish dishes and sauces.

1 Take a pan big enough to hold all the ingredients and put on a medium heat.

2 Heat the oil, then add the prawns and crab body.

3 Sweat off for 8 minutes. Add the vegetables and tomato purée and continue to sweat for a further 8 minutes, then add the wine, water, stock and herbs.

4 Bring to a simmer and skim off any impurities. Continue to cook for 30 minutes, skimming when a scum develops on the top of the nage.

5 When ready, pass the nage into a container through a fine sieve; do not force. Refrigerate and use as required.

Store for up to three days in an airtight container in the refrigerator, or freeze for up to one month.

27 Lobster sauce

Ingredient	Makes 1 litre
Live hen (female) lobster	¾–1 kg
Butter or oil	75g
Onion, roughly cut (mirepoix)	100g
Carrot, roughly cut (mirepoix)	100g
Celery, roughly cut (mirepoix)	50g
Brandy	60ml
Flour	75g
Tomato purée	100g
Fish stock	1.25 litres
Dry white wine	120ml
Bouquet garni	
Crushed clove garlic	Half
Salt	

1 Wash the lobster well.
2 Cut in half lengthwise, tail first, then the carapace.
3 Discard the dark green sac from the carapace, clean the trail from the tail, and remove any spawn into a basin.
4 Wash the lobster pieces.
5 Crack the claws and the four claw joints.
6 Melt the butter or oil in a thick-bottomed pan.
7 Add the lobster pieces and the onion, carrot and celery.
8 Allow to cook steadily without colouring the butter for a few minutes, stirring continuously with a wooden spoon.
9 Add the brandy and allow it to ignite.
10 Remove from the heat, mix in the flour and tomato purée.
11 Return to a gentle heat and cook out the roux.
12 Cool slightly, gradually add the fish stock and white wine.
13 Bring back to the boil, stirring continuously.
14 Add the bouquet garni and garlic and season lightly with salt.
15 Simmer for 15–20 minutes.
16 Remove the lobster pieces.
17 Remove the lobster meat from the pieces.
18 Crush the lobster shells, return them to the sauce and continue simmering for 30 minutes.
19 Crush the lobster spawn, stir into the sauce, reboil and pass through a coarse strainer.

This sauce may be made in a less expensive way by substituting cooked lobster shell (not shell from the claws), which should be well crushed, in place of the live lobster.

Variation: Lobster sauce or coulis
Omit the flour and finish with 250ml double cream.

28 Butter sauce (beurre blanc)

1 Reduce the water, vinegar and shallots in a thick-bottomed pan to approximately 2 tablespoons.
2 Allow to cool slightly.
3 Gradually whisk in the butter in small amounts, whisking continually until the mixture becomes creamy.
4 Whisk in the lemon juice, season lightly and keep warm in a bain marie.

The sauce may be strained if desired. It is suitable for serving with fish dishes.

Variation
This butter sauce can be varied by adding, for example, freshly shredded sorrel or spinach, or blanched fine julienne of lemon or lime.

Ingredient	4 portions	10 portions
Water	125ml	300ml
Wine vinegar	125ml	300ml
Shallot, finely chopped	50g	125g
Unsalted butter	200g	500g
Lemon juice	1 tsp	2½ tsp
Salt and pepper		

29 Butter-thickened sauce for fish (monter au beurre)

This is the traditional way to make a classic à la carte fish sauce.

1 Strain off the liquid in which the fish has been poached.
2 Reduce the liquid to a glaze.
3 Away from the heat, gradually and thoroughly mix in small pieces of butter.
4 If the sauce is to be glazed, some lightly whipped double cream may be added.

Professional tip
Many sauces can be enriched and given a sheen by adding butter in this way at the last moment before serving.

30 Hollandaise sauce

1 Place the peppercorns and vinegar in a small pan and reduce to one third.

2 Add 1 tablespoon of cold water and allow to cool. Add the egg yolks.

3 Put on a bain marie and whisk continuously to a sabayon consistency.

4 Remove from the heat and gradually whisk in the melted butter.

5 Add seasoning and pass through muslin or a fine chinois.

6 Store in an appropriate container at room temperature.

Egg-based sauces should not be kept warm for more than 2 hours. After this time, they should be thrown away. They are best made fresh to order.

Faults

If you add oil or butter when making hollandaise or mayonnaise, the sauce may curdle because the lecithin in the egg yolk has had insufficient time to coat the droplets. This can be rectified by adding the broken sauce to more egg yolks.

Variations
- Mousseline sauce: hollandaise base with lightly whipped cream.
- Maltaise sauce: hollandaise base with lightly grated zest and juice of one blood orange.

Ingredient	Makes 500g
Peppercorns, crushed	12
White wine vinegar	3 tbsp
Egg yolks	6
Melted butter	325g
Salt and cayenne pepper	

31 Béarnaise sauce

1 Place the shallots, tarragon, peppercorns and vinegar in a small pan and reduce to one third.

2 Add 1 tablespoon of cold water and allow to cool. Add the egg yolks.

3 Put on a bain marie and whisk continuously to a sabayon consistency.

4 Remove from the heat and gradually whisk in the melted butter.

5 Add seasoning. Pass through muslin or a fine chinois.

6 Store in an appropriate container at room temperature.

Note: Egg-based sauces should not be kept warm for more than 2 hours. After this time, they should be thrown away. They are best made fresh to order.

Variations

- Choron sauce: add 200g tomato concassé, well dried. Do not add the chopped tarragon and chervil to finish.
- Foyot or Valois sauce: add 25g warm meat glaze. Typically served with steaks.
- Paloise sauce: this sauce is made as for béarnaise using chopped mint stalks in place or the tarragon in the reduction. To finish, add chopped mint instead of the chervil and tarragon.

Ingredient	Makes 500g
Shallots, chopped	50g
Tarragon	10g
Peppercorns, crushed	12
White wine vinegar	3 tbsp
Egg yolks	6
Melted butter	325g
Salt and cayenne pepper	
Chervil and tarragon to finish, chopped	

32 Sabayon sauce

Ingredient	Makes 8 portions
Egg yolks	4
Caster or unrefined sugar	100g
Dry white wine	250ml

1 Whisk egg yolks and sugar in a 1 litre pan or basin until white.
2 Dilute with the wine.
3 Place pan or basin in a bain marie of warm water.
4 Whisk mixture continuously until it increases to four times its bulk and is firm and frothy.

Sabayon sauce may be offered as an accompaniment to any suitable hot sweet, e.g. pudding or soufflé.

It may also be made using milk in place of wine, which can be flavoured according to taste, e.g. with vanilla, nutmeg or cinnamon.

33 Sabayon with olive oil

This sauce is an alternative to hollandaise. It is healthier than hollandaise because it does not contain the cholesterol of a hollandaise made with butter.

Ingredient	4–6 portions	8–12 portions
Crushed peppercorns	6	15
Vinegar	1 tbsp	2½ tbsp
Egg yolks	3	8
Olive oil	250ml	625ml
Salt and cayenne pepper		

1 Place the peppercorns and vinegar in a small sauteuse or stainless steel pan and reduce to one third.
2 Add 1 tablespoon of cold water and allow to cool.
3 Add the egg yolks and whisk over a gentle heat in a sabayon.
4 Remove from heat and cool.
5 Gradually whisk in the tepid olive oil.
6 Correct the seasoning.
7 Pass through a muslin or fine strainer.
8 Serve warm.

Faults

Should the sauce curdle, place a teaspoon of boiling water in a clean sauteuse and gradually whisk in the curdled sauce. If this fails to reconstitute the sauce, then place an egg yolk in a clean sauteuse with a dessertspoon of water. Whisk lightly over gentle heat until slightly thickened. Remove from the heat and gradually add the curdled sauce, whisking continuously. To stabilise the sauce during service, 60ml thick béchamel may be added before straining.

34 Suprême sauce

This sauce can be served hot with boiled chicken, vol-au-vonts, etc. and can also be used for white chaud-froid sauce. The traditional stock base is a good chicken stock.

Ingredients	4 portions	10 portions
Butter or oil	100g	250g
Flour	100g	250g
Stock (chicken, veal, fish or mutton)	1 litre	2.5 litres
Mushroom trimmings	25g	60g
Egg yolk	1	2
Cream	60ml	150ml
Lemon juice	2–3 drops	5–6 drops

1 Melt the fat or oil in a thick-bottomed pan.
2 Add the flour and mix in. Cook to a sandy texture over a gentle heat, without colouring.
3 Allow the roux to cool, then gradually add the boiling stock. Stir until smooth and boiling.
4 Add the mushroom trimmings. Simmer for approximately 1 hour.
5 Pass through a final conical strainer.
6 Finish with a liaison of the egg yolk, cream and lemon juice. Serve immediately.

Professional tip
Once the liaison has been added, do not reboil this sauce.

35 Onion soubise

Ingredient	Makes 300g
Onions, peeled	750g
Sprigs thyme	2
Butter	100g
Chicken stock	250ml
Cream	125ml

1 Slice the onions finely and pick the thyme.
2 Cook together in the butter until golden, then add the stock and reduce by half.
3 Add the cream, bring to the boil and liquidise until smooth.

36 Fresh tomato sauce (cooked)

1 Skin, halve, remove the seeds and chop the tomatoes.
2 Sweat the chopped onion and garlic in the butter.
3 Add the tomatoes and season.
4 Simmer for 15 minutes.
5 Purée in a liquidiser or food processor.
6 Bring to the boil and correct the seasoning.

Professional tip

Fully ripe, well flavoured tomatoes are needed for a good fresh tomato sauce. Italian plum tomatoes are also suitable and it is sometimes advisable to use tinned plum tomatoes if the fresh tomatoes that are available lack flavour and colour.

Variations

Herbs such as rosemary, thyme or bay leaf may be added and shallots used in place of onion.

Asparagus, celery, fennel, mushroom, onion, salsify, spinach, tomato, watercress and yellow pepper are other vegetables that can be seasoned with herbs or spices, then processed, strained and thinned to a suitable consistency for a sauce, or enriched with butter and/or cream.

Some pulse vegetables, e.g. lentils, peas and beans, are suitable for cooking and processing into sauces.

Ingredient	4 portions
Tomatoes, fully ripe	1 kg
Onion, chopped	50g
Clove garlic, chopped	1
Butter	25g
Salt and pepper	
Pinch of sugar	

37 Pesto sauce

Ingredient	Makes 250ml
Fresh basil leaves	100ml
Pine nuts (lightly toasted)	1 tbsp
Garlic (picked and crumbled)	2 cloves
Parmesan cheese (grated)	40g
Pecorino cheese (grated)	40g
Olive oil	5 tbsp
Salt and pepper	

1 Place all ingredients into a food processor and mix to a rough-textured sauce.

2 Transfer to a bowl and leave for at least 1 hour to enable the flavours to develop.

Pesto is traditionally served with large flat pasta called Trenetta. It is also used as a cordon in various fish and meat-plated dishes, e.g. grilled fish, medallions of veal.

Variation
Use flat-leaved parsley in place of basil, and walnuts in place of pine nuts.

38 Roasted pepper oil

▲ Infusing the ingredients

Ingredient	Makes 500ml
Virgin olive oil	500ml
Mirepoix	200g
Red pepper (cleaned and roasted)	2
Yellow pepper (cleaned and roasted)	2
Green pepper (cleaned and roasted)	2
Bay leaves	2
Black peppercorns	8
Garlic cloves	2
Sea salt	½ tsp
Olive oil	1 tsp
White wine vinegar	1 tsp

1 Heat 1 teaspoon of olive oil in a pan.

2 Fry the mirepoix and one of each of the peppers until golden brown.

3 Add 1 bay leaf, the black peppercorns, garlic and the sea salt, and cook for a further 3–4 minutes.

4 Add half the olive oil and bring to simmering point.

5 Take off the heat and allow to cool.

6 When completely cool, pass through a chinois into a clean Kilner jar.

7 Add the white wine vinegar and the remaining olive oil, roasted peppers and remaining bay leaf.

8 For full flavour, leave for at least one month before use. To use, strain the oil, discarding the ingredients.

Recipe supplied by Mark McCann.

39 Thousand Island dressing

Thousand Island dressing is an example of a dressing which has become a standard condiment.

Ingredient	
Mayonnaise	250ml
Tomato ketchup	30ml
White wine vinegar	1 tbsp
Caster sugar	1 tsp
Finely chopped onion	15g
Finely chopped clove of garlic	1
Tabasco sauce	2 drops
Worcester sauce	½ tsp
Gherkins finely chopped	2

1 Mix all the ingredients together; use as a dressing for fish cocktails and salads.

40 Mediterranean dressing

Ingredient	
Vinaigrette	125ml
Feta cheese, diced	25g
Parsley, chopped	¼ tsp
Oregano, chopped	¼ tsp
Plum tomato, skinned, deseeded, diced	1

1 Combine all the ingredients.

41 Spicy honey and mustard dressing

Ingredient		
Olive oil	100ml	
Honey	2 tsp	
Dijon mustard	1 tsp	
Lime juice	1 tbsp	
Zest of lime	1	
Thyme, chopped		

1 Mix together the ingredients, except the thyme.

2 Finish with chopped thyme to serve.

42 Mango and lime dressing

Ingredient		
Mango, peeled, chopped	1	
Zest and juice of lime	1	
Dijon mustard	1 tsp	
Sugar	½ tsp	
Rice vinegar	100ml	
Olive oil	100ml	

1 Mix together the mango, lime zest and juice, mustard and sugar.

2 Add the vinegar and oil.

43 Italian dressing

Ingredient		
Olive oil		100ml
White wine vinegar		2 tbsp
Onion, chopped, blanched		20g
Garlic clove, finely chopped		1
Parsley, chopped		1 tbsp
Oregano, chopped		1 tsp

1 Mix together the ingredients.

44 Green or herb sauce

Ingredient	4 portions	10 portions
Spinach, tarragon, chervil, chives, watercress (mixture)	50g	125g
Mayonnaise	250g	625g

1 Pick, wash, blanch and refresh the green leaves.
2 Squeeze dry.
3 Pass through a very fine sieve.
4 Mix with the mayonnaise.
May be served with cold salmon or trout.

45 Cumberland sauce

Ingredient	4 portions	10 portions
Redcurrant jelly	100ml	250ml
Chopped shallots	5g	12g
Lemon juice	¼	½
Port	2 tbsp	5 tbsp
Juice and zest of orange	1	2
English mustard	¼ level tsp	½ tsp

1 Warm and melt the jelly.
2 Blanch the shallots well and refresh.
3 Add the shallots to the jelly with the remainder of the ingredients, except the orange zest.
4 Cut a little fine julienne of orange zest, blanch, refresh and add to the sauce.

May be served with cold ham.

46 Red onion confit/marmalade

Ingredient	Quantity
Red onions, sliced	1 kg
Butter	50g
Soft brown sugar	50g
Red wine vinegar	250ml
Blackcurrant cordial or red wine (optional)	60ml

1 Slowly sauté the onions in the butter in a thick-bottomed pan.
2 Cook thoroughly but with little or no colour.
3 Add the other ingredients and reduce slowly until slightly thick.
4 Season lightly with salt and milled pepper.
5 When cold, store in covered jars or basins in the refrigerator.

May be served as a garnish/accompaniment to many hot or cold dishes.

47 Yoghurt and cucumber dressing

Ingredient	4 portions	10 portions
Cucumber	100g	250g
Natural yoghurt	125ml	300ml
Salt and pepper		
Chopped mint	¼ tsp	¾ tsp

1 Peel the cucumber, blanch in boiling water for 5 minutes, refresh and drain.
2 Purée the cucumber in a food processor.
3 Add the cucumber to the natural yoghurt, season and finish with freshly chopped mint.

48 Balsamic dressing

Ingredient	Makes 225ml
Balsamic vinegar	50ml
Grape seed/corn oil	50ml
Light olive oil	125ml
Seasoning	

1 Place the vinegar and corn/grape seed oil in a large bowl and whisk to an emulsion.
2 Slowly add the olive oil, about 50ml at a time, bringing to an emulsion at each stage and season.
3 Pour into a jar or bottle and store in the fridge until ready to use. If the mix separates during storage, simply shake the bottle or jar to re-emulsify.

49 Fig and apple chutney

Ingredient	6 portions
Cooking apple, peeled and cut into 2.5cm dice	1
Onion, diced	25g
Dried fig, chopped	50g
White wine vinegar	25ml
English mustard	½ tbsp
Cayenne pepper	pinch
Clove garlic	Half
Sultanas	50g
Sugar	10g

1 Combine all the ingredients in a heavy saucepan.

2 Bring to the boil then lower the heat and simmer for 2 hours until thick.

3 Add a splash of water if the mixture dries out before the 2 hours are up.

4 Leave to cool, and then briefly liquidise the mixture until it is the consistency of jam. Store in the refrigerator.

Test yourself

Sauces

11 State the quality points to look for in a sauce.

12 Why is the use of good quality stocks important when making a sauce?

13 Describe a nage.

14 List three ways of thickening a sauce.

15 Describe what is meant by an emulsified sauce.

16 How can hollandaise be rectified when it has curdled?

17 Name three derivatives of a velouté sauce.

18 Name three derivatives of a brown sauce produced from a jus-lié.

19 What is a coulis?

20 How should cold sauces be stored?

21 How should stocks be finished and stored?

22 List four types of flavoured oils.

Dressings

23 Describe what is meant by a dressing.

24 List three types of dressings.

25 At what temperature should chilled dressings be served?

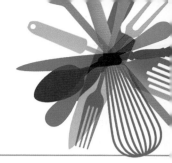

7 Advanced vegetable and vegetarian dishes

This chapter covers the following units:

NVQ:
→ Cook and finish complex vegetable dishes.

VRQ:
→ Advanced skills and techniques in producing vegetable and vegetarian dishes.

Introduction

The aim of this chapter is to develop the necessary advanced skills, knowledge and understanding of the principles in preparing and cooking vegetables to dish specifications. The emphasis in this chapter is to develop precision, speed and control in existing skills and to develop more refined and advanced techniques.

Learning objectives

By the end of this chapter you should be able to:
→ Identify suitable commodities and recipes for vegetarian/vegan diets.
→ Explain how to select the correct type, quality and quantity of vegetables to meet recipe and dish requirements.
→ Describe the composition of different vegetables and factors affecting it, including cooking methods.
→ Describe the cooking requirements for different vegetables.
→ Identify the quality points for a range of vegetable dishes and describe how to achieve the desired outcome.
→ Describe how to store vegetables and vegetable dishes.
→ Demonstrate a range of cooking and preparation techniques using the correct tools and equipment appropriate to the different cooking methods.
→ Describe how to adapt recipes and cooking techniques to maximise nutritional value and healthy options in vegetable and vegetarian professional cookery.
→ Demonstrate professional ability in minimising and correcting faults in complex vegetable and vegetarian dishes.
→ Explain how to carry out different finishing methods and identify relevant sauces and dressings.

Recipes included in this chapter

Vegetables

Vegetables, like fruits, are the edible products of certain plants. Vegetables are savoury rather than sweet and, in most countries, they are associated with poultry, meat or fish as part of a meal or as an ingredient. Some vegetables are botanically classed as fruits – for example, tomatoes are berries and avocadoes are drupes – but are used as vegetables because they are not sweet. People who include plenty of vegetables in their diet are more energetic and do not get tired as easily. A diet rich in fresh vegetables ensures smooth functioning of all bodily processes. As carbohydrates and fat are only present in very small quantities in vegetables, problems of weight gain or obesity can also be controlled with a vegetable diet.

Composition

Most vegetables contain at least 80 per cent water (the remainder is made up of carbohydrate, protein and fat). The water content varies depending on the type of vegetable: squashes contain a high percentage of water; potatoes contain a great deal of starch. Corn, carrots, parsnips and onions contain invert sugars.

Because of the high water content, most fruits and vegetables provide only minerals, certain vitamins and some fibre. However, these additions – particularly of vitamins such as ascorbic acid – can be of immense value in our diet.

A typical plant cell is like a balloon, not blown up with air but with water. The internal pressure in the cell is called **turgor pressure** and can be as high as nine times that of atmospheric pressure. When plant tissues are in a state of maximum water content they are said to be **turgid** or in a complete state of turgor. In a normal cell in full turgor the pressure is equalised by the elasticity of the cell wall.

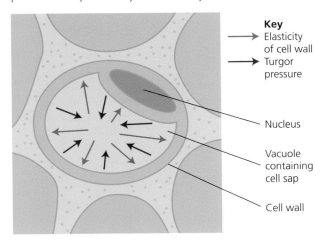

Key
→ Elasticity of cell wall
→ Turgor pressure

Nucleus

Vacuole containing cell sap

Cell wall

▲ Equalised pressures within a plant cell

Once the supply of water is reduced or cut off to the cell, water is gradually lost from the plant by **transpiration** and the turgor pressure cannot be maintained. The elasticity of the cell wall now exceeds the outward turgor pressure and the cell starts to collapse. This is repeated throughout the plant, causing it to wilt. The rate of wilting varies: harvested leafy vegetables such as lettuce, for example, only last short periods before becoming limp.

As plant cells become older, a complex substance called lignin is deposited in the cell wall, thus making it tougher. Water evaporates and sugars become concentrated – old, raw carrots, for example, are sweeter than young ones. Sugars change as soon as the vegetable is separated from the plant. (A good example is corn, which is often rushed straight from the stalk to the pot in order to preserve its taste.)

The cell wall of fruit and vegetable cells is made of carbohydrates, mainly cellulose. Cells are held together by pectins and hemicellulose; these change as the vegetable ripens, so that the cells part easily, giving a softer texture. Starch is the second most common carbohydrate after cellulose and is the main food material in most vegetables.

<table>
<tr><td rowspan="6">**Key term**</td><td>**Pectin** – a complex polysaccharide present in primary cells. Pectin is formed in the middle of plants and it helps bind cells together.</td></tr>
<tr><td>**Polysaccharides** – these are formed from monosaccharide units.</td></tr>
<tr><td>**Protopectin** – substance found in the skins of citrus fruits and apples and can be hydrolysed to form pectin.</td></tr>
<tr><td>**Cellulose** – an insoluble substance; the main constituent of plant cell walls and of vegetable fibres such as cotton. The plant uses this for support in stems, leaves, husks of seeds and bark.</td></tr>
<tr><td>**Hemicellulose** – a polysaccharide similar to cellulose but consists of many different sugar building blocks in shorter chains than cellulose.</td></tr>
</table>

Factors affecting the composition of vegetables:

Origin – where vegetables are grown affects their composition and nutritive value. The type of soil (acid, alkali, rocky, sandy or clay) is a factor.

Seasons – most fruit and vegetables are seasonal and dependent on the climate. The climate and the weather affect the size of fruits and vegetables and the yield per acre. Adverse weather conditions can reduce yields dramatically. Variations in weather conditions may be offset by using protected environments such as polytunnels or heated greenhouses. Good weather conditions can increase yields and produce high quality fruits and vegetables.

Types of vegetables

New vegetable varieties and hybrids, with different flavours and colours, are constantly being developed.

Root vegetables

These are plant roots commonly consumed as vegetables. Remove root vegetables from their sacks and store in bins or racks in a cool, dark place.

Vegetable	Description	Nutritional information	Quality points	Shelf life	Cooking methods
Beetroot	Two main types: round and long. Closely related to sugar beet. Leaves can be cooked like spinach.	High in vitamin A, iron and calcium. Good source of potassium. Helps renew immunity cells and purify blood	Firm to touch. Good, even colour; fresh, earthy aroma	Up to 6 weeks	Steam, boil, bake, roast. Can be used in salads and soups
Carrot	Bright orange colour. Young carrots are tender; older ones should be firm.	Large amounts of carotene and vitamin A, good amount of vitamins B3, C and E. Good quantities of potassium, calcium, iron and zinc	No blemishes, good colour. Young carrots should have intact, feathery tops	Up to 10 days	Boil, steam, roast. Can be used in soups and salads
Cassava	West Indian root used in Caribbean dishes. Native to Brazil. Used to make tapioca.	Good source of vitamins A, C, K, folate	Long, tubular, brown in colour, no blemishes, firm	Up to 2 weeks	Boil, grated
Celeriac	Large, light brown, celery flavour. Knobbly with partly brown skin, crunchy texture if eaten raw.	Vitamin C, potassium, calcium, iron and magnesium	Firm, good colour, no blemishes	Up to 14 days	Boil, steam, roast, bake, deep fried crisps, salads, soups
Ginger	Light brown, knobbly. Used since ancient times for its anti-inflammatory, anti-microbial properties	Source of folate, vitamin C, E, calcium, magnesium, phosphorus	Firm, no blemishes	Up to 2 weeks	Used in relishes, sauces, stir fry

Vegetable	Description	Nutritional information	Quality points	Shelf life	Cooking methods
Horseradish	A strong-flavoured root, often grated.	Source of calcium, iron, magnesium, potassium, phosphorous, sodium.	Flesh should be slightly moist and the skin should be easy to peel	Up to 14 days	Used as a relish, garnish
Mouli or daikon radish (sometimes known as oriental radish)	Smooth-skinned, long, white radish, mild, less peppery than the red radish.	Source of vitamins B1, B2, B3, calcium, vitamin C, iron and magnesium	Free from blemishes, smooth skin	Up to 14 days	Boil, stir fry, salads, pickles
Parsnip	Normally a winter crop, usually at their best after the first frost.	Source of vitamins A, C, calcium, iron and potassium	Firm, pale, ivory colour without any sprouting roots	Up to 10 days	Boil, roast, deep fry, soups
Radish	Peppery, flavour. Pungency not only depends on the varieties but on the soil they are grown in.	Source of vitamins A, C, K and folate	Generally bright red in colour, free from blemishes	Up to 5 days	Salads, crudités
Salsify and scorzonera (black salsify)	Closely related vegetables, members of the same family as dandelion and lettuce. Pale, creamy flesh. Flesh will oxidise quickly so must be kept in acidulated water.	Good source of vitamin C, folate. Scorzonera good source of potassium, magnesium, phosphorus	White and pale brownish skin; scorzonera has a black skin, firm to touch	Up to 10 days	Boil, steam, deep fry and shallow fry
Swede	Swedes were known as turnip-rooted cabbages until 1780s when Sweden began exporting the vegetable to Britain and they became known as swedes. Firm, mottled purple colour.	Good source of calcium and potassium	Firm, unblemished, good shape, not split or damaged	Up to 2 weeks	Boil, purée, roast, soup, steam
Turnip	Round white/green flesh, mild flavour, the leaves can be cooked as turnip tops.	Good source of calcium and potassium	Firm flesh, leaves intact.	Up to 2 weeks	Boil, fry, roast, soups, steam

Tubers

The tuber is the enlarged tip of an underground stem (rhizome). The plant uses this tip to store food.

Vegetable	Description	Nutritional information	Quality points	Storage	Cooking methods
Jerusalem artichoke	Jerusalem artichokes are related to the sunflower and have nothing to do with Jerusalem. One explanation for their name was that they were called girasol, 'Jerusalem', because the yellow flowers turned towards the sun. Small knobbly tubers; a lovely distinct flavour.	Good source of vitamins A, C and folate.	Firm to touch no blemishes	5°C up to 10 days	Boil, roast, soup
Sweet potato	Deep orange colour when cut, skin colour can range from white/pink to reddish yellow brown. Commonly used in African and Caribbean cooking.	Good source of vitamins A, C, K and betaine.	Firm flesh, no sign of woodiness and free from blemishes	5°C for up to 3 weeks	Boil, roast, bake, deep fry
Yam	A staple food for many cultures. Many different varieties, with different shapes, colours from white to yellow to purple. Some wild yams are said to contain a toxic alkaloid dioscorine and histamine (an allergen). This has not been detected in all yams, but for precautions it is not advisable to eat yams raw.	Good source of vitamins A, C, K, folate and choline.	Course skin, firm and unbroken flesh, should be moist	5° up to 3 weeks	Boil, steam, roast, deep and shallow fry

Potatoes

Potatoes are a type of tuber. There are thousands of different varieties grown around the world. In the UK over 450 varieties are grown and used as part of a healthy meal. Some of the most popular potato varieties in the UK are listed below. Potatoes are a seasonal crop and so availability will vary.

Variety	Description	Best for	Season
Annabelle	An oval-shaped, yellow salad potato with a waxy yellow flesh.	Salads, boiling, or for best results steaming	Salad
Cara	White skinned, with creamy flesh, pink eyes and a soft, waxy texture.	Mash, baking	Main
Carlingford	A small firm potato with a fresh-flavoured, waxy texture. It can be eaten hot or cold.	Salads, boiling, steaming	New
Charlotte	A small, deep yellow fleshed potato with a firm waxy texture.	Often used in salads. For boiling, steaming, baking, roasting, sauté	Salad

Variety	Description	Best for	Season
Desirée	Has a red or pink skin with yellow and creamy flesh. It has a deep flavour, making it suitable for many forms of cookery.	Boiling, baking, roasting, chips, sauté, mash, wedges	Main
Duke of York	Has a yellow flesh and loose skin. They are harvested early before maturity for a fresh flavour.	Steaming, boiling	Main
Dunbar Rover	A very rare variety with an oval shape, white skin and snowy white flesh. It has a lovely buttery flavour and fine texture.	Baking, boiling, steaming, roasting, chips	Main
Estima	Has a pale yellow skin, with a firm, moist texture and a mild flavour. It is best for simple recipes.	Mash, especially good for baking	Main
Exquisa	Has a unique teardrop or apostrophe shape. It has a smooth, soft texture and a rich, buttery taste.	Boiling, steaming	Salad
Galante	An early potato with yellow flesh and smooth yellow skin.	Boiling, steaming	Salad
Harmony	A very smooth, round potato with a firm texture. It has a very pale white skin and flesh and is suitable for potato recipes that require wedges.	Mash, baking wedges, boiled, baking jacket potatoes	Main
Kerr's Pink	Rosy-skinned with creamy white flesh, which has a mealy, floury texture when cooked. Kerr's Pink is best for simple recipes.	Boiling, steaming, mash, roasting	Salad
King Edward	A traditional English potato, with some red blushes on the skin. It has a floury texture making it ideal for recipes that include baking and roasting.	ideal for baking and roasting; can be used for boiling, mash, chips	Main
Lady Balfour	Named after Lady Eve Balfour – a pioneer of the UK organic movement. Extremely versatile and can work in many different potato recipes. It has a uniform oval shape, creamy skin and pale tasty flesh.	Baking, mash, roasting, chips, wedges	Main
Maris Peer	A small firm potato with a fresh-flavoured, waxy texture. It can be eaten hot or cold.	Most suitable for boiling or steaming. Can be used for salads, baking, mash, roasting, chips, wedges	Main
Maris Piper	One of the best varieties for chipping, but can also be used in simple recipes like roasting, baking and mashing. It is a white fleshed potato with a dry floury texture and full flavour.	Baking, roasting, chips, wedges	Main
Melody	A smooth skinned potato, with a yellow flesh which is quite dry.	Suitable for all types of cookery but mostly used for baking jacket potatoes or mashing	Main
Red Duke of York	Has a thick red russet skin and soft, floury, yellow flesh with a delicate flavour.	Wedges, roasting, baking	Main
Red King Edward	A rare version of the King Edward, oval in shape with red skin and a white flush. It has a floury texture.	Baking, boiling, roasting, steaming, chips	Main
Romano	Red-skinned, with creamy, mildly nutty flesh and a soft, dry texture. The skin fades to rusty beige during cooking.	Baking, mash, roasting	Main
Saxon	Has white skin and white flesh with a firm and moist texture. It has a creamy flavour.	Chips, boiling, mash, baking, jacket potatoes	Main
Sante	Yellow-skinned with a dry, firm texture.	Boiling, roasting, chips, wedges	Main

Variety	Description	Best for	Season
Vales Sovereign	A relatively new variety and has a great all-round cooking quality with an oval shape, a good texture and cream coloured flesh.	Particularly good for mash, baking, roasting, chips, wedges; also a good baking potato	Main
Vivaldi	A pale yellow potato with a velvety texture. It is particularly good as a boiled potato as it holds its shape whether hot or cold. It has a mild sweet flavour.	Mash, steaming, baking, microwaving; particularly good for boiling	Main
Yukon Gold	Has an attractive smooth yellow skin with very yellow flesh.	Baking, roasting, chips	Main

Professional tips

- In order to get better-browned roast potatoes and oven-fried potatoes, refrigerate the potatoes whole for a day before cutting and cooking them. Chilling converts some of the starch in the potatoes to sugar and enhances browning.
- For potatoes to retain a firm texture in potato salads or during a long cooking process, first cook at a low temperature (54–60°C for 20–30 minutes), drain and let the potatoes cool. This initial low-temperature cooking activates an enzyme in the potatoes' cell walls that prevents the cells from weakening and making them mushy during longer cooking.
- Pre-cooling can also help keep mashed potatoes from becoming gluey because the potatoes firm up before the final cooking and mashing.

Bulbs

Bulbs belong to the Allium family. They are perennial, meaning that they last a long time throughout the year and recur yearly. They have a strong flavour and are pungent and therefore are used to add flavour to many dishes. The pungency varies from variety to variety; the method and style of cooking will also affect their flavour and pungency.

Vegetable	Description	Nutritional information	Quality points	Storage	Cooking methods
Fennel	A swollen leaf base, with a distinctive flavour. Crisp texture.	Good source of vitamins A, C and folate	Good round size, free from blemishes firm to touch	5°C for up to 2 weeks	Used in a variety of dishes, soups, sauces, stir fries etc.
Garlic	An onion-like bulb with a papery skin, inside which are small individually wrapped cloves. Has a pungent distinctive flavour, often crushed and chopped.	Good source of vitamins A and C, magnesium, phosphorus, potassium, sodium and selenium	Tight cloves, firm without sprouting	5°C for up to 2 weeks	Used in a variety of dishes, soups, sauces, stir fries etc. Can be roasted whole.
Leek	Long white stems, bright green leaves. The tougher top leaves are used in stocks and sauces. Summer leeks have a milder flavour than winter leeks.	Good source of vitamins A, C, E and K, folate, choline	The bottom part of the leek should be white, with pale green colour at the top. Check that the white bottom has not gone to seed and become woody and hard	In a dark place at 5°C for up to a week	Boil, braise, soups, sauces

Vegetable	Description	Nutritional information	Quality points	Storage	Cooking methods
Onion	There are varieties with different coloured skins and varying strengths. All have a papery outer skin. There are also very small button or pickling onions. Spanish onions are generally milder than other varieties. Red onions (sometimes called Italian onions) have a milder flavour. Yellow onions are the most pungent of all and are excellent for pickling. White onions, come in various shapes and sizes. The very small white onions called 'Paris silver skin' are used for commercial pickling. Vidalia are popular American onions. Bermuda onions are similar in size to Spanish onions but more squat and have a mild flavour.	Good source of vitamins, A, C and K, folate, pantothenic acid, choline	Papery skin, very dry, moist inside, no sprouting	Onions have a relatively long shelf life. Up to several weeks, if stored in a dark, cool place	Braise, fry, boil, roast
Spring onions	Slim and tiny bulbs. Look like miniature leeks. They may be harvested before they are fully developed or when the bulb is fully formed. Known as scallions in Ireland, cibies (or sibies) in Scotland and gibbons in Wales.	Good source of vitamins A and C, magnesium, phosphorous, sodium, zinc, manganese, selenium	Firm bulb, green tops, fresh and not withered, no discolouration or wilting	5°C for up to 3 days	Salads, stir fries
Shallots	Similar to onions, but a milder flavour. Common varieties: banana, brown English and pink shallot. In delicate culinary dishes, finely chopped shallots will dissolve into the sauce.	Good source of vitamin A	Dry papery skin, with no sprouting	Cool, dry, dark place or refrigerate at 5°C	As a flavouring ingredient for many dishes

Leafy vegetables

These are the leaves of plants, grown using **photosynthesis** – a natural process that uses water and sunlight to produce sugar and oxygen, which takes place in the leaves. These vegetables supply rich protein, fibres and minerals like iron and calcium to our body. They also give us phytonutrients like carotenoids, vitamins A, C and K and folic acid. The fat content of leafy vegetables is very low, so they are ideal for those following a weight-loss regime.

During storage, leafy vegetables lose moisture and vitamin C degrades rapidly. They should be stored for as short a time as possible in a cool place in a container, such as a plastic bag or a sealed plastic container.

Vegetable	Description	Nutritional information	Quality points	Storage	Cooking/ uses
Chicory	Usually white, there are some red varieties. Bitter flavour.	Good source of vitamins A, C and K, folate, choline	Tight close leaves, crisp, firm, no wilting	5°C for up to 3 days	Braise, salads
Chinese leaves	Long white, densely packed leaves, sharp bitter taste on outside leaves, inner leaves are milder.	Good source of vitamins A, C and K, folate	Crisp leaves, no wilting	5°C up to 3 days	Salads, stir fries
Corn salad (lambs lettuce)	Small, tender, dark leaves with a tangy nutty taste.	Good source of vitamins A and C	Crisp leaves, no wilting	5°C up to 3 days	Salads, garnishing
Cress	There are over 15 varieties, many with different flavours.	Very good source of vitamins A, C and K, choline	Fresh green colour, no wilting of stems or leaves	5°C up to 3 days	Salads, garnishing
Lettuce, e.g. cos, rocket, oak leaf, romaine, iceberg, round, radicchio, frisée, endive	There are many varieties, with flavours ranging from peppery to bitter.	Good source of vitamins A, C and K, folate, calcium, iron and potassium	Strong vibrant leaves, crisp, no discoloration or wilting	5°C up to 3 days	Salads, braise
Mustard and cress	Sharp, warm flavour, small leaves, bright green colour.	Good source of vitamins A, C and, B3, iron, sulphur, zinc, folate	Good green colour, no wilting or bruising	5°C up to 3 days	Salads, garnish
Nettles	Young green leaves; once cooked the sting disappears. Nettles contain prostaglandins which have properties that support resistance to inflammation.	Relatively high in protein, calcium, phosphorous, iron, magnesium, beta carotene, vitamins A, C, D and B	No wilting, good colour	5°C up to 4 days	Salads
Sorrel	Tender, dark green leaves with a musky flavour.	Good source of vitamins C and A, iron, calcium, phosphorous, potassium and magnesium.	Dark green, no wilting or bruised leaves	5°C up to 5 days	Soups, salads, stir fries
Spinach	Tender, dark green leaves, mild musky flavour.	Good source of vitamins A, C, K, B, folate, choline, betaine	Dark green, no blemishes or wilting	5°C up to 3 days	Boil, sauté, purée, stir fries, salads

Vegetable	Description	Nutritional information	Quality points	Storage	Cooking/ uses
Pousse	A baby form of spinach, very delicate flavour.	The same as spinach	Crisp leaves, no wilting or bruising	5°C up to 3 days	Boil, sauté, purée, stir fries, salads
Swiss chard	Large, ribbed, slightly curly leaves, flavour similar to spinach, but milder.	Good source of vitamins C, K, B, folates, calcium, sodium, potassium, iron, manganese and phosphorous	Leaves intact, no blemishes, good green colour	5°C 3 to 5 days	Sauté, boil, purée, stir fries
Vine leaves	From grape vines. Distinctive pointed leaves, dark green in colour.	Good source of vitamins A, C and K, folate, choline	No wilting or bruising	5°C up to 2 days	Soups, salads, garnishing, wrapping
Watercress	Long stems with round, dark, tender green leaves and a pungent peppery flavour.	Good source of vitamins A, C and K, choline	No wilting or bruising	5°C up to 2 days	Soups, salads, garnishing

▲ Lettuces: from left to right, iceberg, romaine, rocket, oak leaf, radicchio and frizzy endive

Brassicas

Brassicas are part of the mustard family, also collectively known as cruciferous vegetables, or cabbages. The edible parts of the plants develop above the ground. These vegetables are rich in vitamin C, dietary fibre and calcium. They supply us with a component called sulforaphane, which is an anti-cancer and anti-diabetic compound. Their fat content is low.

Vegetable	Description	Nutritional information	Quality points	Storage	Cooking methods
Broccoflower	A cross between broccoli and cauliflower. Chinese broccoli is a leafy vegetable with slender heads of flower.	Vitamin C, iron, sodium	Good foliage, strong and firm	5°C for up to 7 days	Boil, steam, purée, soups, salads
Broccoli	Varieties include calabrese, white, green, purple sprouting.	Good source of vitamins A, C, E and K, folate, choline; high in calcium	Floral heads should not have any discolouration or yellowing	5°C up to 3–5 days	Boil, steam, stir fry
Brussel sprouts 	Small, green buds, grow on thick stems.	Good source of vitamins A, C, K, folate, choline	Tight closed heads, good green colour, no wilting or blemishes	5°C up to 5 days	Boil, steam, sauté, stir fry
Cabbage 	Three main types: green, white and red. Many varieties of green cabbage available at different seasons of the year. Savoy (shown here) is considered the best of the winter green cabbages. White cabbage is used for coleslaw.	Good source of vitamins A, C, K, folate, choline	Early green cabbage is deep green and loosely formed, later in the season they firm up with solid hearts	Leafy cabbage 5°C up to 5 days. White and red 5°C up to 14 days	Boil, steam, braise, stir fry, pickle, salads
Cauliflower 	Large, immature flower head with small, creamy white buds.	Good source of vitamins B and C; also contains K, magnesium, phosphorus, potassium, iron	Firm, white heads, no browning or yellow spots; fresh leaves	Refrigerate at 3°C up to 5 days	Boil, steam, sauté, stir fry, roast, use as a crudité
Chinese mustard greens	Deep green, mustard flavour.	Good source of vitamin K and folic acid. Includes mainly di-indolyl-methane and sulforaphane. Have proven benefits against cancer by inhibiting cancer cell growth.	Deep green, no bruising or wilting	5°C up to 5 days	Salads, stir fry

Vegetable	Description	Nutritional information	Quality points	Storage	Cooking methods
Kale and curly kale	Thick, green leaves like other brassicas.	Contains health-promoting phytochemicals that protect against cancer.	Bright green leaves with no discoloration or limp leaves	5°C up to 5 days	Boil, steam, stir fry
Pak choi, bok choi	A common ingredient of Chinese cookery, bright green leaves to pale green base.	Vitamins A, C, E, K, folate, choline	No bruising or wilting leaves, a good green colour	5°C up to 3 days	Braise, stir fry
Romanesco	Green and white cross between broccoli and cauliflower.	High vitamin C; also B, K, manganese, magnesium, phosphorus, potassium	Good colour, no yellowing, bruising or wilting	5°C up to 5 days	Boil, steam, stir fry

Pods and seeds

This group of vegetables includes fresh legumes, seed-bearing pods and seeds. All are best eaten young when they are at their sweetest and most tender. Legumes have all the valuable nutrients: proteins, carbohydrates, vitamins and useful minerals like iron, potassium and calcium. They have very high fibre content. Beans are considered as a suitable substitute to meat for vegans as they contain essential proteins similar to those in meat.

Vegetable	Description	Nutritional information	Quality points	Storage	Cooking methods
Broad beans	Pale green, oval-shaped, contained in a thick fleshy pod.	Good source of vitamins A, C, K, B, folate, choline	The outer pod is thick and strong	5°C up to 3 days	Boil, steam, salads
Butter or lima beans	Butter beans are white, large, flat and oval-shaped. Lima beans are smaller.	Good source of vitamins A, C, K, folate, choline	The outer pod is strong and crisp	5°C up to 3 days	Boil, stew, soups, salads
Mangetout (snow peas, sugar peas)	Flat pea pod with immature seeds.	Vitamins C, A, B, calcium, zinc folate, iron	Bright green, crisp	5°C for 1 day	Boil, stir fry, salads
Okra	Curved and pointed seed pods.	Vitamins A, C, E, K, B, folate, choline	Bright green, no bruising	5°C up to 3 days	Boil, fry, soups, braise, curries

Vegetable	Description	Nutritional information	Quality points	Storage	Cooking methods
Peas	Firm, round and sweet when freshly picked, pods are bright green. Frozen peas are usually of superior quality as the peas are picked at the optimum peak when the starch is converted to sugar, then quickly frozen.	Vitamins A, C, E, K, B1, calcium, iron, magnesium, phosphorous, sodium, selenium	Fresh pods, bright green, crisp when broken in half	Frozen – store at -18–20°C for up to 3 months. Fresh – use on day of purchase	Boil, steam
Runner beans	Bright green colour, long and crisp.	Vitamins A, C, D, E, K, folate, choline	Bright green and crisp and snap easily; not too lumpy, indicating older beans	5°C up to 3 days	Boil, steam
Sweetcorn (maize, sudan corn)	The cob is covered with a straw-like leaf, pale green.	Corn is a good source of phenolic flavonoid antioxidant, ferulic acid. Ferulic acid plays a vital role in preventing cancers, is anti-ageing and anti-inflammatory. Vitamins A, B, folates, zinc, magnesium, copper, iron and manganese	Bright yellow cob, no discoloration	5°C up to 5 days	Boil, steam, soups, salads

Stems

Stems support the entire plant and have buds, leaves, flowers and fruit. These vegetables are grown for their edible stems.

Vegetable	Description	Nutritional information	Quality points	Storage	Cooking methods
Asparagus	Three main types: white with creamy stems and a mild flavour; French with violet or bluish tips and a stronger, more astringent flavour; and green with aromatic flavour.	Vitamins A, C, K, folate, choline	Firm and crisp, not damaged or woody	5°C for up to 5 days	Boil, steam, poach, soups, salads

Vegetable	Description	Nutritional information	Quality points	Storage	Cooking methods
Bean sprouts	Slender young sprouts of the germinating soya or mung bean.	Very little nutritious value	No discolouration or wilting	5°C up to 3 days	Stir fry, salads
Cardoon	Longish plant with root and fleshy ribbed stalks similar to celery, leaves are grey/green in colour.	Vitamins B, C, manganese, calcium, iron, selenium, zinc	Heavy leaves without blemishes	5°C for up to 7 days	Braise, salads
Celery	Long-stemmed bundles of fleshy, ribbed stalks, white to light green in colour.	Vitamins A, C, K folate, choline	The leaves should be bright green with no blemishes	5°C for up to 5 days	Braise, soups, stocks, sauces, salads
Chicory	Conical heads of crisp white, faintly bitter leaves.	Vitamins A, C, E, K, folate, choline	White in colour, light yellow tips, no bruising	5°C up to 5 days	Braise, salads
Globe artichoke	Resemble fat pine cones with overlapping fleshy green inedible leaves. All connected to an edible fleshy base.	Vitamins A, C, K, B, folate, choline	Tough green leaves with no discolouration	5°C up to 7 days	Boil in acidulated liquid or steam
Kohlrabi	Stem that swells to turnip shape above the ground.	Vitamins A, C, E, K, B, folate, choline	No blemishes or bruising	5°C up to 10 days	Stews, purée, boil, steam
Samphire	Two types: marsh samphire (also called glass wort or sea asparagus) which grows in estuaries and salt marshes; and white rock samphire (sea fennel) which grows on rocky shores.	Vitamins A, C, calcium, iron	No wilting, good colour	5°C up to 5 days	Boil, steam, stir fry
Sea kale	Delicate white leaves with yellow fruits edged in purple.	Rich in potassium, sulphur, vitamin C	Long tender stalks with fresh leaves at the top. No blemishes and firm to touch	5°C up to 3 days	Boil, braise, salad, sauté

Vegetable fruits and squashes

These are fruits used as vegetables. They are normally grown from a stem and are the fruits with seeds. These vegetables provide antioxidants, phytochemicals, roughage and various useful minerals to our body. Antioxidants present in these vegetables are vital for maintaining good health and preventing diseases like cancer and coronary heart diseases. They also ensure protection of our skin from harmful UV radiations of the sun.

Vegetable	Description	Nutritional information	Quality points	Storage	Cooking methods
Aubergine/egg plant 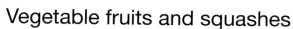	Firm, elongated, varying in size with smooth shiny skins, ranging in colour from purple to red to black. Inner flesh is white with tiny soft seeds.	Vitamins A, C, K B, folate, choline	Firm with no bruising or irregular shape	5°C for 3-5 days	Bake, fry, grill, stuffed
Avocado	Bland, mild, nutty flavour. Green to purple/black skin depending on ripeness.	Good source of Omega 6 poly unsaturated fatty acids, iron, copper, magnesium, manganese, potassium Vitamin As, C, E, K	Check for ripeness: should be soft to touch. No blemishes or bruising	Once ripe use on day of purchase. Overripe very quickly and deteriorate as they ripen	Salads
Chilli	Spicy; varieties range in colour from green to almost chocolate-coloured; wide variation in shape and size.	Rich in vitamin C; also contains A, B, potassium, manganese, iron, calcium	Firm, no blemishes, bright colour, smooth (except wrinkled varieties)	Refrigerate at 3°C up to 5 days	In regional dishes, e.g. Thai; to garnish/finish; in a powder
Christophine (chow-chow, chayote, vegetable pear)	Several varieties, white or green, spiny or smooth-skinned, some pear-shaped. Inside flesh is firm and white. Similar flavour to courgette or marrow.	Vitamins A, B, C, K, folate, amino acids, calcium, zinc, potassium, iron, magnesium	Firm, plump, few wrinkles	Refrigerate at 3°C up to 4 days	Boil, steam, shallow fry, salads, South American dishes
Courgette	Baby marrow, light to dark green in colour.	Vitamins A, C. Rich in flavonoid poly-phenolic anti-oxidants such as carotenes, lutein and zea-xanthin. Helps against ageing and various disease processes in the body	No discolouration or blemishes	5°C for 3-5 days	Boil, steam, braise, sauté, stir fry

Vegetable	Description	Nutritional information	Quality points	Storage	Cooking methods
Cucumber	Long, smooth-skinned fruiting vegetable, ridged and dark green in colour.	Vitamins A, C, K, folate, choline	The flesh should be firm and free from blemishes and bruising	5°C up to 6 days	Salads, braise
Durian	Strong odour. The seeds can be eaten when cooked. Large fruit can grow up to 30cm long. Shape is oblong to round. Husk colour is green to brown. Widely used in South East Asia, known as the king of fruits.	Rich in vitamin C, potassium, good source of carbohydrates, fat and protein	Firm, prickle covered. Tough outer husk	5°C up to 5 days	Used in confectionery, ice cream and milkshakes; can be eaten raw
Marrow	Long, oval shaped, edible gourds with ridged, green skins and bland flavour.	Vitamins A, C, calcium, iron	Firm, no blemishes or bruising	5°C for up to 5 days	Boil, steam, sauté, bake
Pepper	Available in a variety of colours: red, green, orange, white. When green peppers are ripe, they turn to yellow to orange and then to red. They must remain on the plant to do this.	Vitamins A, C, E, K, B, folate, choline	Clean, shiny skin with no bruising or blemishes	5°C for up to 5 days	Salads, braise, stir fry, bake
Pumpkin	Vary in size, orange in colour, round and heavy.	Vitamins A, C, K, B, folate, choline	Firm, tough skin, no bruising	5°C up to 5 days	Soups, salads, purée
Squash	Many varieties: e.g. acorn, butternut, summer crookneck, hubbard, kaboche.	Vitamins A, C, E, K, folate, pantothenic acid	Flesh firm, no uneven colour or bruising	5°C up to 5 days	Soups, purée, boil, steam, fry, bake
Tomato	Many varieties available including cherry, plum, yellow globe, large ridged beef.	Vitamins A, C, K, folate, choline. Antioxidants in tomatoes found to be protective against cancers. Lycopene in tomatoes, a flavonoid antioxidant, prevents skin damage from ultra-violet rays and offers protection against skin cancer	Good colour, red, firm flesh, no cracks or bruising, when cut juicy and sweet	5°C up to 5 days	Grill, stuff, bake, salads, sauces, soups

Fungi and mushrooms

A mushroom is the fleshy, spire-bearing, fruiting body of a fungus. Mushrooms are grown above ground on soil. Mushrooms can make a delicious food but many are poisonous; it is therefore important that every mushroom intended for eating be accurately identified. Mushrooms are a good source of vitamins D, C and B, as well as calcium and magnesium, and especially phosphorus, potassium and selenium.

Vegetable	Description	Quality points	Storage	Cooking methods
Field mushroom	Found in meadows from late summer to autumn, creamy white cup and stalk, strong earthy flavour.	Firm	5°C for up to 3 days	Sauté, grilling, stir fry, soup
Cultivated mushrooms	Button shape, small, succulent, weak in flavour; cup: open or flat, firm.	Firm, no slime	5°C for up to 3 days	Sauté, grilling, stir fry
Ceps (porcini)	Short stalks with slightly raised veins and tubes underneath the cup in which the brown spores are produced.	Firm, no slime	5°C for up to 3 days	Sauté, grill
Chanterelles or girolles	Wild, funnel-shaped, yellow-capped mushroom with a slightly ribbed stalk that runs under the edge of the cap.	Firm, no slime	5°C for up to 3 days	Sauté, grill
Horns of plenty	Trumpet-shaped, shaggy almost black, wild mushroom.	Firm, no slime	5°C for up to 3 days	Sauté, grill
Morels	Delicate wild mushrooms varying in colour from pale beige to dark brown/black.	Spongy, no slime	5°C for up to 3 days	Sauté, grill
Oyster	The mushroom has a pronounced cap.	Creamy gills and firm flesh	5°C for up to 2 days	Stir fry, sauté, grilling
Shitake	Firm texture, harvested from hard wood trees, comes in a range of colours.	Firm, whole, not broken or bruised	5°C for up to 2 days	Stir fry, sauté, grilling
Truffles	Almost all truffles are ectomycorrhizal (symbiotic) and are therefore found close to trees. There are hundreds of species: e.g. the white truffle of the Piedmont region in northern Italy; the black truffle or black Périgord truffle, which is named after the Périgord region in France and grows near oak and hazelnut trees. Truffles have a strong aroma.	Up to 7cm in diameter, both the black and white truffle should have a good colour and size and be firm	Fresh truffles should be consumed within a few days of being unearthed. To store longer, wrap in absorbent paper and store in a dry glass or plastic container at the bottom of a refrigerator	Used in a variety of dishes and as a garnish in sauces and salads

▲ Clockwise from top left: girolles, chestnut mushrooms, hedgehog mushrooms, large cup (Portobello) mushrooms, button mushrooms, brown chanterelles and shitake mushrooms

Key term

Symbiotic – refers to two species existing together, but not necessarily benefitting each other.

Aquatic vegetables

These are edible vegetables which grow in water or partially submerged in water.

Vegetable	Description	Nutritional information	Quality points	Storage	Cooking methods
Water chestnut	This is the common name for a number of aquatic herbs and their nut-like fruit. The best known variety is the Chinese water chestnut also known as the Chinese sedge. Sweet, crunchy flavour with nutty overtones.	Rich source of calcium, folic acid, iron, vitamins A and C	Firm, no blemishes	Unpeeled fresh water chestnuts will keep for up to 2 weeks in a plastic bag in the refrigerator	Edible cooked or raw. Used in Chinese cookery

Edible flowers

These are flowers that can be consumed safely. Edible flowers may be preserved for future use using techniques such as drying, freezing or steeping in oil. They can be used in drinks, jellies, salads, soups, syrups and main dishes.

Flower-flavoured oils and vinegars are made by steeping edible flower petals in these liquids. Candied flowers are crystallised using egg white and sugar. The flowers should be fresh, with no withering and of good colour. Most edible flowers can be stored in polyethylene bags at 5°C for 1 week.

Name	Description	Flavour
Apple	White to pink small radiating petals (can contain cyanide precursors, so eat in moderation)	Floral; garnish or candied
Basil	White, pink or lavender tiny petals	Basil with citrus notes
Borage	Pale blue, triangular petals	Cucumber
Marigold	Golden orange, short, plump petals	Peppery and tangy
Carnation	White to deep red petals	Light clove
Chervil	White, delicate, tiny blossoms	Light anise
Chrysanthemum	White, yellow, orange and red spiky petals	Tangy and peppery
Citrus blossoms (orange, lemon, lime, grapefruit)	White to golden, firm, round petals; often distilled into flavour extract	Citrus
Clover	Pale purple, ball-shaped flower	Sweet anise
Daisy	White or yellow narrow petals	Mildly bitter, edible but not delicious
Dandelion	Yellow, spiky petals	Sweet, honey-like
Hibiscus	White, pink, yellow, or crimson, large, showy, trumpet-shaped blossoms	Citrus and cranberry
Honeysuckle	White to yellow tiny cup-shaped blossoms	Honey
Jasmine	Creamy white tiny blossoms	Sweet floral, very aromatic
Lavender	Pale purple small blossoms	Sweet floral, very aromatic
Lemon verbena	Creamy white tiny blossoms	Citrus
Lilac	Pink to purple tiny blossoms	Slightly bitter, lemony, very aromatic
Nasturtium	White to bright orange trumpet-shaped blossom	Sweet and peppery
Pansy	White, yellow or purple two-tiered blossom	Sweet and mild

Name	Description	Flavour
Peony	White to deep red, soft, rose-like petals	Sweet floral
Rose	All colours, soft, velvety petals	Highly floral and aromatic
Squash blossoms	Yellow to pale orange blossoms	Mild squash flavour, often stuffed and fried
Thyme	Pale purple, tiny blossoms	Mild thyme flavour
Violet	Pale to deep purple, small blossoms	Sweet floral

Health and safety

Some vegetables unfortunately contain components which have adverse effects. For examples, potatoes (particularly green ones) contain alkaloids (an alkali like nitrogen) containing substances such as solanine. High levels of this can be toxic. Some vegetables such as cassava, sweet potato and kidney beans contain small amounts of cyanide. Enzymes naturally present in a product can release this cyanide during storage or cooking and poisoning may occur in undernourished people. The cabbage family contains goitrin which can interfere with the uptake of iodine by the body, causing thyroid problems in susceptible people.

Quality

When choosing vegetables you should avoid those that are limp and wilting, discoloured or damaged by harvesting. Leaf vegetables need careful picking over to avoid serving garden pests on the plate. Vegetables should be prepared for the pot as simply as possible. Wash them just before cooking, don't soak them and, as they lose nutrients through peeling and cutting, they should be peeled only thinly (vitamins are usually found just under the skin).

For more detailed quality points for specific vegetables, refer to the tables above.

Storage

When detached from the tree or dug from the ground, a fruit or vegetable can continue to live and respire, and for some varieties this can be for several months.

The storage life of a plant depends on its chemical composition, resistance to microbiological attack, external temperature, and the pressure of various gasses in the storage atmosphere.

On contact with air, certain vegetables tend to discolour, regardless of whether they are cooked or raw. This is because certain enzymes cause **oxidisation**. **Respiration** is the major process of interest in most post-harvest fruit and vegetables: the rate of respiration indicates the rapidity with which changes in fruit and vegetables are taking place. Fruit and vegetables can be divided into two groups, according to their respiratory behaviour:

- **Climatic fruits** are generally fleshy (for example, banana, mango, pear and avocado) and show a rapid rise in the rate of respiration after harvesting, which leads to ripening and then **senescence** of the fruit. They deteriorate quickly, presenting many problems in storage.
- **Non-climatic fruits** (for example, citrus groups, pineapple, fig and grape) and most vegetables (except tomato) show no rapid increase in respiration, and changes in ripening and maturation are gradual. They keep well for long periods under normal conditions.

Key term

Respiration – the exhalation of energy-rich organic compounds to form simpler compounds and yield energy.

Oxidisation – the interaction of oxygen (in the air) with the vegetable, facilitated by enzymes and causing discolouration and some loss of nutrients.

Senescence – ageing of vegetables, caused by enzyme action. Cell walls degrade, causing loss of texture. Nutrients, colour and flavour are also lost.

Professional tip

Respiration slows down as the temperature of the environment decreases, thus delaying ripening and senescence. However, chilling some fruit – particularly those of tropical origin – below a certain temperature may cause damage. Below 13°C some enzyme systems are inhibited in tropical fruit, while others continue, causing the accumulation of toxic intermediate products.

Storage information for specific vegetables is provided in the tables above, but more generally:

- The fresher the vegetables, the better the flavour. If storage is necessary, then it should be for the shortest time possible.
- Store all vegetables in a cool, dry, well ventilated room at an even temperature of 4°C to 8°C to minimise spoilage. If vegetables are stored at the incorrect temperature, micro-organisms may develop.

- Check vegetables daily and discard any that are unsound.
- Store green vegetables on a well ventilated rack. Green vegetables lose vitamin C quickly if they are bruised, damaged or stored for too long.
- Store salad vegetables in a cool place and leave in their containers.
- Store frozen vegetables at -18°C or below. Keep a check on use-by dates, damaged packaging and any signs of freezer burn. Thaw out correctly and never refreeze once thawed.
- Store raw vegetables and cooked vegetables in separate areas to prevent bacteria passing.

Preservation

Vegetables may be preserved using various methods, including:

- Canning – used for artichokes, asparagus, carrots, celery, beans, peas, tomatoes, mushrooms and truffles.
- Drying – the seeds of legumes (peas and beans) have their moisture content reduced to 10 per cent.
- Freezing – many vegetables, such as peas, beans, sprouts, spinach and cauliflower, are deep frozen.
- Pickling – uses acid (usually as vinegar) to discourage the action of enzymes or micro-organisms. Many pickles have a high salt level which helps the acid in its preservation action. Pickling is also carried out to produce and enhance flavour and taste experiences. The simple method involves cooking the vegetables to a soft texture and soaking them in a weak brine to draw out some of the moisture. This is followed by immersion in vinegar, sometimes sweetened and spiced. In fermented pickles, stronger brine is used to prevent the growth of undesirable bacteria, but the brine is weak enough to allow lactic acid bacteria to grow. Sauerkraut, for example, is made by fermenting cabbage. Acid in the pickles helps to retain vitamin C in products which are high in vitamin C.
- Blanching – this helps to preserve colour, especially in green vegetables.

Professional tip

Some vegetable pigments are lost in the cooking process. Purple broccoli contains both chlorophyll (green) and anthocyanin (purple), but the latter is water soluble, so cooked broccoli always looks green. Red cabbage reacts like litmus paper – it turns blue in the presence of an alkali (the lime in tap water), so you need to add a dash of acid, such as vinegar, to preserve the colour.

Effects of cooking on vegetables

The application of heat, whether by boiling, baking or stir-frying, tenderises the food by weakening the cell walls and extracting water. First, heat denatures the proteins that make up the cell membranes; water leaks from the cells, the tissue loses turgor (rigidity), and the plant becomes wilted and flabby. Heat causes some hemicelluloses in the cell wall to dissolve and the distribution of the pectin substances is altered (soluble pectin increases; insoluble protopectins decrease). Cell walls are substantially weakened and tissue is more tender.

The problem with cooking vegetables is how to make the tissue tender without making it too soft. Usually we take the common-sense approach of sampling the food during cooking and stopping when it is tender, but still firm. In some cases colour can be used to indicate that a vegetable is cooked. Experience and personal taste remain the best guides, but as a general rule, leaf vegetables, with their relatively thin, exposed and delicate layer of tissue, need only a minute or two of heating, while stem and root vegetables may require much longer. All root vegetables (except new potatoes) are started off by cooking in cold, salted water. Vegetables that grow above ground are started in boiling, salted water; this is so that they may be cooked as quickly as possible for the shortest time so that maximum flavour, food value and colour are retained.

Approximate times only are given in the recipes in this chapter for the cooking of vegetables, as quality, age, freshness and size all affect the length of cooking time required. Young, freshly picked vegetables will need to be cooked for a shorter time than vegetables that have been allowed to grow older and that may have been stored after picking.

Advanced techniques for preparing and cooking vegetables

Turning – vegetables should be shaped evenly for cookery and presentation, e.g. in a barrel or olive shape.

Shaping – shaping of vegetables for cooking and presentation, e.g. shaping potatoes into small cakes for fondant potatoes.

Carving – vegetables and fruit can be carved into different shapes and designs. In Thailand this is a great art form.

Turning and shaping potatoes

1 Top and tail the potato

2 Using a turning knife, start to cut

3 Complete the cut

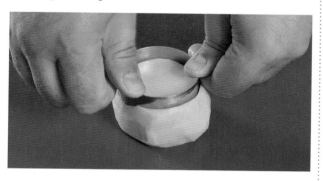

4 Alternatively, to shape a potato into a socle, top and tail it, then press a pastry cutter down into it

5 Lift away the trimmings

6 Shape the edges with a peeler

7 Turned potatoes (left) and a potato socle (right)

Preparing globe artichokes

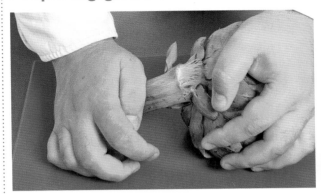

1 Break off the stalk.

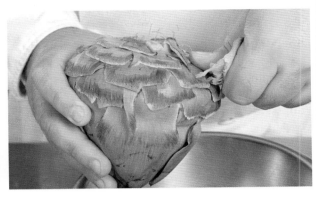

2 Pull off the outer leaves.

3 Trim away any green parts left at the base, then rub with a cut lemon to prevent browning.

4 Cut off the remaining leaves.

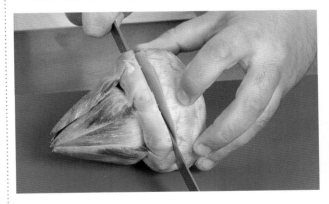

5 Cut open the artichoke.

6 Scoop out the choke (the furry part) before cooking the artichoke.

Video: globe artichokes
http://bit.ly/1i8gGP4

- **Marinating** – vegetables are often marinated before cooking to increase and enhance flavours using scented oils, herbal oils, lemon juice, vinegar and red or white wine.
- **Smoking** – a variety of woods are used (e.g. hickory and oak). They are often enhanced with spices, herbs and other infusions, for example, tea smoking of asparagus, Jerusalem artichokes and fennel.
- **Sous vide** – a method of cooking and processing by which the vegetables are cooked in a vacuum pouch under vacuum at low temperatures, steamed in a water bath and reheated. It is particularly suitable for delicate vegetables such as asparagus. The vegetables retain flavour and colour and will keep refrigerated once cooked in a sealed vacuum pouch for up to five days.
- **Etuvé** – a method of cooking in which vegetables are sweated in butter or vegetable oil, covered, without colouring, to extract flavour. Particularly suitable vegetables are red and white cabbage and peas (French-style peas).
- **Drying** – a modern technique used in professional cookery; it involves drying slices of vegetables for garnish. Vegetables such as tomatoes, celeriac and beetroot are dried, either in a warm oven slowly or in a dehydration machine. Vegetable crisps, parsnips, celeriac and sweet potatoes can be deep fried or dried by baking slowly in an oven.

Vegetarian cooking

People choose to eat vegetarian or vegan food for a variety of reasons, including religious beliefs, ethical and ecological views against meat eating, as well as for health reasons. There are different types of vegetarians, including:

- **Lacto vegetarians** – this group eats milk and cheese, but not eggs, whey or anything that is produced as a result of an animal being slaughtered (meat, poultry, fish or any by-products such as fish oils, rennet and cochineal).
- **Ovo-lacto vegetarians** – eat eggs but otherwise are the same as lacto vegetarians.
- **Demi-vegetarians** – this group usually chooses to exclude red meat, though they may eat it occasionally. White poultry and fish are generally acceptable.
- **Vegans** – avoid all animal products and by-products including milk, cheese, yoghurt, eggs, fish, poultry, meat and honey. They eat only items or products from the plant kingdom.

Vegetarians expect strict working practices when preparing vegetarian dishes. All utensils, equipment and working surfaces must be kept separate from other food which is non-vegetarian. If using the same working area, this must be cleaned thoroughly before any vegetarian food preparation. A separate colour-coding system has been developed to separate all vegetarian equipment from other kitchen equipment.

Excluded items in a vegetarian menu

- **Rennet or pepsin-based cheeses** – rennet is an enzyme from the stomach of a newly killed calf; pepsin is from pigs' stomachs. Replace these items with approved 'vegetarian' cheeses or non-rennet cheeses such as cottage and cream cheese (cheese is still not suitable for vegans, unless made with soya milk). Check to ensure ready-made vegetarian dishes include vegetarian cheese.
- **Battery farmed eggs** – these may have been fed fish meal; also strict vegetarians consider the battery rearing of hens is cruel. Replace these items with free-range eggs. Strict vegetarians will eat only free-range eggs – this is impractical for most manufacturers and caterers so clear labelling is important so as not to mislead.
- **Whey** – a by-product of cheese-making and therefore may contain rennet. Whey may be found in biscuits, crisps and muesli.
- **Cochineal (E120)** – made from the cochineal beetle. This is often present in glacé cherries and mincemeat. Choose an alternative red colouring.
- **Alcohol** – some wines or beers may be 'fined' using isinglass (a fish product) or dried blood. Some ciders contain pork to enhance their flavour. Check with the wholesaler or manufacturer if in any doubt before use in cooking or serving to a vegetarian.
- **Meat or bone stock and animal-based flavourings** – stock for soups or sauces; animal-based flavourings for savoury dishes. Replace these items with stock made from vegetables or yeast extract or bought vegetable stock cubes/bouillon. Alternative flavourings include soy sauce, miso, Holbrook's Worcester sauce (which is anchovy free).
- **Animal fats** – includes suet, lard or dripping and ordinary white cooking fats; some margarines contain fish oil. Bought-in pastry may contain lard or fish oil margarine. Replace these items with Trex and Pura white vegetable fats or 100 per cent vegetable oil margarines. Suenut or Nutter (available from health food stores) and white Flora can also be used.

- **Oils containing fish oil** – replace these items with 100 per cent vegetable oil (sunflower, corn, soya, ground nut, walnut, sesame, olive) or mixed vegetable oil. Fish oils may well be 'hidden' in margarine and products such as biscuits, cakes and bought-in pastry items. Check suitability before using.
- **Setting agents** – gelatine, aspic, block or jelly crystals (for glazing, moulding and in cheesecakes and desserts). Some yoghurt is set with gelatine, as are some sweets, particularly nougat and mints. Replace these items with agar agar, Gelozone, and apple pectin. Other gums are also used as substitutes, for example, alginates (produced from seaweed), other pectins, xanthan gum and the tree gums tragacanth and carob bean gum.
- **Animal fat ice cream** – replace animal fat ice cream with vegetable fat ice cream.

Professional tip

Never claim that a food or dish is vegetarian if you have used a non-vegetarian ingredient. If you knowingly mislead a customer into believing you have a suitable vegetarian choice on offer, you can be prosecuted under the Trade Descriptions Act 1968/Food Safety Act 1990.

Professional tip

Many additives contain meat products. Look on the label for 'edible fats', 'emulsifiers', 'fatty acids' and the preservative E471. The safest course is to ask the manufacturer of any bought-in product you wish to use.

Protein and vegetarian diets

In a meat-eater's diet, meat, poultry and fish provide a considerable amount of the daily protein intake. Any vegetarian or vegan meal must therefore contain an adequate source of protein to replace meat protein.

Protein is made up of amino acids. Human protein tissue and animal proteins contain all the 'essential' amino acids (the body can manufacture the non-essential amino acids). Vegetable proteins, however, contain fewer of the essential amino acids. Vegetarians therefore need to combine various plant proteins in one dish or meal to provide the equivalent amino acid profile of animal protein. This is known as **protein complementing**. The lack of some amino acids in one plant is compensated for by another. For example, putting together 60 per cent beans (or other pulses) or nuts with 30 per cent grains or seeds and grains and 10 per cent green salad or vegetables makes the ideal combination.

Sources of vegetarian protein

The best sources of vegetarian protein are cheese, eggs, milk and textured vegetable protein (TVP), followed by tofu, soya beans and all other pulses and nuts. Seeds, cereals (preferably wholegrain), millet, wheat, barley and oats are also good sources of protein for vegetarians. Vegans exclude cheese, eggs and milk as a source of protein and substitute with soya milk, soya cheese and soya yoghurt. All other items above are acceptable.

Fifty grams of meat is equivalent in protein to the following (all cooked weights):

- 75g fish
- 50g (hard cheese)
- 100g soft or cottage cheese
- 2 eggs
- 100g nuts
- 50g peanut butter
- 150g pulses (lentils, peas, beans)
- 75g seeds.

Leaves, stems, buds and flowers are almost all water and have insignificant protein content. A dish such as ratatouille, made from 'water' vegetables, is suitable only as a side dish. 'Vegetable' curries, hotpots and similar dishes should all include a recognisable and good vegetable protein source.

Textured vegetable protein (TVP)

TVP, also known as soy protein, is a deflated soy flour product, a by-product of extracting soya bean oil. It is quick to cook and has a protein content similar to meat. TVP is usually made from high-protein (50%) soy flour or concentrate but can also be made from cotton seeds, wheat or oats. It is processed into various shapes, chunks, flakes, mince. After processing, it is fried and rehydrated before use at a 2:1 ratio. Soy protein TVP can also be used as a low-cost/high-nutrition extender in meat and poultry products.

Myco-protein

Myco-protein is also known as fungi protein. Quorn is the leading brand of myco-protein and is used as an alternative to meat.

1 Anna potatoes

1 Grease an Anna mould using hot oil.

2 Trim the potatoes to an even cylindrical shape.

3 Cut into slices 2mm thick.

4 Place a layer of slices neatly overlapping in the bottom of the mould, season lightly with salt and pepper.

5 Continue arranging the slices of potato in layers, seasoning in between.

6 Add the butter to the top layer.

7 Cook in a hot oven (210–220°C) for 45 minutes to 1 hour, occasionally pressing the potatoes flat.

8 To serve, turn out of the mould and leave whole or cut into four portions.

Ingredient	4 portions	10 portions
Oil		
Peeled potatoes	600g	1.5 kg
Salt and pepper		
Butter	25g	60g

Energy	Cal	Fat	Sat fat	Carb	Sugar	Protein	Fibre
688 kJ	159 kcal	5.4g	3.3g	25.8g	0.9g	3.2g	2.0g

Finely slice the potatoes

Layer the potato slices in the mould

Carefully remove from the mould

2 Potato blinis

Ingredient	4 portions	10 portions
Dry mash	500g	1.25 kg
Crème fraiche	125g	300g
Flour	25g	60g
Whole eggs	4	10
Yolks	2	5
Seasoning		

Energy	Cal	Fat	Sat fat	Carb	Sugar	Protein	Fibre
1474 kJ	352 kcal	21.7g	11.1g	28.1g	1.6g	13.1g	1.9g

1 Mix the mash with the crème fraiche and the flour.
2 Separate the eggs and add the yolks to the mash.

3 Whip the egg whites to a snow, carefully fold into the mash, check for seasoning and allow to rest for 1 hour.

4 Heat a little oil in a non-stick pan and place a small amount of mix in the pan (approximately 1 tablespoon). Turn over when it is golden brown.

3 Potato cakes with chives

Ingredient	4 portions	10 portions
Large potatoes	4	10
Egg yolks	2	5
Butter	50g	125g
Chives, chopped	50g	125g
Salt and pepper		

Energy	Cal	Fat	Sat fat	Carb	Sugar	Protein	Fibre
1181 kJ	282 kcal	14.2g	2.1g	34.6g	1.4g	6.0g	2.8 g

1 Bake the potatoes in their jackets.
2 Halve and remove the potato from the skins.
3 Mash with the yolks and butter.
4 Mix in the chopped chives and season.
5 Mould into round cakes, 2cm diameter.
6 Lightly flour and shallow fry to a golden colour on both sides.

4 Sauté potatoes with onions and mixed herbs

Ingredient	4 portions	10 portions
Potatoes, peeled, sliced approx. 0.5cm thick	600g	1.5 kg
Onion, shredded	200g	500g
Duck fat or vegetable oil	100g	250g
Parsley, chopped	½ tsp	1½ tsp
Chervil, chopped	½ tsp	1½ tsp
Thyme, chopped	½ tsp	1½ tsp

Energy	Cal	Fat	Sat fat	Carb	Sugar	Protein	Fibre
525 kJ	124 kcal	2.2g	0.7g	28.2g	4.8g	3.2g	2.9g

1 Mix the sliced potato and onion together and lightly season with salt and pepper.
2 Place the duck fat in a deep oven-proof dish and put into a pre-heated oven at 180°C or heat on top of the stove.
3 When the fat is very slightly smoking, place the potatoes and onion in the dish and place into the oven.

4 After approximately 10 minutes, turn the potatoes to colour evenly.
5 After approximately 20 minutes, add the chopped thyme and continue to turn the potatoes.
6 Cook until tender.
7 Finish and serve with chopped parsley and chervil.

5 Potatoes cooked in milk with cheese

Ingredient	4 portions	10 portions
Potatoes	500g	1.25 kg
Milk	250ml	600ml
Salt and pepper		
Grated cheese (Gruyère, Emmental, Cheddar or Parmesan)	50g	125g

Energy	Cal	Fat	Sat fat	Carb	Sugar	Protein	Fibre
730 kJ	173 kcal	5.6g	3.4g	24.3g	3.6g	7.8g	2.2g

1 Slice the peeled potatoes to ½cm thick.

2 Place in an ovenproof dish and just cover with milk.

3 Season, sprinkle with grated cheese and cook in a moderate oven (190°C) until the potatoes are cooked and golden brown.

6 Braised (fondant) potatoes with thyme

Ingredient	4 portions	10 portions
Potatoes	500g	1.25 kg
White stock	375ml	1 litre
Powdered thyme, pinch of		
Butter or margarine	50g	125g
Salt and pepper		

Energy	Cal	Fat	Sat fat	Carb	Sugar	Protein	Fibre
802 kJ	192 kcal	10.9g	6.5g	21.9g	0.9g	3.0g	2.2g

1 Trim and cut the potatoes to an even size.

2 Place in a dish, half cover with the stock to which the thyme has been added.

3 Brush with melted butter and season.

4 Cook in a hot oven (230°C), brushing occasionally with melted butter.

5 When ready, the potatoes should have absorbed the stock and be golden brown in colour and cooked through.

Note: As thyme is a strong, pungent herb, it should not be used in excess.

7 Potatoes baked in stock with cheese and garlic

Ingredient	4 portions	10 portions
Potatoes, peeled	400g	1kg
Stock	375ml	1 litre
Egg	1	2–3
Grated cheese	50g	125g
Clove of garlic, crushed and chopped	1	2
Butter or margarine	50g	125g

Energy	Cal	Fat	Sat fat	Carb	Sugar	Protein	Fibre
1082 kJ	260 kcal	16.4g	9.5g	21.9g	0.9g	7.4g	2.2g

1 Thinly slice the potatoes.

2 Mix the stock, egg and grated cheese in a basin.

3 Butter an earthenware dish and add the potatoes, stock and garlic.

4 Sprinkle with more grated cheese and a little melted butter.

5 Bake in a moderate oven (190°C) until the potatoes are cooked and golden brown.

8 Deep-fried vegetables in tempura batter

Ingredient	4 portions	10 portions
Mangetout, topped and tailed	12	30
White button mushrooms, halved	100g	250g
Carrot, cut in matchstick pieces	100g	250g
Sweet potato, peeled and sliced thinly	100g	250g
Batter	**Makes enough for 10 portions**	
Flour	350g	
Cornflour	100g	
Baking powder	50g	
Egg	3	
Water (sparkling), ice cold	625ml	

1 Flour the vegetables, then shake off excess flour.

2 Sift the flour, cornflour and baking powder. Add the egg and enough water to give a light batter consistency.

3 Pass the vegetables through batter, remove excess, deep-fry in hot oil (175°C). Drain and serve immediately.

Professional tip

To create a lighter, thinner, crispier crust for tempura or other fried foods, add 1 teaspoon (5ml) baking soda per 250g of flour in the batter.

9 Celery royale

Ingredient	4 portions	10 portions
Butter	25g	75 g
Celery, finely chopped	100g	250g
Béchamel sauce, cold	2 tbsp	6 tbsp
Egg	1	3
Egg yolks	3	6
Seasoning		

Energy	Cal	Fat	Sat fat	Carb	Sugar	Protein	Fibre
588 kJ	142 kcal	13.6g	6.6g	0.8g	0.8g	4.4g	0.4g

1 Cook the celery in the butter until soft. Purée in a food processor.

2 Add the béchamel sauce and the beaten egg and egg yolks. Season.

3 Place in greased individual moulds or a dariole mould.

4 Place in a bain marie in the oven at 180°C or a combination oven injected with steam for 25 to 30 minutes.

Variations

- Asparagus royale: use asparagus instead of celery.
- Leek royale: use leeks instead of celery.
- Green pea royale: use 100g of pea purée instead of the celery, and white stock with a pinch of sugar instead of béchamel sauce.
- Tomato royale: as for green pea royale, using 200g of tomato purée made with fresh tomatoes.
- Carrot royale: cook 75g carrots in the butter and then purée them. Add 2 tablespoons double cream to the béchamel sauce, with 1 egg and 2 egg yolks.

Note: Royales are cut into various shapes and used as a garnish for soups, salads and other dishes.

Professional tip

A royale must be completely cold before shaping, to make it easy to handle.

10 Vegetable soufflé

Ingredient	4 portions	10 portions
Eggs	3–4	7–10
Seasoned vegetable purée, stiff	400g	1 kg
Double cream	12g	5 tbsp

1 Separate the eggs and stiffly beat the whites.
2 Mix the egg yolks and cream into the vegetable purée.
3 Fold in the beaten egg whites.
4 Prepare the soufflé moulds with butter and flour. Place the mixture into the moulds.
5 Bake in a hot oven at 220°C until set.

11 Grilled vegetables

1 Wash, peel and trim the vegetables. Discard the pith and seeds of the peppers.
2 Dry well, brush with olive oil, season lightly.
3 Grill the vegetables with the most tender last.
4 Serve as a first course accompanied by a suitable sauce, e.g. spicy tomato sauce, or as a vegetable accompaniment to a main course.

Professional tip

Tender young vegetables can be cooked from raw and are best cooked on an under-fired grill or barbecue.

Variation

If vegetables other than baby ones, e.g. carrot, turnip, parsnip, are grilled, then they first need to be cut into thick-ish slices and parboiled until half-cooked, drained well, dried, then brushed with oil and grilled. They can be served plain, sprinkled with chopped, mixed herbs, or with an accompanying sauce.

Ingredient	4 portions	10 portions
Small asparagus stems	8	20
Red onion, thickly sliced	8 slices	20 slices
Long red or orange peppers	2	5
Tomatoes	4	10
Olive oil		

Energy	Cal	Fat	Sat fat	Carb	Sugar	Protein	Fibre
308 kJ	73 kcal	1.1g	0.1g	13.0g	11.7g	3.8g	6.4g

12 Braised chicory

Ingredient	4 portions	10 portions
Fish or chicken stock	200ml	500ml
Chicory, medium heads	8	20
Fresh lemon juice	3 tbsp	8 tbsp
Caster sugar	3 tbsp	8 tbsp
Sea salt and freshly ground black pepper		
Butter	25g	60g

Energy	Cal	Fat	Sat fat	Carb	Sugar	Protein	Fibre
476 kJ	114 kcal	6.3g	3.7g	17.8g	13.5g	1.1g	1.8g

1 Have the stock ready and set aside.

2 Trim the chicory of any bruised outside leaves, then trim the ends and use a small, sharp knife to remove the bitter core at the base of each head.

3 Bring a pan of water to the boil and add the lemon juice, 1 tablespoon of the sugar and salt to taste. Blanch the chicory for 8–10 minutes and drain well.

4 Drain all the liquid from the chicory and, in a large frying pan, heat the butter and brown the chicory on all sides, deglaze with a little stock and simmer for a few minutes, basting the chicory at all times.

5 Arrange the heads in a single layer on a platter, sprinkle with the remaining sugar, and season with salt and pepper. Leave to cool for about 10 minutes.

13 Braised red cabbage with apples, red wine and juniper

Ingredients	4 portions	10 portions
Red cabbage, finely shredded	400g	1.25 kg
Vegetable oil	2 tbsp	5 tbsp
Cooking apples, peeled, cored and diced	200g	500g
Red wine	250ml	600ml
Salt and pepper		
Juniper berries	12	30

Energy	Cal	Fat	Sat fat	Carb	Sugar	Protein	Fibre
368 kJ	88k cal	5.8g	0.6g	8.2g	7.8g	1.3g	3.3g

1 Blanch and refresh the cabbage.

2 Heat the oil in a casserole, add the cabbage and apples, and stir.

3 Add the wine, seasoning and juniper berries.

4 Bring to the boil, cover and braise for approximately 40–45 minutes until tender.

5 If any liquid remains when cooked, continue cooking uncovered to evaporate the liquid.

14 Leaf spinach with pine nuts and garlic

Ingredient	4 portions	10 portions
Spinach	1 kg	2.5 kg
Pine nuts	125g	300g
Oil or butter	1 tbsp	2–3 tbsp
Garlic clove, chopped	1	2–3
Salt and pepper		

Energy	Cal	Fat	Sat fat	Carb	Sugar	Protein	Fibre
614 kJ	149 kcal	12.5g	1.1g	3.0g	2.8g	6.0g	3.4g

1 Cook the spinach for 2–3 minutes and drain well.

2 Lightly brown the pine nuts in oil, add garlic and sweat for 2 minutes.

3 Add coarsely chopped spinach and heat through over a medium heat.

4 Correct seasoning and serve.

15 Salsify with onion, tomato and garlic

Ingredient	4 portions	10 portions
Salsify	400g	1.25 kg
Margarine, oil or butter	50g	125g
Onions, chopped	50g	125g
Clove of garlic, crushed and chopped	1	2–3
Tomatoes, skinned, deseeded and diced	100g	250g
Tomato purée	25g	60g
White stock	250ml	600ml
Seasoning		
Parsley, chopped		

Energy	Cal	Fat	Sat fat	Carb	Sugar	Protein	Fibre
558 kJ	135 kcal	10.7g	6.5g	13.2g	4.1g	2.1g	3.8g

1 Wash and peel the salsify, cut into 5cm lengths. Place immediately into acidulated water to prevent discoloration.

2 Place salsify into a boiling blanc or acidulated water with a little oil and simmer until tender, approximately 10–40 minutes. Drain well.

3 Melt the margarine, oil or butter, add the onion and garlic. Sweat without colour.

4 Add the tomatoes, and tomato purée, cook for 5 minutes.

5 Moisten with white stock, correct seasoning.

6 Place the cooked and well drained salsify into the tomato sauce.

7 Serve in a suitable dish, sprinkled with chopped parsley.

Butter was used for the nutritional analysis.

16 Stir-fried bok choi

Ingredient	4 portions	10 portions
Baby bok choi	4 bunches	10 bunches
Fresh ginger, grated	1 tsp	2½ tsp
Soy sauce	2 tbsp	5 tbsp
Sugar	1 tsp	2½ tsp
Sesame oil	1 tsp	2½ tsp
Vegetable oil	2 tbsp	2½ tbsp
Water	1 tbsp	2½ tbsp

Energy	Cal	Fat	Sat fat	Carb	Sugar	Protein	Fibre
210 kJ	51 kcal	3.7g	0.4g	3.1g	3.0g	1.2g	1.6g

1 Wash the baby bok choi and drain. Separate the stalks and leaves.
2 Split the bok choi lengthways into halves or quarters.
3 Heat the vegetable oil in a wok. Add the ginger and stir-fry for about 30 seconds.
4 Add the bok choi. Quickly stir-fry for 1 minute.
5 Add the soy sauce, sugar and stir-fry for a further 1 minute.
6 Add the water. Simmer for 1 minute.
7 Stir in the sesame oil, garnish with sesame seeds and serve.

There are more than 20 varieties of bok choi available in Asia.

17 Gribiche

Ingredient	4 portions	10 portions
Egg, hard-boiled, finely grated	100g	225g
Capers, chopped	40g	100g
Shallot, finely diced	20g	50g
Muscatel vinegar	5g	12g
Salt	4g	10g
Parsley, chopped	4g	10g
Crème fraiche	50g	125g

Energy	Cal	Fat	Sat fat	Carb	Sugar	Protein	Fibre
389 kJ	94 kcal	7.9g	4.2g	1.6g	0.7g	4.2g	0.8g

1 Combine all the ingredients. Store until required.
2 Serve on toast as an appetiser or canapé.

Peas were used instead of capers in the nutritional analysis.

18 Bulgur wheat and lentil pilaff

Ingredient	4 portions	10 portions
Green lentils	100g	250g
Bulgur wheat	100g	250g
Coriander, ground	1 tsp	2½ tsp
Cinnamon, ground	1 tsp	2½ tsp
Olive oil	1 tbsp	2½ tbsp
Streaky bacon, finely chopped	200g	500g
Onion, finely chopped	50g	125g
Garlic clove, finely chopped	1	3
Cumin seeds	1 tsp	2½ tsp
Salt and pepper		
Parsley, finely chopped	2 tbsp	5 tbsp

Energy	Cal	Fat	Sat fat	Carb	Sugar	Protein	Fibre
1399 kJ	336 kcal	19.2g	5.3g	25.0g	0.9g	17.2g	0.2g

1 Soak the lentils in cold water for 1 hour. Soak the bulgur wheat in boiling water for 15–20 minutes. Drain both.

2 Place the lentils in a saucepan with the coriander and cinnamon and approximately half of the water – enough to cover the lentils. Bring to the boil and simmer until the lentils are cooked and the liquid has been absorbed.

3 Heat the olive oil in a pan and fry the bacon until crisp. Remove and drain.

4 Add the onion and garlic to the pan. Add a little more oil if required and sweat for approximately 10 minutes until the onion is slightly brown.

5 Stir in the cumin seeds and cook for 1 minute. Return the bacon to the pan.

6 Stir the drained bulgur wheat into the cooked lentils, then add the mixture to the frying pan. Season with salt and pepper, stir in the parsley and serve.

For a vegetarian dish, replace the bacon with tomato.

19 Galette of aubergines with tomatoes and mozzarella

Ingredient	4 portions	10 portions
Aubergines	400g	1 kg
Onions, finely chopped	100g	250g
Garlic, crushed and chopped	2	5
Vegetable oil		
Low-fat yoghurt	500ml	1¼ litres
Cornflour	2 tsp	5 tsp
Coriander, finely chopped	1 tsp	2½ tsp
Sugar	1 tsp	2½ tsp
Plum tomatoes	400g	1kg
Buffalo mozzarella	150g	375g
Seasoning		
Coriander, to garnish		

Energy	Cal	Fat	Sat fat	Carb	Sugar	Protein	Fibre
940 kJ	224 kcal	9.6g	6.2g	20.7g	17.0g	15.0g	3.4g

1   Slice the aubergines thinly. Place on a tray and sprinkle with salt. Leave for 1 hour.

2   Dry the aubergines in a cloth. Quickly fry in oil until golden brown on both sides. Remove from pan, drain well.

3   Sweat the finely chopped onion and garlic in oil without colour.

4   Mix together the yoghurt, cornflour, coriander and sugar, add to the onion and garlic. Bring to the boil and season lightly. Remove from heat.

5   Blanch and peel the tomatoes, chop into ½cm slices.

6   Arrange 4 large slices of aubergine on a greased baking sheet. Spread the yoghurt mixture on each slice, and top with tomato and mozzarella. Continue to build layers of aubergine, tomato and mozzarella. Finish with a layer of yoghurt mixture.

7   Bake in a hot oven (220°C) for 20 minutes until golden brown. Serve immediately, garnish with coriander.

20 Spinach subric

Ingredient	4 portions	10 portions
Spinach	750g	1.5 kg
Butter	25g	50g
Béchamel sauce (thick)	125mls	250mls
Eggs	1	2
Egg yolks	2	4
Nutmeg	¼ tsp	½ tsp
Vegetable oil	75mls	150mls
Salt and pepper		

Energy	Cal	Fat	Sat fat	Carb	Sugar	Protein	Fibre
1041 kJ	251 kcal	21.0g	8.4g	10.5g	2.8g	5.7g	0.3g

1   Remove the stalks from the spinach, wash well, cook in a small amount of boiling water for approximately 5 minutes, refresh, drain and squeeze well.

2   Place in a food processor and coarsely purée.

3   Heat the butter in a suitable pan, heat the spinach and cook to dry out the water.

4   Add the thick béchamel, the egg and egg yolks, season and add the grated nutmeg and mix well.

5   Heat the oil in a shallow pan.

6   Carefully spoon the spinach mixture into the oil, cook on both sides until golden brown.

7   Drain the subrics and serve immediately as an appetiser or as a vegetarian dish. Serve with a suitable sauce, e.g. cream sauce, and suitable leaves.

For a healthier version, in place of cream, use fromage frais or yoghurt.

21 Spinach, ricotta and artichoke filo bake with cranberries

Ingredient	4 portions	10 portions
Spinach	400g	1 kg
Ricotta	500g	1.25 kg
Olive oil	1 tbsp	2–3 tbsp
Onion, sliced	100g	250g
Salt, freshly ground black pepper		
Parsley, fresh, chopped, to taste		
Filo pastry	275g	700g
Artichokes, tinned, drained	200g	500g
Cranberries, frozen	500g	1.25 kg
Butter	25g	60g

Energy	Cal	Fat	Sat fat	Carb	Sugar	Protein	Fibre
2191 kJ	523 kcal	25.2g	12.9g	51.2g	14.7g	24.3g	7.5g

1 Cook, refresh and drain the spinach, then chop finely.

2 Break down the ricotta and mix with the spinach.

3 Sauté the onion in the olive oil without colour.

4 Add to the ricotta and spinach with the chopped parsley. Season to taste.

5 Line a lightly buttered flan dish with 3 layers of filo pastry, leaving overhang.

6 Fill with spinach mixture and press drained artichokes evenly around the dish.

7 Top with cranberries.

8 Gather in the overhanging filo pastry, adding more layers to cover centre.

9 Lightly pull up the edges of the pastry, keeping the leaves loose, brush with butter and bake at 180°C for 35 minutes approximately.

This recipe was contributed by Gary Thompson.

22 Stuffed vegetables

Certain vegetables can be stuffed and served as a first course, as a vegetable course and as an accompaniment to a main course.

The majority of vegetables used for this purpose are the bland, gourd types, such as aubergines, courgette and cucumbers, in which case the stuffing should be delicately flavoured so as not to overpower the vegetable.

Below are some of the more popular types of vegetable used for this purpose and the usual type of stuffing in each case. There is, however, considerable scope for variation and experimentation in any of the stuffings.

Artichoke bottoms

Use duxelle stuffing; serve with a cordon of thin demi-glace or jus-lié flavoured with tomato.

Aubergine

The cooked, chopped flesh is mixed with one of the following to make the stuffing:

- Cooked, chopped onion, sliced tomatoes and chopped parsley.
- Duxelle, sprinkled with fresh breadcrumbs, grated cheese – and gratinated.
- Cooked, chopped onion, garlic, tomato concassé, parsley, breadcrumbs – and gratinated.
- Diced or minced cooked mutton, cooked, chopped onion, tomato concassé, cooked rice and chopped parsley.

Serve with a cordon of tomato sauce or coulis.

Mushrooms and ceps

Use duxelle stuffing.

For stuffed ceps, forest style, use equal quantities of duxelle stuffing and sausage meat (or omit).

Stuffed cabbage

Use veal, pork or chicken stuffing and serve with pilaff rice.

Cucumber

This can be prepared in two ways.

1 Peeled, cut into 2cm pieces, the centres hollowed out with a parisienne spoon and then boiled, steamed or cooked in butter.

2 The peeled, whole cucumber is cut in halves lengthwise, the seed pocket scooped out and the cucumber cooked by boiling, steaming or in butter.

Suitable stuffings can be made from a base of duxelle, pilaff rice or chicken forcemeat, or any combination of these.

To stuff the cucumber pieces, pipe the stuffing from a piping bag and complete the cooking in the oven. When the whole cucumber is stuffed, re-join the two halves, wrap in muslin and braise.

Lettuce

Stuff with two parts chicken forcemeat and one part duxelle, and braise.

For a vegetarian version, stuff with duxelle only.

Turnips

1 Peel the turnips, remove the centre almost to the root and blanch the turnips.

2 Cook and purée the scooped-out centre and mix with an equal quantity of potato purée.

3 Refill the cavities and gently cook the turnips in butter in the oven, basting frequently.

Turnips may also be stuffed with cooked spinach, chicory or rice.

More stuffings for vegetables

- Stuff with pilaff rice, varied if required with other ingredients, e.g. mushrooms, tomatoes or duxelle.
- Use duxelle stuffing with garlic and diced ham. Serve with a cordon of demi-glace flavoured with tomato.
- Use chopped, hard-boiled egg bound with thick béchamel, grated cheese and gratinated. Serve with a cordon of light tomato sauce.
- Use scrambled egg, mushrooms and diced ham, sprinkled with breadcrumbs fried in butter.
- Use risotto with tomato concassé. Coat with thin tomato sauce.
- Use cooked tomato concassé, chopped onion, garlic and parsley, bound with fresh breadcrumbs – gratinate. May be served hot or cold.
- Use pilaff rice in which dice of tomato and red pepper have been cooked. Cook gently in the oven and sprinkle with chopped parsley.

23 Tofu and vegetable flan with walnut sauce

4 Sprinkle over the herbs. Cook for 1 minute. Drain vegetables and allow to cool.

5 Warm the milk to blood heat. Whisk the egg and tofu in a basin, add seasoning then gradually incorporate milk. Whisk well.

6 Fill the flan case with the drained vegetables and add the tofu and milk mixture.

7 Bake for 20 minutes approximately at 180°C. Serve with walnut sauce (see below).

Walnut sauce

Ingredient	4 portions	10 portions
Onion, finely chopped	100g	250g
Garlic clove, chopped	1	2–3
Walnut oil	50g	125g
Brown sugar	10g	25g
Curry powder	25g	60g
Lemon, grated zest and juice	1	2–3
Peanut butter	25g	60g
Soy sauce	1 tsp	2–3 tsp
Tomato purée	25g	60g
Vegetable stock	375ml	900ml
Seasoning		
Walnuts, very finely chopped	100g	250g
Arrowroot	10g	25g

1 Fry the onion and garlic in the walnut oil, add the sugar and cook to a golden-brown colour.

2 Add the curry powder, cook for 2 minutes.

3 Add zest and juice of lemon, peanut butter, soy sauce and tomato purée. Mix well.

4 Add vegetable stock, bring to boil, simmer for 2 minutes, season.

5 Add chopped walnuts.

6 Dilute the arrowroot with a little water and gradually stir into sauce. Bring back to the boil stirring continuously. Simmer for 5 minutes.

7 Correct seasoning and consistency.

Tofu and vegetable flan

Ingredient	4 portions	10 portions
Shortcrust pastry	150g	375g
Sunflower oil	4 tbsp	10 tbsp
Carrots, diced	50g	125g
Mushrooms, sliced	50g	125g
Celery, diced	50g	125g
Broccoli florets, blanched and refreshed	100g	250g
Fresh chopped basil	3g	7.5g
Dill weed, chopped	3g	7.5g
Milk, skimmed	125ml	300ml
Egg	1	2–3
Tofu	200g	500g
Seasoning		

Energy	Cal	Fat	Sat fat	Carb	Sugar	Protein	Fibre
2359 kJ	568 kcal	45.8g	9.3g	25.8g	5.6g	15g	2.5g

1 Line 18cm flan ring(s) with shortcrust pastry and bake blind for approximately 8 minutes in a preheated oven at 180°C.

2 Heat the sunflower oil in a sauté pan, add the carrots, mushrooms and celery, gently cook for 5 minutes without colouring.

3 Add the broccoli, cover and cook gently until just crisp, stirring frequently and adding a little water if the mixture begins to dry.

24 Tomato and lentil dahl with roasted almonds

Ingredient	4 portions	10 portions
Vegetable oil	2 tbsp	5 tbsp
Butter or margarine	25g	60g
Onions, finely chopped	50g	125g
Carrots, finely chopped	1	2
Garlic cloves	3	7
Cumin seeds, crushed and chopped	2 tsp	5 tsp
Yellow mustard seeds	2 tsp	5 tsp
Root ginger, grated	1 tsp	2½ tsp
Ground turmeric	2 tsp	5 tsp
Chilli powder	1 tsp	2½ tsp
Garam masala	1 tsp	2½ tsp
Red lentils	200g	500g
Water	400ml	1 litre
Coconut milk	400ml	1 litre
Tomatoes, peeled, deseeded and finely diced	5	12
Seasoning		
Lime juice	2 limes	5 limes
Coriander, fresh, chopped	4 tbsp	10 tbsp
Almonds, flaked	50g	125g

1 Heat the oil and butter in a suitable saucepan. Add the onion and sweat for 5 minutes without colour. Add the carrot, garlic, cumin and mustard seeds, and ginger. Cook for a further five minutes.

2 Stir in the ground turmeric, chilli powder and garam masala and cook for 1 minute. Stir well.

3 Add the lentils, water, coconut milk and tomatoes.

4 Season and bring to the boil. Simmer for approximately 45 minutes, stirring occasionally.

5 Stir in the lime juice and fresh coriander. Cook for a further 15 minutes until the lentils are tender.

6 Serve sprinkled with coriander and flaked almonds.

Energy	Cal	Fat	Sat fat	Carb	Sugar	Protein	Fibre
1819 kJ	434 kcal	24.5g	5.4g	41.2g	12.3g	17.8g	0.6g

25 Vegetable, bean and saffron risotto

Ingredient	4 portions	10 portions
Vegetable stock	185ml	1 litre
Saffron	5g	12g
Butter	50g	125g
Onion, chopped	25g	60g
Celery	50g	125g
Short-grain rice	100g	250g
Small cauliflower	1	2–3
Vegetable oil	4 tbsp	150ml
Large aubergine	1	2–3
Haricot beans, cooked	100g	250g
Peas, cooked	50g	125g
French beans, cooked	50g	125g
Tomato sauce made with butter and vegetable stock	250ml	600ml
Parmesan cheese, grated	25g	60g

Energy	Cal	Fat	Sat fat	Carb	Sugar	Protein	Fibre
2017 kJ	485 kcal	35.8g	8.6g	4.5g	34.8g	5.6g	4.5g

1 Infuse the vegetable stock with the saffron for approximately 5 minutes by simmering gently, while maintaining the quality of stock.

2 Melt the butter, add the onion and celery and cook without colour for 2–3 minutes. Add the rice.

3 Cook for a further 2–3 minutes. Add the infused stock and season lightly. Cover with a lid and simmer on the side of the stove.

4 While rice is cooking, prepare the rest of the vegetables. Cut the cauliflower into small florets, wash, blanch and refresh, quickly fry in the sunflower oil in a sauté pan. Add the aubergines cut into ½cm dice and fry with the cauliflower. Add the cooked haricot beans, peas and French beans.

5 Stir all the vegetables together and bind with tomato sauce.

6 When the risotto is cooked, serve in a suitable dish. Make a well in the centre. Fill the centre with the vegetables and haricot beans in tomato sauce.

7 Sprinkle the edge of the risotto with grated Parmesan cheese to serve.

Note: This dish is suitable for vegetarians if the Parmesan cheese is omitted.

26 Vegetable curry using textured vegetable protein

Ingredient	4 portions	10 portions
Textured vegetable protein (TVP) chunks	100g	250g
Vegetable stock	500ml	1.25 litres
Onion, finely chopped	50g	125g
Garlic, crushed and chopped	2	5
Vegetable oil	2 tbsp	5 tbsp
Curry powder	1 tbsp	2½ tbsp
Garam masala	¼ tsp	¾ tsp
Fresh ginger, grated	½ tsp	1¼ tsp
Cooking apple, peeled and finely chopped	75g	180g
Carrots, cut into ½cm dice	50g	125g
Celery, cut into ½cm dice	50g	125g
Mushrooms, sliced	100g	250g
Plum tomatoes (canned are suitable in their juice)	400g	1 kg
Tomato purée	2 tsp	5 tsp
Water or vegetable stock	280ml	700ml

1 Soak the TVP in the vegetable stock for 10 minutes. Drain.

2 Fry the onion and garlic in the vegetable oil until cooked and slightly coloured.

3 Add the TVP, curry powder, garam masala and ginger. Stir well.

4 Add the apple, carrot, celery, mushrooms, chopped tomatoes and tomato purée. Simmer for 2 minutes.

5 Add the water or vegetable stock. Simmer for approximately 15–20 minutes.

> **Professional tip**
>
> This curry may be slightly thickened with arrowroot or cornflour.

Energy	Cal	Fat	Sat fat	Carb	Sugar	Protein	Fibre
698 kJ	167 kcal	12.0g	1.4g	9.2g	6.9g	6.4g	0.3g

27 Vegetable gougère (filled choux ring)

Ingredient	4 portions	10 portions
Choux pastry		
Water	250ml	625ml
Sunflower margarine	100g	250g
Strong flour	125g	300g
Eggs (medium)	4	10
Gruyère cheese, diced	75g	180g
Seasoning		

Energy	Cal	Fat	Sat fat	Carb	Sugar	Protein	Fibre
3557 kJ	855 kcal	61.5g	19.3g	47.4g	1.7g	30.7g	0.2g

1 Make the choux pastry, cool and add the finely diced Gruyère cheese.

2 With a 1cm plain tube, pipe individual rings approximately 8cm in diameter onto a very lightly greased baking sheet.

3 Brush lightly with eggwash and relax for approximately 15 minutes.

4 Bake in a preheated oven at 190°C for 20–30 minutes.

5 When cooked, place on individual plates. Fill the centre with a suitable filling.

Possible fillings
- ratatouille
- stir-fry vegetables
- cauliflower cheese
- button mushrooms in a tomato and garlic sauce
- leaf spinach with chopped onions in a béchamel sauce
- button mushrooms and sweetcorn in a béchamel yoghurt sauce.

28 Leek roulade with ricotta corn filling

Ingredient	4 portions	10 portions
Leek, large, finely shredded	1	3
Clove of garlic, chopped	1	3
Butter	75g	180g
Plain flour	100g	250g
Milk	250 mls	625 mls
Eggs, separated	4	10
Grated Parmesan cheese	30g	75g
Ricotta corn filling		
Ricotta cheese	100g	250g
Sweetcorn, cooked	250g	625g
Chives, fresh chopped	1 tbsp	2½ tbsp

Energy	Cal	Fat	Sat fat	Carb	Sugar	Protein	Fibre
2211 kJ	529 kcal	32.1g	16.1g	38.5g	7.6g	24.0g	0.6g

1 Line a Swiss roll tin with silicone paper approximately 25cm × 30cm.

2 Melt 50g butter in a saucepan and sweat the leek and garlic, stir for about 5 minutes.

3 Melt the remaining butter in a saucepan, add the flour, stir for 1 minute, gradually adding the boiling milk. Add the egg yolks and leek mixture.

4 Place this mixture in a suitable bowl.

5 Beat the egg whites until soft peaks, fold carefully into the leek mixture.

6 Spread this mixture onto the Swiss roll tin. Cook in a hot oven 180°C–200°C for approximately 10 minutes or until risen and golden brown.

7 When cooked, carefully turn out onto baking paper sprinkled with Parmesan. Remove the silicone paper.

8 To make the ricotta filling, place the ricotta cheese in a bowl, add the sweetcorn and chives and mix well.

9 Spread the roulade with the ricotta filling.

10 Carefully roll the roulade neatly.

11 Serve sliced with a suitable sauce, e.g. tomato and basil.

29 Ratatouille pancakes with a cheese sauce

Ingredient	4 portions	10 portions
Pancake batter		
Flour, sifted	100g	250g
Skimmed or whole milk	250ml	625ml
Egg	1	2–3
Salt, pinch of		
Butter, melted	10g	25g
Ratatouille		
Courgettes, diced	200g	500g
Aubergines, diced	200g	500g
Red pepper, diced	50g	125g
Green pepper, diced	50g	125g
Tomatoes, concassée	200g	500g
Onion, chopped	50g	125g
Clove garlic, chopped	1	2
Vegetable oil	50ml	125ml
Cheese sauce		
Skimmed or whole milk	500ml	1.25 litres
Butter or vegetable oil	50g	125g
Flour	50g	125g
Onion, studded with clove	1	2–3
Parmesan, grated	25g	60g
Egg yolk	1	2–3
Seasoning		

Energy	Cal	Fat	Sat fat	Carb	Sugar	Protein	Fibre
2398 KJ	571 kcal	35.8g	6.5g	46.1g	19.0g	19.6g	6.5g

Skimmed milk was used for the nutritional analysis.

1. Infuse the milk for the sauce with the studded onion.

2. Make pancakes: combine the sifted flour, eggs, milk and salt, and whisk well. Strain, then add the melted butter. Pour a thin layer of the mixture into the bottom of the pan, cook for a few seconds, turn and cook on the other side. Repeat for each pancake.

3. Make a ratatouille: cook the onions gently in the oil for 5–7 minutes, without colour. Add the garlic, courgette, aubergine and peppers, season lightly, cover and cook for 4–5 minutes, tossing occasionally. Add the tomato and cook for 20–30 minutes or until tender.

4. For the sauce, first make a white roux with the butter and flour. Gradually add the boiling infused milk. Clean round the saucepan and simmer for 20 minutes. Strain through a conical strainer into a clean pan. Add seasoning, grated Parmesan and egg yolk, over heat but without reboiling. Finally, season with cayenne pepper.

5. Fill the pancakes with the ratatouille, roll up and serve on individual plates or a service dish. Coat with cheese sauce and sprinkle with grated Parmesan. Finish by gratinating under the salamander.

30 Vegetarian strudel

Ingredient	4 portions	10 portions
Strudel dough		
Strong flour	200g	500g
Salt, pinch of		
Sunflower oil	25g	60g
Egg	1	2–3
Water at 37°C	83ml	125ml
Filling		
Cabbage leaves, large	200g	500g
Vegetable oil	4 tbsp	150ml
Onion, finely chopped	50g	125g
Cloves garlic, chopped	2	5
Courgettes	400g	1 kg
Carrots	200g	500g
Turnips	100g	250g
Tomato, skinned, deseeded and diced	300g	750g
Tomato purée	25g	60g
Toasted sesame seeds	25g	60g
Wholemeal breadcrumbs	50g	125g
Basil, fresh, chopped	3g	9g
Seasoning		
Eggwash		

Energy	Cal	Fat	Sat fat	Carb	Sugar	Protein	Fibre
2117 kJ	504 kcal	27.6g	4.0g	54.1g	10.5g	14.3g	9.7g

1 To make the strudel dough, sieve the flour with the salt and make a well.

2 Add the oil, egg and water, and gradually incorporate the flour to make a smooth dough; knead well.

3 Place in a basin, cover with a damp cloth; allow to relax for 3 minutes.

4 Meanwhile, prepare the filling: take the large cabbage leaves, wash and discard the tough centre stalks, blanch in boiling salted water for 2 minutes, until limp. Refresh and drain well in a clean cloth.

5 Heat the oil in a sauté pan, gently fry the onion and garlic until soft.

6 Peel and chop the courgettes into ½cm dice, blanch and refresh. Peel and dice the carrots and turnips, blanch and refresh.

7 Place the well drained courgettes, carrots and turnips into a basin, add the tomato concassé, tomato purée, sesame seeds, breadcrumbs and chopped basil, and mix well. Season.

8 Roll out the strudel dough to a thin rectangle, place on a clean cloth and stretch until extremely thin.

9 Lay the drained cabbage leaves on the stretched strudel dough, leaving approximately a 1cm gap from the edge.

10 Place the filling in the centre. Eggwash the edges.

11 Fold in the longer side edges to meet in the middle. Roll up.

12 Transfer to a lightly oiled baking sheet. Brush with the sunflower oil.

13 Bake for 40 minutes in a preheated oven at 180–200°C.

14 When cooked, serve hot, sliced on individual plates with a cordon of tomato sauce made with vegetable stock.

Test yourself

1 Name five categories of vegetables.

2 Name four nutrients found in vegetables.

3 How should each of the following vegetables be stored?

  (a) bean sprouts

  (b) potatoes

  (c) onions

  (d) fennel

  (e) broad beans

  (f) mangetout

  (g) mushrooms

4 State the key stages in the preparation, cooking and serving of whole globe artichokes.

5 Name a sauce suitable to serve with:

  (a) globe artichokes

  (b) asparagus.

6 Explain what is meant by blanching.

7 State how vegetable preparation and refreshing can be made healthier.

8 State two ways in which the composition of each of the following can be affected when they are coming to the end of their growing season:

  (a) leeks

  (b) asparagus.

9 Explain what happens to Vitamin C in vegetables in storage and when they are cooked.

10 Describe the effect that boiling can have on the volume of each of the following:

  (a) spinach

  (b) leaf vegetables, in terms of cellulose and fibre structure

  (c) tubers, in terms of starch.

8 Advanced meat dishes

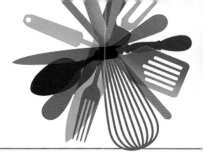

This chapter covers the following units:

NVQ:
→ Prepare meat for complex dishes.
→ Cook and finish complex meat dishes.

VRQ:
→ Advanced skills and techniques in producing meat dishes.

Introduction
The aim of this chapter is to enable you to develop the necessary advanced skills, knowledge and understanding of the principles in preparing and cooking meat to dish specifications. The chapter will help you to develop precision, speed and control in existing skills and develop more refined and advanced techniques.

Learning objectives
By the end of this chapter you should be able to:
→ Identify suitable commodities and recipes for meat dishes and dietary requirements.
→ State factors affecting the composition of meat.

→ Describe how the composition of different meats influences the choice of processes and preparation methods.
→ Describe the range of products available after the dissection of a carcass.
→ Explain how to select the correct type, quality and quantity of meat to meet recipe and dish requirements.
→ Describe how to store meat and meat dishes.
→ Demonstrate a range of cooking techniques using the correct tools and equipment appropriate to the different cooking methods.
→ Describe the methods for producing fine and coarse forcemeats.
→ Compare the effects of different preservation methods for meat.
→ Describe how to adapt recipes and cooking techniques to maximise nutritional value and healthy options in meat dishes and professional cookery.
→ Demonstrate professional ability in minimising faults in complex meat dishes.

Recipes included in this chapter

Origins

From the earliest times, man has been a carnivore, with meat providing so much protein and essential vitamins to the diet. Although the importance of meat in the diet has been reappraised – whether its drawbacks in terms of causing high cholesterol and its high price outweigh its value as a protein provider in our diets, or in the case of vegetarians, ethical considerations – meat is still the most expensive item on the budget and a great deal of thought should be given to selecting meat and using it wisely.

Butchers' meat today is largely a product of selective breeding and feeding techniques, whereby animals are reared carefully to reach high standards that meet the specific needs of the customer – that is to say, for lean and tender meat. Modern cattle, sheep and pigs have been bred to be well-fleshed yet compact in stature in comparison to some of the older or rarer breeds of the past, some of which are now sourced by chefs for specific types of cooking or menus.

Composition

Meat can be defined as the skeletal muscle of an animal that has been bred for the table. It is classed as either red or white. The most commonly used meats are beef, pork and lamb/mutton, but there are other types of meat that appear on menus in some establishments, including goat, veal and venison (farmed).

Meat is natural and therefore not a uniform product, varying in quality from carcass to carcass, while flavour, texture and appearance are determined by the type of animal and the way it has been fed. Flavour is not found only in meat that possesses a proportion of fat, although fat does give a characteristic flavour to meat and helps to keep it moist during roasting. Neither is the colour any guide to quality. Consumers are inclined to choose light coloured meat – bright red beef, for example, as seen on some supermarket shelves – because they think and believe that it will be fresher than the darker red meat. Freshly butchered beef is bright red because of the pigment in the tissues, **myoglobin**, which is affected by the oxygen in the air. After several hours, the colour changes to dark red or brown as the pigment is further oxidised to become **metamyoglobin**. The colour of fat can also vary from almost white in lamb to bright yellow in some beef. Colour again is affected by the feed, by the breed and to a certain extent on the time of the year.

The most useful guide to tenderness and quality is through knowledge of the cuts/joints of meat and their location on the carcass; this will give an indication of the fibre make-up of the muscles. In the main, the leanest and tenderest cuts – the prime cuts – tend to come from the hind quarter of the animal; the coarse cuts tend to be from the hardest working areas of the animal – the neck, legs and fore quarters – the parts of the animal that tend to have had the most muscular exercise and where the fibres have become hardened. Slower cooking methods are generally used for these areas to help break down the connective tissue. The meat from young animals is generally more tender.

Quality

When sourcing meat, there are a number of considerations for the chef, including origin, breed, cut/joint, and slaughter method (including religious requirements, such as kosher butchery in Judaism and halal butchery in Islam), as well as the factors of feed, well-being (organics) and period of ageing. All of these will affect the price per kilo.

Through good sourcing and knowledge of the commodity, it is possible to purchase quality meats even on a restricted budget and produce quality dishes. As a guide the following should be considered:

- **Cost** – Consider using the less utilised cuts of meat from the higher quality animal. An innovative and skilled chef using good preparation and presentation skills could make good use of these cuts.
- **Waste or trim** – When buying butchered meat ensure you order correctly to a specification in terms of weight, cut etc. Use correct cooking methods to ensure correct yield in terms of joints.
- **Butchery** – Consider buying by carcass and butchering in the kitchen, which can help to reduce add-on costs from the butcher, although you need to ensure you have the skills available to make the most of the entire carcass.

Chefs in all areas of catering need to be aware of customer interest in the source of products. For some people, this includes traceability and impact on resources. This has led to an increase in chefs who link up with specific suppliers or butchers. The provenance and history as well as the appropriate quality tags, e.g. BPEX, EBLEX, Red Tractor and Freedom Foods, will all influence the purchase decisions for the caterer.

> **Take it further**
>
> For more information on quality tags, see the following:
> - BPEX: www.bpex.org.uk
> - EBLEX (the organisation for the beef and sheep industry): www.eblex.org.uk
> - Red Tractor: www.redtractor.org.uk
> - Freedom Food: www.freedomfood.co.uk

Key term

Myoglobin – pigment in the tissues of mammals which gives meat its bright red colour.

Metamyoglobin – created when myoglobin is oxidised. Changes the colour of meat to dark red or brown.

Types of meat

Beef

Beef is the meat of domesticated cattle. The cuts used in catering vary considerably from the very tender fillet steak to the tougher parts like the shin. There are 17 primary cuts from a side of beef, split between hind and fore quarter or front and back of the animal, each one composed of muscle, fat, bone and connective tissue. The least developed muscles, usually from the inner areas, can be roasted or grilled, while leaner and more sinewy meat is cut from the more highly developed external muscles. Exceptions are rib and loin cuts which come from the immobile external muscles.

Quality points

Lean meat should be bright red with small flecks of white fat (marbling). The fat should be firm, brittle in texture, creamy white in colour, and odourless. Meat that has traceable origin back to the farm is a legal requirement in the UK.

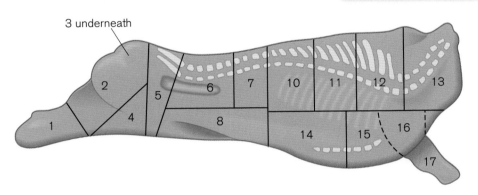

▲ A side of beef – see Table 8.1 for explanation of the numbers

Table 8.1 Joints, uses and weights of beef joints

Joint	Use	Average weight
	Hindquarter	85–90 kg
Fillet	Leanest and most tender, grilling, frying, roasting, whole or as steaks	3–3.5 kg
Sirloin (6)	Boneless steak, which is lean grilling, frying and roasting, whole or as steaks	12–14 kg
Wing rib (7)	Normally last 3 ribs, roasting on or off the bone	5–6 kg
Thin flank (8)	Also known as the skirt, can be gristly, it is usually stewed or minced for cooking	8–10 kg
Thick flank (4)	Coarse joint that can be braised as a piece	10–11 kg
Shin (1)	Stewing although most frequent use is in making consommé as a mince or fine cut	6–7 kg
Topside (2)	Braising, stewing or slow roasting cooking method (second class roasting)	9–10 kg
Silverside (3)	Coarse joint, stewing or traditional salting and boiled (boiled salt beef)	12–14 kg
Rump (5)	Roasting or cut to steaks and frying	10–11 kg
	Forequarter	75–80 kg
Middle rib (11)	Second class roasting joint that can also be braised	8–8.5 kg
Fore rib (10)	Prime roasting or frying as rib eye steak	6–6.5 kg
Chuck rib (12)	Stewing or braising, remove gristle and sinew	12–13 kg
Sticking piece (13)	Lean and high in flavour, stewing and mincing	6–7 kg
Shank (17)	As for shin, mince, stew or use to flavour consommé	5–5.5 kg
Leg of mutton cut (LMC) (16)	Braising or stewing, can be trimmed to make paupiettes	7–8 kg
Brisket (15)	Boiling or slow pot-roasting bring out the best flavour	14–16 kg
Plate (14)	Trim fat and cut for dice or mince for stewing	8–8.5 kg

▲ A boned shin of beef

▲ Rolled silverside of beef

▲ Fore rib of beef

Larding a fillet of beef

1 Use strips of larding bacon or pork fat.

2 Insert the bacon lengthways using a larding needle.

3 Repeat across the meat.

Tying and rolling a sirloin

1 Lift the fat away from the sirloin, keeping it attached along one edge.

2 Remove the membrane before putting the fat back in place.

3 Roll up the sirloin and tie securely.

Veal

Veal is the meat of young calves (normally 12 to 24 weeks old) and is making a comeback onto menus. Because of the young age of the animal, the flesh is normally a pale pink colour and it is often considered a white meat.

The joints of veal are shown in the diagram and Table 8.2. Veal is butchered in a similar way to lamb and many joints share the same names.

A saddle of veal consists of the two loins. It is usually divided into the two loins, cut into chops or boned and cut into noisettes or escalopes.

The mignon or filet mignon is the fillet that is attached to the underside of the loin. It is used for escalopes, stir fries, veal stroganoff and other sauté dishes.

The cushion of veal corresponds to the topside in an adult animal, and the under cushion to the silverside. The mignon is equivalent to beef fillet.

Quality points

- Veal should be pale pink and firm.
- Cut surfaces should be moist, bones should be pinkish white and porous with a small amount of blood in their structure.
- The fat should be firm and pinkish white.
- The kidney should be firm and well covered in fat.
- The liver and sweetbreads from veal are still considered to be the best offal dishes.

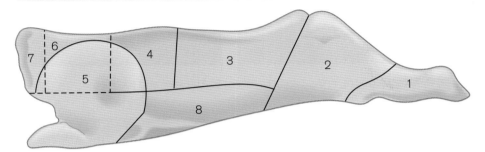

▲ A side of veal – see Table 8.2 for explanation of the numbers

Table 8.2 Joints, uses and weights of veal

Joint	Approx. weight (kg)	Uses
Knuckle (1)	2	Osso bucco, sauté, stock
Leg (2)	5	See below
Loin (3)	3.5	Roast, fry, grill
Best end (4)	3	Roast, fry, grill, cutlets
Shoulder (5)	5	Braise, stew
Neck end (6)	2.5	Stew, sauté
Scrag (7)	1.5	Stew, stock
Breast (8)	2.5	Stew, roast
Joints of the leg		
Cushion/nut	2.75	Roast, braise, sauté, escalopes
Under cushion/under nut	3	
Thick flank	2.5	
Knuckle	2.5	Osso bucco, sauté

Preparation of veal escalopes

1 Slice the joint into escalopes.

2 Bat out each escalope with a cutlet bat.

3 Pané if required.

Boning a loin of veal

1 The loin of veal.

2 Using a boning knife, carefully remove the fillet.

3 The joint and the fillet.

4 Bring the loin down from the rib to the backbone, and trim the rib bones.

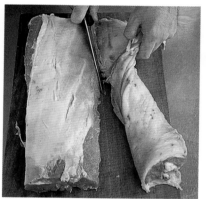

5 Trim the excess fat and sinew from the meat.

6 The fillet, boned loin and bones (which can be used for stock).

Lamb and mutton

Lamb is the flesh of young sheep, normally under one year old. Mutton is the older animal, which gives slightly larger joints and it tends to need longer cooking due to the development of the muscles.

- Spring lamb – normally in its first 6 months but often under 4 months.
- Lamb – up to 1 year old.
- Hogget – is normally over one year old but hasn't developed the flavour of mutton.
- Mutton – normally over 2 years old with a strong, well developed flavour.

The breed and location (upland or lowland) of where the animal has been reared will have a bearing on the price.

Quality points

- Good quality lamb should have a fine, white fat, with pink flesh where freshly cut. In mutton the flesh has a deeper colour.
- Lamb has a very thin parchment-like covering on the carcass, known as the fell or bark, which is left on for roasting as this helps maintain shape during cooking.

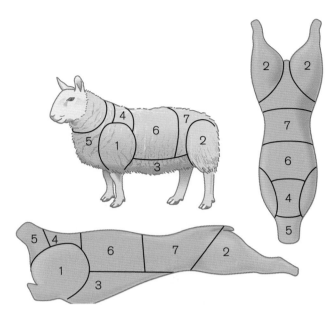

▲ A carcass of lamb, viewed from different angles – see Table 8.3

Table 8.3 Joints, uses and weights of lamb

Joint	Uses	Approx. weight
Whole carcass		18–21 kg approx.
Shoulders (1)	Roasted, boned and stuffed	2 × 2.5 kg
Legs (2)	Roasting with the knuckle end often braised (shank)	2 × 2.5 kg
Breasts (3)	Boned, rolled and slow roasted	2 kg
Middle neck (4)	Neck fillet can be grilled, otherwise braising	2 kg
Scrag end (5)	Stewing or broths, traditional Irish stew	1 kg
Best end (6)	Roasting as small joint, grilled as cutlets	2 kg
Saddle (7)	Roasting as small joint, frying, grilling	3 kg

▲ A shoulder of lamb

▲ A shoulder of lamb that has been boned and rolled

▲ A pair of best ends of lamb

▲ Lamb chops: from left to right, two rosettes, a valentine and two noisettes (all cut from the boned-out loin) and a Barnsley chop or double loin chop (cut across the saddle, on the bone)

Tunnel boning and trimming a leg of lamb

1 Use a boning knife to scrape back the flesh and expose the knuckle bone. Cut flesh away from the bone up to the joint, then cut through the joint.

2 Remove the knuckle bone, reshape the joint and tie it neatly.

Boning and tying a saddle of lamb

1 Remove the aitch (pelvic bone).

2 Remove excess fat and then the sinew.

3 Trim the sinew.

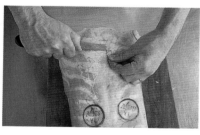

4 Cut off the flaps, leaving about 15cm each side so that they will meet in the middle under the saddle.

5 Trim skin and excess fat from the back.

6 Fold the edges in neatly.

7 Score the skin neatly and tie.

Pork

Pigs may have been the first animals to be domesticated and used as a food source. Pigs are easy to breed and tend to reach a slaughter weight within 6 months. Pork is widely used around the world, although it is unacceptable to certain religious groups for a variety of reasons.

Pigs will eat almost anything so the farmer has to regulate their diet to produce the right combination of fatness, weight and carcass quality. There is a variety of pigs now available for the table; the breed and location will have an influence on the price and eating quality of the pork.

Quality points

- Lean flesh of pork is usually pale pink.
- The fat is normally white, firm, smooth and not excessive. Bones are usually small, fine and pinkish.
- The quality of the skin or rind depends on the breed.
- Suckling pig weighs 5–9 kg dressed and is usually roasted whole.
- Boars are male pigs but also the term for wild pigs which are now farmed in some areas. The flesh is normally darker and richer in flavour.

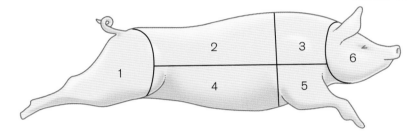

▲ A side of pork – see Table 8.4 for explanation of numbers

Table 8.4 Cuts, uses and weights of pork

Whole side of pork		16–18 kg
Head (6)	Boiled for brawn although cheeks can be braised	4 kg
Spare rib (3)	Minced for sausages, slow roasting, grilling, frying	2 kg
Loin (2)	Roasting on or off bone, cut to chops, grilling, frying	5 kg
Leg (1)	Roasting, muscles used to make escalopes, frying	4 kg
Belly (4)	Sausages, although now popular slow braised or roasted	3 kg
Shoulder (5)	Boned and slow roasted, diced for stewing	3 kg
Trotters	Boiled and then added to brawn, produces gelatinous stock	0.5 kg

The leg of pork can be broken down into the chump, silverside, topside and thick flank. The hock, or knuckle, is at the base of the leg.

▲ Loin of pork

▲ Leg of pork, boned

▲ Pig's trotters

Bacon

Bacon is cured pork from a pig specifically bred to produce bacon, with a shape, size and fat ratio that yield the most economic bacon joints.

The meat is dry cured and smoked or wet cured by soaking in brine. Soaked bacon may then be smoked; if not smoked it has a milder flavour but doesn't keep as long as smoked versions.

This process was used to store meat and there are a number of variations with different flavours from herbs, spices and salt; the time taken for curing also affects the flavour of the meat.

Quality points

- The flesh should be firm, pink, with no sign of stickiness or unpleasant smell.
- The rind should be thin and smooth.
- The fat layer should be white, smooth and not excessive in proportion to the lean meat.

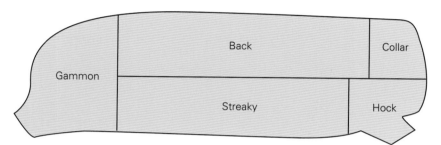

▲ A side of bacon

Table 8.5 Joints and uses of bacon

Joint	Approx. weight (kg)	Uses
Collar	4.5	Boil, grill
Hock	4.5	
Back	9	Grill, fry
Streaky	4.5	
Gammon	7.5	Boil, grill, fry

Offal

Offal is the name given to parts taken from the inside of the carcass; these include liver, kidney, heart, sweetbread and tripe. Other edible parts of the carcass, which include oxtail, tongue, head and sometimes hooves, are sometimes included under this term. Fresh offal should be purchased as required. It can be refrigerated in clean, covered conditions at around 0°C for no more than 7 days. Frozen offal should be kept frozen until required.

Quality points

- Offal products should not have an unpleasant smell.
- Liver should be moist and smooth. It should not contain an excessive number of tubes.
- Kidneys should be supplied with the suet (casing of fat) left on, to prevent them drying out. Remove the suet before cooking. Both suet and kidneys should be moist.
- Hearts should not be too fatty or contain too many tubes. When cut, they should be moist, not sticky.
- Sweetbreads should be fleshy, a good size and creamy white in colour.
- Tongues must be fresh. There should not be an excess of waste at the root end.
- Heads should be fresh, with plenty of flesh. They should not be sticky.
- Oxtails should be lean (not too much fat), with no stickiness.

For examples of offal preparation, see Recipes 7, 8, 13, 14 and 16.

Advanced techniques for preparing and cooking meat

Butchery

There are two main types of butchery that are used in the preparation of meat into joints or cuts for cooking.

- **Traditional butchery** – this follows the bone structure and keeps muscles with different characteristics in one joint, which gives a variety of eating qualities and textures as in a traditional leg of lamb that is cut direct from the back bone and then the pelvic bone (aitch) is split to give 2 legs.
- **Seam butchery** – this follows the muscle formation and the natural seams of meat.

The advantage of the seam method is that it is believed to produce less waste and is more consistent in terms of eating quality.

Classification and joints

For economic reasons in terms of saving both labour and storage space, most caterers tend to purchase meat pre-jointed and fully prepared for cooking rather than by the carcass. Knowledge of where the specific cuts come from on the animal helps chefs in deciding the cooking method to use in relation to that cut.

Cooking

Meat is an extremely versatile commodity that can be cooked in a variety of ways, and matched with practically any vegetable, fruit and herb. Raw meat is difficult to chew because of the muscle fibre or collagen. Cooking will break the muscle fibre or proteins down to make them more edible (mincing – as in steak tartare – can have the same effect). When you cook meat, the protein gradually coagulates as the internal temperature increases. At 69°C, coagulation is complete, the protein begins to harden and further cooking makes the meat tougher.

The cut or joint will impact on the cooking method chosen. The time and temperature of cooking affect the way the meat will taste.

Since tenderness combined with flavour is the aim in meat cookery, much depends on the ratio of time and temperature. In principle, slow cooking retains juices and produces a more tender result than does fast cooking at high temperatures. There are, of course, occasions when high temperatures are essential. For instance, you need to grill a steak under a hot flame for a very limited time in order to obtain a crisp, brown surface and a pink juicy interior – using a low temperature would not give the desired results. But in potentially tough cuts such as breast or where there is a quantity of connective tissue as in a neck of lamb, then a slow rate of cooking converts the tissue to gelatine and helps make the meat more tender. Meat containing bone will take longer to cook as bone is a poor conductor of heat. Tough or coarse cuts of meat should be cooked by braising, pot-roasting or stewing. Marinating in a suitable marinade, such as wine or wine vinegar, helps to tenderise the meat and imparts additional flavour.

Meat bones are useful for giving flavour to soups and stocks, especially beef ones with plenty of marrow. In fact, marrow can be used to finish sauces or poached and used as a garnish. Veal bones are gelatinous and help to enrich and thicken soups and sauces as in *osso bucco*. Animal fat can be rendered down for frying or used as an ingredient in its own right as in suet or lard for pastries.

Larding and barding

Larding and barding are techniques for keeping meat moist and succulent during cooking. Larding involves inserting strips of fat into the lean meat (see page 238). Barding lays fat, or bacon, over meat; this is usually used for poultry or game.

Mincing

Mincing is the technique of processing meat by chopping it finely, usually using a machine. Other ingredients may be added to the meat.

Forcemeat

Forcemeat is a combination of meat with other ingredients such as fat, herbs or spices. It is used as a stuffing, or in sausages, pâtés and terrines.

Coarse forcemeat is usually minced, and fine forcemeat puréed.

Storage

Meat is a highly perishable and potentially high-risk commodity that can develop bacteria and moulds quickly, therefore it needs to be stored appropriately. This includes temperature and coverings. It is usual to store raw meat between 1 and 4°C.

- Raw and cooked meat need to be separated in storage to avoid cross-contamination risks.

- Hanging carcass – store at 1°C chill temperature for 7–28 days, dependent on animal.
- Chilled meat must be used by the 'use by' date.
- Chilled cooked meat should be stored at or below 5°C.
- Vacuum-packed meat – store below 3°C, ensure good circulation and that pouches are not punctured during storage until required for use. There are a number of kitchens that use their own vacuum-packing process. It is recommended that separate machines are used for cooked and raw products.
- Temperatures of chillers and freezers should be monitored regularly to ensure correct operation.
- Thaw out frozen meats correctly when necessary and never refreeze them once they have thawed out.

Preservation

There are a number of preservation methods that are used for extending the potential shelf life of meat or for the addition of flavour and tenderising meat prior to serving. These include:

- Canning – certain meats are preserved in tins – e.g. stewing steak, corned beef and even steak and kidney pies. These are all in a cooked state which need reheating or can be eaten cold.
- Drying – strips or pieces of meat are hung to dry. More popular in hotter countries; an example is Bresaola.
- Salting or dry curing – where meat is smothered in salt to help preserve the meat; used in some Spanish hams.
- Pickling or wet curing – where meat is soaked in brine for a period of time, as for gammons or salt beef.
- Smoking – traditionally where meat was hung in chimney spaces to dry out; wood smoking contains elements that help preserve as well as add flavour. Hot smoking helps to cook the meat at the same time, whereas cool smoking adds flavour and preserves. Typical woods used are hickory, oak and apple.
- Marinating – steeping of meat in a marinade to tenderise prior to cooking, normally contains something acidic, spicy and/or oil.
- Freezing – small carcasses such as lamb or mutton are often frozen without too much impact on flavour, as in NZ lamb, whereas beef is frozen pre-jointed (PJ). Meat should be frozen quickly to ensure smaller ice crystals are formed in the flesh. It should be covered to minimise the risk of freezer burn.

1   Slow-cooked beef fillet with onion ravioli

Ingredient	4 portions	10 portions
Beef		
Vegetable oil		
Centre cut fillet	1 × 600g	2 × 750g
Thyme, strands of	3	7
Garlic clove, thinly sliced	1	2
Ravioli pasta		
Flour	550g	1.4 kg
Egg yolks	5	12
Eggs	4	10
Chicken mousse		
Breast of chicken	300g	1 kg
Salt to taste		
Cream	200ml	500ml
Ravioli mix		
Chicken mousse (see above)	300g	750g
Lyonnaise onions	100g	250g
Parsley	25g	60g
Sherry vinegar	50ml	125ml
Seasoning		
To finish		
Haricots verts, cooked	200g	500g
Button onions, cooked	12 (100g)	30 (250g)
Sherry jus	200ml	500ml
Picked lemon thyme	1 tsp	2 tsp

Energy	Cal	Fat	Sat fat	Carb	Sugar	Protein	Fibre
5119 kJ	1221 kcal	52.5g	25.3g	116.4g	9.1g	77.2g	6.9g

For the beef

1 Preheat the oven to 59°C.
2 Heat a pan with a little corn oil and carefully put the beef fillet in it, browning on all sides. Add thyme and garlic at the end (this operation should take no more than 2 minutes).
3 Remove from the pan and allow to cool.
4 Wrap the fillet in foil and place in an oven already pre-set at between 55 and 60°C (the theory behind this cooking is that, for a medium rare 'doneness', the core temperature will be between 57 and 59°C; therefore, to achieve this preferred cooking degree throughout the fillet, the oven should be set at between 55 and 60°C).
5 It will take approximately 50–60 minutes for the temperature to penetrate to the core of the fillet. This will then last for an extra 1–1½ hours after this time (obviously the longer in the oven, the more it will dry out).
6 When ready to serve, remove from the oven and re-seal the fillet in a hot pan – this should take no more than 30 seconds. There is no need to rest the meat as the proteins will not have shrunk to a degree that requires it to be rested.

For the mousse

1 Blitz the chicken for 1 minute with the salt.
2 After standing for 30 seconds, add the cream.
3 Pass through a fine sieve and reserve.

For the ravioli pasta

1 Place the flour in a food processor.
2 Whisk the eggs together and pass through a chinois to get rid of any membrane.
3 Slowly incorporate the egg mix into the flour.
4 When all the liquid is used, take out of the food processor.
5 Work together on the bench, as though working bread dough, for 5 minutes.
6 Rest for 1 hour before rolling out to your required thickness.

For the ravioli mix

1 Place the chicken mousse, Lyonnaise onions and parsley in a bowl. Mix well.

2 Add the sherry vinegar and check the seasoning.

3 Weigh out into 25g balls and reserve until you need to make the ravioli.

To make the ravioli

1 Roll the pasta very thinly and cut into discs approximately 12cm in diameter.

2 Cover the 8 discs in cling film to prevent them drying out.

3 Lay a disc on the workbench and dab with a little water.

4 Place in the centre a ball of mousse and top with another disc ensuring that all the air is removed and an even shape is formed.

5 When all ravioli are complete, blanch in boiling water for 2 minutes then arrest the cooking with iced water.

6 Drain, store and cover in cling film to prevent drying.

To finish

1 Reheat the haricots verts, button onions and sauce, keeping them all warm. Meanwhile, reheat the ravioli in boiling water for 2 minutes. Remove and allow to drain. Carve the beef into equal slices (2 per portion).

2 Place the button onions and haricots verts in the centre of the plate, top with the beef, then the ravioli on top of the beef. Finish with the sauce and serve garnished with lemon thyme.

Professional tip

A modern cooking method is adopted here. The risk involved in cooking at high heat for a short period of time and getting the degree of cooking correct has always been an issue, however with this method the beef can stay in the oven for up to 2 hours and still be able to be served due to the protein shrink temperature.

2 Tournedos Rossini (modern)

Ingredient	4 portions	10 portions
Tournedos		
Oil	50ml	125ml
Tournedos of beef (approx. 200g each)	4	10
Girolles, washed and prepared	500g	1¼ kg
Butter	50g	125g
Slices white bread, without crusts and trimmed to the size of the beef	2	5
Garlic cloves, thinly sliced	1½	3
Slices foie gras (approx. 30g each)	4	10
Madeira sauce (3 parts reduced veal stock to 1 part Madeira)	150ml	375ml
Truffle, sliced	1	2
Madeira jelly (see below) cut into 1cm dice	100g	250g
Thin slices Parma ham (baked in an oven until crisp)	4	10
Chervil		
Seasoning		

Ingredient	4 portions	10 portions
Jelly		
Madeira	125ml	300ml
Port	100ml	250ml
Brandy	50ml	125ml
Agar agar (powdered)	2.1g	5.25g

Energy	Cal	Fat	Sat fat	Carb	Sugar	Protein	Fibre
3532 kJ	848 kcal	59.4g	21.5g	17.9g	3.2g	57.3g	2.2g

For the jelly

1 Mix all the ingredients together.

2 Place in a pan and bring to the boil, whisking for 1 minute.

3 Pour into a shallow container and allow to set at room temperature.

4 Store in the refrigerator until ready.

For the tournedos

1 In a small frying pan, add the oil and heat it, lightly season the tournedos and seal well in the pan. Cook until desired degree of cooking is achieved. Remove from the pan and rest on a wire rack in a warm place.

2 In the same pan, add the girolles and butter and bring to a foam. Carefully remove the girolles from the butter using a slotted/perforated spoon. Keep warm.

3 In the same pan add the bread slices and cook until golden and crisp. Remove and put them on the draining tray next to the beef.

4 Remove the butter from the pan, leaving a small amount of residue in the bottom. Heat up and then place in the garlic slices. Cook gently for 1–2 minutes.

5 Add the foie gras slices and cook gently.

6 Meanwhile, reduce the Madeira by one third, then add the reduced veal stock.

To finish

1 Place the tournedos on top of the bread croute, then the crisp ham slice on top of the beef with the foie gras topping the Parma ham.

2 Finally, arrange the girolles and jelly around the plate with the beef stack in the centre. Drizzle the sauce over and around, and finish with freshly sliced truffle and chervil.

Liver paté was used instead of foie gras and mushrooms instead of truffles for the nutritional analysis.

> **Variations**
> For the classical presentation:
> - Use grilled field mushrooms, each one slightly larger than the steak, rather than fried girolles.
> - Omit the ham.
> - Top with the grilled mushroom and dress with a red wine jus or shallot sauce, rather than jelly.
>
> Why not try a pork variation and use fillet, with a roast apple croute, and top with black pudding instead of foie gras: 'pork Rossini'?

Whisk the jelly

Fry the girolles in foaming butter

Place the meat on the bread

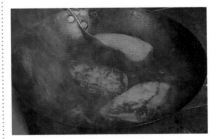
Fry the foie gras

3 Slow-cooked sirloin with Lyonnaise onions and carrot purée

Ingredient	4 portions	10 portions
Beef		
Sirloin, denuded with fat tied back on	1.2 kg	3 kg
Seasoning		
Oil	50ml	125ml
Garlic clove, sliced	1	2
Sprigs thyme	1	2
Bay leaves	1	2
Lyonnaise onions		
Onions	200g	500g
Seasoning		
Carrot purée		
Medium-sized carrots	600g	1½ kg
Star anise	1	2
To serve		
Jus de viande (meat juice)	150ml	375ml
Sprigs of chervil		

Energy	Cal	Fat	Sat fat	Carb	Sugar	Protein	Fibre
1949 kJ	467 kcal	25.3g	6.9g	16.0g	14.0g	44.8g	4.3g

Professional tip

This recipe uses the same slow cooking method used in recipe 1 for beef fillet, but a slightly higher temperature is needed because there is a little more collagen in sirloin than in fillet.

For the beef

1 Preheat the oven to 180°C. Season the beef and heat the oil in the pan. Add the garlic, thyme, bay leaves and the beef.

2 Place the beef in the oven for 15 minutes. Remove, and turn the oven down to 69°C. When the oven has reached this new temperature, return the beef to it for a further 1 hour 10 minutes.

3 While the beef is cooking, make the carrot purée and the Lyonnaise onions (see below) and keep warm.

For the onions

1 Finely slice the onions and put them into a large induction pan while cold.

2 Put on medium heat and season.

3 When the onions are starting to colour, turn down and cook slowly for approximately 2 hours.

4 Cool and refrigerate.

For the carrot purée

1 Peel the carrots and juice just over half of them into a small pan.

2 Cut the remaining carrots into equal slices of about 1cm and place into the carrot juice.

3 Boil the carrots, ensuring that you scrape down the sides of the pan.

4 For the last 8 minutes of cooking, before all the liquid has completely evaporated, drop in the star anise. Pass, retaining the juice.

5 Remove the star anise pod(s) and blitz the purée for 7 minutes, adding the retained juice.

To finish

1 When the beef is cooked, remove from the oven and carve evenly. Place a portion of carrot purée and Lyonnaise onions on each plate. Top with the beef and pour over the jus de viande (meat juice), garnish with sprigs of chervil and serve.

4 Braised short rib with horseradish couscous

Ingredient	4 portions	10 portions
Rib meat		
Rib meat off the bone	400g	1 kg
Vegetable oil	50ml	125ml
Onion, cut into medium mirepoix	125g	300g
Carrot, cut into medium mirepoix	125g	300g
Garlic clove	1	2
Sprigs of thyme	1	2
Red wine	400ml	1 litre
Brown stock	300ml	750ml
Sherry vinegar	25ml	60ml
Couscous		
Water	300ml	750ml
Olive oil	50ml	125ml
Salt and cayenne pepper		
Couscous	150g	375g
Fresh horseradish, grated	20g	50g
Lemon, juice of	Half	1
To finish		
Salad rocket	260g	625g
Shaved Parmesan	100g	250g
Vinaigrette	50ml	125ml

Energy	Cal	Fat	Sat fat	Carb	Sugar	Protein	Fibre
3019 kJ	726 kcal	44.91g	11.2g	46.2g	6.6g	36.3g	2.1g

Professional tip

Ribs are full of flavour due to the amount of collagen in them – the animal breathes continuously, working the muscle group a great deal, hence the amount of flavour. This is a very versatile piece of meat.

For the rib meat

1 Preheat the oven to 130°C. Trim any excess fat from the meat.

2 Heat the oil in a heavy casserole and add the rib meat, mirepoix, garlic and thyme. Cook for 5–6 minutes until brown.

3 Add the wine, bring to the boil and simmer until reduced by half. Add the stock, then cover with foil and cook in the oven for 2 hours.

4 Remove from the oven and leave the meat to cool in the liquor. Remove the ribs and set aside. Strain the sauce into a clean pan and bring to the boil.

5 Simmer until it has reduced to a thick sauce, but be careful not to over-reduce.

6 Meanwhile, trim any elastin or connective tissue from the rib meat, being careful to leave it whole and keep warm.

7 The sauce may need to be adjusted with the vinegar to cut through the richness – be careful not to add too much as you want only an undertone of vinegar.

For the couscous

1 Bring the water, oil, salt and cayenne pepper to the boil.

2 Place the couscous in a bowl with the finely grated horseradish and pour on the liquid. Place cling film tightly over the top and leave for 5 minutes.

3 Remove the cling film and add a little more oil to help free the grains.

4 Allow to infuse.

5 Finish with lemon juice and check the seasoning.

To finish

1 Mix the rocket, Parmesan and vinaigrette together and check the seasoning.

2 Place a mould of couscous on the plate and divide the rib meat equally between the plates.

3 Pour over the sauce, top the ribs with the salad and serve.

5 Bresaola (cured silverside)

Ingredient	4 portions	10 portions
Beef silverside	400g	1 kg
Coarse salt	120g	250g
Branches of rosemary (each about 15–23cm)	1	2
Sprig of thyme	1	1
Bay leaves	1	2
Garlic clove, crushed	1	3
Black peppercorns	3	6
Juniper berries, crushed	3	6
Orange, zest of	1	3

Energy	Cal	Fat	Sat fat	Carb	Sugar	Protein	Fibre
1395 kJ	332 kcal	10.2g	4.0g	0.2g	0.2g	43.3g	0.0g

1 Trim all the surface fat and silverskin off the joint. Don't try to remove the single vein of silverskin running through the centre of the muscle – the meat will fall apart if you do.

2 Make up the cure by combining all the other ingredients. You can use a grinder.

3 Rub the joint with half of the cure. (Place the remainder in a vacuum bag or jar and seal for later use.)

4 Place the meat into a vacuum or freezer bag and seal it. Place into the fridge and leave to marinate for one week, turning daily.

5 After the first week, take the meat out of the bag, dry it with paper towels and then rub it with the second half of the cure. Place it into a clean bag and seal. Return to the fridge and marinate for a second week.

6 After the second week, take the meat out of the bag, remove any remaining cure and pat dry with paper towels. Tie two pieces of string vertically around the meat, then tie a series of butcher's knots horizontally around it. Wrap it in clean muslin (to absorb any moisture that may seep out). Label clearly with the date and weight. Hang the muslin-wrapped meat in a place that is cool but not too dry.

7 Check it regularly during hanging for any unpleasant odour and weigh it carefully. The bresaola is ready after about three weeks and when it has lost about 30% of its weight. A small amount of white mould usually appears on the surface during hanging. If there is a great deal of mould, wash it off with a clean piece of muslin soaked in vinegar.

8 To serve, slice paper thin. Serve plain or dressed with a little olive oil and lemon juice. Accompany with a rocket salad and shaved Parmesan.

Note

Bresaola della Valtellina takes its name from the famous geographical district in which it was first produced. Since ancient times, techniques for preserving meat by salting and drying have been known. The use of such techniques in the Valtellina district of Italy is noted in writings dating back as far as 1400.

Bresaola is made from raw beef that has been salted and naturally aged. The meat, which is eaten raw, has a delicate flavour and a capacity to melt in the mouth that is highly appreciated by consumers.

6 Roast wing rib with Yorkshire pudding

Ingredient	10 portions
Beef	
Piece wing rib of beef	1 × 2 kg
Beef dripping	25g
Yorkshire pudding	
Eggs	2
Milk	200ml
Ice cold water	100ml
Plain flour	110g
Gravy	
Carrots	50g
Onion	50g
Red wine	200ml
Plain flour	30g
Beef stock	300ml
Prepared English mustard or horseradish sauce, to serve	

(Mirepoix spans Carrots and Onion rows.)

Energy	Cal	Fat	Sat fat	Carb	Sugar	Protein	Fibre
3185 kJ	758 kcal	31.04g	13.0g	31.6g	4.5g	90.0g	1.6g

For the Yorkshire pudding

1 Place the eggs, milk and water in a bowl and combine well with a whisk.

2 Gradually add the flour to avoid lumps and whisk to a smooth batter consistency.

3 Place in the refrigerator overnight to rest (this will give you a better lift in the oven).

4 Preheat the oven to 180°C.

5 Heat oil in a Yorkshire pudding tray by placing a small amount in the bottom of each well, and place the tray in the oven for 5 minutes.

6 Carefully fill the wells on the tray to two-thirds full and return to the oven.

7 When the puddings have risen and are golden brown, remove from the oven and keep warm.

For the beef

1 Preheat the oven to 195°C.

2 Place the dripping in a heavy roasting tray and heat on the stove top.

3 Place the beef in the tray and brown well on all sides.

4 Place in the oven on 195°C for 15 minutes then turn down to 75°C for 2 hours.

5 Remove and allow to rest before carving.

For the gravy

1 Remove the beef. Place the tray with the fat, sediment and the juice back on the stove.

2 Add the mirepoix and brown well.

3 Add the red wine and reduce by two-thirds.

4 Mix the flour and a little stock together to form a viscous batter-like mix.

5 Add the stock to the roasting tray and bring to the boil.

6 Pour in the flour mix and whisk into the liquid in the tray.

7 Bring to the boil, simmer and correct the seasoning.

8 Pass through a fine strainer or chinois and retain for service.

To complete

1 Slice the beef and warm the Yorkshire puddings, serve with the gravy, horseradish and mustard.

Professional tip

This dish would work well with most vegetables or potatoes. As an alternative, why not add slightly blanched root vegetables to the roasting tray at the start of the beef cooking, remove and reheat for service? They will get maximum flavour from the beef and juices.

7 Braised oxtails with garlic mash

1 Preheat the oven to 200°C.

2 Separate the trimmed tails between the joints and season with salt and pepper.

3 In a large pan, fry the tails in the dripping until brown on all sides, then drain in a colander.

4 Fry the chopped carrots, onions, celery and leeks in the same pan, collecting all the residue from the tails.

5 Add the chopped tomatoes, thyme, bay leaf and garlic and continue to cook for a few minutes.

6 Place the tails in a large braising pan with the vegetables.

7 Pour the red wine into the first pan and boil to reduce until almost dry.

8 Add some of the stock then pour onto the meat in the braising pan and cover with the remaining stock.

9 Bring the tails to a simmer and braise in the pre-heated oven for 1½–2 hours until tender.

10 Lift the pieces of meat from the sauce and keep to one side.

11 Push the sauce through a sieve into a pan, then boil to reduce it, skimming off all impurities, to a good sauce consistency.

12 While the sauce is reducing, quickly cook the diced garnish: carrot, onion, celery and leek in 1 tablespoon of water and a little butter for 1–2 minutes.

13 When the sauce is ready, add the tails and vegetable garnish, and simmer until the tails are warmed through.

14 Add the diced tomato, and spoon into hot bowls allowing 3 or 4 oxtail pieces per portion.

15 Sprinkle the oxtail with chopped parsley and serve with the garlic mash.

Ingredient	4 portions	10 portions
Oxtail		
Oxtails, trimmed of fat	2	5
Seasoning		
Beef dripping	100g	250g
Carrots, chopped	225g	550g
Onions, chopped	225g	550g
Celery sticks, chopped	225g	550g
Leeks, chopped	225g	550g
Tomatoes, chopped	450g	1 kg
Sprig of thyme	1	2
Bay leaf	1	2
Garlic clove, crushed	1	3
Red wine	570ml	1.5 litres
Veal/brown stock	2.25 litres	6 litres
Garnish		
Carrot, finely diced	100g	250g
Onion, finely diced	100g	250g
Celery sticks, finely diced	75g	185g
Small leeks, finely diced	75g	185g
Tomatoes, skinned, deseeded and diced	4	10
Cooked mashed potato with garlic	500g	1.25 kg
Fresh parsley, chopped	1 heaped tbsp	3 tbsp

Energy	Cal	Fat	Sat fat	Carb	Sugar	Protein	Fibre
2457 kJ	587 kcal	27.0g	12.2g	41.7g	20.4g	46.5g	8.68g

Professional tip

Choose oxtails that clearly have plenty of flesh around the bone: one complete oxtail will serve two people.

Alternatively, oxtail is good with haricot or cannellini beans, which seem to absorb a great deal of the flavour.

8 Pickled ox tongue

1 Wash the ox tongues and place in 2 litres of water with the salt and star anise for 3 hours.

2 Roast the mirepoix in a heavy-duty pan until golden brown.

3 Add the red wine vinegar and bring to the boil.

4 Add the chicken stock and red wine.

5 Pour onto the ox tongues and cook for 3 hours.

6 Remove the tongues from the liquid and pass the liquor.

7 Peel off the skin and store in retained liquor until cold. Remove from the liquid and wrap tightly in cling film.

Data on saturated fats was estimated based on other data about ox tongue.

Uses

Ox tongue can be sliced thinly when cold and served with pickled beetroot salad or, alternatively, diced and put through a meat sauce for either fish or meat preparations.

The texture of ox tongue is quite spongy so when using warm in certain dishes, something with a crisp, crunchy texture should be added to the dish to balance out the plate.

Ingredient	4 portions	10 portions
Ox tongues	1 kg	2.5 kg
Rock salt	200g	600g
Star anise	2	3
Mirepoix		
Carrot	250g	625g
Onions	250g	625g
Leeks	250g	625g
Celery	250g	625g
Red wine vinegar	500ml	1.25 litres
Chicken stock	1 litre	2.5 litres
Red wine	500ml	1.25 litres

Energy	Cal	Fat	Sat fat	Carb	Sugar	Protein	Fibre
2128 kJ	513 kcal	41.8g	16.9g	0.0g	0.0g	34.1g	0.0g

Note

Tongue has been cooked, pressed, pickled and canned since before the Second World War and remains an under-utilised product.

9 Stuffed saddle of lamb with apricot farce

Ingredient	4 portions	10 portions
Saddle, boned	1	2
Pig's caul/crepinette (optional)	200g	500g
Oil	50ml	125ml
Lamb jus	200ml	500ml
Farce		
Best lamb mince	250g	625g
Dried apricots, chopped	100g	250g
Chives, chopped	1 tsp	2 tsp
Seasoning		
Fondant potato		
Medium potatoes, preferably Maris Pipers	4	10
Vegetable oil	50ml	125ml
Garlic cloves, split	2	5
Butter	150g	375g
Water or white stock	100ml	250ml
Seasoning		
Roast vegetables		
Butter	50g	125g
Oil	50ml	125ml
Medium carrots, peeled and cut into 4	2	5
Medium parsnips, peeled and cut into 4	2	5
Leeks, trimmed of green and root, cut into 10cm lengths	2	5

Energy	Cal	Fat	Sat fat	Carb	Sugar	Protein	Fibre
4086 kJ	983 kcal	70.2g	27.8g	40.8g	19.6g	49.2g	8.0g

For the lamb

1 Preheat the oven to 180°C.

2 With the lamb on the chopping board, fat side down, bat out the fat flanks either side.

3 Mix all the farce ingredients and place in the lamb's centre cavity. Wrap over the fat flanks, encasing the farce and creating a tight cylinder shape.

4 Open up the crepinette onto the board and lay the lamb inside. Tightly double-wrap the lamb with crepinette and tie in four equal sections (tie tightly, but not overtight, as during cooking the lamb will expand and either burst from the wrap or split the string).

5 Heat the oil in a thick-bottomed pan and seal the lamb saddle all over until golden. Place in the oven for 20 minutes on 180°C, then reduce the temperature to 65°C for a further 45 minutes.

6 Check if cooked by inserting a probe in the centre, which should read 55–59°C.

For the fondant potato

1 Preheat the oven to 170°C.

2 Peel the potatoes and cut into slabs approximately 4cm thick.

3 Using a round pastry cutter, cut the slabs of potato into 5cm rounds and trim off the sharp edges.

4 In an ovenproof saucepan, heat the vegetable oil over a medium-high heat.

5 Add the potatoes and garlic; brown the potatoes on one side, taking care not to scorch them. Turn the potatoes over when golden brown and add the butter, water and seasoning.

6 Bring to a simmer, then transfer to the oven for 12–15 minutes or until the centre of the potatoes is soft. Remove from the oven and leave to soak up the butter for about 1 hour.

7 If the liquid in the pan has evaporated and only butter is left as the cooking medium, top up the pan with hot liquid and bring back to an emulsion.

To finish

1 Start to cook the fondant potatoes (see above).

2 Start to cook the lamb (see above).

3 Meanwhile, place the butter and oil for the vegetables in a thick pan, and heat and cook the vegetables until golden on the stove.

4 Drain into colander and keep warm.

5 Remove the lamb from the oven when ready and allow to rest for 10 minutes.

6 Remove the string and place on a carving board.

7 Meanwhile, arrange the fondants and roast vegetables on the plate. Slice the lamb into four equal pieces and place neatly over the vegetables.

8 Finish the dish by coating the lamb and around with lamb jus.

Tripe was used for pig's caul, for the nutritional analysis.

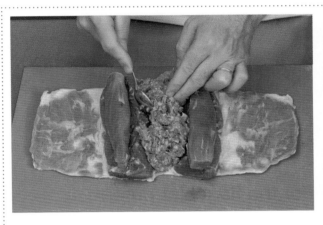
Place the farce into the cavity

Wrap over the flanks

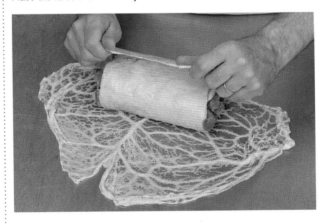
Wrap in crepinette

Variation

Omit the dried apricots for mint and peas, then substitute the roast vegetables and potato for a light couscous salad and wilted greens.

10 Roast rump of lamb, flageolets purée and balsamic dressing

Ingredient	4 portions	10 portions
Lamb		
Lamb rumps off the bone (150–160g each), trimmed	4	10
Salt and pepper		
Flageolets purée		
Dried flageolet beans	225g	560
Chicken stock	1 litre	2.5 litres
Smoked back bacon or pancetta trimmings	100g	250g
Carrots	50g	125g
Shallots	1	3
Garlic cloves	2	5
Sprig of thyme	1	3
Salt	10g	25g
Fondant potato		
Medium potatoes, preferably Maris Pipers	4	10
Vegetable oil	50ml	125ml
Garlic cloves, split	2	5
Butter	150g	375g
Water	100ml	250ml
Seasoning		
Balsamic dressing		
Garlic clove	1	3
Redcurrant jelly	2 tsp	5 tsp
Lemon oil	1 tsp	3 tsp
Good-quality balsamic vinegar	2 tsp	5 tsp
Thyme leaves	1 tsp	3 tsp
Dried tomato fillets, diced	6	15
Lamb jus	300ml	750ml
To finish		
Spinach, washed and picked	200g	500g

Energy	Cal	Fat	Sat fat	Carb	Sugar	Protein	Fibre
3870 kJ	930 kcal	61.4g	22.9g	43.3g	10.6g	53.2g	18.7g

For the flageolet purée

1 Soak the beans overnight. Next day, rinse in cold water and place in a saucepan over medium heat with the stock, bacon, carrot, shallot, garlic and thyme.

2 Simmer for 20 minutes, then add the salt and cook the flageolets for a further 30 minutes until tender.

3 If the liquid starts to reduce too much, top up with some water.

4 Leave the beans to cool in the cooking liquid. When cold, strain off and reserve the liquid. Remove and discard the garlic. Place the beans and other ingredients in a food processor and liquidise until very smooth, adding the retained cooking liquor if the consistency appears to be too thick.

5 When the beans are the consistency of soft mashed potato, keep warm until ready to serve.

For the fondant potato

1 Pre-heat the oven to 170°C.

2 Peel the potatoes and cut into slabs approximately 4cm thick.

3 Using a round pastry cutter, cut the slabs of potato into 5cm rounds and trim off the sharp edges.

4 In an ovenproof saucepan, heat the vegetable oil over a medium-high heat.

5 Add the potatoes and garlic, brown the potatoes on one side, taking care not to scorch them. Turn the potatoes over when golden brown and add the butter, water and seasoning.

6 Bring to a simmer, then transfer to the oven for 12–15 minutes or until the centre of the potatoes is soft. Remove from the oven and leave to soak up the butter for about 1 hour.

7 If the water in the pan has evaporated and only butter is left as the cooking medium, top up the pan with hot water and bring back to an emulsion.

For the balsamic dressing

1 In a small saucepan, combine all the ingredients for the dressing and bring to a simmer.

2 Leave to simmer for 1–2 minutes until the flavour has infused fully.

3 Cover with cling film and retain for serving.

To complete

1 Pre-heat the oven to 180°C.

2 Place a large frying pan on the stove over a medium-high heat and add the oil.

3 Season the lamb with salt and pepper and place in the pan, fat-side down.

4 Cook until golden brown on the fatty side, then turn over and seal the lean side.

5 When all the lamb is sealed, place it on a wire rack over a baking sheet and cook in the oven for 12–13 minutes, turning occasionally.

6 Test that the lamb is cooked to a core temperature of 57–59°C.

7 Meanwhile, drain the potatoes and reheat in the oven until warm.

8 Reheat the flageolets and, when the lamb is cooked, remove it from the oven and leave to rest in a warm place for 7–10 minutes.

9 In a saucepan, wilt the spinach in a little butter, then drain and place on the serving plates. Top with the fondant.

10 Warm the dressing through.

11 Spoon on the flageolet purée.

12 Carve the lamb and serve with the fondant potato.

13 Pour the sauce over, ensuring each plate has an equal serving.

Roast lamb and flageolet purée are synonymous with each other in classic French recipes.

Broad beans were used instead of flageolet beans for the nutritional analysis.

Professional tips

You may need to soak the beans overnight.

The fondant potatoes need to be made 1 hour before serving to ensure maximum flavour.

11 Pot roast chump of lamb with root vegetables

Ingredient	4 portions	10 portions
Beef dripping	100g	250g
Lamb chump, trimmed and boned	400g	1 kg
Small whole onions, peeled	400g	1 kg
Small carrots, peeled	400g	1 kg
Celery sticks, cut in three	400g	1 kg
Swede, peeled and cut into chunks	200g	500g
Field mushrooms	100g	250g
Hot stock	275ml	700ml
Clarified butter	200ml	500ml
Bay leaves	1	3
Sprig of thyme		
Butter	25g	60g
Flour	25g	60g
Seasoning		

Energy	Cal	Fat	Sat fat	Carb	Sugar	Protein	Fibre
3231 kJ	781 kcal	66.1g	37.7g	24.4g	16.7g	23.7g	6.3g

1 Pre-heat the oven to 140°C.

2 Melt the dripping in a thick cooking pot and when it's hot, put in the lamb and sear and brown it all over, then transfer it to a plate. Next, lightly brown the onions, carrots, celery and swede, and remove them temporarily to the plate.

3 Empty all the fat from the pot, then replace the lamb chump and arrange the vegetables and mushrooms around the meat. Add the hot stock, clarified butter, bay leaves and thyme, and a little salt and pepper. Cover with foil and a tightly fitting lid and, as soon as you hear simmering, place in the centre of the oven and leave for about 15 minutes.

4 When ready, place the meat and vegetables on a warmed serving dish, then skim off the fat. Bring the liquid to the boil and boil briskly until reduced slightly. Mix the butter and flour to a paste, then add this to the liquid and whisk until the sauce thickens. Serve with the meat and some sharp English mustard.

Professional tip

This method of cookery traps in flavour, keeps the meat moist and makes a sauce while doing so. The root vegetables in this will give the meat flavour, which makes them a suitable garnish with the lamb.

12 Fillet and rack of lamb with asparagus risotto

Ingredient	10 portions
Fillet or loin of lamb (about 70g per portion once trimmed)	1 kg
Rack of lamb (6 bones)	2
Asparagus	1 kg
Red wine	250ml
For the risotto	
Cream	2 tbsp
Butter	25g
Double cream	200ml
Lemon juice	1
Olive oil	
Shallots, finely chopped	2
Arborio rice	250g
Asparagus stock (cooking liquor)	750ml
Crème fraiche	150ml
To finish	
Purée of asparagus with cream	
Lamb stock, good	250ml

Energy	Cal	Fat	Sat fat	Carb	Sugar	Protein	Fibre
2521 kJ	610 kcal	52.7g	27.5g	7.7g	3.6g	22.5g	0.3g

Asparagus risotto

1 Trim the bottoms off the asparagus. Blanch and then set aside 10 spears for garnish.

2 Chop and cook the remaining asparagus and trimmings. Retain the liquor after cooking for use as the stock in the risotto. Put to one side.

3 Purée the cooked asparagus. Pass through a fine sieve. Add 2 tablespoons of cream and season. Put to one side.

4 Sweat the chopped shallots in butter until soft. Add the risotto rice. Then add slowly the asparagus stock and cook the rice until al dente (approximately 12 minutes)

6 Finish with crème fraiche and a little of the asparagus purée.

The lamb

1 Trim the lamb joints.

2 Poach the loin of lamb in red wine and lamb stock.

3 Pan fry the rack of lamb, then finish in the oven, cooking to the desired degree.

4 Slice both joints once cooked.

Service

1 Place a portion of risotto in the centre of each plate. Garnish with asparagus purée with cream.

2 Place three slices of loin on top, and slices of rack at the side.

3 Garnish with asparagus and finish with lamb jus.

Professional tip

The lamb joints can be bought already prepared.

13 Lamb's kidneys with juniper and wild mushrooms

Ingredient	4 portions	10 portions
Lamb's kidneys	12	30
Shallots, chopped	25g	60g
Juniper berries, crushed	12	30
Gin, marinated with the berries for one day	60ml	150ml
White wine	125ml	300ml
Strong lamb stock	500ml	1.25 litre
Selected wild mushrooms	50g	125g
Large potatoes	2	5
Oil and butter to sauté		

Energy	Cal	Fat	Sat fat	Carb	Sugar	Protein	Fibre
821 kJ	193.9 kcal	3.5g	1.1g	18.0g	1.2g	23.7g	0.4g

Variation

Finely shred the potatoes into matchsticks on a mandolin, dry in a clean cloth, season and cook in a ring mould as a fine potato cake.

1 Remove the fat and thin film of tissue covering the kidneys.

2 Season well and sauté in a hot pan, keeping them pink. Remove and keep warm.

3 Add the shallots and some crushed juniper berries to the pan, flambé with a little gin, pour in the white wine and reduce well. Add the lamb stock and reduce by half. Pass and finish with butter.

4 Prepare the mushrooms and sauté in hot oil, adding butter to maintain their earthy flavour, then keep warm.

5 Bake the potatoes in the oven until cooked. Scoop out the flesh, pass through a sieve and add a little butter. Mould into cakes, pass through flour and shallow fry on both sides. (They may also be passed through beaten egg and breadcrumbs before frying.)

6 To serve, place the potato cake in the centre of a serving dish, slice the kidneys and arrange attractively in a circle on the potato. Garnish the kidneys with the wild mushrooms, serve the sauce separately and serve immediately.

14 Lamb's liver flavoured with lavender and sage, served with avocado and sherry sauce

Ingredient	4 portions	10 portions
Lamb's liver	400g	1 kg
Milk	250ml	625ml
Honey	50g	125g
Sage, bunch, chopped	1	2
Lavender, bunch	1	2
Garlic cloves	2	4
Sesame oil	25ml	50ml
Avocados	1	2
Potato	1 kg	2½ kg
Sauce		
Baby onions	200g	500g
Garlic clove, crushed	1	1
Sesame oil	25ml	50ml
Sherry	50ml	125ml
Veal stock	250ml	625ml
Unsalted butter	50g	125g

Energy	Cal	Fat	Sat fat	Carb	Sugar	Protein	Fibre
2123 kJ	509 kcal	38.62g	13.3g	17.6g	15.6g	24.0g	2.0g

1 Remove skin and arteries from the liver and place on one side.

2 Mix the milk, honey and half the chopped sage with the lavender and uncrushed garlic cloves.

3 Place the liver in the milk mixture and leave for 24 hours.

4 For the sauce, peel the baby onions, blanch them, then refresh.

5 Sauté the crushed garlic in sesame oil with the baby onions. Add the sherry and reduce by half.

6 Add the veal stock and reduce this by two-thirds. Take off heat and cool slightly.

7 Whisk in the unsalted butter and season.

8 Heat sesame oil for the liver, remove the liver from the marinade and drain. Season liver with salt, pepper, remaining sage and any remaining lavender.

9 Sauté liver lightly until pink.

10 To serve, pour the sauce onto the plate, place the liver on top. Serve with potato purée with diced avocado. Garnish with the onions from the sauce. Finish with sprigs of lavender.

15 Rendang kambang kot baru (lamb in spicy coconut sauce)

Ingredient	4 portions	10 portions
Peanut oil	35ml	125ml
Shallots	25g	60g
Garlic clove, finely chopped	1	2½
Candlenuts or brazils or macadamia nuts, ground	50g	125g
Turmeric powder	5g (1 tsp)	12g (2½ tsp)
Coriander powder	5g (1 tsp)	12g (2½ tsp)
Lemon grass stalk, finely chopped	1	2–3
Lamb loin chops	4 × 150g	10 × 150g
White cabbage leaves, blanched	4	10
Spicy coconut sauce (see below)	500 ml	1.25 litres

Energy	Cal	Fat	Sat fat	Carb	Sugar	Protein	Fibre
3099 kJ	749 kcal	68.7g	23.8g	3.1g	1.8g	30.1g	0.2g

1 Place the peanut oil, shallots, garlic and ground nuts in a basin. Sprinkle with turmeric and coriander. Add the lemon grass and season. Mix to a thick paste. Add oil until a spreadable mixture is obtained.

2 Spread mixture onto the lamb, marinate for 2 hours in a refrigerator.

3 Quickly fry the lamb on both sides for 1 minute.

4 Wrap each chop in a blanched cabbage leaf.

5 Place the lamb in a suitable pan for braising, cover with hot spicy coconut sauce. Place a lid on the pan and braise in a moderate oven (180°C) for approximately 10 minutes.

6 Serve with stir-fried vegetables, flavoured with turmeric.

Spicy coconut sauce (saus rendang)

Ingredient	Makes 500ml
Peanut oil	35ml
Shallots, finely chopped	50g
Ginger, finely chopped	12g
Garlic clove, finely chopped	1
Greater galangal, peeled and finely chopped	12g
Candlenuts, brazil nuts or macadamia nuts	25g
Lemon grass stalk, chopped	1
Kaffir lime leaves	4
Turmeric powder	2 g
Coriander powder	2g (½ tsp)
Red chilli juice	375ml
Coconut milk	250ml
Salt	

1 In a saucepan, heat the peanut oil and sauté the shallots, ginger, garlic and greater galangal for approximately 5 minutes, until they are a light-brown colour.

2 Add the nuts, lemon grass, lime leaves, turmeric and coriander. Continue to sauté for a further 2 minutes.

3 Add the chilli juice and coconut milk and season. Bring to the boil and simmer for approximately 6–10 minutes, stirring continuously.

4 Remove the lime leaves.

5 Liquidise and pass through a strainer, use as required.

Fresh chillies were used instead of chilli juice for the nutritional analysis.

> **Professional tip**
>
> Greater galangal is less aromatic and pungent than lesser galangal. It is always used fresh in Indonesia. If unavailable, use fresh ginger but double the amount.

16 Sautéed veal kidneys with shallot sauce

Ingredient	4 portions	10 portions
Veal kidneys, free from fat and cut into individual nodules	250g	625g
Shallots, sliced	50g	125g
Butter	75g	180g
White wine vinegar	50ml	125ml
Cream	250g	625g
Tarragon, chopped	1 tsp	2 tsp
Vegetable oil	1 tbsp	3 tbsp
Brandy	75ml	180ml

Energy	Cal	Fat	Sat fat	Carb	Sugar	Protein	Fibre
2212 kJ	536 kcal	53.1g	31.5g	2.2g	2.0g	12.1g	0.2g

For the sauce

1 Place a hot pan in the middle of the stove for the kidneys.
2 Sweat the shallots in butter in a pan and add the white wine vinegar. Reduce by half.
3 Add cream and bring to the boil. Reserve to finish.

To sauté the kidneys

1 Place the vegetable oil and the kidneys in the hot pan.
2 Caramelise to a golden-brown colour.
3 Turn over and remove the pan from the stove.
4 Let the residual heat carry on cooking for 3–4 minutes.

To finish

1 Remove the kidneys from the pan, drain off the liquid, return the pan to the stove and de-glaze with the brandy.
2 Reduce slightly and add the shallot sauce.
3 Bring to the boil, add the tarragon, return the kidneys to the pan and serve.

This dish would traditionally be served with sautéed potatoes and haricots verts, but due to the versatility of the kidneys, pretty much most things will suit (excluding salad).

Ox kidneys were used instead of veal for the nutritional analysis.

> **Professional tip**
>
> The offal in veal has a subtle flavour due to the age of the animal, and pairing it with the shallots here offers an undertone of sweetness to the slightly bitter note of the kidney.

17 Osso bucco (braised shin of veal)

1 Pre-heat the oven to 180°C.

2 Season the flour and use to coat the meat well on both sides.

3 Heat the butter and oil in a casserole, add the veal and fry, turning once, until browned on both sides. Add the wine and cook, uncovered, for 10 minutes. Blanch, peel and chop the tomatoes, and add along with the stock and herbs.

4 Cover and cook in the centre of the oven until the meat is very tender and falls away from the marrow bone in the middle.

Delicious served with sauté potatoes or with a risotto alla Milanese.

Note

Part of the attraction of this dish is the marrow found in the bones. Traditionally, osso bucco is served with a gremolata, which is a combination of chopped parsley, garlic and lemon zest that is added to the dish at the very end. It has been omitted from this recipe, offering you a simple base.

Ingredient	4 portions	10 portions
Salt and ground pepper		
Plain flour	45g	112g
Thick slices of veal shin on the bone	4 × 200g	10 × 200g
Butter	50g	125g
Oil	2 tbsp	5 tbsp
White wine	150ml	375ml
Plum tomatoes	450g	1.125 kg
Light veal or chicken stock	300 ml	750 ml
Sprigs of parsley and thyme		
Bay leaf	1	1

Energy	Cal	Fat	Sat fat	Carb	Sugar	Protein	Fibre
1732 kJ	413 kcal	19.7g	8.5g	12.5g	3.9g	47.3g	1.5g

18 Veal chops with cream and mustard sauce

Ingredient	4 portions	10 portions
Veal chops	4	10
Butter or oil	50g	125g
Dry white wine	125ml	300ml
Veal stock	125ml	300ml
Bouquet garni		
Salt and pepper		
Double cream	60ml	150ml
French mustard, to taste		
Parsley, chopped		

Energy	Cal	Fat	Sat fat	Carb	Sugar	Protein	Fibre
1390 kJ	331 kcal	12.74g	7.3g	29.2g	4.5g	26.7g	2.6g

1 Shallow-fry the chops on both sides in hot butter or oil, pour off the fat.

2 Add white wine, stock, bouquet garni and season lightly; cover and simmer gently until cooked.

3 Remove chops and bouquet garni, reduce liquid by two-thirds, then add cream, the juice from the chops and bring to boil.

4 Strain the sauce, mix in the mustard and parsley, correct seasoning, pour over chops and serve.

19 Slow roast belly of pork, squash fondant, wilted spinach and spicy liquor

Ingredient	4 portions	10 portions
Pork belly	550g	1.4 kg
Leeks	100g	250g
Carrot	100g	250g
Butter	50g	125g
Squash	400g	1 kg
Spinach	100g	250g
Marinade		
Lime	1	3
Coriander leaves	25g	60g
Ginger	10g	25g
Soy sauce	25mls	60g
Cooking liquor		
Lime	1	3
Coriander leaves	25g	60g
Ginger root	10g	25g
Soy sauce	25ml	60ml
Light brown stock	500ml	1.25 litres

Energy	Cal	Fat	Sat fat	Carb	Sugar	Protein	Fibre
2144 kJ	516 kcal	38.9g	16.7g	13.1g	8.6g	29.9g	4.4g

1 Trim the pork belly and marinade it with lime juice, ginger, coriander and soy sauce, overnight if possible.

2 Make the cooking liquor by combining the ingredients. Add all the vegetables except the squash and spinach. Heat it up.

3 Pre-heat the oven to 175°C.

4 Put the trimmed and tied belly into the hot cooking liquor and cover with foil. Cook slowly for up to 3 hours: after 1 hour, turn the oven down to 160°C, and after 2 hours, remove the foil.

5 Prepare the squash and cook it fondant style (see fondant potato, page 215). Wilt the spinach.

6 Remove the pork from the liquor once cooked. Reduce the liquor and pass.

7 Plate with squash in centre of plate, with spinach, and a portion of belly pork on top. Drizzle with the reduced cooking liquor.

20 Roast shoulder of pork with crackling and apple sauce

Ingredient	4 portions	10 portions
Pork		
Pork shoulder joint	450g	1–1.5 kg
Olive oil to rub on joint		
Fine sea salt and freshly ground black pepper		
Bramley apple sauce		
Bramley cooking apples	200g	500g
Butter	10g	25g
Caster sugar	1 tbsp	3 tbsp
Gravy		
Plain flour	1 tsp	2 tsp
Meat or vegetable stock	100ml	450ml

Energy	Cal	Fat	Sat fat	Carb	Sugar	Protein	Fibre
1711 kJ	410 kcal	28.6g	9.6g	29.1g	11.7g	10.9g	0.8g

1 Preheat the oven to 180°C.

2 Rub the pork skin all over with kitchen paper. Leave for half an hour for the skin to dry (if the skin is moist it will not make crackling). Check the skin is evenly scored. If it is not, make further cuts in the flesh with a large, very sharp knife.

3 Brush the skin very lightly with oil. Sprinkle the skin with a thin, even layer of salt and a little pepper. Set the joint in a roasting tin and place in a preheated oven for 30 minutes, then reduce the temperature to 160°C for a further 1 hour, 20 minutes.

4 Meanwhile, make the Bramley apple sauce. Cut the apples into quarters using a small, sharp knife. Peel, core and slice the quarters then place in a pan with 3 tablespoons of cold water and bring to the boil. Reduce the heat to medium, cover the pan with a lid and cook for 6–8 minutes, until the apples are soft and pulpy.

5 Remove the apples from the heat and beat with a wooden spoon until smooth, then beat in the butter and sugar. If the sauce is too thin, return it to the heat and cook gently, stirring until it thickens slightly. Transfer to a serving bowl.

6 When the pork is cooked, remove from the oven and rest. Cover loosely with foil and leave for 15 minutes while you make the gravy. Using a large spoon, remove as much surface fat from the pan juices as you can. Place the roasting tin on the hob and reheat the juices. Remove from the heat and stir in the flour. Return to the hob and cook gently for 2 minutes. Gradually add the stock, stirring all the time until the gravy is slightly thickened. Simmer for 5 minutes. Pass and check seasoning.

7 Using a sharp carving knife and a fork to steady the meat, remove the crackling from the joint and place on a board. Cut the crackling into pieces (you can do this with kitchen scissors). Carve the pork into thick slices and serve each portion with some crackling, gravy and a generous spoonful of apple sauce.

Separate nutrition figures were used for the pork crackling and the meat, based on estimated serving sizes.

> **Professional tip**
>
> As this is a traditional roast, most seasonal vegetables will go with it.
>
> How to get the best crackling on roast pork is the subject of much debate in the kitchen. The secret of success is a good layer of fat beneath the rind. Also, the rind should be scored evenly all over. It helps if you choose a larger joint like this shoulder so there is more time in the oven to develop crisp crackling.

Test yourself

1   State the different categories for lamb and mutton.

2   List the quality points that you would check when you take delivery of lamb.

3   What are the main types of butchery used in preparing cuts for cookery?

4   In preparing a leg of lamb for roasting, what are the two methods that could be used to remove the bones?

5   List four preservation methods that could be used for meat and give an example of each.

6   Explain the main benefit gained from hanging meat.

7   List at least four factors that have an effect on the quality and composition of meat.

8   Name five cuts or joints from beef and indicate a cookery method you would consider for each.

9   What are the main benefits to resting a piece of meat after roasting, prior to carving?

10 State at least two safety points to consider when boning a shoulder of pork.

11 List four types of offal that can be used from pork and give a dish or garnish recommendation for each.

12 State two ways in which marrow from beef bones can be used in dishes.

13 Describe the process or practices that a chef can use to maximise yield from joints of beef.

14 State the potential hazards that may be increased from undercooked meat.

15 Describe the preparation of an 'entrecote steak' from a whole off-the-bone sirloin.

9 Advanced poultry and game dishes

This chapter covers the following units:

NVQ:
→ Prepare poultry for complex dishes.
→ Prepare game for complex dishes.
→ Cook and finish complex poultry dishes.
→ Cook and finish complex game dishes.

VRQ:
→ Advanced skills and techniques in producing poultry and game dishes.

Introduction

The aim of this chapter is to allow you to develop the necessary advanced skills, knowledge and understanding of the principles in preparing and cooking poultry and game to dish specifications. The emphasis in this chapter is to develop precision, speed and control in existing skills and develop more refined and advanced techniques.

Learning objectives

By the end of the chapter you should be able to:
→ Identify suitable commodities and recipes for poultry and game dishes and dietary requirements and explain the differences between different types of poultry and game.
→ Explain how to select the correct type, quality and quantity of poultry and game to meet recipe and dish requirements.
→ Describe the factors affecting the composition of meat and how it influences the choice of processes and preparation methods.
→ Identify appropriate seasons for game.
→ Describe how to store and preserve poultry and game as well as dishes containing poultry and game.
→ Demonstrate a range of cooking techniques using the correct tools and equipment appropriate to the different cooking methods.
→ Describe how to adapt recipes and cooking techniques to maximise nutritional value and healthy options in meat dishes and professional cookery.
→ Demonstrate professional ability in minimising faults in complex poultry and game dishes.

Recipes included in this chapter

Poultry

Origins

The word poultry describes all domestic birds that are bred especially for the table. It covers chickens, turkeys, ducks, geese, guinea fowl, squab, and recently quail has been added to this list. Over the years poultry farming has developed enormously and today poultry is the most popular meat from any animal or bird. In Britain alone some 800,000 tons of poultry are eaten each year – this includes free-range, organic and battery-reared chickens.

Over recent years there has been an increase in customers expecting kitchens to provide information on the provenance of food, particularly with an indication of the type of poultry it is using. More chefs are turning to free-range or organic labels as well as corn-fed chicken, as they offer more depth in flavour also. The use of specific breeds of poultry is more common with turkey and duck, as the particular breed gives a distinctive flavour and, for some, a link to the local producer is important.

Composition

The flesh of poultry is more easily digested than that of butchers' meat. It contains a good level of protein but is low in fat, which makes it a natural choice for some customer groups.

Nowadays poultry is readily available – fresh and frozen products are sold, with suppliers selling as either oven-ready or fresh (which will have the entrails still in them).

Generally a larger bird gives better value because of the meat-to-bone ratio, which tends to be higher in larger birds. As well as being available whole (which allows the chef to make use of the carcass as well as the flesh), it can be bought as portions – legs, drumsticks, breasts or crowns. Chicken, duck and turkey are also available as smoked products which can be incorporated into menus.

The lighter or white meat is found on the breast and is more tender than the dark or leg meat due to the work the leg muscles perform. Leg meat has more fat than the breast meat – even in the intensive rearing systems, the birds tend to spend lots of time on their legs developing the darker meat.

Factors affecting composition:
- A bird's origin (for example, the region of origin, or whether the bird was free range or organic) affects the composition of the flesh and its tenderness and flavour.
- The breed of bird affects its size, shape and quality. New breeds are introduced frequently, and many have been bred for flavour.
- Age has an impact: young birds are the most tender. Older birds have a stronger flavour and are tougher, requiring longer cooking.
- The type of feed given to birds may affect flavour, colour and quality; for example, the flesh of corn-fed chickens has distinctive yellow colouring.
- Seasonality has little effect on the composition of the flesh. Some birds (e.g. goose) are unavailable at some times of year.
- Slaughtering and processing methods have an effect: minimising stress for the birds during slaughtering better results. Processing establishes the ch... the muscle constituents and how they int... the muscles, affecting meat quality.

Chicken

Within the UK around 90 per cent of the chicken used is battery reared and is probably the most popular on menus. It comes in a variety of types suitable for different dishes and products.

Quality points

- Fresh chicken should have a plump breast, a pliable breast bone and firm flesh.
- The skin should be white with a faint bluish tint and unbroken.
- The legs should be smooth with small scales and spurs if the feet are still attached.

▲ Chickens, before and after preparation

▲ A corn-fed chicken – notice the yellower flesh

Table 9.1 Types of chicken

Type	Age and use	Average weight
Poussin	Up to 6 weeks, grilling or frying	250–400g
Spring chicken	6–8 weeks, grilling, frying, roasting	1–1.25 kg
Broiler chicken	3 to 4 months, roasting or portions	1–2 kg
Medium to large	3–6 months, roasting	2–4 kg
Capon (castrated cockerel)	Up to 7 months, pot roast	2–4 kg
Boiling	Up to 10 months, boiling	2–4 kg

Professional tip

Chicken is a delicate meat as most birds are under 6 months old when they are slaughtered. Chefs need to be considerate when cooking, especially the breast, as a high heat would render the fibres tough and dry. The correct method of cookery is important to ensure the quality of the finished dish.

Faults

This chicken looks done on the outside, but when it is cut open, you can see it is not cooked through.

Suprêmes

The suprême is the wing and half the breast of a chicken with the trimmed wing bone attached. The white meat of one chicken yields two suprêmes.

Use a chicken weighing between 1¼ and 1½ kg.

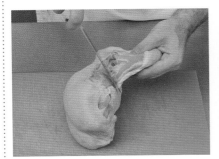

1 Cut off each leg of the chicken.

2 Remove the wishbone.
3 Scrape the wing bone bare adjoining the breasts.
4 Cut off the winglets near the joints, leaving 1½ to 2cm of bare bone attached to the breasts.

5 Cut the breasts close to the breast bone and follow the bone down to the wing joint. Cut through the joint.

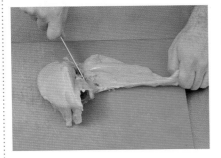

6 Lay the chicken on its side and pull the suprêmes off, using the knife to assist.
7 Lift the fillet from each supreme and remove the sinew from it.
8 Make an incision lengthways along the thick side of each supreme, open it and place the fillet inside.

9 Close, lightly flatten with a bat moistened with water and trim if necessary.
10 Suprêmes can be poached or shallow-fried. For shallow frying, they may be prepared as for a sauté of chicken, or stuffed and floured or crumbed.

Chicken for sauté

Chickens between 1¼ and 1½ kg may be cut into 10 pieces for sauté, yielding 4 portions. The pieces may be prepared on the bone, or skinned and boned for easier eating. Boning the chicken slightly increases shrinkage, so the portions look smaller; it also takes longer to prepare.

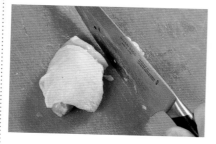

1 Remove the winglets.

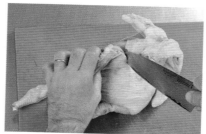

2 Remove the legs, cutting around the oyster.

3 Cut off the feet.

4 Separate the thighs from the drumsticks.

5 Trim the drumsticks neatly.

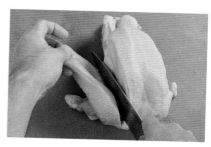

6 Remove each breast.

7 Separate the wings from the breasts and trim them.

8 Cut into the cavity, splitting the carcass (this may be used for stock).

9 Cut each breast in half.

The winglets, giblets and carcass are used to make chicken stock.

Chicken is prepared in the same way for a fricassée, blanquette or pie.

Recipes for chicken sauté include sauté chicken chasseur (see Recipe 6) or variations using different herbs (e.g. tarragon or rosemary), wines and garnishes (e.g. wild mushrooms).

Chicken mousse, mousselines and quenelles

Chicken mousse, mousseline or quenelles are smooth, light dishes, easy to digest. They are made from a mixture known as chicken forcemeat or farce. Adjust sizes according to the requirements of the dish, whether a light first course, a main course or a garnish.

Chicken mousse (Recipe 4) is cooked in a mould but turned out for service, so the basic mixture must be fairly firm or the mousse will break up.

Making mousselines

1 Mould the forcemeat using two tablespoons, dipping the spoons in hot water to prevent the mixture sticking to them.

2 Place the mousselines into a buttered shallow dish and carefully cover with chicken stock.

3 Cover with buttered paper and cook gently in the oven.

4 Serve two per portion.

Quenelles are also shaped with two spoons, like mousselines; the size can be varied by using different-sized spoons, e.g. teaspoons if the quenelles are to be used in vol-au-vents. Pea-sized quenelles can be formed by piping, instead of moulding; these are used to garnish soups.

Mousses, mousselines and quenelles may also be made from other foods, for example, ham, hare, partridge or pheasant.

Turkey

Most turkeys are battery farmed, although there are a number of free-range options available around the UK. These also include specialist local breeds as in Norfolk turkeys – Blacks or Bronze birds that have a broader breast.

Quality points
- The breast on whole turkeys should be large, the skin undamaged and with no signs of stickiness.
- The legs of younger birds are black and smooth, the feet supple with a short spur.

Turkeys can vary in weight from 3½ kg to 20 kg. Allow 200g raw weight per portion.

Clean and truss a turkey in the same way as a chicken. Always remove the wishbone and draw the sinews out of the legs before trussing.

When cooking a large turkey, the legs may be removed, boned, rolled, tied and roasted separately from the rest of the bird. This will reduce the cooking time and enable the legs and breast to cook more evenly.

Goose and duck

The type and breed of duck or goose will influence the price and yield. There are three main types of duck on the UK market:
- Aylesbury – a roasting bird.
- Barbary – normally the duck that is sold portioned with well-formed breast.
- Gressingham – normally older, so develops a better flavour than the other two.

Goose is seasonal, with availability from Michaelmas (29 September) until Christmas.

Quality points
A duck or goose should be free from bruising, feathers and blemishes. It should have a pleasant smell, be moist but not sticky; the skin should be pale and undamaged.

The average goose weighs between 5 kg and 6 kg. A gosling weighs between 2½ kg and 3½ kg.

Clean and truss a goose in the same way as a chicken.

Roast goose (see Recipe 19) is traditionally served with sage and onion dressing and apple sauce. Other dressings include peeled apple quarters with stoned prunes or peeled chestnuts.

Guinea fowl

When plucked, these grey and white feathered birds resemble chickens with slightly darker flesh. Guinea fowl needs to be cooked carefully so as not to dry the flesh out.

Quality points
Normally around 10–15 weeks' old at slaughter, the quality points for chicken also apply for guinea fowl.

▲ A turkey

▲ Turkey prepared for cooking

▲ A goose

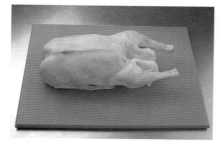

▲ Goose prepared for cooking

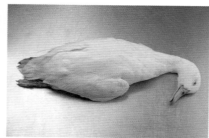

▲ A duck

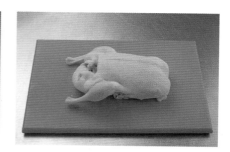

▲ A Gressingham duck prepared for cooking

▲ Guinea fowl, before and after preparation

Offal

- Poultry is often delivered with the heads on. Only the neck part is used, to make stock. The gizzard may be used in stock too.
- Chicken hearts are used for stock or stuffing.
- Chicken liver is used for pâté. Goose liver is used for foie gras.
- Chicken kidneys are used for stock.
- Game bird livers and kidneys can be used in the same way as those of chickens, as long as the bird has not hung for a long time, making it putrid inside.
- The cockscomb may be used as a garnish, but this is not common.

Advanced preparation and cooking techniques

Butchery

The preparation of poultry and game in the kitchen may involve the breaking down of whole birds or small carcasses in relation to using the most appropriate cooking method. This may involve boning, rolling, stuffing and tying. The inclusion of fat, for example, barding (laying over or round) or larding (inserting in), helps to keep prime pieces moist.

Classification and joints

For economic reasons of saving both labour and storage space, most caterers tend to purchase poultry and game pre-jointed and fully prepared for cooking rather than by the carcass. Knowledge of the specific birds and cuts from larger animals helps chefs in deciding the cooking method to use in relation to that cut.

Drawing and washing

This is the removal of the entrails from the inside of poultry and game birds. To remove, make a small lateral incision into the backside (vent) of the bird, then insert a forefinger and middle finger and roll them around the cavity of the bird to loosen the membrane that holds the innards in. When loose, remove from the bird by pulling through the vent and discard; ensure all the innards are removed, then wash and dry the bird well.

Leg removal and crowning

A crown is useful for roasting when only a few portions are required or the legs are not wanted.

Preparing a crown

1 Cut off each leg.

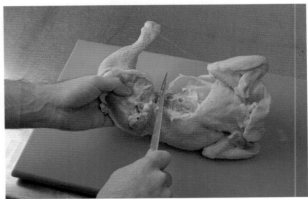

2 Sever the leg at the joint with a sharp tap of the knife.

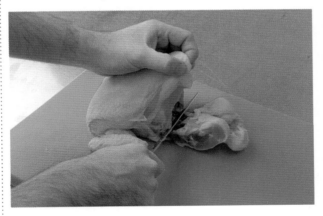

3 Cut off the front of the bird, leaving the crown.

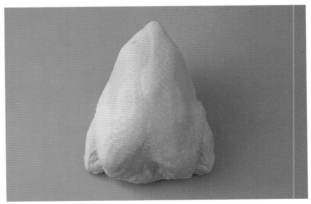

4 A prepared crown.

Video: preparing a duck for roasting
http://bit.ly/1kedUYT

Preparation of ballotines

Ballotines are boned-out, stuffed legs of poultry, usually chicken, duck or turkey.

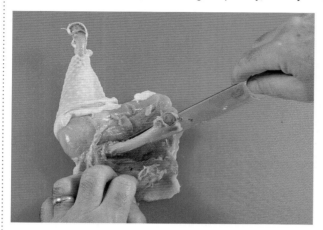

1 Cut the leg open around the bone.

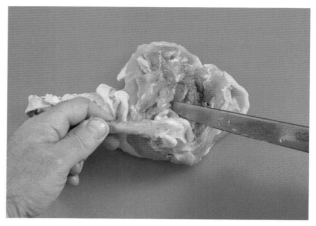

2 Start to separate the knuckle from the meat. (Scrape the bone.)

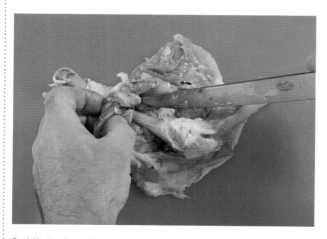

3 Lift the knuckle away (tunnel boning).

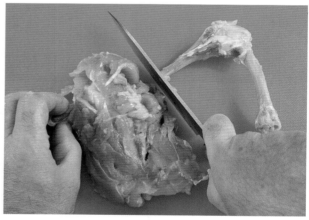

4 Cut the bone below the knuckle.

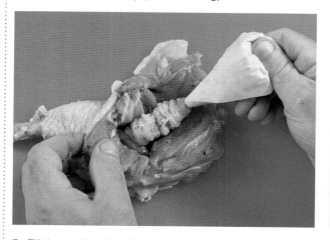

5 Fill the cavity of the thigh and drumstick with stuffing.

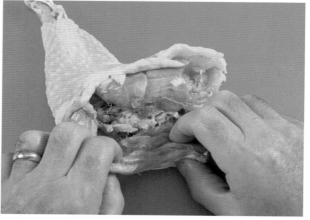

6 Close up in a neat shape and secure with string.

Stuffing

In the interests of food safety, stuffing should only be used in small birds such as poussin, or for ballotines, where a savoury stuffing is required. For larger birds, where stuffing is normally considered to be part of the dish (e.g. turkey and stuffing), then it is safer to cook separately.

Cooking

Much depends on the age and type: young birds can be grilled, roasted or fried, whereas older birds tend to be pot roasted or casseroled (the slower cooking methods), for example, coq au vin.

Sous vide

This is a method of cooking and processing which is cooking in a vacuum pouch under vacuum at low temperatures, steaming in a water bath and reheating. For meat, this is normally individual portions and may include sauce, liquor or garnish. The objective is to rationalise kitchen procedures at busy times without having a detrimental effect on the quality of the individual dishes prepared in this way.

Storage

Fresh, uncooked poultry should be used within two to three days of purchase, provided it has been kept in a refrigerator. Prepared game and poultry must be covered at all times and kept below other food items to ensure that no cross-contamination occurs. If poultry is purchased frozen, it should be removed from the freezer, placed directly in the fridge on a drip tray, covered and allowed to defrost thoroughly in the fridge.

Health and safety

To help minimise the risks from bacterial infection from poultry, chefs need to ensure that they put in place measures that prevent cross-contamination from raw products by following correct hygiene processes.

- Raw and cooked meat need to be separated in storage to avoid cross-contamination risks.
- Frozen poultry should be fully defrosted in the fridge.
- Chilled poultry must be used by the 'use by' date.
- Chilled, cooked poultry should be stored at or below 5°C.

- Vacuum-packed poultry should be stored below 3°C – ensure good circulation and that pouches are not punctured during storage until required for use. There are a number of kitchens that use their own vacuum-packing process – it is recommended that separate machines are used for cooked and raw products.
- Follow good practice in labelling and stock rotation (see Chapter 2).
- Temperatures of chillers and freezers should be monitored regularly to ensure correct operation.
- Thaw out frozen meats correctly when necessary and never refreeze them once they have thawed.

Preservation

There are a number of preservation methods that are used for extending the potential shelf life of poultry and game, or for the addition of flavour and tenderising prior to serving. These include:

- **Canning:** certain poultry is preserved in tins; these are all in a cooked state which need reheating or can be eaten cold.
- **Drying:** strips or pieces of poultry or game that is hung to dry; more popular in hotter countries.
- **Salting or dry curing:** meat is smothered in salt to help preserve the flesh.
- **Pickling or wet curing:** meat is soaked in brine for a period of time. Poultry may be cured in a brine before smoking.
- **Smoking:** traditionally the meat was hung in chimney spaces to dry out. Wood smoking contains elements that help preserve as well as adding flavour; hot smoking helps to cook the flesh at the same time whereas cool smoking adds flavour and preserves. Typical woods used are hickory, oak and apple.
- **Marinating:** steeping of poultry or game in a marinade to tenderise prior to cooking; the marinade normally contains something acidic, spicy and/or oil.
- **Confit:** cooking and then storing in duck or goose fat; the traditional version is 'confit duck'.
- **Freezing:** whole or in portions, meat should be frozen quickly to ensure smaller ice crystals are formed in the flesh. Poultry and game should be covered to minimise the risk of freezer burn.

Regional recipe - goose liver, heart and kidney parfait

Contributed by City College Norwich

Ingredient	
Goose livers	3
Goose hearts	3
Goose kidneys	6
Brandy	25ml
Madeira	25ml
Goose back fat	
Shallots, chopped	30g
Tarragon	5g
Egg white	50g
Cream	30ml
Seasoning	
Truffle oil	2ml

1 Marinate the offal (hearts, livers and kidneys) in the brandy and Madeira overnight.

2 Spread some of the goose fat out on cling film and allow to set – it may need to be almost frozen before it can be cut to shape to line the mould.

3 Sauté the offal in hot goose fat, then place on one side.

4 Sweat the shallots and tarragon in the same pan.

5 Blend the offal and shallots through the food processor until smooth. Add the egg whites and cream. Season to taste and add truffle oil.

6 Line a mould or terrine with the set goose fat. Fill with the parfait mix.

7 Cook in a water bath in the oven at 150°C for 20 minutes. Alternatively, place in a vacuum pouch (low setting) and cook in a water bath at 55°C for 40 minutes.

8 Slice and serve on a puff pastry base. Garnish with pickled vegetables and micro lettuce.

1 Chestnut and apple forcemeat

Ingredient	4 portions	10 portions
Eating apples, peeled and cored	200g	450g
Tinned whole unsweetened chestnuts, roughly mashed	350g	900g
Salted or fresh belly of pork, cut into small dice	60g	175g
Shallots, finely chopped	1	2
Garlic cloves, crushed	1	2
Parsley, finely chopped	1 tbsp	3 tbsp
Egg	1 yolk	1

1 Gently stew the apples, covered, in 1 tablespoon of water until reduced to a pulp.

2 Add to the roughly mashed chestnuts, pork, shallots, garlic and parsley, and bind with the egg.

Use with turkey, chicken, duck, pheasant or guinea fowl.

2 Foie gras stuffing

1 Sauté the goose liver and shallots in butter in a small sauté pan for 2 minutes. Scrape into a mixing bowl.

2 Boil the port in the sauté pan until reduced to 2 tablespoons. Add to the mixing bowl.

3 Add the foie gras to the bowl and mix well until everything is incorporated. If the mixture is too wet for easy stuffing, beat in the breadcrumbs.

4 Season to taste with the allspice, thyme, salt and pepper. Place into oiled foil and shape into a neat roll. Cook in the oven for 30 minutes towards the end of the goose roasting time. Remove from the foil, cut into thick slices and serve with the goose.

Ingredient	4 portions	10 portions
Goose livers, finely chopped (approx. 200g each)	1	2
Shallots, finely chopped	1 tbsp	2 tbsp
Butter	5g	10g
Port	75ml	150ml
Foie gras, chopped, or liver pâté	55g	110g
Allspice	Pinch	Pinch
Thyme	Pinch	Pinch
Salt and pepper		

3 Pork, sage and onion forcemeat

Ingredient	10 portions
Large onion, finely chopped (approx. 100 g)	1
Dried sage	1 heaped tsp
White breadcrumbs	4 heaped tbsp
Boiling water	2–3 tbsp
Sausage meat	900g
Salt and freshly ground black pepper	To taste

1 Mix the onion, sage and breadcrumbs in a large bowl, then add the boiling water and cool thoroughly.

2 Work the sausage meat into it and season.

Use with turkey, chicken, pork, veal, duck, pheasant, goose or guinea fowl.

4 Chicken mousse

Ingredient	10 quenelles (5 portions)
Chicken breast	350g
Whipping cream	250ml
Salt and pepper to season	To taste

Energy	Cal	Fat	Sat fat	Carb	Sugar	Protein	Fibre
3056 kJ	294 kcal	54.3g	13.0g	3.4g	1.4g	58.6g	0.0g

1 Roughly dice the chicken and place in the freezer for 10 minutes to chill.

2 When the chicken is cold, place in a food processor and blend slightly, then season with a little salt.

3 Pour in the cold cream gradually until all of it has emulsified with the chicken.

4 Form the mousse into quenelles using 2 tablespoons. Dip the spoons into water after each quenelle. Place on a tray lined with cling film and chill.

5 Poach in a little simmering water, adjust the seasoning, drain and store well covered with cling film in the refrigerator. Serve with a suitable sauce or as an accompaniment to another dish.

Variation: unmoulded chicken mousse
● Fill a buttered mould with the uncooked mixture. (Individual or 2- or 4-portion moulds can be used.)
● Cook in a bain marie in a moderate oven.
● Turn out of the mould for service.

5 Chicken soufflé

Ingredient	4 portions	10 portions
Raw chicken, without skin or sinew	250g	600g
Butter	50g	125g
Thick velouté	250ml	600ml
Salt and pepper		
Eggs, separated	3	8

Energy	Cal	Fat	Sat fat	Carb	Sugar	Protein	Fibre
397 kJ	94.6 kcal	5.1g	3.2g	0.4g	0.4g	11.7g	0.0g

1 Finely dice the chicken and cook in the butter.

2 Add to the velouté and purée in a food processor. Check seasoning.

3 Beat the yolks into the warm mixture.

4 Fold in the stiffly beaten whites carefully.

5 Place into individual buttered moulds or one mould.

6 Bake at 170°C for approximately 15 minutes (25–30 for a large mould) and serve.

7 Serve a suitable sauce separately, for example, mushroom or suprême sauce.

6 Sauté chicken chasseur

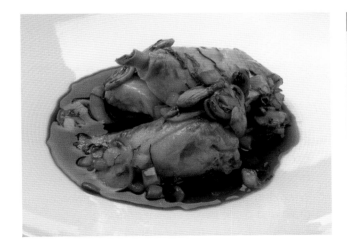

Ingredient	4 portions	10 portions
Butter or oil	50g	125g
Salt, pepper		
Chickens, each 1.25–1.5 kg, cut for sauté	1	2½
Shallots, chopped	10g	25g
Button mushrooms, washed and sliced	100g	250g
Dry white wine	3 tbsp	8 tbsp
Jus-lié, demi-glace or reduced brown stock	250ml	625ml
Tomato concassé	200g	500g
Parsley and tarragon, chopped		

Energy	Cal	Fat	Sat fat	Carb	Sugar	Protein	Fibre
2430 kJ	579 kcal	45.8g	20.7g	2.1g	1.6g	37.6g	1.5g

1. Place the butter or oil in a sauté pan on a fairly hot stove.
2. Season the pieces of chicken and place in the pan in the following order: drumsticks, thighs, wings and breast.
3. Cook to a golden brown on both sides.
4. Cover with a lid and cook on the stove or in the oven until tender. Dress neatly in a suitable dish.
5. Add the shallots to the sauté pan, rubbing them into the pan sediment to extract the flavour. Cover with a lid and cook on a gentle heat for 1–2 minutes.
6. Add the washed, sliced mushrooms and cover with a lid. Cook gently for 3–4 minutes, without colour. Drain off the fat.
7. Add the white wine and reduce by half. Add the jus-lié, demi-glace or reduced stock.
8. Add the tomatoes. Simmer for 5 minutes.
9. Correct the seasoning and pour over the chicken.
10. Sprinkle with chopped parsley and tarragon and serve.

7 Grilled chicken suprêmes with asparagus and balsamic vinegar

Ingredient	4 portions	10 portions
Asparagus pieces	20	50
Banana shallots, finely diced	2	5
Vinaigrette	20ml	50ml
Lemon, juice of	1	3
Chicken suprêmes	4 × 150g	10 × 150g
Vegetable oil	50ml	125ml
Salt and pepper		
Aged balsamic vinegar	20ml	50ml

Energy	Cal	Fat	Sat fat	Carb	Sugar	Protein	Fibre
1329 kJ	318 kcal	17.0g	2.4g	3.5g	2.9g	37.8g	1.2g

1. Cook the prepared asparagus in boiling, salted water for approximately 4 minutes, refresh in ice water.
2. Sweat the shallots without colour until soft, drain and mix with the vinaigrette and lemon juice to form a dressing.
3. Place the chicken on an oiled tray and season with salt and pepper.
4. Place the chicken on a pre-heated grill and grill gently for 3–4 minutes either side, ensuring an even bar mark on the sides.
5. Warm the asparagus through in boiling water, drain and place in the shallot dressing.
6. Take the cooked chicken and place on a plate or serving dish, lay the asparagus on top, finish with the dressing and balsamic vinegar.
7. Serve immediately.

▲ The bar marks created by grilling the chicken

8 Fricassée of guinea fowl with wild mushrooms

Ingredient	4 portions	10 portions
Guinea fowl	1 × 1.3 kg	2 × 1.6 kg
Salt and pepper		
Butter	50g	125g
Flour	35g	90g
Chicken stock	500ml	1.25 litres
Egg yolks	1–2	3–5
Cream	4 tsp	10 tsp
Wild mushrooms, cooked and chopped	200g	500g
Parsley, chopped	2 tsp	5 tsp

Energy	Cal	Fat	Sat fat	Carb	Sugar	Protein	Fibre
2537 kJ	606 kcal	28.3g	11.6g	7.5g	0.7g	80.7g	0.9g

1 Cut the guinea fowl as for sauté, and season with salt and pepper.

2 Place the butter in a sauté pan. Heat gently. Add pieces of bird. Cover with a lid.

3 Cook gently on both sides without colouring. Mix in the flour.

4 Cook out carefully without colouring. Gradually mix in the stock.

5 Bring to the boil and skim. Allow to simmer gently until cooked.

6 Mix the yolks and cream in a basin (liaison).

7 Pick out the guinea fowl into a clean pan.

8 Pour a little boiling sauce onto the yolks and cream, and mix well.

9 Pour all back into the sauce, combine thoroughly but do not re-boil.

10 Correct the seasoning and pass through a fine strainer.

11 Add the cooked wild mushrooms, mix and incorporate well.

12 Pour over the guinea fowl and reheat without boiling.

13 Serve sprinkled with chopped parsley. The dish may also be garnished with heart-shaped croutons, fried in butter.

9 Confit chicken leg with leeks and artichokes

1 Gently heat the confit oil, add the garlic, bay and thyme.

2 Put the chicken legs in the oil and place on a medium to low heat, ensuring the legs are covered.

3 Cook gently for 3–3½ hours.

4 To test if the legs are cooked, squeeze the flesh on the thigh bone and it should just fall away.

5 When cooked, remove the legs carefully and place on a draining tray.

6 Heat the vegetable oil in a medium sauté pan, add the artichokes and leeks, colour slightly and then add the brown chicken stock.

7 Reduce the heat to a simmer and cook for 4–5 minutes; meanwhile place the confit leg on a baking tray and place in a pre-heated oven at 210°C; remove when the skin is golden brown (approximately 5 minutes), taking care as the meat is delicate.

8 Place the chicken, leeks and artichokes in a serving dish or on a plate, check the leeks and artichokes are cooked through, and bring the stock to a rapid boil, working in the butter to form an emulsion.

9 Add the chopped chives to the sauce and napé over the chicken leg.

Ingredient	4 portions	10 portions
Confit oil	1 litre	2.5 litres
Garlic cloves	4	10
Bay leaf	1	3
Sprig of thyme		
Chicken legs	4 × 200g	10 × 200g
Vegetable oil	50ml	125ml
Globe artichokes, prepared, cooked and cut into quarters	4	10
Whole baby leeks, blanched	8	20
Brown chicken stock	250ml	625ml
Butter	50g	125g
Chives, chopped	1 tbsp	3 tbsp
Seasoning		

Professional tip

Confit oil is 50/50 olive oil and vegetable oil infused with herbs, garlic, whole spice or any specific flavour you wish to impart into the oil; then, through slow cooking in the oil, the foodstuff picks up the flavour.

This dish utilises the by-product of the chicken crown; it is not only very cost-effective but has great depth of flavour due to the work the muscle group has done.

Immerse the chicken legs in confit oil, ensuring they are covered, and cook

Check whether the chicken is done using the squeeze test

Remove from the oven when the chicken is golden brown

Add butter to the stock to form an emulsion

10 Pan-fried chicken suprêmes with lemon, capers and wilted greens

1 Take the suprêmes, pané evenly and set aside.

2 Slice each lemon evenly into 4, removing seeds.

3 To a pre-heated pan, add the oil and gently place in the chicken.

4 Cook gently on an even heat for approximately 6–8 minutes both sides until the breadcrumbs are an even golden brown. Use a probe to check the core temperature; if it has not reached 75°C, finish for a few minutes in the oven.

5 Reheat the greens in the melted butter and season lightly.

6 Place the greens in the centre of the plate and top with a chicken suprême.

7 Place 2 slices of lemon on the chicken and sprinkle the capers on and around to finish.

Ingredient	4 portions	10 portions
Chicken breast suprêmes	4 × 150g	10 × 150g
Pané à l'anglaise (flour, eggwash and breadcrumbs)	250g	600g
Lemons, peeled and pithed	2	5
Vegetable oil	50ml	125ml
Mixed greens, e.g. spinach, bok choi, prepared, blanched	250g	625g
Butter	50g	125g
Salt and pepper		
Baby capers	2 tbsp	5 tbsp

Variations

This dish will work equally well with an escalope of chicken, turkey, veal or pork, halving the cooking time. Or why not pané using ground porridge oats for a healthier option?

Energy	Cal	Fat	Sat fat	Carb	Sugar	Protein	Fibre
2393 kJ	571 kcal	28.5g	9.4g	35.3g	2.3g	45.3g	1.9g

11 Poached guinea fowl with Muscat grapes

Ingredient	4 portions	10 portions
Chicken stock	2 litres	5 litres
Guinea fowl crowns	2 (300g)	5 (750g)
Vegetable oil	50ml	125ml
Button onions	12	30
Reduced chicken stock	250g	625g
Natural yoghurt	100g	250g
Juice of lemon	1	3
Cooked salsify batons	200g	500g
Salt and pepper		
Peeled Muscat grapes	250g	625g

Energy	Cal	Fat	Sat fat	Carb	Sugar	Protein	Fibre
1273 kJ	305 kcal	19.2g	3.5g	22.2g	16.0g	14.9g	3.9g

1 Bring the stock up to a simmer; place the crowns in the stock and simmer for 11 minutes.

2 Meanwhile, place the oil in a medium pan and heat to a moderate heat. Place the onions in the pan and colour slightly. Add the reduced chicken stock and cook the onions gently.

3 When the onions are cooked, put to one side and retain. Remove the guinea fowl crowns and rest for 2–3 minutes, breast-side down.

4 Remove the breasts from the crown and check if cooked; if not quite cooked place back in the cooking liquor for 1–2 minutes. Then place the breasts on a plate or in a serving dish and keep warm.

5 Add the yoghurt, lemon juice and salsify to the onions with the reduced stock. Warm slightly – do not boil – adjust the seasoning, add the peeled grapes and pour over the guinea fowl. Garnish with microherbs.

12 Guinea fowl *en papillote* with aromatic vegetables and herbs

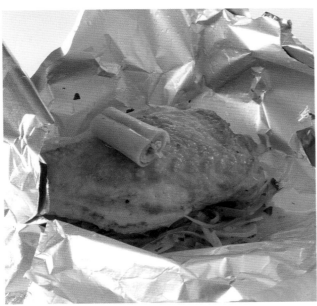

Ingredient	4 portions	10 portions
Guinea fowl breasts	4 × 150g	10 × 150g
Butter	50g	125g
Shallots, chopped	25g	65g
Onions, sliced	25g	65g
Carrots, cut into julienne	25g	65g
Leeks, cut into julienne	25g	65g
Sprigs of lemon thyme	3	7
Sticks of lemon grass	1	2.5
Oil	25g	65g
Dry white wine	100ml	250ml
Seasoning		
Coriander, chopped	15g	35g

Energy	Cal	Fat	Sat fat	Carb	Sugar	Protein	Fibre
1540 kJ	370 kcal	23.8g	9.2g	1.8g	1.4g	37.2g	0.5g

1 In a pre-heated pan seal off the guinea fowl breasts in butter.

2 Sweat the shallots, onions, carrots, leeks and herbs in half the butter without colour. Add the wine and allow to reduce.

3 Season and add the coriander.

4 Cut greaseproof paper or aluminium foil into large heart shapes, big enough to hold one breast each. Oil the paper or foil liberally.

5 Place a small pile of the vegetable and herb mix to one side of the centre of the paper or foil. Place a breast on top and a piece of lemon grass on the breast, and cover with a little of the wine mix. Fold or pleat the paper or foil tightly. Place on an oiled tray in a hot oven (240°C) until the bag expands. Cook for 8–10 minutes. Serve immediately.

Partridge was used instead of guinea fowl for the nutritional analysis.

Reduce the cooking liquid

Place the ingredients on the foil, topping with lemon grass

Fold over the foil

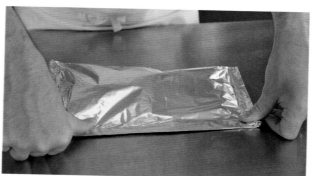

Seal the package

13 Roast Gressingham duck with jasmine tea and fruit sauce

Ingredient	4 portions	10 portions
Sultanas	100g	250g
Hot jasmine tea	200ml	500ml
Oven-ready Gressingham ducklings, wishbones removed, wings and neck cut short (reserve for sauce)	1 × 1.6 kg	2 × 2 kg
Water	700ml	1.75 litres
Unsalted butter	10g	25g
Caster sugar	10g	25g
Orange juice	150ml	375ml
Soy sauce	1 tsp	3 tsp
Cherry brandy	100ml	250ml
Jasmine tea leaves	10g	25g
Lime, juice of	One quarter	1
Salt and pepper		
Duck fat	1 tbsp	3 tbsp

Energy	Cal	Fat	Sat fat	Carb	Sugar	Protein	Fibre
2045 kJ	487 kcal	18.6g	6.6g	32.9g	18.6g	40.5g	0.6g

Steps 1–9 can be prepared 24–48 hours before the dish is to be served and can be stored, ready to use.

1 Soak the sultanas in hot jasmine tea for 24 hours.

2 Chop the wings and neck into small pieces. Place in a large pan and caramelise.

3 Drain, cover with water, bring to the boil and skim.

4 Pass through a fine sieve and boil to reduce to 200ml. Keep warm.

5 In a separate pan, melt the butter and sugar together to make a caramel. Add the orange juice, soy sauce, cherry brandy and the reduced duck stock, boil and reduce by half.

6 Bring to simmering point, add the jasmine tea leaves and leave to infuse for 1 minute.

7 Add the lime juice, taste and correct seasoning with salt and pepper.

8 Strain through a fine sieve to remove the tea leaves. Reserve.

9 With the point of a very sharp knife, score very light criss-cross lines across the breasts and legs (so the fat will run out easily).

10 Cut the legs off the ducks just above the joint.

11 Melt the fat in a large roasting tray on the top of the cooker and colour the ducklings for about 5 minutes each side and for 3 minutes on each breast.

12 Season with salt and pepper, then roast in the oven – on one breast for 10 minutes, then on the other for another 10 minutes, then a further 15 minutes on their backs (a total of 35 minutes).

13 Remove from the oven and discard the fat. Let the ducklings rest on their breasts for 5–10 minutes. Carve and serve.

14 Add the sultanas to the sauce, pour over the carved duck and serve with fondant potatoes and stir-fried vegetables.

14 Garbure of duck

1 Place all ingredients in a pan (except the potatoes and parsley) and cook slowly for 1 hour, stirring frequently.

2 After 1 hour add the potatoes and cook for a further 25 minutes.

3 Remove the bouquet garni, add the parsley and work in the butter on a high heat.

4 Serve in a bowl with traditional country French bread, toasted.

Ingredient	4 portions	10 portions
Duck legs confit, with skin and bone removed	4 × 150g	10 × 150g
Small green cabbage	half	1
Small onion	half	1
Napoli salami in thin batons	40g	100g
Chicken stock	200ml	500ml
Haricots blanc, half-cooked	25g	65g
Flageolet, half-cooked	25g	65g
Butter	75g	185g
Leek, in rounds	1	3
Bouquet garni	small	large
Potatoes, cut into 1cm dice	250g	625g
Chopped parsley	2 tsp	6 tsp

Professional tip

This recipe can also double up as a secondary protein element on most duck dishes. It is a cost-saving addition as you will be able to reduce the amount of prime breast meat that is needed.

15 Confit duck leg with red cabbage and green beans

Ingredient	4 portions	10 portions
Confit oil	1 litre	2.5 litres
Garlic cloves	4	10
Bay leaf	1	3
Sprig of thyme	1	2
Duck legs	4 × 200g	10 × 200g
Butter	50g	125g
Green beans, cooked and trimmed	300g	750g
Braised red cabbage (see page 219)	250g	625g
Seasoning		

1 Gently heat the confit oil, add the garlic, bay leaf and thyme.

2 Put the duck legs in the oil and place on a medium to low heat, ensuring the legs are covered.

3 Cook gently for 4–4½ hours (if using goose, 5–6½ hours may be needed).

4 To test if the legs are cooked, squeeze the flesh on the thigh bone and it should just fall away.

5 When cooked, remove the legs carefully and place on a draining tray.

6 When drained, put the confit leg on a baking tray and place in a pre-heated oven at 210°C; remove when the skin is golden brown (approximately 9–10 minutes), taking care as the meat is delicate.

7 Heat the butter in a medium sauté pan and reheat the green beans.

8 Place the braised cabbage in a small pan and reheat slowly.

9 Place the duck leg in a serving dish or plate along with the red cabbage and green beans.

Professional tip

Confit duck legs can be prepared up to three or four days in advance. Remove them carefully from the fat they are stored in, clean off any excess fat and place directly into the oven. This is a great timesaver in a busy service.

16 Pan-fried breast of duck with vanilla and lime

Ingredient	4 portions	10 portions
Duck		
Barbary duck breasts, off the bone	4 × 150g	10 × 150g
Vanilla sauce		
Vanilla pods	3	7
Oil, to brown mirepoix		
Mirepoix	200g	500g
Port wine	25ml	60ml
Armagnac	10ml	25ml
Red wine	75ml	185ml
Brown stock reduced to glaze	250ml	625ml
Good-quality vanilla extract	25ml	65ml
To finish		
Spinach, picked and washed	200g	500g
Butter	20g	50g
Limes, cut into segments	3	8

Energy	Cal	Fat	Sat fat	Carb	Sugar	Protein	Fibre
1611 kJ	387 kcal	24.5g	6.9g	4.3g	3.9g	31.7g	2.4g

For the vanilla sauce

1 Cut the vanilla pods lengthways with a sharp knife and scrape the seeds from the centre; retain seeds for later use.

2 In a saucepan, heat a little oil and cook the mirepoix until slightly golden brown. Add the vanilla pod husks and cook for 2–3 minutes to extract the flavour.

3 Add the wine and Armagnac and stir to the boil. Pour in the brown stock and simmer for 45 minutes.

4 Add the vanilla extract and pour the sauce through a fine strainer into a clean pan. Remove the pod husks, wash them off and retain for another use.

5 Add the seeds from the pod to the sauce, whisking to disperse them evenly.

For the duck

1 Trim any sinew and excess fat from the meat-side of the breast.

2 Turn the breast with the flesh-side facing down and, using a sharp knife, score the fat, working up the breast, being careful not to cut too far into the breast and expose the meat.

To complete

1 Heat a heavy-based pan on the stove over a medium heat. Place the duck breasts in the pan fat-side down and allow the fat to render into the pan, creating a cooking medium for the duck.

2 Cook, turning once, for 6–7 minutes and ensure that the skin achieves a golden-brown colour and that the fat is rendered from the breast.

3 When the duck is cooked pink, remove from the pan and leave to rest for 5–6 minutes in a warm place.

4 Warm the sauce, wilt the spinach in some butter and season to taste. Drain the spinach well and place a thin line of it down the centre of the plate. Cut each duck breast in two diagonally, giving two good strips.

5 Place on the spinach, garnish the dish with the lime segments, pour over the sauce and serve.

Professional tips

● The lime and vanilla in this dish cut through the richness of the duck.

● The sauce in this recipe can be made in advance if needed and stored for later use. If vanilla pods are not available, increase the amount of vanilla extract.

● In this and most other duck recipes, the breast is taken off the bone before cooking to ensure that all the fat is rendered down to give a crispy skin. This also helps to time the cooking perfectly, which is difficult when cooking the bird whole because the ratio of bone to meat to fat varies from duck to duck.

17 Ballotine of duck leg with black pudding and apple

1 Bone out the duck leg as for ballotine (see page 278).

2 Combine all ingredients well, taking care not to overwork the mousse as splitting may occur.

3 Fill the duck cavity with the mousse and carefully wrap in foil, ensuring the mousse is well encased in the duck leg.

4 Place the leg in a pre-heated oven at 180°C for 25 minutes, remove and allow to rest for a further 5 minutes.

5 Remove from the foil and place in a hot pan with oil and cook until golden brown, rest for a further 3 minutes.

6 Slice 3 pieces and retain the drumstick. Serve (e.g. with braised red cabbage scented with orange and bay leaf, or with beetroot, carrot and leek).

Variation

This recipe can be used with a chicken ballotine, but the cooking time should be reduced by 5–10 minutes.

Ingredient	4 portions	10 portions
Large duck legs	4 × 200g	10 × 200g
Good-quality black pudding cut into ½cm dice	200g	500g
Dried apple chopped into ½cm dice	75g	185g
Chives, chopped	1 tsp	3 tsp
Chicken mousse (see recipe 4)	300g	750g

Energy	Cal	Fat	Sat fat	Carb	Sugar	Protein	Fibre
2529 kJ	604 kcal	34.3g	14.4g	19.2g	11.9g	56.1g	2.5g

18 Confit duck leg rillette

Ingredient	4 portions	10 portions
Duck confit legs, skin and bone removed	4 × 200g	10 × 200g
Shallots, cooked and chopped	200g	500g
Parsley, chopped	15g	40g
Cognac	15ml	35ml
Orange zest (optional)	1	3
Freshly ground black pepper		
Sea salt flakes		
Fat reserved from the confit	30g	75g

1   Place the duck confit meat in the oven to warm through, which will allow the food processor to break down the meat more easily.

2   Combine all the ingredients, except the fat, in the bowl of an electric mixer fitted with a dough hook. Beat at low speed for about 1 minute, or until everything is well mixed. Or use a food processor, taking care not to purée the mixture or let it turn into a paste. The texture should be like finely shredded meat.

3   Place in ramekins and press down lightly, ensuring an even top. Pour over the reserved confit fat ensuring that it covers the meat to form a seal when set.

4   Use immediately (if doing so, serve without topping off with the oil), or cling film well and store in the refrigerator for up to 1 week.

5   Serve with a chutney or a relish as an hors d'oeuvre (starter).

Mix together until the correct texture is achieved, as shown here

Place in ramekins and spoon fat over the meat

19 Roast goose with citrus fruits

Ingredient	4 portions	10 portions
Goose, oven-ready, crowned	1 (4 kg in weight)	2 (5 kg each)
Orange	1	3
Lemon	1	3
Lime	1	3
Salt and pepper		
Chicken stock	400ml	1 litre
Confit goose legs (see Recipe 15, Confit duck legs, for instructions)	2	4

This recipe makes use of the whole goose, with each muscle group cooked in the way that gives it the most flavour and best texture.

1 Pre-heat the oven to 210°C.

2 Lightly score the fat on the crown in a harlequin style at 3mm intervals, being careful not to penetrate the meat.

3 Zest and juice the oranges, lemons and limes and mix the zest and juice together. Season.

4 Place the shells of the fruit inside the goose carcass, ensuring it is tightly packed.

5 Place a cooking wire over a drip tray and place the goose on top. Pour the zest and juice mix over the breasts and rub it into the incisions in the fat.

6 Place the tray with the goose into the pre-heated oven.

7 Bring the stock to the boil. When the goose has been in the oven for 10 minutes, carefully open the oven door and pour the hot stock over the goose – this will supercharge the cooking, releasing more fat from the breasts. Close the oven door and turn it down to 175°C.

8 Roast for ¾ of an hour, basting at 20-minute intervals with the fat and stock in the roasting tray.

9 Add the confit legs to the oven and continue roasting and basting for 30 minutes.

10 Remove the legs and crown from the oven, placing the crown breast-side down to retain as much moisture as possible.

11 Remove the breasts and carve into portions. Include both breast and leg meat in every serving. Accompany with fondant potatoes, glazed roast carrots and roast gravy.

20 Roast turkey

Ingredient	
Turkey, with legs on	4–5 kg
Sea salt and freshly ground black pepper	
Unsalted butter, melted	625g

Energy	Cal	Fat	Sat fat	Carb	Sugar	Protein	Fibre
1076 kJ	257 kcal	9.6g	3.2g	0.0g	0.0g	42.0g	0.0g

1 Adjust an oven rack to its lowest position and remove the other racks in the oven. Pre-heat to 165°C. Remove turkey parts from neck and breast cavities and reserve for other uses, if desired. Dry bird well with paper towels, inside and out. Salt and pepper inside the breast cavity.

2 Set the bird on a roasting rack in a roasting pan, breast-side up, brush generously with half the butter and season with salt and pepper. Tent the bird with foil.

3 Roast the turkey for approximately 2 hours, depending on size. Remove the foil and baste with the remaining butter. Increase the oven temperature to 220°C and continue to roast until an instant-read thermometer registers 74°C in the thigh of the bird, about 45 minutes more.

4 Remove turkey from the oven and set aside to rest for 15 minutes before carving. Carve and serve with roast gravy, cranberry sauce, bread sauce, and either or both sausage meat and chestnut dressing and parsley and thyme dressing. Chipolata sausages and rolled rashers of grilled bacon may also be served.

Professional tip

The secret to keeping turkey moist is to baste as much as you can and, when the turkey is cooked, place it on its breast, breast-side down, allowing all the cavity juices to penetrate the meat.

Whole roasted turkeys may be required for presentation, especially for buffets. However, for bulk production, the legs may be removed before cooking, boned and stuffed. Cooking the legs and the crown (or just the breast) separately gives a greater yield and less waste.

21 Steamed and pan-fried turkey fillet wrapped in basil leaves and Parma ham, with chargrilled Mediterranean vegetables, mushroom ratatouille and tomato coulis

1 Trim and remove sinews from the turkey, wrap in basil leaves and Parma ham. Wrap in cling film and chill.

2 Prepare vegetables for chargrilling, cut into pieces 60 to 70cms in length, thickness of a finger. Brush with oil and seasoning.

3 Prepare vegetables for ratatouille, cut into pieces no bigger than 20cms, add trimmings from courgette, aubergine and red pepper. Heat olive oil and sweat vegetables adding a third of the tomatoes, season and add half of remaining basil leaves.

4 Sweat carrot, garlic and shallot in a pan with a little butter, add tomatoes and purée and cook down, season and add stock allow to simmer for 30 minutes. Remove from heat and strain, keep warm for service.

5 Blend the remaining basil leaves with a little olive oil for drizzling on the dish.

6 Chargrill the vegetables on a hot grill to form lines.

7 Steam (or poach) the turkey for 12–15 minutes, unwrap then fry in a hot pan to crisp the Parma ham.

8 To serve, place ratatouille in centre of plate, place chargrilled vegetables crossed on top. Slice turkey at an angle and place over vegetables. Drizzle plate with basil oil and serve tomato coulis over the turkey. Garnish with a deep-fried bay leaf.

Ingredient	4 portions	10 portions
Turkey fillets	4 × 150g	10 × 150g
Slices of Parma ham	4	10
Bacon trimmings	70g	175g
Unsalted butter	100g	250g
Olive oil	50ml	125ml
Basil	10g	25g
Chargrilled vegetables		
Courgettes	2	5
Aubergine	1	3
Red pepper	1	3
Ratatouille		
Mushrooms	100g	250g
Red pepper	1	3
Red onion	1	3
Tomato coulis		
Plum tomatoes	500g	1.25 kg
Carrot	1	3
Shallot	1	3
Clove garlic	1	3
Tomato purée	15g	40g
Vegetable stock	250ml	625ml

Energy	Cal	Fat	Sat fat	Carb	Sugar	Protein	Fibre
2687 kJ	646 kcal	42.2g	35.5g	16.3g	14.1g	51.2g	7.0g

Game

Origins

The word game is used for culinary purposes to describe animals or birds that are hunted for food, although many animals categorised as game are now being bred or domesticated (for example, squab (pigeon), ducks and venison). Wild animals, because of their diet and general lifestyle, contain a different set of enzymes than poultry. These are the enzymes that start to soften the tissue while game is hung, softening the proteins and adding the characteristics of the 'gamey' flavour.

Composition

Game is less fatty than poultry or meat and is easier to digest (although the flesh of wild waterfowl tends to be a little oily). When selecting game it is important to know its life age and how long it has hung for as this will determine the cooking method. With the larger game animals such as venison, a rule of thumb is to break carcasses down in the same way as for lamb.

Types of game

Game is split into two main categories: feathered and furred. All require some hanging before preparation. Normally birds are not plucked or drawn before hanging, whereas furred game is hung with the skin on but gutted.

Game is a seasonal product that has some restrictions on availability depending on the type, although some traditional game is now available for longer periods through modern farming and rearing practices.

Table 9.2 shows the seasons for game.

Quality points

- Venison joints should be well fleshed and have a dark brownish red colour.
- The ears of hares and rabbits will tear easily when they are younger.
- Game birds' breasts should be plump, legs smooth and beaks should break easily.
- If in feather, then the plumage should be soft and quill feathers should be pointed.

Table 9.2 Seasonality of game

	Jan	Feb	Mar	Apr	May	Jun	Jul	Aug	Sep	Oct	Nov	Dec
Furred												
Hare	Available	Available	Available	Available	Available	Available	Available	At best	At best	At best	At best	At best
Rabbit	Available	Available	Available	Available	Available	Available	Available	Available	Available	Available	Available	Available
Venison	Available	Available	Available	Available	Available	Available	Available	Available	Available	Available	Available	Available
Feathered												
Goose (wild)	Available								At best	At best	At best	At best
Goose (farmed)									At best	At best	At best	At best
Grouse								12th Available	Available	Available	Available	Available
Mallard (wild duck)	Available								Available	Available	Available	Available
Moorhen									Available	Available	Available	Available
Partridge (English grey leg)	At best								Available	Available	Available	Available
Partridge (French red leg)	At best								Available	Available	Available	Available
Pheasant	At best									At best	Available	Available
Pigeon (farmed)	Available	Available	Available	At best	At best	At best	At best	At best	At best	At best	Available	Available
Pigeon (English wood)	Available	Available	Available	Available	At best	At best	At best	At best	At best	At best	Available	Available
Quail	Available	Available	At best	At best	At best	At best	At best	At best	At best	Available	Available	Available
Snipe	At best							12th Available	Available	Available	Available	Available
Teal	Available								Available	Available	Available	Available
Woodcock	At best									At best	Available	At best

Key

Available ▮ At best ▯

Feathered game

Table 9.3 shows the uses and quality points for feathered game.

Table 9.3 Uses and quality points for feathered game

Bird	Quality points	Hanging time	Uses
Pheasant	Look for young birds: flexible beak, pliable breast bone, underdeveloped spurs or none. Last large feather in the wing is pointed		Roast, braise, pot roast
Partridge	Look for young birds: smooth legs, flexible beak, pliable breast bone, underdeveloped spurs or none	3 to 5 days	Roast, braise
Woodcock	Soft, supple feet; clean mouth and throat; fat and firm breast	3 to 4 days	Roast
Snipe			Roast or use in steak puddings or pies
Wild duck	Soft and pliable beak and feet		Roast (slightly underdone), braise
Teal			Roast, braise
Grouse	Look for young birds: soft, downy plumes on the breast and under the wing; pointed wings; rounded, soft spur knob (not hard and scaly)		Roast (slightly underdone) and serve hot or cold
Quail	Plump, with firm white fat		Stuff, roast, spit roast, casserole, poach in rich chicken/veal stock
Pigeon	Plump; mauve-red flesh; pinkish claws		

▲ Male and female pheasants, and one that has been prepared for cooking

▲ Red-legged and English partridges, shown whole and after preparation

▲ Grouse, before and after preparation

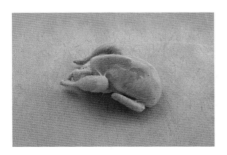

▲ A quail

▲ Wood pigeons, before and after preparation

Woodcock and snipe

Woodcock has a distinctive flavour, and leaving the entrails in during cooking will accentuate this. The vent must be checked carefully for cleanliness.

Snipe resemble woodcock but are smaller. The flavour can be accentuated in the same way.

▲ A snipe

Wild duck

The most common wild duck is the mallard, which is the ancestor of the domestic duck. Teal is a smaller species.

It is particularly important that water birds, like wild duck and teal, are only eaten in season; out of season the flesh is coarse and has a fishy flavour.

▲ A wild duck

Furred game

Hare

Hare should be used when young (2½ to 3 kg in weight). In a young animal it should be possible to take the ear between the fingers and tear it quite easily, and the hare lip, which is clearly marked in older animals, should be only faintly defined.

Hare should be hung for about a week before cleaning.

Rabbit

Rabbits are available wild or farm-reared. They have shorter ears, legs and bodies than hares.

Butchering a rabbit

1 Carefully remove the fur.
2 Cut an incision along the belly.
3 Remove the intestines.
4 Clean out the forequarter, removing all traces of blood.

Rabbits can be cut up in different ways depending on the size of rabbit and the dish, for example:
- The saddle can be removed and left whole for roasting, braising and pot roasting.
- The legs, forelegs, forequarter (well trimmed) and saddle may be cut into two pieces each (or three for a large saddle).
- The two nuts of meat can be removed from the saddle. Cut half through each one lengthwise and carefully beat out with a little cold water to form escalopes.

▲ A rabbit and a hare

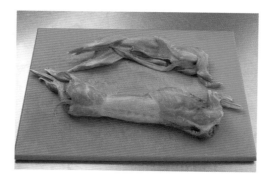

▲ Rabbit and hare after skinning

Venison

Venison is the meat of the red deer, fallow deer or roebuck. The meat of the roebuck is considered to have the best and most delicate eating quality.

Joints of venison should be well fleshed and a dark brownish-red colour.

After slaughter, carcasses should be hung well in a cool place for several days.

▲ A side of venison

The prime cuts are the legs, loins and best ends. Joints of venison are usually marinated before cooking. Methods of preparation include:

- Bone and roll the shoulder for roasting (only shoulders from young animals will be suitable).
- Cut chops from the loin and cutlets from the best end and trim well. Shallow fry, if the meat is tender, or braise.
- Cut steaks or escalopes from the boned-out nuts of meat from the loins. Trim well and thin slightly with a meat bat. Escalopes can be quickly shallow-fried.
- Cut up the shoulders of older animals for stewing or braising.

A spicy, peppery sauce is usually offered. De-glaze the cooking pan with stock, red wine, brandy, Madeira or sherry and use this in the sauce. Possible extra ingredients include cream, redcurrant jelly, cooked beetroot or sliced mushrooms.

Possible accompaniments for venison include a purée of lentils, dried beans or root vegetables, or braised red cabbage.

Wild boar

Wild boar are best slaughtered between 12 and 18 months old, weighing 60 kg to 75 kg on the hoof, and in late summer when the fat content is lower. The meat is hung for 7 to 10 days.

Buy from suppliers who use breeding stock as pure as possible and who allow the boars to roam freely and forage for food.

Prime cuts (leg, loin, best end) can be marinated and braised.

Young boar, slaughtered at up to 6 months, are sufficiently tender to cook as noisettes and cutlets or roast as joints.

Hanging game

Game bought through a main dealer will probably have been hung correctly prior to delivery to the kitchen. The general rules are to hang in a cool, dry, airy place, protected from flying insects to prevent infestation. In the UK it is common to hang birds from the head in pairs known as a **brace**, whereas rabbits, hares and other game are hung by the rear legs and the head hanging down.

Preparation techniques

Game is often purchased from specialist suppliers, and is usually supplied prepared for cooking.

Plucking and skinning

Plucking is the removal of the feathers. Most game is supplied plucked, or with the fur removed, to observe hygiene regulations and reduce the risk of contamination.

Some types of game may need to be skinned.

Boning, trussing and tying

Excess bones are usually removed from game before cooking. This makes carving easier.

Game birds are trussed like chickens, except woodcock and snipe, which are trussed using their beaks.

Game joints, like other meats, often need to be tied to improve their shape and make carving easier.

Cooking

Game meat responds best to roasting – young game birds in particular should be roasted and it is traditional to leave them unstuffed. Due to the low fat content, it is traditional to cover (or bard) game with fat or bacon to help keep moisture in during the cooking process.

Older or tougher cuts of game such as the haunch of venison should be casseroled or used in pies and terrines. Marinating in oil vinegar or wine with herbs and spices helps to make the meat more tender as well as enhance the flavour.

22 Game farce

Ingredient	4 portions	10 portions
Butter	50g	125g
Game livers	100g	250g
Onion, chopped	25g	60g
Thyme, sprig		
Bay leaf	1	2–3
Salt, pepper		

Health and safety !

Traditionally, the livers were kept underdone, but this is not advisable in a modern kitchen due to the risk of food poisoning, particularly *campylobacter*.

1 Heat half the butter in a frying pan.

2 Quickly toss the seasoned livers, onion and herbs in the butter, browning well but keeping underdone. Pass through a sieve or mincer.

3 Mix in the remaining butter. Correct the seasoning.

4 Place in an oven-proof dish and oven cook in a bain marie (a roasting tray half-filled with water) at 180°C for 25 minutes. Check with a probe that the centre has reached 75°C.

23 Roast pigeon with red chard, celeriac and treacle

Ingredient	4 portions	10 portions
Celeriac, peeled and cut into 2½ cm dice	1	2
Milk and water, to cook the celeriac		
Squab pigeon crowns (350–400g per crown)	4	10
Butter, to place in the squab cavity	50g	125g
Parsnip batons	12	30
Butter, to cook the parsnips	75g	185g
Vegetable stock	200ml	500ml
Reduced brown stock	100ml	250ml
Red chard, picked and washed	200g	500g
Black treacle	5g	15g

Energy	Cal	Fat	Sat fat	Carb	Sugar	Protein	Fibre
2735 kJ	655 kcal	40.6g	18.4g	6.2g	3.3g	66.9g	2.1g

1 In a saucepan, cook the chopped celeriac until soft in a mixture of half milk and half water.

2 Drain off all the liquid, then purée the celeriac in a food processor.

3 When smooth, place in a clean pan and return to the stove. Cook until thick, letting as much moisture evaporate as possible.

4 Preheat the oven to 180°C. Adjust the seasoning to taste.

5 Roast the squab on the bone for 3 minutes on its back and 3 minutes on its breast.

6 Remove and rest with the butter evenly placed in the cavity to add extra moisture to the bird.

7 Cook the parsnips in butter and vegetable stock, place to one side and keep warm.

8 Wilt the chard, season and drain.

9 Warm the celeriac purée and place on the plate with the chard and the parsnip batons.

10 Carefully remove the two breasts from the bone, place on the chard and liberally drizzle the sauce over the breasts and the plate.

11 If red chard is unavailable you can substitute with spinach or bok choi.

12 Reduce brown stock by half again, add treacle sauce and plate.

Chicken was used for the saturated fat analysis.

24 Pot au feu of pigeon

Ingredient	4 portions	10 portions
Squabs, legs removed	6 (approx. 2 kg)	10 (approx. 3.5 kg)
Carrot, peeled and cut into 4 laterally	1	3
Celery stick, cut into 4	1	3
Baby turnips	8	20
Medium turnips, peeled and blanched	2	4
Leek, washed, cut in rounds	1	3
Small shallots, peeled and left whole	8	20
Smoked streaky bacon, rind removed	100g	250g
Chicken/game stock	400ml	1 litre
Bouquet garni with 4 black peppercorns and 1 clove of garlic	1	3
Salt		

Energy	Cal	Fat	Sat fat	Carb	Sugar	Protein	Fibre
1545 KJ	367 kcal	11.2g	3.44g	11.12g	9.98g	56.2g	4.79g

1 Place the squab legs in a large casserole and arrange the vegetables tightly in one layer with the legs, add the bacon, then cover with the stock to about 4cm above the ingredients (you may need to top this up with water).

2 Add the muslin-wrapped bouquet garni. Season with salt and bring to a gentle boil.

3 Skim off any impurities, cover with a lid (leaving a small gap) and simmer gently for 50 minutes.

4 Skim off any fat or impurities, then add the squab breasts and cook for a further 12 minutes.

25 Sauté of pigeon salad with rocket, Parmesan and beetroot

Ingredient	4 portions	10 portions
Squab crowns (350–400g each)	4	10
Oil	50ml	125ml
Pigeon/game glaze (reduced stock from pigeon bones)	50ml	125ml
Balsamic dressing (see page 186)	50ml	125ml
Salt and pepper		
Cooked beetroot batons	100g	250g
Mixed salad leaves	200g	500g
Rocket, washed and picked	250g	600g
Shaved Parmesan	50g	125g

Energy	Cal	Fat	Sat fat	Carb	Sugar	Protein	Fibre
2059 kJ	492 kcal	27.7g	6.3g	5.0g	4.8g	56.2g	1.5g

1 Remove the breasts from the crowns and cut each breast into three.

2 Heat the oil in a pan, place in the squab and sauté for 2–3 minutes with a little colour, leaving the meat pink.

3 Remove from the pan and place on a draining wire in a warm place. Add the squab glaze and dressing to the pan, removing all sediment.

4 Season the squab before presenting on the plate.

5 Pour into a small dish and allow to cool slightly. Arrange the squab and beetroot in the centre of a plate, dress the mixed leaves, rocket and Parmesan and place carefully on the squab. Finish the dish with the remaining jus/dressing.

Professional tip

This is a suitable summer dish as the squabs could even be chargrilled or cooked on a barbecue for greater depth of flavour. Alternatively, this dish would be just as effective with partridge.

26 Roast breast of pheasant with vanilla and pear

Ingredient	4 portions	10 portions
Cold water	250ml	625ml
Caster sugar	25g	70g
Salt	25g	70g
Large pheasant breasts, removed from the bone with skin and wing tip attached	4	10
Unsalted butter	70g	175g
Shallots, diced	4	9
Large vanilla beans, scraped to gather the seeds	1	3
Dry white wine, preferably Chardonnay	150ml	375ml
Pear/apple cider	300ml	750ml
Whipping cream	300ml	750ml
Preserved ginger, finely chopped	15g	35g
Freshly ground black pepper		
Salt		
Dry red wine	250ml	625ml
Honey	75g	185g
Coriander seeds, toasted and finely crushed	10g	25g
Pears (preferably red), halved, cored and sliced into 1cm thick slices	2	5
Caster sugar		
Sprigs of chervil		

Energy	Cal	Fat	Sat fat	Carb	Sugar	Protein	Fibre
2830 kJ	680 kcal	47.1g	24.5g	27.9g	27.2g	39.0g	2.2g

To cure the pheasant

1 In a large bowl combine the water, 25g/62.5g each of sugar and salt, mixing to dissolve.

2 Add the pheasant breasts, cover with plastic wrap and refrigerate overnight.

To make the sauces

1 In a large sauté pan, heat 1 or 2½ tablespoons (depending on portion size) of the butter over medium to medium-high heat. Add the shallots, cooking until tender (about 3 minutes). Add the vanilla bean and seeds, white wine, cider and cream. Bring to a simmer, cooking until the liquids are reduced and thickened to sauce consistency (about 10 minutes). Add the ginger, season with salt and pepper, keep warm.

2 In another saucepan, combine the red wine and the honey. Bring to a simmer over high heat, cooking until reduced to coat the back of a spoon (about 15 minutes). Reserve.

To cook the pheasants

1 Pre-heat the oven to 190°C.

2 In a large sauté pan, add 2 or 5 tablespoons butter and melt over a high heat. Season the pheasant with salt, pepper and coriander. Add the pheasant skin-side down, cooking until browned and well seared (about 5 minutes). Turn over and transfer the pan to the lower rack of the oven. Cook until just about medium (about 6 to 8 minutes, depending on the size of the pheasant breast). Carefully remove the hot pan from the oven. Allow to rest for a couple of minutes before cutting.

3 In another large pan, heat 1 or 2½ tablespoons of the butter over high heat. Add the pears, cooking until they just begin to soften slightly (about 2 minutes) Add 2 or 5 tablespoons of sugar, cooking and stirring occasionally until browned on the edges (about 4 minutes). Remove the pears and keep warm.

To serve

1 Position the pear slices in the centre of the plate. Slice the pheasant on a bias to yield 4 or 6 thin broad slices. Spoon the vanilla, ginger and red wine sauce over and around the pheasant and on the plate. Garnish with the vanilla bean and sprigs of chervil over the pheasant. Serve immediately.

Variations
This dish will work equally well with corn-fed chicken or guinea fowl. Again, as for Recipe 16, vanilla extract can be used instead of vanilla pods.

27 Poached pheasant crown with chestnuts and cabbage

1 Bring the stock up to a simmer and place the crowns in it; simmer for 11 minutes.

2 Meanwhile, place the butter in a medium pan on a moderate heat and melt. Add the cabbage and bacon lardons and 100 ml of reduced chicken stock.

3 When the cabbage and bacon have formed an emulsion with the butter and stock, put to one side and retain.

4 Remove the pheasant crowns and rest for 2–3 minutes breast-side down.

5 Remove the breasts from the crown (see page 277) and check if cooked; if not quite cooked place back in the cooking liquor for 1–2 minutes. Then place the cooked breasts in a serving dish and keep warm.

6 Add the cooked chestnuts to the remaining reduced chicken stock and bring to the boil, adding the cream. At this point add the chopped parsley and check the seasoning – if the sauce appears to be rich this can be modified with a little lemon juice. You should not be able to taste the lemon but using it allows the deep flavours of the sauce to come through.

7 Place the cabbage mix in the centre of the plate and top this with the pheasant; finish with the sauce over and around, ensuring an even distribution of chestnuts.

Ingredient	4 portions	10 portions
Chicken stock	2 litres	5 litres
Pheasant crowns	2	5
Butter	50g	125g
Savoy cabbages, cored, shredded and blanched	2	5
Cooked bacon lardons	75g	187g
Reduced chicken stock	250ml	625ml
Cooked chestnuts	175g	437g
Double cream	50ml	125ml
Parsley, chopped	5g	12.5g
Salt and pepper		
Lemons, juice of	1	2

Energy	Cal	Fat	Sat fat	Carb	Sugar	Protein	Fibre
2417 kJ	581 kcal	37.4g	17.5g	22.7g	9.4g	39.7g	5.7g

28 Roast partridge

Ingredient	4 portions	10 portions
Grey-legged partridges (approx. 400g each), oven-ready, with livers	4	10
Unsalted butter	20g	50g
Groundnut oil or vegetable oil	20g	50g
Salt		
Freshly ground pepper		
Roasting juices from veal, pork or beef	70ml	175ml
Water	50ml	125ml

Energy	Cal	Fat	Sat fat	Carb	Sugar	Protein	Fibre
2495 kJ	596 kcal	26.7g	7.6g	0.3g	0.1g	88.5g	0.0g

1 Shorten the wings and sear the partridges under a flame to remove the feather stubs. Remove any trace of gall from the liver. Wash briefly, pat dry and reserve.

2 In a roasting tray sear the wing bones and the partridges in butter and oil for 2 minutes on each side and 2 minutes on the breast (6 minutes in total) until they are brown.

3 Season with salt and pepper, and roast in the pre-heated oven for 5–6 minutes, according to the size of the partridges.

4 Remove from the oven and place the partridges on a cooling wire, breast-side down, cover loosely with aluminium foil and allow to rest.

5 In the same roasting tray, fry the livers in the remaining butter and oil for 1 minute, and reserve. Spoon out the excess fat and add the roasting juices and water to the winglets, bring to the boil then simmer for 5 minutes.

6 Taste and season with salt and pepper, then strain through a fine sieve.

7 Serve with roasted seasonal vegetables (e.g. roast parsnips, carrots and braised cabbage). Suggested accompaniments: bread sauce, roast gravy, watercress and game chips. The fried livers may be puréed and served on croutes of toasted bread as an accompaniment.

29 Braised partridge with cabbage

Ingredient	4 portions	10 portions
Old partridges	2	5
Lard, butter, margarine or oil	100g	250g
Savoy cabbage	400g	1.25 kg
Belly of pork or bacon (in the piece)	100g	250g
Carrot, peeled and grooved	1	2–3
Studded onion	1	2–3
Bouquet garni		
White stock	1 litre	2.5 litres
Frankfurter sausages or pork chipolatas	8	20

Energy	Cal	Fat	Sat fat	Carb	Sugar	Protein	Fibre
3137 kcal	750 kJ	57.6g	14.0g	11.5g	6.2g	47.8g	1.2g

Older or red-legged partridges are suitable for this recipe.

1 Season the partridges, rub with the fat, brown quickly in a hot oven and remove.

2 Trim the cabbage, remove the core, separate the leaves and wash thoroughly.

3 Blanch the cabbage leaves and the belly of pork for 5 minutes. Refresh and drain well to remove all water. Remove the rind from the pork.

4 Place half the cabbage in a deep ovenproof dish; add the pork rind, the partridges, carrot, onion, bouquet garni, the remaining fat, and stock, and season lightly.

5 Add the remaining cabbage and bring to the boil, cover with greased greaseproof paper or foil and a lid, and braise slowly until tender, approximately 1–2 hours.

6 Add the sausages halfway through the cooking time by placing them under the cabbage.

7 Remove the bouquet garni and onion, and serve everything else, the pork and carrot being sliced.

Nutritional data is based on 100g cooked meat and oil.

30 Quail with pomegranate and orange

1 Preheat the oven to 180°C.

2 Heat a medium saucepan and add the oil and butter.

3 Season the quails and place in the pan, backs down, colour evenly all over, place in a roasting tray and cook for 6 minutes.

4 Meanwhile, segment the oranges and set aside.

5 Remove the seeds from the pomegranate and also set aside.

6 When cooked, remove the quails from the oven and rest for 2–3 minutes.

7 Remove the quail breasts from the bone and slice into 4 pieces.

8 On the centre of the plate combine the dressed leaves, beans and oranges with a dessertspoon of vinaigrette.

9 Place the quail around the leaves, finish with the pomegranate seeds and serve immediately.

Partridge was used in place of quail for the saturated fat analysis.

Ingredient	4 portions	10 portions
Oil	10ml	25ml
Butter	50g	125g
Salt and pepper		
Oven-ready quails (approx. 75g each)	8	20
Oranges	2	5
Pomegranate	1	2
Mixed leaves	200g	650g
Split cooked green beans	100g	250g
Vinaigrette		

Energy	Cal	Fat	Sat fat	Carb	Sugar	Protein	Fibre
1118 kJ	267 kcal	13.3g	2.6g	8.24g	8.0g	29.2g	2.5g

31 Pot-roasted quail with roast carrots and mashed potato

Ingredients	4 portions	10 portions
Quails (approx. 75g each)	8	20
Pancetta, thinly sliced	12 (approx. 250g)	30 (approx. 625g)
Fresh sage leaves	12	30
Vegetable oil	15ml	50ml
Butter	20g	50g
Salt and freshly ground pepper		
Carrots, peeled and cut into quarters	3 (100g)	8 (250g)
Oil	30ml	80ml
Dry white wine	125ml	310ml
Red wine	125ml	310ml
Brown stock	250ml	625ml
Unsalted butter	50g	125g
Portions of mashed potato	4 (200g)	10 (500g)
Chives, chopped	1 tsp	2 tsp

Energy	Cal	Fat	Sat fat	Carb	Sugar	Protein	Fibre
2839 kJ	680 kcal	41.89g	16.61g	9.99g	2.47g	66.1g	1.15g

1 Wash the quails thoroughly inside and out, then place them in a large colander to drain for at least 20 minutes; pat the quail dry.

2 Stuff the cavity of each bird with 1 slice of pancetta and 1 sage leaf.

3 Put the oil in a large thick-bottomed roasting pan on high heat. When the fat is hot, add all the quails in a single layer and cook until browned on one side, gradually turning them, and continue cooking until they are evenly browned all over.

4 Lightly sprinkle the quails with salt and pepper, then add the carrots and cook for a couple of minutes until a slight colour appears.

5 Add the wine and brown stock then turn the birds once, let the wine bubble for about 1 minute, then lower the heat to moderate and partially cover the pan. Cook the quail until the meat feels very tender when poked with a fork and comes away from the bone (approximately 35 minutes).

6 Check from time to time that there are sufficient juices in the pan to keep the birds from sticking; if this does occur, add 1 to 2 tablespoons of water at a time. When the quails are done, transfer them to a warmed tray and reserve.

7 Turn up the heat and reduce the cooking juices to a glaze – enough to coat all the birds, scraping the bottom of the pan with a spoon to loosen any cooking residues.

8 Add the butter and whisk in to form an emulsion; at this point, if the sauce splits or is too thick, add a little water and reboil.

9 Remove the carrots from the pan and place neatly on the plate with the potato purée.

10 Pass the juices, then pour them over the quail, sprinkle with chopped chives and serve immediately.

Partridge was used in place of quail for the saturated fat analysis, and streaky bacon in place of pancetta throughout.

The quail may be boned out and some suitable stuffing added, before cooking.

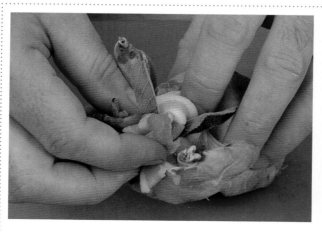

Stuff the quail

Add the stock

32 Rabbit saddle stuffed with liver

Ingredient	4 portions	10 portions
Saddle		
Long saddles of rabbit (approx. 400g each), boned, livers retained	4	10
Spinach leaves	100g	250g
Thin slices Parma ham	9	23
Lentil sauce		
Brown chicken stock	300ml	750ml
Cold butter, diced	40g	100g
Plum tomatoes, cut into concassé	3	8
Cooked Umbrian or Puy lentils	30g	75g
Chives, chopped	1 tsp	4 tsp
Sherry vinegar, to taste		
To serve		
Spinach, picked, washed and wilted	300g	750g

Energy	Cal	Fat	Sat fat	Carb	Sugar	Protein	Fibre
1393 kJ	333 kcal	18.2g	2.7g	6.6g	2.5g	36.0g	2.7g

For the rabbit

1 Split the two natural halves of the rabbit liver with a sharp knife.

2 Keeping them as whole as possible, remove any sinew. Wrap the livers in the spinach leaves.

3 Trim off any excess fat from the belly flaps of the two loins of the boned saddle.

4 Lay the spinach-wrapped livers in the cavity of the two loins of the boned saddle.

5 Lay 2 pieces of the Parma ham on a 30cm-square sheet of kitchen foil and place the rabbit saddle with the livers facing upwards on the ham.

6 Roll into a sausage shape, twisting the ends of the foil to ensure a tight parcel.

7 Repeat with the remaining rabbits and place in the fridge overnight to rest.

For the lentil sauce

1 Place the chicken stock in a saucepan, bring to the boil and simmer until the volume has reduced to 150ml.

2 Add the butter to the reduced stock and whisk to an emulsion, add the tomatoes, lentils and chives, and adjust the seasoning to taste. Finish with the sherry vinegar so you can just taste the acid in the background.

To complete

1 Preheat the oven to 180°C. Place the rabbit saddles on a baking tray and roast for 17 minutes, turning after 10 minutes.

2 Remove the rabbit from the oven and rest for 3–4 minutes in the foil.

3 Remove the foil from the rabbit and slice each saddle into 4 equal pieces.

4 Lay the sliced rabbit on wilted spinach and garnish the dish with the lentil sauce. Serve immediately.

33 Braised baron of rabbit with olives and tomatoes

Ingredient		4 portions	10 portions
Farm-raised rabbit barons (approx. 750–800g each), including bones and trim for gravy		2	5
Carrot		1	3
Onion		1	3
Celery stick		1	3
Olive oil		90ml	225ml
Balsamic vinegar		15ml	40ml
Caster sugar		10g	25g
Dry white wine		750ml	1875ml
Butter, to brown the meat		125ml	310ml
Basil leaves	'Mediterranean influence'	12	30
Black olives		32	80
Sun-dried tomato pieces		8	20
Salt and pepper			

Energy	Cal	Fat	Sat fat	Carb	Sugar	Protein	Fibre
2365 kJ	568 kcal	40.98g	8.42g	5.47g	4.75g	44.6g	0.87g

1 Well ahead of time, prepare the rabbits; cut the rabbits across at the point where the ribs end, and chop the forequarters into small pieces.

2 Cut the vegetables into a mirepoix.

3 In a large saucepan, brown the bones and mirepoix in 2 tablespoons (or 5 tablespoons, depending on portion size) of the olive oil.

4 Add the vinegar and sugar, and toss to coat. Cook until light brown.

5 Pour over almost all the white wine, reserving about 150ml (or 375ml, depending on portion size) for deglazing the roasting pan later. Boil hard to reduce until the liquid has a syrupy consistency.

6 Just cover with cold water, return to the boil and skim.

7 Simmer for 1½ hours.

8 Pass the resulting stock into a bowl, wash out the saucepan and return the stock to it.

9 Bring back to the boil, skim again and, once more, return to a slow simmer until reduced by half. Reserve.

10 Cut the rabbit legs into two (thigh and drumstick), and the rack into two.

11 In a large saucepan, brown the meat cuts in foaming butter; when golden brown, add the finished stock and cook for a further 1½ hours (approximately).

12 When cooked, pass the sauce through a fine strainer, bring to the boil and reduce by half.

13 While reducing, put the rabbit into a casserole/serving dish, julienne the basil and remove the stones from the olives.

14 When the sauce is reduced by half, add the tomatoes, olives and basil. Correct the seasoning, pour over the rabbit and serve.

Note

A baron is the rear end of the rabbit – the saddle and the two hind legs.

Variation

The 'Mediterranean influence' can be omitted, substituted with a British theme – woodland mushrooms, parsnips – and served with braised cabbage.

34 Jugged hare

1 Remove as much blood as possible from the cavity (the game supplier should have retained the blood once the hare had been killed and butchered). Reserve the blood in the fridge well covered.

2 Remove the legs front and back, trim off the rib cage and cut the long saddle into 4 pieces.

3 Mix all the marinade ingredients together and boil for 5 minutes. Allow to cool to room temperature.

4 Once room temperature is achieved, pour the marinade over the hare pieces and place in the fridge for 24 hours.

5 Pre-heat the oven to 125°C, drain the rabbit pieces out of the marinade and separate from the mirepoix.

6 Place a little of the rendered duck fat or lard in a thick-bottomed pan and brown the hare pieces well. Place in a casserole along with the vegetable mirepoix from the marinade and the flour and cook for 2 minutes.

7 Add the marinating liquid and the rest of the ingredients (redcurrant jelly, seasoning), mix well and bring to the boil. Place in the oven with a tight-fitting lid for 3 hours until tender.

8 When cooked remove from the oven and strain the cooking liquor, placing the cooked hare in a serving dish. Discard the vegetables and bring the liquor up to the boil, simmer and skim if necessary.

Ingredient	4 portions	10 portions
Hare		
Hare (approx. 2½ kg each), skinned, whole	1	3
Rendered duck fat/lard	25g	65g
Plain flour	20g	50g
Cold water	400ml	1 litre
Redcurrant jelly	2 tsp	5 tsp
Salt and pepper		
Marinade		
Red wine	750ml	1875ml
Mirepoix	250g	625g
Black peppercorns	10	25
Bay leaves	2	53
Juniper berries, crushed	3	8

Energy	Cal	Fat	Sat fat	Carb	Sugar	Protein	Fibre
1029 kJ	246 kcal	10.3g	5.7g	6.2g	1.7g	32.3g	0.2g

9 Take the retained blood and mix with a little red wine vinegar and water so that it is less solidified. Add to the sauce and heat but do not boil as this will split the mix.

10 Pour over the hare pieces and serve with mashed potato and heirloom carrots or other seasonal winter vegetables.

> **Professional tip**
>
> Traditionally, fried heart-shaped croutons, with the points dipped in the sauce, and chopped parsley are used as a garnish. Once the *liaison au sang* (blood thickening) is added to the sauce, it must not be boiled as this will coagulate the blood and the sauce will appear to be split.

35 Pot roast rack of venison with buttered greens and Merlot sauce

Ingredient	4 portions	10 portions
Venison		
Venison rack, bones cleaned and trimmed	2kg	5kg
Vegetable oil	50ml	125ml
Small mirepoix	250g	625g
Clarified butter	100g	250g
Garlic cloves	2	5
Sprig of thyme	1	3
Salt and pepper		
Merlot sauce		
Oil	100g	200g
Venison trimmings and bones	450g	1kg
Cracked pepper	½ tsp	1½ tsp
Bay leaf	1	3
Sprig thyme	1	2
Carrot, peeled and roughly chopped	1	3
Onion, peeled and roughly chopped	0.5	2
Garlic cloves, split	2	5
Merlot wine	330ml	800ml
Chicken stock	1.5 litres	3.75 litres
Greens		
Butter	50g	125g
Spinach leaves	250g	500g
Spring cabbage, cut into 1½cm strips (blanched for 2 minutes and refreshed in an ice bath)	1	2
Escarole, stalks removed	1	3

For the venison

1 Pre-heat the oven to 180°C. Trim the venison so that the bones rise 5cm above the meat. Tie the venison with kitchen string at intervals along the joint, tying 3 pieces of string between each bone.

2 Place a large pan with a tight-fitting lid over a high heat, add the oil and seal the venison, turning until it is a light golden colour all over.

3 Transfer the meat to a tray and cook the mirepoix in the same way.

4 Add the venison back to the pan, placing it on top of the mirepoix, and cover in the clarified butter. Add the garlic, thyme and seasoning. Place in the oven for 20–25 minutes until medium rare (the residual heat will cook it further).

5 Remove from the oven and set aside to rest for 10 minutes.

Energy	Cal	Fat	Sat fat	Carb	Sugar	Protein	Fibre
2984 kJ	717 kcal	50.0g	23.2g	10.4g	9.2g	60.2g	5.4g

For the Merlot sauce

1 In a large saucepan, heat the oil over a medium heat. Working in batches to prevent steaming and give a good colour, cook the venison trimmings until brown, add the cracked pepper, bay leaf and thyme. Remove from the pan and set aside in a bowl.

2 Reduce the heat and, in the same pan, cook the carrots, onion and garlic for about 10 minutes or until brown. Return the meat to the pan and stir well.

3 Raise the heat and, when the pan is quite hot, add the wine. Bring to the boil.

4 Boil rapidly until the volume of liquid has reduced by half. Add the chicken stock, return to the boil, then lower the heat right down and cook the sauce for about 1 hour.

5 Stir every 10 minutes to prevent sticking and skim off any sediment that rises to the surface.

6 When the sauce has reduced to approximately 400ml (for 10 portions, 1 litre), pour it through a fine strainer into a clean pan.

7 Bring to the boil and simmer until the volume of liquid has reduced to 200ml/500ml, giving a rich plum-coloured sauce.

To complete

1 Add the butter to a pan, place in the greens and heat through, warm the sauce and then slice the venison.

2 Lay the buttered greens in the centre of the plate with the venison on top and finish with the Merlot sauce.

> **Professional tip**
> The Merlot sauce can be used for other rich meat dishes. It can be made two or three days ahead and stored in an airtight container in the refrigerator.

36 Medallions of venison with red wine, walnuts and chocolate

Ingredient	4 portions	10 portions
Vegetable oil	50ml	125ml
Trimmed venison loin	750g	1.8 kg
Salt and pepper		
Merlot sauce (see Recipe 35)	200ml	500ml
Broken walnut pieces	40g	100g
70 per cent bitter chocolate in small broken pieces	50g	125g

Energy	Cal	Fat	Sat fat	Carb	Sugar	Protein	Fibre
1834 kJ	439 kcal	27.0g	8.2g	8.4g	8.2g	43.7g	0.7g

1 Pre-heat the oven to 190°C. Heat the oil in a heavy frying pan and seal the venison loin, add salt and pepper, place on a baking tray and put in the oven, and cook for 6–8 minutes medium rare or according to taste.

2 Meanwhile, warm the sauce and add the broken walnuts.

3 Slice the venison loin equally, nap over the sauce and sprinkle liberally with chocolate pieces.

This method of sprinkling the chocolate on afterwards allows you to taste the sauce and the chocolate separately.

37 Roast grouse

Ingredient	4 portions	10 portions
Grouse		
Young grouse (approx. 750g each), wings removed	4	10
Hearts and livers from the grouse		
Sage leaves	4	10
Butter	20g	50g
Salt and pepper		
Stock		
Oil	30ml	80ml
Grouse wings		
Giblets		
Large onion	1	2
Celery stick	1	3
Carrot	1	3
Red wine	250ml	625ml
Water		
Thyme, sprig of		
Bay leaf	1	3
Toast		
Small slices of bread	16	40
Duck fat		
To serve		
Bread sauce		
Watercress, bunches	1	2
Game chips		

Energy	Cal	Fat	Sat fat	Carb	Sugar	Protein	Fibre
3874 kJ	916 kcal	23.9g	8.6g	104.2g	7.5g	77.9g	3.5g

1 Trim the wings from the young grouse, draw and reserve the livers, hearts and giblets. Season birds liberally inside and out, and put the livers and hearts back in with a sage leaf and a knob of butter. Pre-heat the oven to 210°C.

2 Brown the wings and the remaining giblets with the diced onion, celery and carrot. Deglaze pan with red wine, cover with water and simmer for 30 minutes with the thyme and bay leaf. Strain and reserve.

3 Meanwhile, fry the bread in duck fat. Make bread sauce and pick through watercress.

4 Seal the birds in a little hot oil until golden. Then place the birds in the oven and roast. Cook until a probe inserted between the leg and the cavity reads 75°C, or until the juices run clear.

5 Remove from the pan and leave the birds to rest for 10 minutes.

7 Put the roasting tray over a flame, add a splash of brandy and the stock, and allow to bubble down to a thin gravy.

8 To serve, cook the livers and hearts, mash these up and serve on the toast. Serve with roasted vegetables, game chips, bread sauce, gravy and a watercress garnish.

Professional tip

Good-quality grouse that has not been over-hung is essential for this dish. The meat will already have a strong pungent flavour embedded in it from the birds' diet of heather.

Test yourself

Poultry

1 State the different types of chicken and give a dish recommendation for at least two types.
2 List the quality points that you would check when you take delivery of fresh chicken.
3 Describe the process for preparing duck legs for confit.
4 List at least four factors that have an effect on the quality and composition of poultry.
5 What are the risks associated with poor defrosting process for poultry?
6 Explain the term 'sauté' in relation to cooking chicken.
7 List the potential health benefits of including poultry in the diet.
8 State the potential hazards that may be increased from undercooked poultry.

Game

9 In preparing a game bird for roasting, what techniques are used to keep the meat moist?
10 List four preservation methods that could be used for poultry or game and give an example of each.
11 Explain the main benefit gained from hanging game prior to cooking.
12 List five game birds that are common on menus in the UK.
13 State two ways in which fat may be added to game dishes and give an example of a dish incorporating the technique.
14 What is the significance of 'The Glorious Twelfth' in the UK game season?

10 Advanced fish and shellfish dishes

This chapter covers the following units:

NVQ:
→ Prepare fish for complex dishes.
→ Prepare shellfish for complex dishes.
→ Cook and finish complex fish dishes.
→ Cook and finish complex shellfish dishes.

VRQ:
→ Advanced skills and techniques in producing fish and shellfish dishes.

Introduction

Fish and shellfish are among the most important, useful and interesting range of ingredients available to chefs. Their types are wide-ranging and the diversity is extended further by seasonality and the effects that weather conditions can have on sea fishing. Consideration must also be given to the possibilities of over-fishing and the effect that has on fish stocks and the sustainability of fish.

Learning objectives

By the end of this chapter you should be able to:
→ List the classifications of different types of fish and shellfish.
→ Describe the quality points for a range of fish and shellfish.
→ Describe the ways that fish are caught and kept in best condition.
→ Discuss ways of sustaining fish stocks and use of fish farming.
→ Identify the correct storage requirements for fish and shellfish.
→ Describe the range of products available after filleting.
→ Explain the effects that cooking has on fish, including how the composition of different fish and shellfish affects the cooking method.
→ Compare the effects of different preservation methods for fish and shellfish.
→ Select, prepare and cook a range of fish and shellfish by different methods.

Recipes included in this chapter

Fish

Fish are vertebrates, which means they have a spine. They are divided into two primary groups: flat and round. They can then be categorised further into:

- **Pelagic** fish – swim in mid-depth or shallow waters. They are usually round, oily fish such as mackerel and herring.
- **Demersal** fish – live in deeper waters and tend to feed from the bottom of the sea. They are almost always white fish and can be round such as cod, whiting and haddock or flat such as plaice, turbot or sole.

Table 10.1 Some popular flat fish and their uses

Fish	Size	Description	Uses
Halibut	Up to 3m long; 20–50 kg	Long and narrow; brown with some darker mottling on the upper side; superior flavour	Whole, in fillets or in tronçons Poach; grill; steam; fry; smoke
Dover sole		Oval; rough skin; one dark and one light side; superior texture and flavour to lemon sole	Whole or in fillets Poach; grill; steam; fry
Lemon sole		Similar to Dover sole but the dark side is yellow/brown and the flesh is softer; the price tends to be lower	
Plaice		White on one side; grey/brown with orange spots on the other	Whole or in fillets Poach; grill; steam; fry (popular as deep-fried goujons)
Turbot	3.5–4 kg	Large; diamond shaped; dark, knobbly skin	Whole, in fillets or in tronçons Poach; grill; steam; pan fry

▲ Halibut

▲ Dover sole

▲ Lemon sole

▲ Plaice

▲ Turbot

Table 10.2 Some popular round fish and their uses

Fish	Size	Description	Uses
Bass	Typically 30cm long, but may grow to 60cm	Silver/grey back and white belly; smaller fish may have black spots; soft white flesh; good flavour; best when very fresh; widely available farmed, or wild bass are available at a higher cost	Whole or in fillets Stuff and bake; poach; grill; steam; pan fry
Grey mullet	Typically 30cm long; 500g	Scaly, streamlined body; colour ranges from silver/grey to blue/green; deep sea or offshore mullet has firm, moist flesh with a fine flavour	Whole, in fillets or in darnes Stuff and bake; grill; pan fry; poach; steam
Cod	Typically 100–150cm long; 3–10 kg	The most popular white fish in the UK; sustainable stocks off Scotland and Iceland but there is a need for caution about over-fishing	Traditional fish and chips; steam; poach; pan fry; grill; bake; fish cakes; fish pie
Grouper	Typically fished at 2–10 kg	Varieties include brown, brown spotted, golden and red speckled; part of the sea bass family; pale pink flesh changes to off-white when cooked; pleasant, mild flavour	Similar to cod
Haddock	Typically 0.5–2 kg	Looks like cod but with a 'thumb mark' on the side	Similar to cod; very popular when smoked

Table 10.3 Some popular oily fish and their uses

Fish	Description	Uses
Tuna	Dark red flesh turns lighter when cooked; firm texture; good flavour; dries out if overcooked so usually served 'medium rare'; often canned	Pan fry; grill
Sardines and pilchards	Small fish of the herring family; often canned	Grill or pan fry fresh sardines
Mackerel	Torpedo-shaped; blue/grey skin; deteriorate quickly so use when fresh	Whole or in fillets Pan fry; grill; smoke; pickle
Herring	9–16cm long; rich in Omega-3 fatty acids; stocks are suffering from over-fishing; use from sustainable sources	Whole or in fillets Poach; fry; grill; rollmops (pickled, filleted and rolled around pickled cucumber)
Eels	Live in fresh water, including UK rivers and fish farms; hundreds of species, most common are conger eel and common eel; up to 1 metre long; usually kept alive until needed	Fry; grill; steam; stew; smoke
Salmon	Very popular; cheaper farmed than wild; anadromous (born in fresh water but migrating to the sea and back); often canned	Many different methods; fish cakes; fish pie

▲ Cod

▲ Sardine

▲ Herring

▲ Haddock (smoked)

▲ Mackerel

▲ Salmon

Table 10.4 Examples of exotic fish

Fish	Description	Uses
Barramundi	Native to Australia; freshwater fish; mature quickly, so very suitable for farming; firm, white flesh; mild flavour	Grill; pan fry
Grouper (hammour)	Often over 1 metre long; popular in Australasia and Asia; stocks are very depleted	Any cooking method used for fish
Tilapia	Found in fresh and seawater; often farmed; thick scales; firm, white flesh	Whole or in fillets Pan fry; grill; deep fry
Parrot fish	Brightly coloured skin; soft, delicate flesh; found in warm waters; usually sold whole and ungutted; scale before use, but skin can be left on	Larger fish can be filleted Grill; pan fry; bake; barbecue; well-suited to Asian flavours such as chilli and lemongrass

Freshwater fish

Fish from freshwater rivers and lakes also make a significant contribution to fish stocks and can add variety to menus. Farming of freshwater fish has made them more readily available and has lowered the price. Some popular freshwater fish are listed in Table 10.5.

Table 10.5 Some popular freshwater fish and their uses

Fish	Description	Uses
Zander (sander, sandre)	Very popular in France but stocks have dwindled; firm, fine-textured, white flesh; delicate flavour	Pan fry; grill; poach; steam; serve with beurre blanc
Trout	Usually a freshwater fish but some varieties are anadramous (see Table 10.3); rainbow trout is native to the USA, brown trout to the British Isles; often farmed because easy to breed	Grill; pan fry; steam; poach; smoke
Bream	Freshwater bream are often farmed, and relatively inexpensive; sea bream is considered superior; mild flavour	Grill; bake; fry; steam
Pike	Long, green/brown body with lighter green flecks; long jaws; white flesh; mild taste; farmed so available all year; pin boning is essential	Mousse-based recipes; bake; fry; poach

Quality points

Whole fish:
- eyes clear and bright – not sunken
- gill should be bright red
- scales should be firmly attached to the skin and no areas where scales are missing
- moist with a slippery feel
- fresh looking skin with a natural colour
- tails and fins should feel stiff and not be broken, flesh should feel firm
- should have a fresh sea aroma with no unpleasant or ammonia smell.

Fish fillets:
- fresh appearance and smell, with firm flesh
- neatly prepared, trimmed with no ragged edges
- white fish fillets should have a white translucent colour with no discolouration

- prepared to specification, for example, fillets of the same weight, skin on or off, pin bones removed from round fish.

Smoked fish:
- glossy, fresh appearance
- flesh firm and not sticky
- pleasant smoky aroma.

Frozen fish:
- should be at the required temperature of -18°C or below
- packaging should be intact sufficient to protect the fish from damage
- there should be no freezer burn; this shows as dull white, dry patches on the fish.

Table 10.6 Seasonality of fish

	Jan	Feb	Mar	Apr	May	Jun	Jul	Aug	Sep	Oct	Nov	Dec
Bream					*	*	*					
Brill			*	*	*							
Cod			*	*								
Eel												
Mullet (grey)			*	*								
Gurnard												
Haddock												
Hake			*	*	*							
Halibut				*	*							
Herring												
John Dory												
Mackerel												
Monkfish												
Plaice			*	*								
Red mullet												
Salmon (farmed)												
Salmon (wild)												
Sardines												
Sea bass				*	*							
Sea trout								*	*	*	*	
Skate												
Sole (Dover)	*											
Sole (lemon)												
Trout												
Tuna												
Turbot				*	*	*						
Whiting												

Key

Available

At best

* Spawning and roeing – this can deprive the flesh of nutrients and will decrease the yield.

Storage of fish

Spoilage of fish is mainly caused by the actions of enzymes and various bacteria. Enzymes in the intestines of living fish help to convert food to body tissue and energy. When the fish dies, the enzymes continue working and start to break down the actual fish flesh. Bacteria are present on the skin and in the intestines of the fish and cause no problems while the fish is alive but, along with the enzymes, will start to break down the fish flesh once it is dead. Generally these bacteria will not harm humans but rapidly cause an unpleasant smell and the eating quality will be reduced.

Careful storage directly after the fish is caught is very important and fish are usually placed into temperature-controlled storage very quickly. This may be iced water containers with a mixture of ice and sea water for small vessels, or complex temperature-controlled storage equipment on large vessels.

Once delivered, the higher the temperature the more quickly the breakdown will occur so fish refrigerators need to be kept below 4°C and where possible at a temperature of 1–2°C. Because of the enzymes and bacteria in the intestines, gutted cleaned fish will keep a little longer than those that are not gutted.

Different types of fish will have different storage methods:

- **Fresh fish** – once caught will keep for between 6–8 days as long as properly refrigerated. Because the chef will probably not know when the fish were actually caught, it is advisable to use on the day of delivery, but fish can usually be stored overnight. Remove the fish from the ice/delivery boxes, rinse and pat dry with kitchen paper and store on trays covered with cling film at the bottom of the refrigerator.
- **Ready-to-eat, cooked fish** – fish such as hot smoked salmon or mackerel, cooked prawns or crab should be stored, covered with their packaging or cling film, in the refrigerator on shelves above raw food items to avoid the possibility of cross-contamination. They may also be stored in refrigerators with other cooked items
- **Frozen fish** – must be stored at or below -18°C and defrosted overnight or for several hours at the bottom of the refrigerator in a deep tray covered with cling film. Do not defrost fish by placing it in water as this affects the flavour and texture of the fish and valuable water-soluble nutrients may be lost. Do not re-freeze fish once it has defrosted.
- **Smoked fish** – should be stored in the refrigerator and well wrapped to prevent the strong odour penetrating other foods. Some smoked fish such as haddock may be dyed before smoking so need to be kept away from other foods to prevent the transfer of colour. For more information on the smoking process, see page 327.

Preparing and cooking fish

Fish needs to be prepared with care and skill to avoid undue wastage. Even with skilled preparation, a flat fish can lose 60 per cent of its original weight in preparation due to removing fillets from the bone, loss of head, tail and fins, gutting and skinning. This is one of the reasons that fish may remain a relatively expensive menu item.

Gutting a red mullet

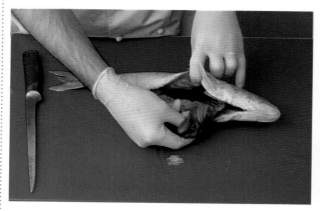

1 Cut from the vent to two-thirds along the fish.
2 Draw out the intestines with the fingers or, in the case of a large fish, the hook handle of a utensil such as a ladle.

3 Ensure that the blood lying along the main bone is removed, then wash and drain thoroughly.

Cutting

Fish are cut in various ways and the cuts used will depend on: the type of fish, its size, the portion size required, the cooking method to be used and the requirements of the finished dish.

Filleting

This involves removing the fish flesh neatly from the bone down the length of the fish. A flat fish will give four fillets, two from each side of the fish on either side of the spine. A round fish will give two fillets either side of the spine. Depending on the size of the fish, the fillets may then be cut further.

Skinning and trimming a Dover sole

1 Score the skin just above the tail.

2 Hold the tail firmly, then cut and scrape the skin until sufficient is lifted to be gripped.

3 Pull the skin away from the tail to the head. Both the black and white skins may be removed in this way.

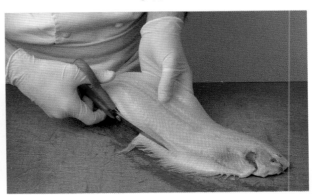

4 Trim the tail and side fins with fish scissors.

Filleting a round fish – salmon

1 Remove the head and clean thoroughly.

2 Remove the first fillet by cutting along the backbone from head to tail. Keeping the knife close to the bone, remove the fillet.

3 After both fillets have been removed, remove the rib cavity bones and trim the fish neatly.

Filleting a flat fish – turbot

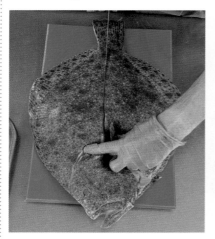

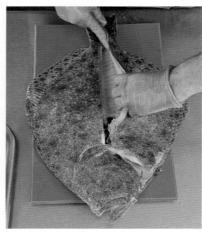

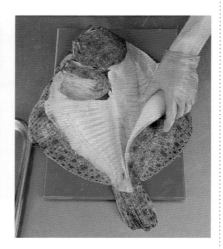

1 Using a filleting knife, make an incision from the head to the tail. Cut around the gill and backbone.

2 Remove the first fillet, holding the knife almost parallel to the surface and keeping the knife close to the bone.

3 Repeat for the second fillet.

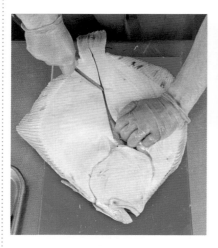

4 Turn the fish over and repeat, removing the last two fillets.

5 Hold the fillet firmly at the tail end. Cut the flesh as close to the tail as possible, as far as the skin. Keep the knife parallel to the work surface, grip the skin firmly and move the knife from side to side to remove the skin.

6 Trim the fillets neatly.

Pin boning

Pin bones are the bones on larger fish that go into the muscles cross-wise. In a fillet of fish you will feel the ends of the bones. The larger pin bones are where the fillet joined and they get smaller towards the tail.

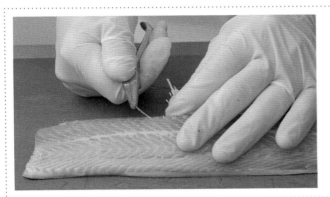

1 To find the pin bones, place the fillet skin side (or where the skin was removed) down on the board. Run the fingers along the length of the upper side and you will feel the tips of the pin bones in the thicker parts of the fish.

2 Using fish pliers (tweezers), gently pull the bones out until they are all removed. Feeling the surface of the fish is the best way to check that all the bones have been removed.

Stuffing

Stuffing (forcemeat or farce) may be used with fish to:
- fill a cavity where the fish was gutted
- add flavour, colour, texture or interest to the fish dish
- as part of the requirements of the dish, as in paupiettes
- as an ingredient of a dish such as a terrine.

The stuffing (forcemeat or farce) used can vary greatly in flavour and texture depending on the dish requirements and the cooking method and presentation style. It will depend on whether the stuffing is required to add flavour or to enhance the natural fish flavours; the stuffing should never overpower the fish. The texture may vary, depending on the dish requirements. For example, a stuffing used for a whole trout may have visible chopped vegetables or nuts and could be quite coarse, but the stuffing for sole paupiettes would be much smoother and more delicate.

Stuffing a whole round fish

1 Scale, trim and gut the fish.

2 Remove the spine, snipping the top with scissors. Clean the cavity.

3 Using a piping bag, fill the cavity with the required stuffing.

4 Secure the cut edges of the fish in place before cooking. (Fish may also be wrapped in foil before cooking.)

Cuts of fish

- **Darnes** – steaks cut through the bone of a round fish such as salmon or cod.
- **Tronçon** – steaks cut through the bone of a large flat fish such as turbot.
- **Suprêmes** – a neat cut taken from a large fillet with the skin and bones removed. For example, salmon, cod, turbot and brill.
- **Goujons** – skinned fillets, usually from a flat fish such as plaice or sole, cut diagonally into strips approximately 8cm × 0.5cm.
- **Paupiettes** – fillets of fish such as sole or plaice spread with a suitable stuffing and rolled.
- **Plaited (en tresse)** – thin strips are cut down the length of a fillet of fish then three strips are plaited together and neatly arranged before cooking.

▲ Cuts of fish: (clockwise from top left) a darne, a tronçon, a fillet, paupiettes, a suprême and goujons

Other preparation techniques

- **Marinating** – coat the fish in a marinade (a mixture of ingredients such as wine, vinegar, oils, herbs and spices) for the required amount of time before cooking. This adds flavour, moisture and may change texture.
- **Coating** – fish may be coated in a number of ways before cooking. Coatings include flour/egg/breadcrumbs (pané), batters, milk/flour. Coatings add texture, flavour and colour, protect the fish from excessive heat such as in frying and prevent cooking oil penetrating the fish; they also help to prevent the fish breaking up when cooking.
- **Topping and covering** – fish may be topped or covered before cooking with flavoured crumb mixtures, pastry, filo, pancetta or other charcuterie items, foil or parchment paper.

Effects of cooking

Muscle fibres in fish are much shorter than in meat and the connective tissue collagen dissolves easily during cooking, allowing fish to cook quickly. Possibly the biggest challenge in fish cookery is to avoid over-cooking which spoils the texture and the flavour of the fish.

As fish cooks, proteins in the muscle fibres coagulate, the flesh changes from translucent to opaque and the flesh becomes firmer. When the collagen softens on heating, its structure changes and turns to gelatine so the fish separates easily into flakes. The processes of coagulation and collagen softening happen simultaneously and at much lower temperatures than with meat. This is why fish is easily overdone. This softening and easy breakdown also makes fish easy to digest

White fish may absorb the fats and oils they are cooked in and this is increased when the fish is coated in breadcrumbs or batter that also absorb the fats and oils, changing the nutritional content of fish. Oily fish do not usually absorb fats or oils during cooking but the existing oils in their structure may liquefy, making the flesh delicate and easy to separate. There will be additional fats and/or oils on the surface of the fish or on any coatings after cooking in fat (e.g. pan frying).

When poaching fish, up to 50 per cent of water-soluble minerals may be lost.

Cooking methods

A wide variety of cooking methods can be used for fish as long as it is borne in mind that the cooking time is likely to be short and over-cooking will impair quality. Suitable methods for cooking fish include poaching, steaming, en papillote, baking (occasionally referred to as roasting), braising/stewing, deep and shallow frying, grilling and griddling. Fish stocks and soups should be boiled and simmered. Any fish or shellfish cooked by poaching can also be steamed.

The composition of individual fish will affect the cooking methods to be used. For example, the fine structure and delicate nature of sole and plaice need a cooking method that is fast and gentle to avoid spoiling the texture and flavour. Therefore these fish would be cooked quickly by steaming, poaching, grilling or frying. Eel, tuna and monkfish have a stronger and firmer texture so could be included in stewed and braised recipes, as well as the faster cooking methods.

Preservation of fish

Fish deteriorates rapidly from the time it is caught and ways have been sought to make fish keep longer. Preservation methods include:

- **Salting** and **drying** such as dried salt cod.
- **Curing**, for example, herrings cured in salt.
- **Pickling** – herrings may be filleted, rolled and pickled in vinegar with herbs and spices,
- **Smoking** – the fish is usually gutted then soaked in a strong salt solution (brine); sometimes a dye is added to improve colour. It is then drained and hung on racks in a kiln and exposed to smoke for five or six hours. Fish can be cold smoked (at 24°C, which smokes but does not cook the fish) or hot smoked (at 82°C; the fish is lightly cooked at the same time).
- **Canning** – oily fish such as salmon, tuna and sardines are preserved in cans surrounded by oil, brine, water or a sauce.

Because of modern **chilling** and **frozen** storage, the above methods used to preserve fish are no longer strictly necessary. However, the taste, texture and characteristics given to the fish by these methods became popular and add interest and variety to menus. These methods may also be used where there is limited refrigerator/freezer space or for 'emergency use'. Some of these preservation methods allow for interesting cold fish dishes to be placed on menus. Fish may also be given a slightly longer shelf life by the use of vacuum packaging and modified atmosphere packaging (see Chapter 2).

Sustainability of fish stocks

The amount of fish available around the UK and especially in the North Sea has become a matter of concern as stocks of fish such as cod have become seriously depleted. There are controls already in place to help prevent further depletion of fish stocks such as restrictions on:

- fishing methods
- the type of fishing equipment used
- certain fishing areas, especially in spawning seasons
- the amounts and sizes of fish caught.

Organisations such as the Marine Conservation Society, the World Wildlife Fund and the Food and Agriculture Organization of the UN are keen to address the current problems by encouraging consumers to use fish that are plentiful and not at risk of depletion. Chefs can play a major role in the interesting and creative use of these fish and in educating consumers in varieties that may not previously have been popular. Table 10.7 shows the fish we should use more of and those that should be avoided.

Table 10.7 Fish we should use more of and avoid

Use more of:	Reduce use or avoid using:
Coley	Atlantic blue-fin tuna
Dab	Atlantic cod
Gurnard	Atlantic halibut
Mackerel	Atlantic salmon
Megrim	European eel
Oysters	King/tiger prawns
Pollock	Sturgeon and caviar
Sardines	Skates and rays
Sea bream	Swordfish
Tilapia	Turbot

Source: Marine Conservation Society

The fishing quota system controls the amounts and types of fish that may be caught and landed in specific areas. This is intended to protect fish stocks and was agreed by governments, those in the fishing industries and marine biologists.

Professional tip

Elvers (glass eels) are very small immature eels that are occasionally popular on menus but are expensive and are sometimes scarce so fishing them may be discouraged.

Table 10.8 The main ways that fish are caught and their impacts

Method	How it is done	Advantages	Disadvantages
Line fishing	Lines are dropped into the sea, usually from boats. Sometimes each line has multiple hooks and in some areas very long lines are used.	• Smaller quantities fished so not likely to threaten fish stocks. • Can be more selective about the fish you want to catch by selection of areas and types of hooks and lines. • Not as complex as other methods. • Reaches the consumer very quickly. • No problem with 'by catch' i.e. catching different species unintentionally. • A tagging scheme monitors where and when fish were caught, methods used, quality and sustainability. • Quality is considered better – whiter, firmer flesh because fish are not piled on top of each other as they are on trawlers.	• More time and labour intensive so fish are more expensive. • Can be hazardous because boats may need to be taken from shore, often in poor weather conditions. • Availability of required fish may be unpredictable. • Any lost lines and fishing hooks can damage fish on the sea bed.
Netting	Large nets are fixed in a specific area and are left for a set length of time before pulling in with the fish catch. Some nets are dragged in from the edges, efficiently trapping the fish.	• Catches a wide variety of fish passing through the netted area. • Once set up is efficient to operate. • Mesh sizes of nets can be specified to avoid netting very small fish. • Supplies can be predicted because specific areas can be netted to catch the required shoals of fish.	• Initial set-up can be time consuming and nets need to be frequently checked and kept in good repair. • Large catches contribute to overfishing and stock depletion. • Species not intended to be caught, such as dolphins and turtles, can get into the nets. • Not able to fully select the type of catch. • Fixing the nets can damage the seabed.
Baiting	Attracting fish with bright lights or to a cage with a bait.	• Over fishing and depletion of stock is unlikely as catches are usually in small numbers. • Catch tends to be undamaged.	• Can be a small return for the amount of effort needed.
Trawling	Trawlers are boats that have large nets on long 'arms' either side of the boat. The bottom of the nets are weighted or on wheels so drag along the seabed which disturbs and helps to catch fish such as cod and sole. The equipment can be complex and sophisticated to find fish shoals and pinpoint the best fishing areas and the type of sea bed. Inspectors may be on board larger trawlers.	• Allows for efficient fishing and large catches. • Not as dangerous for fishing crew as smaller boats may be. • Modern trawlers can be energy efficient. • Spacious and well equipped enough to allow for some fish preparation, packaging and even freezing at sea.	• All fish within the trawled area are caught, some of which cannot be sold as food or may be endangered species. • Dragging the nets along the sea beds can damage eco systems.

The farmed fish industry

Fish farming is often referred to as **aquaculture**. It has come about because of world population growth and predicted further growth, leading to the need for sustainable food supplies, particularly given many fish stocks are now depleted. With aquaculture increasing in popularity, it is very important that steps are taken to ensure that as many farmed fish as possible are reared under the highest standards of animal welfare.

There are many species of fish used in aquaculture and these are farmed in both freshwater and seawater. The most common system consists of fish being hatched in production units then transferred to sea or river cages of progressively larger sizes as they grow. The majority of UK aquaculture is in Scotland and mainly for Atlantic salmon, though other farmed fish species are now becoming available.

Table 10.9

Advantages of aquaculture	Disadvantages
High yields are possible in a relatively small area	High densities of fish in a small area makes them prone to stress-related disease and infection
Relieves pressure on over-fished and wild stock and helps to meet global demands	Disease can be released into water and can affect wild fish stocks
Reduces the cost of some popular fish such as salmon	Overcrowding of fish can lead to them developing toxic amounts of mercury in their bodies
Food is always available so fish grow and mature quickly	Mercury can be released into the water so can pollute water surrounding fish farms
Fish are not prey to natural predators and the environment such as temperature and oxygen levels can be selected	Some consider the quality of the fish is not as high as in wild varieties
No accidental netting of endangered species as occurs with netting and trawling	Treatments such as antibiotics and pesticides can remain in the fish and also be released into surrounding waters
Provides a reliable and sustainable source of fish	
Breeding can be carried out to size and weight specifications	

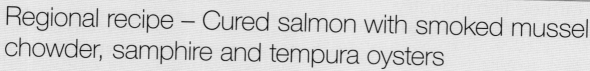

Regional recipe – Cured salmon with smoked mussel chowder, samphire and tempura oysters

Contributed by Jason Sant, Colchester Institute

The quantities in this recipe are for a starter.

	4 portions	10 portions
Samphire	200 g	500g
Cured salmon		
Salmon, skinned and pin boned	300 g	750g
Maldon sea salt	50 g	125g
Muscovado sugar	50 g	125g
Fennel seeds	15 g	40g
Lemon, zest and juice	1	3
Lime, zest and juice	1	3
Orange, zest and juice	1	3
Dill, chopped		
Tempura oysters		
Colchester native oysters	8	20
Self-raising flour	50 g	125g
Cornflour	50 g	125g
Egg yolk	1	3
Sparkling water	150ml	375ml
Ice cubes	A few	A few
Vegetable oil		
Smoked mussel chowder		
Mussels	1 kg	2.5 kg
Butter	50 g	125g
Potatoes, two chopped roughly, one diced neatly for garnish	3	8
Leek, white only	½	2
Shallot	1	3
Garlic clove	1	3
Celery sticks	2	3
White wine	250 ml	625ml
Double cream	175 ml	440ml
Liquid hickory smoke		
Fish stock		
Sea vegetables to garnish		

Energy	Cal	Fat	Sat fat	Carb	Sugar	Protein	Fibre
4225 kJ	1011 kcal	54.2g	49.4g	57.3g	20.2g	68.9g	4.4g

1 Toast the fennel seeds in a dry pan, then crush them. Mix with the salt, sugar, zest, dill and some of the juice to make a paste.

2 Spread the mixture over the fish, wrap in cling film and leave for at least 4 hours.

3 Cook three-quarters of the mussels in the white wine until opened, drain (keeping the liquor) then remove from the shells.

4 In a pan, sweat off the vegetables until tender, then add the mussel cooking liquor and reduce down.

5 Add some fish stock and cook for about 10 minutes, or until the potatoes are soft.

6 Add the mussels and cream. Blend until you have achieved a smooth sauce. Add the liquid smoke to taste, using a smoke gun, and then finish with a little of the citrus juice.

7 Open the oysters (keeping the juice).

8 Make the tempura batter by whisking the ingredients together until the batter coats the back of a spoon.

9 Blanch the diced potatoes and refresh.

10 Wash the salmon and pat dry. Portion (about 75 g each) then place in a vacuum pack bag with a drizzle of citrus juice and seal. Place the fish in a water bath at 52°C for 10 minutes.

11 Meanwhile, warm the sauce and add the remaining quarter of the mussels, finishing with the oyster juice. Do not allow to boil.

12 Blanch the samphire in boiling water and then toss it in a little butter.

13 Dip the oysters in flour and then in the tempura batter. Deep fry at 180°C for about a minute. Drain

14 To serve, place the samphire in the bottom of a warm bowl. Drain the fish and place on top. Add the sauce around. Garnish with the diced potatoes, deep-fried oysters and sea vegetables. Place a glass cloche over the bowl and fill with smoke from the smoke gun; the bowl is removed at the table in front of the guest.

Asparagus was used in place of samphire for the nutritional analysis.

1 Crispy seared salmon with horseradish foam and caviar

1 Place salmon skin-side down in a hot pan with a little vegetable oil and cook on a medium heat until two-thirds of the salmon is cooked.

2 For the foam, sweat the shallots and thyme in half the butter, adding white wine after 2 minutes and reduce by half.

3 Add the cream and horseradish, bring to the boil and infuse for 15 minutes off the heat.

4 Pass through a fine chinois and work in the other half of the butter and the lemon juice while the mix is hot (this will stop it from splitting). Place into a pressurised container used to produce foams.

5 Wilt the spinach in a little butter, add the asparagus to reheat and arrange neatly in the centre of each serving dish.

6 Place the seared salmon skin-side up on the asparagus and spinach, and finish with a quenelle of caviar. Add the foam to the plate at the last minute.

Note: Foams are aerated sauces produced using either a siphon or a blender.

Variations
This dish can be adapted in many ways.
● Substitute the caviar with avruga caviar to save the expense.
● Alternatively, if cost is not an issue, use smoked salmon (a smaller portion) to replace the caviar, and sear that in the same way. The horseradish will be a great foil for this.

Ingredient	4 portions	10 portions
Salmon fillet steaks, skin on, scaled (140g per steak)	4	10
Vegetable oil	50ml	125ml
Baby spinach, washed	400g	1 kg
Butter		
Asparagus spears, blanched	12	30
Garlic cloves, chopped	1	3
Caviar (optional)	50g	125g
For the foam		
Shallots, sliced	2	5
Thyme sprigs	1	3
Butter	80g	200g
White wine	60ml	150ml
Double cream	60ml	150ml
Horseradish, grated	20g	50g
Lemons, juice	1	3

Energy	Cal	Fat	Sat fat	Carb	Sugar	Protein	Fibre
2209 kJ	533 kcal	41.7g	18.3g	4.8g	4.1g	34.7g	3.5g

2 Grey mullet en papillote with green chutney

Ingredient	4 portions	10 portions
Grey mullets (about 500g each), scaled and gutted	2	5
Turmeric powder	½ tsp	1 tsp
Salt	½ tsp	1 tsp
Lemon, juice	1	3
Banana leaves	1	3
Green pepper, cut into sprigs to garnish		
Red chillies, cut into julienne, to garnish		
For the chutney		
Fresh green peppercorns, stripped from the stem	4 tbsp	10 tbsp
Fresh coriander bunches	½	1
Green chillies, deseeded	2	5
Garlic cloves, peeled	5	12
Cumin powder	½ tsp	1½ tsp
Lemon, juice	½	1½
Salt, to taste		

Energy	Cal	Fat	Sat fat	Carb	Sugar	Protein	Fibre
1274 kJ	302 kcal	10.3g	2.8g	2.3g	0.5g	51.0g	0.7g

1. Cut 3 gashes on either side of the fish; rub in the turmeric powder, salt and lemon juice to coat the fish. Set aside for at least 1 hour.
2. Grind or process all the chutney ingredients (not too finely), incorporating the lemon juice and a pinch of salt.
3. Smear the fish with the chutney, making sure it goes into the gashes and the inside cavity.
4. Parcel the fish in two pieces of banana leaf cut to the appropriate size and moistened with vegetable oil. You can secure the parcels with string or with toothpicks (alternatively parcel the fish in oiled kitchen paper or foil).
5. Bake for 15 minutes in an oven heated to 180°C.
6. Serve hot in the banana leaf with tomato rougail as an accompaniment, and garnish with the green pepper and julienned red chillies.

Professional tips

- Kitchen paper or foil may be used instead of banana leaves.
- Sea bass may be used as a substitute for grey mullet.
- Rougail is tomato concassée cooked with garlic, ginger, thyme and parsley.

This recipe was contributed by Mehernosh Mody.

3 Salmon 'mi cuit' with horseradish sauce and green vegetables

Ingredient	4 portions	10 portions
Salmon fillet pieces (about 120g), trimmed of skin and grey fat	4	10
For the cooking oil		
Corn oil	1 litre	3 litres
Star anise	2	5
Bay leaves	2	5
Vanilla pods, used	3	7
Peppercorns	20	50
For the horseradish sauce		
Mashed potato (dry)	200g	500g
Double cream	60ml	150ml
Butter	60ml	150ml
Salt, pepper		
For the greens		
Green cabbage, blanched	100g	250g
Baby bok choi, blanched	4	9
Spinach, washed, picked	200g	500g
Butter	50g	125g

Energy	Cal	Fat	Sat fat	Carb	Sugar	Protein	Fibre
3279 kJ	792 kcal	70.0g	24.6g	13.0g	4.1g	28.4g	3.5g

For the cooking oil

1 Place all the ingredients in pan and heat slowly to about 80°C. Leave to infuse for 1 hour.

2 Remove from the heat and leave at room temperature for at least 24 hours to take on more of the flavour.

For the horseradish and potato sauce

1 Place the potato (retaining 10 per cent), horseradish and other ingredients in a saucepan over a low heat until the consistency resembles a thick, puréed soup. If not, adjust by adding more cream or potato.

2 Adjust the seasoning to taste and keep warm.

To complete

1 Heat the infused cooking oil to 40–43°C, using a digital probe to maintain the temperature and moving the pan on and off the stove.

2 When the oil is at the required temperature, place the salmon pieces in it and cook for 40 minutes. When done, the flesh will still be pinky-orange inside, but do not let this put you off as this means it is cooked perfectly.

3 Remove and drain.

4 In the meantime, reheat the greens in the butter and place in the centre of the plate.

5 Spoon the sauce around and serve immediately.

> **Professional tip**
>
> The term 'mi cuit' is directly translated as 'just cooked' and that is what this salmon is, due to the temperature of the oil and setting temperature of the fish proteins in the salmon. The cooking medium must not go too far above the protein setting temperature if the flesh is to remain soft as the proteins will harden, making the fish tough.
>
> This process works best with organic salmon as the environment it swims in is less claustrophobic, and free from pesticides and bacteria – this is important because of the low cooking temperature of the fish. The process of supply, handling and serving are all crucial to the safe production of this dish: adopt a safe critical path and the dish will be safe to eat.

4 Baked stuffed sardines

1 Allow 3 sardines per portion. Slit the stomach openings of the sardines and gut.

2 From the same opening, carefully cut along each side of the back bones and remove by cutting through the end with fish scissors.

3 Scale, wash, dry and season the fish.

4 Stuff the fish. A variety of stuffings can be used (see below).

5 Place the stuffed sardines in a greased ovenproof dish.

6 Sprinkle with breadcrumbs and oil.

7 Bake in hot oven, 200°C, for approximately 10 minutes and serve.

Possible stuffings

- cooked chopped spinach with cooked chopped onion, garlic, nutmeg, salt, pepper
- fish forcemeat
- thick duxelle.

Professional tip

Herring, mackerel, sea bass and trout can also be prepared and cooked in this way, and there is considerable scope for flair and imagination in the different stuffings and methods of cooking the fish.

5 Whole sole grilled with traditional accompaniments

Ingredient	4 portions	10 portions
Whole sole, white and black skin removed	4	10
Butter for grilling	200g	500
Seasoning		
Parsley butter	100g	250g
Lemons, peeled and cut as required	1	3

Energy	Cal	Fat	Sat fat	Carb	Sugar	Protein	Fibre	Sodium
3058 kJ	740 kcal	65.4g	39.1g	1.0g	0.9g	36.9g	0.3g	0.7g

1 Ensure the fish is clean of roe, scales and skin.

2 Place on a buttered grilling tray and rub soft butter into the flesh.

3 Season and place under the grill.

4 When the butter starts to brown slightly, remove from the grill and turn the fish over carefully, using a roasting fork or a long pallet knife.

5 With a spoon, baste the flesh of the uncooked side and continue cooking. The tail end will cook faster than the head end, therefore the less hot area towards the front of the grill is where the tail should be cooked.

6 To check whether the fish is done, place your thumb just behind the gill area and you should feel the flesh ease away from the bone.

7 Finish with parsley butter and a wedge of lemon.

> **Professional tip**
>
> This is a classic recipe using slip, Dover or lemon sole. There is no need to modernise it.
>
> Sole is a delicate fish: be careful not to over-cook it.

6 Deep poached hake, cockles and prawns

Ingredient	4 portions	10 portions
Onion, finely chopped	100g	250g
Oil	1 tbsp	2 tbsp
Fish stock	250ml	600ml
Parsley, chopped	1 tbsp	2 tbsp
Hake steaks or fillets (150g)	4	10
Cockles, shelled	8–12	20–30
Prawns, shelled	8–12	20–30
Salt, pepper		
Eggs, hard-boiled, coarsely chopped	2	5
Parsley, chopped (to garnish)	½ tsp	1 tsp

Energy	Cal	Fat	Sat fat	Carb	Sugar	Protein	Fibre
975 kJ	233 kcal	9.5g	1.76g	2.14g	1.54g	34.7g	0.39g

1 Lightly colour the onion in the oil, add the fish stock and parsley and simmer for 10–15 minutes.

2 Place the fish in a shallow ovenproof dish, and add cockles and prawns.

3 Pour on the fish stock and onion; season lightly.

4 Poach gently, and remove any bones and skin from the fish.

5 If there is an excess of liquid, strain and reduce.

6 Serve coated with the unthickened cooking liquor, sprinkled with the egg and parsley.

7 Braised tuna Italian style

Ingredient	4 portions	10 portions
Tuna pieces (each 600g)	1	3
Shallots, chopped	100g	250g
Mushrooms, chopped	200g	500g
White wine	125ml	300ml
Fish stock	125ml	300ml
Marinade		
Lemon, juice	1	2–3
Olive oil	60ml	150ml
Onion, sliced	100g	250g
Carrot, sliced	100g	250g
Bay leaf	1	
Thyme, salt, pepper		

Energy	Cal	Fat	Sat fat	Carb	Sugar	Protein	Fibre
1557 kJ	373 kcal	22.3g	4.02g	6.36g	4.97g	37.2g	1.86g

1 Mix together the ingredients for the marinade and marinate the pieces of fish for 1 hour.

2 Remove, dry well and colour in hot oil.

3 Place in braising pan, add shallots and mushrooms.

4 Cover with a lid, cook gently in oven for 15–20 minutes.

5 Add white wine and fish stock, cover, return to oven.

6 Braise gently for approximately 45 minutes until cooked.

7 Carefully remove the fish, correct the seasoning of the liquid (which may be lightly thickened with beurre manié, if required) and serve.

Variations

- Other ingredients that may be used when braising tuna include tomatoes, garlic, basil and vinegar.
- Slices of tuna can also be shallow-fried or cooked meunière with or without the meunière variations.

8 Whole sea bass baked with Swiss chard, spinach and scallops

Ingredient	4 portions	10 portions
Sea bass	1.6 kg (after preparation, about 880g)	4 kg (2.2 kg)
Lemon butter	30g	75g
Swiss chard or spinach	65g	160g
Scallops	280g	700g

Energy	Cal	Fat	Sat fat	Carb	Sugar	Protein	Fibre
1754 kJ	419 kcal	19.0g	9.0g	2.8g	0.4g	59.3g	0.5g

1 Remove all the fins with scissors and scale thoroughly.

2 Wash the fish quickly and dry it thoroughly.

3 Remove the head with two neat cuts at 45° behind the gills.

4 Cut down the back bone about 2cm from the tail on both sides.

5 Cut down to the rib cage.

6 After you reach the rib cage, go behind it and down to the flesh towards the tail.

7 Cut along the rib bones without breaking the flesh and trim right down to the tail. Trim by the belly.

8 Take the guts out and wash and dry the fish thoroughly.

9 Take out the pin bones.

10 Remove any pieces of silver skin from inside the belly.

11 Trim about 1cm of flesh from the body where you made the first incision, so that when you tie it, it comes together.

12 Brush the inside of the fish with softened lemon butter and line it with spinach.

13 Lay the scallops in the centre and fold the spinach over so they are completely covered.

14 Bring the two sides of the bass together. Tie in the middle first, then either side of the middle.

15 At the head end, double-loop the string, hook it under the gills and tie it very tightly to close up the hole.

16 Tie the rest of the bass up, leaving about 3–4mm between each tie.

17 Leave it in a refrigerator for 24 hours to firm up, before use.

18 Place in an oven at 170°C and cook for 40 minutes. Allow to rest for 10 minutes before serving. Serve with a lemon butter sauce and chopped chives.

Fillet the fish and trim the scallops

Stuff the fish and tie securely

9 Salmon fishcakes

Ingredient	4 portions	10 portions
Salmon fillet – skinned and boneless	400g	1 kg
Cooked mashed potato (make sure this is fairly dry)	225g	560g
Spring onions	3	8
Flat leaf parsley	10g	25g
Dill	Sprig	Sprig
Butter	50g	125g
Juice of lemon	1	2
Crème fraiche	1 tbsp	2½ tbsp
Thai fish sauce (*nam pla*) – optional	Few drops	1 tsp
Eggs	1	2
For pané		
Plain flour, seasoned		
Eggs		
Breadcrumbs		

Energy	Cal	Fat	Sat fat	Carb	Sugar	Protein	Fibre
1821 kJ	438 kcal	30.2g	13.6g	18.0g	1.5g	24.5g	1.6g

1 Place the salmon in an oiled roasting tin, season with salt and pepper, dot with butter and squeeze the lemon juice over.

2 Bake for approximately 7 minutes at 200°C.

3 Allow the salmon to cool a little then flake into bite sized pieces.

4 Chop the spring onions and herbs.

5 Add the salmon to the potato, herbs, spring onion, beaten egg and crème fraiche. Add a little *nam pla* if using and season.

6 Divide the mixture, allowing two pieces per portion. Form into neat, even-sized cake shapes (a ring mould could be used), place on a tray lined with cling film and chill thoroughly for several hours.

7 Coat with seasoned flour, egg and breadcrumbs (pané), chill well again or the formed cakes can be frozen.

8 Cook in a deep fryer at 180–190°C as required, drain on kitchen paper.

9 Serve with a suitable sauce or salsa and/or mixed salad leaves.

10 Gravlax

Ingredient	4 portions	10 portions
Middle-cut, fresh, descaled raw salmon	750g	1.75 kg
Dill bunches, washed, chopped	1	2
Caster sugar	25g	60g
Salt	25g	60g
Peppercorns, crushed	1 tbsp	2 tbsp

Energy	Cal	Fat	Sat fat	Carb	Sugar	Protein	Fibre
1417 kJ	339 kcal	19.3g	3.3g	6.3g	6.3g	35.6g	0.1g

1 Cut the salmon lengthwise and remove all the bones.

2 Place one half, skin-side down, in a deep dish.

3 Add the dill, sugar, salt and peppercorns.

4 Cover with the other piece of salmon, skin-side up.

5 Cover with foil, lay a tray or dish on top, and evenly distribute weights on the foil.

6 Refrigerate for 48 hours, turning the fish every 12 hours and basting with the liquid produced by the ingredients. Separate the halves of salmon and baste between them.

7 Replace the foil, tray and weights between basting.

8 Lift the fish from the marinade, remove the dill and seasoning, wash, dry and top with chopped dill.

9 Place the halves of salmon on a board, skin-side down.

10 Slice thinly, detaching the slice from the skin.

11 Garnish gravlax with lemon and serve with mustard and dill sauce.

Gravlax can be served in appetisers, canapés, buffets, starters and open sandwiches.

11 Arabic-style stuffed red snapper

Ingredient	4 portions	10 portions
Red snapper	4	10
Red chillies	12	30
Red onions	300g	750g
Turmeric powder	12g	25g
Salt	1 tsp	2 tsp
Vegetable oil	4 tbsp	10 tbsp
Red chillies to garnish	4	10
Spring onions to garnish	8	20
Fresh coriander sprigs to garnish	4	10

1 Prepare the red snapper by removing the scales, eyes and gills.

2 Gut the fish by making a cut from the stomach to the head, remove all the innards. Clean under running cold water to remove debris.

3 Across the back make three incisions (ciseler).

4 Remove the stalk and seeds from the first lot of chillies and chop them.

5 Finely chop the onions.

6 In a pestle, place the chillies, onions, turmeric powder and salt, blend to a paste. (Alternatively use a food processor.)

7 Rub the fish with the paste inside and out.

8 Heat some oil in a suitable wok or pan, and gently fry the fish for 4–5 minutes on each side.

9 Serve on a suitable plate garnished with chilli, spring onion and coriander. Serve with plain boiled rice.

Note: Other fish that may be used include mackerel, trout and red mullet.

Energy	Cal	Fat	Sat fat	Carb	Sugar	Protein	Fibre
1210 kJ	292 kcal	21.5g	2.7g	7.9g	6.2g	17.7g	1.6g

12 Roast salt cod and clam chowder

To finish		
Heads escarole leaves, picked and stalks removed *or*	1	3
Spinach	200g	500g
Chopped chervil		

Energy	Cal	Fat	Sat fat	Carb	Sugar	Protein	Fibre
2759 kJ	664 kcal	46.5g	21.5g	18.8g	7.1g	48.3g	2.2g

For the salt cod

1 Place the skinned and trimmed cod fillet on a tray. Rub it all over with the spices and sprinkle evenly with the salt (note: if the fish tapers at one end, reduce the amount of salt placed on this area).

2 Wrap in cling film and place in the fridge. After 2 hours, turn the cod fillet over and return to the fridge for 1 hour.

3 Fill a clean sink half full with cold water. Unwrap the cod, rinse off the salt in the sink and leave the cod to soak for 30 minutes.

4 Remove from the water and dry well.

For the clams

1 Take a large saucepan with a tight-fitting lid and place over a medium heat with the shallots and butter, cook for 1 minute without letting the shallots colour.

2 Add the washed clams, shake the pan, then add the wine and place the lid on the pan immediately.

3 Leave the clams to steam for 1–2 minutes so that they open and exude an intense liquor.

4 Remove the lid and make sure all the clams are open. Remove the pan from the heat and discard any with closed shells.

5 Place a colander over a large bowl and pour the contents of the pan into the colander, reserving the liquor for the chowder.

6 Allow the clams to cool. Pick out the meat and discard the shells. Store the clam meat in an airtight container in the fridge until you are ready to serve the chowder.

Ingredient	4 portions	10 portions
Salt cod		
Cod fillet, skinned and trimmed	500g	1¼ kg
Ground cumin	2 tbsp	5 tbsp
Five-spice powder	1 tbsp	3 tbsp
Sea salt	200g	500g
Clams		
Shallots, finely diced	2 medium	5 medium
Butter	50g	125g
Clams, shells tightly closed	500g	1¼ kg
White wine or dry vermouth	20ml	50ml
Chowder		
Vegetable oil	50ml	125ml
Smoked bacon, cut into 1cm dice	50g	125g
Medium onion, cut into 1cm dice	1	3
Medium carrot, cut into 1cm dice	1	3
Garlic cloves, finely chopped	2	5
Celery sticks, cut into 1cm dice	1	3
Medium potato, peeled and cut into 1cm dice	1	3
Medium yellow pepper, cut into 1cm dice	1	3
Chicken stock	1 litre	2½ litres
Whipping cream	100ml	250ml
Butter	50g	125g
Salt and pepper		

For the chowder

1 In a large saucepan, heat the oil. When hot, add the bacon and cook for about 5 minutes until crisp and brown.

2 Using a perforated spoon, transfer the bacon onto kitchen paper to drain. Add the onion, carrot, garlic, celery and potato to the saucepan, reduce the heat to medium-low and cook the vegetables for 3–4 minutes without colouring.

3 Add the peppers and cook for 5 minutes. Pour in the reserved liquor from the clams and the chicken stock.

4 Bring to the boil and simmer for 10 minutes or until the volume of liquid has reduced by about half.

5 Add the cooked bacon and cream, then bring to the boil and reduce for a further 2 minutes until the soup thickens slightly.

6 Just before serving, whisk in the butter.

To complete

1 Remove the cod fillet from the fridge and cut into equal portions.

2 Wilt the escarole/spinach in a large pan, drain and set aside, keeping warm.

3 Heat a non-stick pan on the stove with 50ml vegetable oil, place in the cod portions and cook for 2 minutes until golden brown on one side, turn and remove the pan from the heat, allowing the cod to finish cooking off on the stove in the residual heat of the pan.

4 Add the clams to the hot chowder and stir for 30 seconds until they are reheated.

5 Place the escarole/spinach in a small ball in the centre of the serving bowls, lay the cod on top of this and then add the chowder. Garnish with the chervil.

> **Professional tip**
>
> For this recipe the cod should be salted in advance – salting the day before use and storing in the refrigerator will yield a better result.

13 Baked pollock with Moroccan spices

Energy	Cal	Fat	Sat fat	Carb	Sugar	Protein	Fibre
2346 kJ	563 kcal	32.8g	4.7g	2.3g	0.6g	65.3g	0.6g

Ingredient	4 portions	10 portions
Pollock fillets	4 × 150g	10 × 150g
For the spice dressing		
Ground cumin	1 tsp	2½ tsp
Paprika	½ tsp	1 tsp
Chilli powder	½ tsp	1¼ tsp
Garlic cloves, crushed and chopped	4	10
Lemon, juice	2	5
Olive oil	200ml	500ml
Coriander, chopped	5 tbsp	7 tbsp
Salt	Pinch	pinch

1 Place all the spice dressing ingredients in a food processor. Blitz the ingredients.

2 Marinate the fish for 30 minutes in half of the dressing.

3 Place the fish in a suitable roasting tray. Cook in the oven for approximately 20 minutes at 200°C.

4 When the fish is cooked, brush it with the remainder of the dressing. Serve with steamed couscous or braised rice.

14 Steamed grouper with soy sauce and ginger

Ingredient	4 portions	10 portions
Grouper, whole	600g	1½ kg
Ginger, fresh, finely sliced	50g	125g
Spring onions, finely shredded	4	10
Vegetable oil	3 tbsp	7½ tbsp
Sauce		
Soy sauce	2 tbsp	5 tbsp
Water	180ml	450ml
Thai fish sauce	1 tsp	2½ tsp
Sugar	1 tsp	2½ tsp

Energy	Cal	Fat	Sat fat	Carb	Sugar	Protein	Fibre
1205 kJ	289 kcal	17.1g	2.2g	3.6g	2.7g	30.2g	0.2g

1 To make the sauce, bring the water to the boil and add the soy sauce, fish sauce and sugar. Stir well, then remove from the heat.

2 Steam the fish, spring onions and ginger in a suitable container such as an oriental steamer. Steam for approximately 8–10 minutes until cooked.

3 Serve the fish with the spring onions and ginger. Garnish with fresh coriander.

4 Heat the oil until slightly smoking, and drizzle it over the fish. Pour the sauce over the fish.

15 Eel with white wine, horseradish and parsley

Ingredient	4 portions	10 portions
Onion, chopped	50g	125g
Butter	100g	250g
Prepared eels	600g	1½ kg
White wine	125ml	300ml
Bouquet garni	1	2
Potatoes, diced	200g	500g
Whipping cream	100ml	250ml
Freshly grated horseradish	1 tsp	3 tsp
Parsley, chopped	½ tsp	1 tsp
Oil for frying		

Energy	Cal	Fat	Sat fat	Carb	Sugar	Protein	Fibre
2770 kJ	669 kcal	58.0g	30.2g	10.6g	2.0g	26.9g	0.9g

1 Sweat the onions in the butter for 4–5 minutes without colour.

2 Add the eel to the pan and seal well.

3 Add the white wine, bring to the boil, adding the bouquet garni and simmer for 15 minutes.

4 Add the potato and cook for a further 20 minutes until the eel is tender.

5 Remove the eel and potatoes and keep warm, pass the stock off into a clean pan and reduce to sauce consistency.

6 Add the cream, horseradish and chopped parsley, and bring to the boil.

7 Check the consistency and seasoning, correct if necessary.

8 Add the cooked eels to the pan of sauce, coat well and place in a serving dish.

Professional tip

This is a traditional dish but still a classic.

An eel is made up of lateral muscle groups either side of the backbone that work to move the fish around. Eels need a longer cooking time than other fish and shellfish species.

16 Fish en papillote

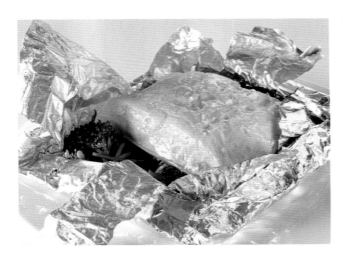

This method of cookery is fresh-tasting and suitable for most fish or shellfish. This recipe outlines the technique.

1 The fish should be portioned, free from bones and may or may not be skinned.

2 Garnish with a fine selection of vegetables chosen from carrots, leeks, celery, white mushrooms and wild mushrooms; a small amount of freshly chopped herbs may be added as desired.

3 Moisten with a little dry white wine, then seal the foil parcel and bake for 15–20 minutes (size and fish dependent).

4 Serve with an appropriate sauce (e.g. white wine).

Place the fish and vegetables onto the foil

Seal the foil package

Make sure it is tightly sealed on all sides

17 Bouillabaisse

Energy	Cal	Fat	Sat fat	Carb	Sugar	Protein	Fibre
2881 kJ	689 kcal	42.3g	8.5g	4.8g	2.1g	67.6g	0.5g

Note

This is a thick, full-bodied fish stew – sometimes served as a soup – for which there are many variations. When made in the south of France, a selection of Mediterranean fish is used. If made in the north of France, the recipe given here is typical.

Ingredient	4 portions	10 portions
Assorted prepared fish, e.g. red mullet, whiting, sole, gurnard, small conger eel, John Dory, crawfish tail	1½ kg	3¾ kg
Mussels (optional)	500g	1¼ kg
Chopped onion or white of leek	75g	180g
Garlic, crushed	10g	25g
White wine	125ml	300ml
Water	500ml	1¼ litres
Tomatoes, skinned, deseeded, diced	100g	250g
or		
Tomato purée	25g	60g
Pinch of saffron	1	2
Bouquet garni (fennel, aniseed, parsley, celery)	1	2
Olive oil	125ml	300ml
Chopped parsley	5g	12g
Salt and pepper		
Butter ⎤ Beurre manié	25g	60g
Flour ⎦	10g	25g
French bread		

1 Clean, descale and wash the fish. Cut into 2cm pieces on the bone; the heads may be removed. Clean the mussels if using, and leave in their shells.

2 Place the cut fish, with the mussels and crawfish on top, in a clean pan.

3 Simmer the onion, garlic, wine, water, tomato, saffron and bouquet garni for 20 minutes.

4 Pour on to the fish, add the oil and parsley, bring to the boil and simmer for approximately 15 minutes.

5 Correct the seasoning and thicken with the beurre manié.

6 The liquor may be served first as a soup, followed by the fish accompanied by French bread that has been toasted, left plain or rubbed with garlic.

Professional tip

If using soft fish, e.g. whiting, add it 10 minutes after the other fish.

Cut the fish on the bone, into even-sized pieces

Add the liquid to the fish pieces

Thicken with beurre manié

18 Fish soup with rouille

1 Heat the olive oil in a very large, heavy-bottomed saucepan and add the onion, leek, fennel and celery. Cook over a medium to low heat until the vegetables are soft but not coloured.

2 Add the garlic and tomatoes and cook for about 10 minutes, until the tomatoes are soft.

3 Add the fennel seeds, saffron, orange rind, tomato purée, peppercorns and fish trimmings and cover with the water. Bring to the boil then simmer for about 40 minutes, stirring often.

4 Strain the cooking liquor into another large pan. Press the vegetable mixture to get out as much flavour as possible. Discard the bones and vegetables.

5 Bring the liquor up to simmering point and poach the fish in it for about four minutes. Leave the fish to cool in the liquid a little, then purée in a blender. Taste for seasoning and adjust if necessary.

6 To make the rouille, put the garlic into a pestle and mortar with some salt and grind to a purée. The salt acts as a good abrasive. Transfer to a bowl and mix in the yolks, then start adding the oil drop by drop, beating all the time (use a whisk or a spatula). The mixture should thicken as you add the oil. Stir in the cayenne. Add the tomato purée, then lemon juice to taste. Add more lemon juice or cayenne if you want.

7 Serve the soup hot, offering grated gruyère and baguette croûtes, and rouille on the side. Garnish the bowl with a small piece of poached fish, and flat parsley.

Ingredient	4 portions	10 portions
Olive oil	2 tbsp	5 tbsp
Onions, finely chopped	1	3
Leeks, sliced	1	3
Fennel bulb, chopped	½	1
Celery stick, chopped	½	1
Garlic cloves, finely chopped	1	2
Tomatoes, chopped	200g	500g
Fennel seeds	2 seeds	¼ tsp
Good pinch of saffron stamens		
Broad strip of orange rind	1	1
Tomato purée	1 tsp	½ tbsp
Peppercorns	3	8
Fish trimmings and bones, including heads, washed, with the eyes and gills removed	800g	2kg
Water	1 litre	2.4 litres
Fish fillets (e.g. bream, bass, haddock, mullet or gurnard), skinned, cut into chunks	180g	450g
Grated gruyère and baguette croûtes, to serve		
Rouille		
Garlic cloves, chopped	1	3
Salt		
Egg yolks, pasteurised	1	2
Olive oil	60ml	150ml
Cayenne pepper	¼ tsp	½ tsp
Tomato purée	2 tsp	4 tsp
Lemon juice	¼ tsp	½ tsp

Energy	Cal	Fat	Sat fat	Carb	Sugar	Protein	Fibre
2792 kJ	665 kcal	31.8g	5.6g	75.7g	9.5g	24.0g	7.9g

19 Fish forcemeat or farce

Salmon, sole, trout, brill, turbot, halibut, whiting, pike and lobster can all be used for fish forcemeat in the preparation of, for example, mousse of sole, mousselines of salmon, quenelles of turbot, all of which would be served with a suitable sauce (white wine, butter sauce, lobster, shrimp, saffron and mushroom).

Ingredient	4 portions	10 portions
Fish, free from skin and bone	300g	1 kg
Salt, white pepper		
Egg whites	1–2	4–5
Double cream, ice cold	250–500ml	600–1¼ litres

Energy	Cal	Fat	Sat fat	Carb	Sugar	Protein	Fibre
2403 kJ	582 kcal	54.4g	33.5g	1.7g	1.7g	21.4g	0.0g

1 Process the fish and seasoning to a fine purée.
2 Continue processing, slowly adding the egg whites until thoroughly absorbed.
3 Pass the mixture through a fine sieve and place into a shallow pan or bowl.
4 Leave on ice or in refrigerator until very cold.
5 Beating the mixture continuously, slowly incorporate the cream.

6 When half the cream is incorporated, test the consistency and seasoning by cooking a teaspoonful in a small pan of simmering water. If the mixture is very firm, a little more cream may be added, then test the mixture again and continue until the mixture is of a mousse consistency.
7 Form the mixture into mousses, mousselines or quenelles (see below) to cook.

Mousse of fish forcemeat

1 Butter the inside of individual or large moulds. Fill with forcemeat.
2 Cook in a bain marie in a moderate oven or in a low-pressure steamer.
3 Turn out of the mould for service.

Professional tips

As mousses are turned out of the mould for service, the mixture should not be made too soft, otherwise they will break up.

It is sounder practice to use individual moulds because for large moulds the mousse needs to be of a firmer consistency to prevent it collapsing.

Mousselines

1 Mould the forcemeat using two tablespoons, dipping the spoons frequently into boiling water to prevent the mixture sticking.
2 Place the mousselines into shallow buttered trays and cover with salted water or fish stock.
3 Cover with buttered greaseproof paper and cook gently in the oven or steamer.

Shellfish mousselines are best cooked in shallow individual moulds because of their looser texture.

Quenelles

Quenelles are made in various shapes and sizes as required:

- moulded with dessert or teaspoons
- piped with a small plain tube.

They are cooked in the same way as mousselines.

▲ Shaping quenelles of fish forcemeat

Variations

- For a coarser forcemeat with a different mouth feel, mince the fish using a coarse mincer, rather than puréeing it.
- When making lobster forcemeat, use raw lobster meat and ideally some raw lobster roe as this gives authentic colour to the mousse when cooked.
- For scallop forcemeat, use cooked scallops. In order to achieve sufficient bulk, it is sometimes necessary to add a little of another fish, e.g. whiting, sole, pike.

20 Smoked haddock paté

Ingredient	4 portions	10 portions
Smoked haddock	570g	1.4 kg
Fresh bay leaves	2	2
Milk	1 litre	2.2 litres
Double cream	140ml	350ml
Juice and zest of lemon	1	2
Butter, melted (plus a little extra)	75g	150g
Horseradish	1 tbsp	2 tbsp
Worcestershire sauce	½ tbsp	1 tbsp
Ground white pepper	½ tsp	1 tsp
Cayenne pepper	½ tsp	1 tsp
Salt		

Energy	Cal	Fat	Sat fat	Carb	Sugar	Protein	Fibre
2117 kJ	507 kcal	35.9g	22.0g	12.3g	11.7g	35.1g	0.3g

1 Place the haddock and bay leaves in a saucepan and cover with milk.

2 Bring to the boil, then simmer over a low heat for 8–10 minutes.

3 Drain in a colander, then place the haddock in a bowl.

4 Remove the skin and any bones. Allow to cool.

5 Transfer to a food processor with the remaining ingredients, season with salt, and blend to a coarse paste.

6 Transfer to a ramekin or terrine lined with cling film. Pour over enough melted butter to cover.

7 Refrigerate overnight, then serve in the ramekins or turn out from the terrine and slice very carefully. Serve with toast.

21 Fish soufflé

Haddock, sole, salmon, turbot, lobster, crab, etc. can all be used for soufflés.

1 Cook the fish in the butter and process to a purée.
2 Mix with the béchamel, pass through a fine sieve and season well.
3 Warm the mixture and beat in the egg yolks.
4 Carefully fold in the stiffly beaten egg whites.
5 Place into individual buttered and floured soufflé moulds.
6 Bake at 220°C for approximately 14 minutes; serve immediately. A suitable sauce may be offered, e.g. white wine, mushroom, shrimp, saffron, lobster.

Ingredient	4 portions	10 portions
Raw fish, free from skin and bone	300g	1 kg
Butter	50g	125g
Thick béchamel	250ml	600ml
Salt and cayenne pepper		
Eggs, separated	3	7

Energy	Cal	Fat	Sat fat	Carb	Sugar	Protein	Fibre
1278 kJ	308 kcal	21.1g	12.0g	5.6g	3.2g	23.9g	0.2g

> **Professional tip**
>
> If a large mould is used, increase the cooking time. The use of an extra beaten egg white will increase the lightness of the soufflé. A pinch of egg white powder added before whipping will strengthen the foam.
>
> Lobster soufflés can be cooked and served in the cleaned half shells of the lobsters.

22 Pike sausages (*cervelas de brochet*)

Ingredient	4 portions	10 portions
Pike meat	200g	500g
Egg white	1	3
Double cream	½ litre	1.4 litres
Salt, white pepper		
Sausage skins	100g	250g

Energy	Cal	Fat	Sat fat	Carb	Sugar	Protein	Fibre
3135 kJ	760 kcal	77.8g	47.3g	2.4g	2.4g	12.9g	0.0g

1　Prepare mousseline mixture as in Recipe 19.

2　Place sausage skins in water; then hang up, knot one end.

3　Using a piping bag, stuff the skins with mousse, being careful not to force it, then knot the other end with a piece of string.

4　Divide sausage into sections by loosely tying with string.

5　Gently poach the sausages in water at 82°C for 15 minutes.

6　Once cooked, carefully remove the sausages and allow to drain for 1–2 minutes.

7　With a sharp knife, remove the sausage skins carefully so as not to spoil the shape, drain well on a clean serviette and serve with a suitable sauce, e.g. tomato, and garnish.

The number of sausages produced will vary according to the size required.

Prepare a fish mousseline

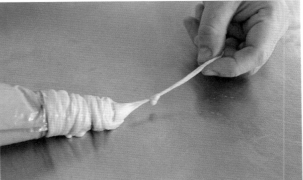

Pipe the mousse into the sausage skin

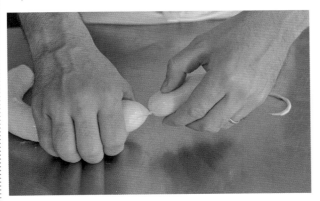

Divide into sections

Remove the skin after cooking

Variations

As with meat sausages, the variations of fish sausages (also called *boudin*) that can be produced are virtually endless. Almost any type of fish or shellfish can be used, and the fish may be chopped or minced instead of making up a firm mousseline mixture as above. The filling can also be a combination of two or more fish, and additional ingredients can be added (e.g. dry duxelle, brunoise of skinned red peppers, a suitable chopped herb such as dill, chervil, parsley and/or a touch of spice).

Video: fish sausages
http://bit.ly/NVVAZt

23 Bengali-style tilapia

Ingredient	4 portions	10 portions
Tilapia fillets	4 × 100–150g	10 × 100–150g
Turmeric	½ tsp	1¼ tsp
Ground coriander	½ tsp	1¼ tsp
Cumin	¼ tsp	1 tsp
Curry powder	1 tsp	2½ tsp
Water	2 tbsp	5 tbsp
Onion, finely shredded	50g	125g
Vegetable oil	50ml	125ml
Kachumbar		
Onion, finely shredded	1	2½
Plum tomatoes, chopped	2	5
Green chilli, finely chopped	1	2½
Fresh coriander, chopped	1 tbsp	2½ tsp

Energy	Cal	Fat	Sat fat	Carb	Sugar	Protein	Fibre
1131 kJ	271 kcal	15.3g	2.0g	6.1g	4.5g	28.1g	2.0g

1 Place the tilapia fillets in a suitable dish.

2 Mix the turmeric, coriander, cumin, curry powder and water in a bowl and pour over the fillets of fish. Leave to marinate.

3 Heat the oil in a frying pan and sweat the shredded onion until slightly coloured.

4 Add the fish fillets to the pan. Cook until golden brown on both sides.

5 Prepare the kachumbar by combining all the ingredients in a bowl. Season.

6 Serve the fish on a plate with the kachumbar on the side. Garnish with fresh limes if desired.

24 Thai-style shallow-fried fish

Ingredient	4 portions	10 portions
Fish fillets (e.g. sole, lemon sole, plaice), skinned	4 × 150g	10 × 150g
Marinade		
Garlic clove, chopped	3	6
Ginger, grated	2 tbsp	6 tbsp
Yellow bean sauce	1 tbsp	2 tbsp
Black bean paste	1 tbsp	2 tbsp
Light soy sauce	4 tbsp	10 tbsp
Thai fish sauce	2 tbsp	5 tbsp
Toasted shrimp paste	1 tsp	2 tsp
Palm sugar	2 tsp	5 tsp
Fish stock	150ml	425ml
Onion, finely chopped	1 small	2
Red chilli, chopped	1½	3
Lemon grass stalk, finely chopped	1	3

1 To make the marinade, combine all the ingredients. Rub into both sides of the fish, and leave it to stand for at least 1 hour in the chiller.

2 Remove the fish from the marinade and allow it to drain.

3 Transfer the remaining marinade to a small saucepan. Simmer to reduce to a sauce-like consistency.

4 Heat enough oil to shallow fry the fish in a suitable pan over medium heat, and slide the fish into the hot oil. Fry until the fish is cooked through, turning once.

5 Serve on a bed of boiled rice, pouring the reduced sauce over the fish. Garnish with lime, coriander and a julienne of lime zest.

Energy	Cal	Fat	Sat fat	Carb	Sugar	Protein	Fibre
1145 kJ	271 kcal	3.7g	0.5g	16.3g	9.5g	44.1g	1.4g

25 Steamed fish with Thai flavours

Ingredient	4 portions	10 portions
Fish fillets (e.g. grouper, tilapia)	4 × 150g	10 × 150g
Lemon grass stalk, finely sliced	4	10
Bird's eye chillies (*chili padi*), chopped	2	5
Red chillies, finely chopped	2	5
Lime, zest	1	3
Kaffir lime leaves, finely sliced	2	5
Dried tamarind skin (*asam keping*)	2	5
Oyster sauce	2 tbsp	5 tbsp
Thai fish sauce to taste	1 tsp	3 tsp
Lime juice to taste	1 tsp	3 tsp
Sugar to taste	1 tsp	3 tsp
Salt to taste		
Cloves of garlic, sliced	2	5
Spring onions, cut into julienne	15g	35g
Coriander, chopped	15g	35g
Cooking oil	1 tbsp	3 tbsp
Sesame oil	1 tbsp	3 tbsp

1 Place the fish in the centre of a baking dish. Place all the ingredients except the coriander, spring onion, garlic and oils on top of the fish.

2 Steam the fish for about 10–12 minutes.

3 Sprinkle the spring onion and coriander on the fish.

4 Heat up the cooking oil and sesame oil in a frying pan. Fry the garlic. Spoon the oil mixture over the fish to serve.

Note

Tamarind is a tropical fruit with an acid taste.

26 Steamed sea bass with asparagus and capers

Ingredient	4 portions	10 portions
Sea bass		
Sea bass fillets (approx. 160 g each, cut from a 2–3 kg fish) skin on, scaled and pin-boned	4	10
Court bouillon	1.5 litre	3.25 litres
Salt and pepper		
Caper dressing		
Fine capers	1 tbsp	3 tbsp
Aged balsamic vinegar	5 tbsp	12 tbsp
Lemon oil	5 tbsp	12 tbsp
Fennel cream		
Fennel bulbs	1	3
Garlic clove	1	3
Vegetable oil	50ml	125ml
Fish stock	100ml	250ml
Whipping cream	100ml	250ml
Butter	50g	125g
Garnish		
Asparagus spears	8	20
Extra fine green beans, blanched	100g	250g

Energy	Cal	Fat	Sat fat	Carb	Sugar	Protein	Fibre
3356 kJ	812 kcal	73.7g	24.6g	3.0g	2.6g	33.4g	1.7g

For the caper dressing

1 In a small bowl, combine all the dressing ingredients, adjust the seasoning to taste and set aside until serving.

For the fennel cream

1 Remove the root and stalks from the fennel bulbs and trim off any blemishes on the outer leaves.

2 Finely chop the fennel and crush the garlic. Heat the oil in a saucepan, add the fennel and garlic and cook slowly over a moderate heat for 6–7 minutes without letting them colour.

3 Add the fish stock, raise the heat under the pan and boil until reduced by half.

4 Pour in the cream, return to the boil and simmer until reduced by half.

5 Remove the pan from the heat and allow to cool slightly.

6 Transfer the mixture to a food processor and purée until fine. Return the sauce to a clean pan and set aside in a warm place – you need to whisk in the butter just before serving.

To complete

1 Ensure the sea bass is free of bones and scales.

2 Bring the court bouillon to a simmer and place the bass fillets in the bouillon.

3 Place the pan to one side, away from the heat, and allow to cook through with residual heat.

4 Meanwhile, reheat the asparagus and beans in a little butter.

5 Drain the fish, place it on the beans and asparagus on suitable plates.

6 Whisk the butter into the fennel cream, and serve on the side or pour over the fish. Garnish the dish with the caper dressing.

27 Roast monkfish with pancetta and white bean cassoulet

Ingredient	4 portions	10 portions
Paprika	10g	25g
Flour	40g	100g
Monkfish tail, boned, skinned and trimmed	1 large (1.5 kg)	3 large (1.5 kg each)
Lemons, juice of	1	3
Cassoulet		
Vegetable oil	60ml	150ml
Carrot, onion and celery brunoise	100g	250g
Pancetta, skinned and cut into lardons	200g	500g
Fish stock	20ml	50ml
White beans, cooked	200g	250g
Cream	100ml	250ml
Wholegrain mustard	1 tsp	3 tsp
Chives, chopped	1 tbsp	3 tbsp
Pickled–braised fennel		
Medium fennel bulbs	2	5
Vegetable oil	4 tsp	10 tsp
Carrots, peeled	50g	125g
Onions, peeled	50g	125g
Garlic clove, split	1	3
White wine vinegar	75ml	200ml
Chicken stock	200ml	500ml
Butter	50g	125g

Energy	Cal	Fat	Sat fat	Carb	Sugar	Protein	Fibre
3635 kJ	870 kcal	53.6g	41.2g	23.5g	5.2g	74.5g	6.4g

For the pickled-braised fennel

1 Remove the tops and root of the fennel, halve and shred the bulb finely.

2 Heat the oil in a saucepan and add all the vegetables. Cook for 5–6 minutes without letting them colour.

3 Add the vinegar and cook for a further 2 minutes, then add the chicken stock and simmer until half the liquid is evaporated and the fennel is tender.

4 Remove and discard the carrot and shallot, and keep the fennel mix warm.

For the white beans

1 Heat the vegetable oil in a thick-bottomed pan, add the brunoise and cook without colour for 2 minutes.

2 Add the pancetta and cook for a further 2 minutes, and then add the fish stock and the cooked beans.

3 Bring to the boil and reduce to a simmer for 5 minutes, ensuring the bacon flavour penetrates the beans.

4 Add the cream, reduce to a semi-thick sauce and keep warm.

To complete

1 Pre-heat the oven to 180°C.

2 Mix the paprika and flour together, and roll the monkfish well in the mix.

3 Heat the oil in a thick-bottomed pan that can go into an oven comfortably.

4 Place the monkfish in the pan and brown well on all sides, place in the oven for 12 minutes, turning and basting every 2–3 minutes.

5 Once cooked, allow to rest for a further 3 minutes.

6 Meanwhile bring the beans and fennel up to a simmer, adding the mustard and chives to the beans and the butter to the fennel.

7 Serve the fish on a bed of braised fennel, surrounded by the cassoulet. Garnish with chervil.

28 Oven-baked marinated cod with bok choi

Ingredient	4 portions	10 portions
Cod fillet (approx. 175g each) with skin off	4	10
Baby bok choi	4	10
Red pepper, cut into julienne	1	3
Yellow pepper, cut into julienne	1	3
Red onion, thinly sliced	1	3
Bean sprouts	100g	250g
Sesame oil	1 tbsp	3 tbsp
Vegetable oil	2 tbsp	5 tbsp
Soy sauce	1 tbsp	3 tbsp
Coriander, chopped	1 tbsp	3 tbsp
Marinade		
Soy sauce	100ml	250ml
Sesame oil	50ml	125ml
Rice wine	100ml	250ml
Black bean paste	100g	250g

Energy	Cal	Fat	Sat fat	Carb	Sugar	Protein	Fibre
1676 kJ	402 kcal	22.9g	3.1g	12.6g	10.9g	36.9g	2.8g

For the marinade

1 Mix all the ingredients together to a smooth consistency.

2 Place the cod fillets in the marinade for 12 hours.

3 After that, remove the fillets and wash off the excess marinade (it is not essential to get it all off).

For the cod

1 Pre-heat the oven to 180°C.

2 Put the cod on a lightly oiled baking tray and place in the oven.

3 Meanwhile heat the oil in a wok (if not available, use a heavy cast pan to retain the heat).

4 Place all the remaining ingredients in the wok/pan, excluding the soy and coriander.

5 Cook for 2–3 minutes until cooked but with a bite. Retain and keep warm.

6 Check the cod; it should take between 5 and 6 minutes according to thickness and, if timed well, will be ready when the vegetables are cooked.

7 Place the vegetables in the centre of the serving plate/dish and top with the cod. Any excess juices from the baking tray or the wok/pan may be poured around the fish, then serve.

Oil for stir frying was not included in the nutritional analysis.

Try something different
There is an eastern influence here, which can be adapted to suit a more European palate by changing the marinade to one using lemon and garlic, and serving spinach, green beans and even olives to accompany the fish.

29 Fillets of fish in white wine sauce

Ingredient	4 portions	10 portions
Fillets of white fish (e.g. sole, plaice)	400–600g	1–1.5 kg
Butter, for dish and greaseproof paper		
Shallots, finely chopped and sweated	10g	25g
Fish stock	60ml	150ml
Dry white wine	60ml	150ml
Lemon, juice of	Quarter	Half
Fish velouté	250ml	625ml
Butter	50g	125g
Cream, lightly whipped	2 tbsp	5 tbsp

Energy	Cal	Fat	Sat fat	Carb	Sugar	Protein	Fibre
1421 kJ	342 kcal	24.0g	12.8g	5.8g	0.9g	25.9g	0.2g

1 Skin and fillet the fish, trim and wash.
2 Butter and season an earthenware dish.
3 Sprinkle with the sweated chopped shallots and add the fillets of sole.
4 Season, add the fish stock, wine and lemon juice.
5 Cover with buttered greaseproof paper.
6 Poach in a moderate oven at 150–200°C for 7–10 minutes.
7 Drain the fish well; dress neatly on a flat dish or clean earthenware dish.
8 Bring the cooking liquor to the boil with the velouté.
9 Correct the seasoning and consistency, and pass through double muslin or a fine strainer.
10 Mix in the butter then, finally, add the cream.
11 Coat the fillets with the sauce. Garnish with *fleurons* (puff pastry crescents).

Variations

Add to the fish before cooking:
- fish *bonne-femme* – 100g thinly sliced white button mushrooms and chopped parsley
- fish *bréval* – as for *bonne-femme* plus 100g diced, peeled and deseeded tomatoes.

Alternatively, poach the fillets in white wine and serve with beurre blanc (see page 175).

Shellfish

Shellfish live mainly on or near sea beds. Similar fishing methods are used as for other fish species (trawling, netting, lines/hooks and cages). Some may be collected and hand-picked by divers and fish farming methods are also used. Mussels are frequently farmed attached to ropes that are then pulled in to harvest them. Work is being completed to ensure sustainability of shellfish stocks with fishing methods, limited fishing areas and information for consumers to help with maintaining stocks of shellfish.

Shellfish are all invertebrates, which means that they do not have an internal skeleton. The main groups are:
- **Molluscs** – have either an external hinged double shell (referred to as **bivalves** – examples are mussels and scallops), or a single univalve spiral shell (for example, winkles and whelks). They may also have soft bodies such as squid and octopus.
- **Crustaceans** – have tough protective outer shells and also have flexible joints to allow for quick movement as can be seen in crab and lobster.
- **Cephalopods** – the main edible cephalopods are squid (sometimes called calamari), cuttlefish and octopus. They are technically molluscs, but look very different and tend to be cooked differently. They need to be cooked very quickly or have a long slow cooking otherwise they can be tough and chewy.

Types and varieties of shellfish

Cockles – small bivalves with cream-coloured shells measuring approximately 2–3cm. Cockles live in sand so need thorough cleaning. They are washed in several changes of cold water then left in salt water until needed. Cook by steaming, in boiling salted water or on a preheated griddle, until the shell opens. Cockles are used in soups, salads, in fish dishes or on their own. They can replace mussels in any mussel recipe.

Clams – another small bivalve that opens on cooking. They have a mild sweet flavour and are very popular in USA for 'clam bakes' and for clam chowders. They are also frequently served with pasta. Razor clams have a long razor-like shape.

▲ Clams

Mussels – larger bivalves with a distinctive oval shape, blue/black shell and a good distinctive flavour. They are popular served on their own with the sauce they were cooked in or as a component of other dishes. Discard any mussels with broken shells and any that do not open after cooking.

▲ Mussels

Scallops – bivalves with a fan-shaped shell. They vary in size from 15cm for great scallops, around 8cm for bay scallops, and queen scallops that are the size of cockles. Inside is the round white flesh and the orange coral which is the roe and is often discarded. Scallops are popular and prices tend to remain high. Prise the two halves of the shell apart and cook by steaming, poaching, frying or grilling.

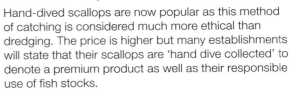

▲ Scallops

> **Professional tip**
>
> Hand-dived scallops are now popular as this method of catching is considered much more ethical than dredging. The price is higher but many establishments will state that their scallops are 'hand dive collected' to denote a premium product as well as their responsible use of fish stocks.

Oysters – highly regarded bivalves with a number of different species available. Oysters available around British coastlines are categorised as native, flat or rock oysters with Colchester and Whistable being significant oyster areas. Traditionally oysters are eaten raw so it is essential they are very fresh and cleaned well. Preparation usually involves prising the two halves of the shell apart with a small, pointed oyster knife. The oysters are then served with lemon and red pepper. Store oysters in their delivery boxes, in the refrigerator, covered with damp cloths or kitchen paper. The shells should be tightly closed or should close when tapped. Discard any that do not close.

Whelks and winkles – look a bit like snails and are usually available in the shell, ready cooked.

Lobster – remains very popular and commands high prices. They have large front claws that are bound or taped on live lobsters to prevent injury to the handler. Lobsters are dark blue/black when live, turning to bright red on cooking. There are some classic dishes for lobster such as Newburg and Thermidor but lobster simply cooked by boiling, steaming or grilling is probably the most popular. Lobster is served both hot and cold with a variety of sauces and the meat is also used on canapés, in soups, salads and composite fish dishes. Because lobster deteriorates very quickly, it is best bought live and cooked as needed.

▲ Lobster

> **Professional tip**
>
> There is much differing opinion about how live lobsters should be killed and one of the following is usually used:
> - placing the live lobster into a large pan of rapidly boiling water which kills it quickly
> - freezing the live lobster for a few minutes to make it drowsy then boiling as above
> - plunging the point of a sharp knife between the eyes to kill the lobster
> - use of a 'stun box' which kills the lobster quickly with an electrical charge.

Crab – best bought live and cooked as needed. Crabs should feel heavy for their size. The two types of flesh from a crab are very different. There is the soft brown, well flavoured flesh under the main shell and the firm white mild flavoured flesh in the claws. Serve as dressed crab on a buffet or with salad, in hors d'oeuvres, canapés, sandwiches, soups and fish cocktails.

▲ Crab

Shrimps – very small and pink/brown in colour. Used for garnishes, cocktails, salads, in soups and sauces, omelettes, canapés and potted shrimps.

Prawns – larger than shrimps and can be very large (such as tiger prawns). Cooked by boiling, steaming, grilling, frying and they have a wide range of culinary uses.

Scampi and Dublin Bay prawns – scampi are found in Mediterranean seas and the similar Dublin Bay prawn is found around Scottish coasts. They look like small lobsters and only the tail flesh is used.

Crayfish –the fresh water variant of the lobster and very popular on cold buffets, especially where lobster is used. They are about 8cm in length, dark brown or grey, turning red on cooking.

▲ Crayfish

Crawfish – similar to lobster, usually between 1–2 kg in weight, without the large claws, all the flesh is in the tail. Crawfish are killed in the same way as lobster, are brick red in colour when cooked and are popular on buffets because of their size and appearance.

Squid – popular in Mediterranian cuisine and available all year, squid has increased in popularity in the UK in recent years. Squid have a firm texture and a strong flavour, and weigh between 100g and 1kg. In cleaning and preparation, the ink sac and transparent cartilage must be removed. The tubular body can be cut into rings and shallow or deep fried, steamed or poached. The unsliced body can be stuffed and steamed or baked; whole squid can be cut and opened out flat and then grilled, pan-fried or griddled.

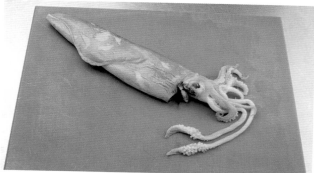

▲ Squid

Cuttlefish – are similar to squid but can be tougher, and can be hard to find because most UK catches go to France and Spain. They typically weigh between 200–450g and inside the cuttlefish body there is the hard white bone, known as the cuttlebone, sometimes used in bird cages.

Octopus – are larger than squid and cuttlefish, often between 225–500g. Octopus available in UK is often frozen (this is thought to help tenderise it). Mediterranean chefs will often tenderise octopus by beating it with a cutlet bat or against rocks. Either cook very quickly or stew slowly.

Choosing and buying shellfish

Shellfish are prized for their unique flavours and their tender, fine-textured flesh. They spoil rapidly due to their protein structure of certain amino acids that encourage bacterial development. To ensure freshness, buy shellfish live and cook them when needed. Purchase of all kinds of live shellfish is now possible due to globalisation and rapid transport systems which has increased trade and demand for shellfish.

When choosing shellfish:

- Shells should not be cracked or broken.
- Shells of oysters and mussels should be tightly shut or should do so when tapped. Discard any that remain open.
- Lobsters, crabs and prawns should have good colour and feel heavy for their size.
- Lobsters and crabs should have all of their limbs and claws.

Table 10.5 Seasonality of shellfish

	Jan	Feb	Mar	Apr	May	Jun	Jul	Aug	Sep	Oct	Nov	Dec
Crab (brown cock)												
Crab (spider)												
Crab (brown hen)												
Clams												
Cockles												
Crayfish (signal)												
Lobster												
Langoustines												
Mussels												
Oysters (rock)												
Oysters (native)												
Prawns												
Scallops												
Squid												

Key

Available	
At best	

Preparing shellfish

Opening and cleaning a scallop

  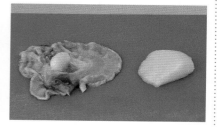

1 Using a small knife with a firm blade, insert the point into the place where the shell hinges and prise the shell open a little.

2 Run a flexible filleting knife over the flat side of the shell to release the scallop. Open the shell fully. Remove the scallop from the bottom shell.

3 Pull off the 'frill', and black stomach sac, leaving just the white scallop flesh and the orange coloured coral. Rinse thoroughly in cold water.

Shelling (shucking) oysters

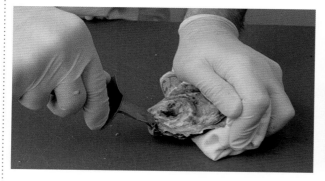 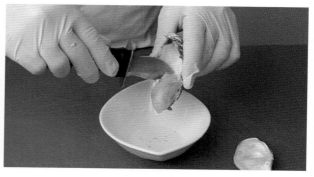

1 Hold the oyster with a thick oven cloth to protect your hand. With the oyster in the palm of your hand, push the point of the knife about 1cm deep into the 'hinge' between the lid and the body of the oyster.

2 Once the lid has been penetrated, push down. The lid should pop open.

3 Lift up the top shell, cutting the muscle attached to it.

4 Remove any splintered shell from the flesh and solid shell.

Preparing squid

1 Pull the head away from the body, together with the innards.

2 Taking care not to break the ink bag, remove the long transparent blade of cartilage (the backbone or quill.

3 Cut the tentacles just below the eye and remove the small round cartilage at the base of the tentacles.

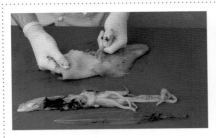

4 Scrape or peel off the reddish membrane that covers the pouch, rub with salt and wash under cold water.

5 Discard the head, innards and pieces of cartilage. Cut up the squid as required.

Preparing clams

To ensure freshness, the shells of clams should be tightly shut. They can be steamed and poached like mussels.

Clams should be soaked in salt water for a few hours so that the sand in which they exist can be ejected.

Clams can be cooked with lemon juice, au gratin (fresh breadcrumbs, chopped garlic, parsley, melted butter), in pasta, stir-fry and fish dishes as garnishes and/or a component of a sea food mixture, and as a soup (clam chowder). Certain types can be prepared and served raw.

Other preparation methods

- **Trimming** – many shellfish need some trimming to remove inedible parts and to improve their appearance. For the parts that need trimming, see the individual instructions in this chapter.
- **Cutting** – because of the small sizes involved, many shellfish need no cutting at all, only trimming or maybe shelling. However, cutting will be needed for items such as octopus, squid, lobster and where shell fish have been made into dishes such as terrines.
- **Marinating** – octopus, cuttlefish or large prawns may be marinated before cooking. This adds flavours and can tenderise.
- **Coating** – shellfish may be coated before frying and the usual coatings are flour/egg/breadcrumbs (pané), batter or a thin pastry such as filo.
- **Blending** – with shellfish, blending usually refers to soups that are blended until smooth and for producing terrines, mousseline, paté, stuffing or farce.

Cooking shellfish

The muscle makeup of shellfish is very different to that of meat. When cooked, the connective tissue becomes very fragile; the muscle fibres are shorter than in meat and the fat content relatively low. Shellfish need very little cooking and should just be cooked to the point where the proteins coagulate. Cooking beyond this stage causes the flesh to dry out, resulting in a tough, dry texture and loss of flavour. Shellfish are well known for their dramatic colour changes on cooking from blue/grey to vibrant pink, orange or red. This is because they contain red and yellow pigments called carotenoids bound to molecules of protein. Once heat is applied, the bonds are broken and the vibrant colours are shown.

Storage

As soon as shellfish are removed from their natural environment some spoilage will occur. After they are caught, they are often stored in temperature-controlled sea water tanks with aeration systems through the water. Some larger seafood establishments have similar systems to store their shellfish but these tend to be costly. Best practice is to arrange for daily fresh deliveries, store in a fish refrigerator below 4°C and preferably between 1–2° C. Make sure the shellfish remain moist during storage. Once cooked, either serve immediately or store for a short time at the required refrigerator temperature. Always store cooked shellfish on a top shelf, well away from raw products or keep in a cooked food refrigerator.

Shellfish such as lobsters can also be blanched quickly to remove shell and membrane but will still need to be stored as a raw product because they need further cooking.

Allergies

Some people can develop severe allergic reactions to certain shellfish. It is important to supply accurate information to those requesting allergy information. Pre-packaged shellfish must state all traces of crustacean shellfish in the package but do not have to list molluscs or cephalopods. Take note of 'hidden' shellfish sources such as stocks, nages and soups.

30 Sauté squid with white wine, garlic and chilli

Ingredient	4 portions	10 portions
Squid, cleaned	600g	1½ kg
Vegetable oil	60ml	150ml
Garlic cloves, crushed	2	5
Sprigs of parsley, chopped	3–4	7–8
Red chilli pepper, seeds removed, finely chopped	1	3
White wine	60ml	150ml
Fish stock	60ml	150ml

Energy	Cal	Fat	Sat fat	Carb	Sugar	Protein	Fibre
1084 kJ	260 kcal	17.6g	2.4g	2.3g	0.3g	23.4g	0.2g

1 Cut the squid into halves and then into thick strips.

2 Place a pan containing the vegetable oil on the hottest point on the stove.

3 Place the squid in the pan and sauté quickly (this will not take long – the squid will toughen if cooked for too long).

4 Add the garlic, chopped parsley and the chilli. Toss the squid around the pan, working in all the flavours.

5 Add the wine and stock, quickly bring to the boil, check the seasoning and serve.

Professional tip

The texture of squid is unlike that of other species. The flesh is very high in protein and dense, giving it that 'rubbery' texture when overcooked. Cook quickly and over a high heat.

Cut the body of the squid in half

Cut into strips

Cook the squid with the other ingredients

31 Crab tartlets or barquettes

Ingredient	4 portions	10 portions
Shallot, finely chopped, cooked in oil or butter	100g	250g
Raw mushroom, finely chopped	200g	500g
White wine	30ml	125ml
Crab meat, cooked	200g	500g
Salt and cayenne pepper		

Energy	Cal	Fat	Sat fat	Carb	Sugar	Protein	Fibre
730 kJ	175 kcal	8.7g	2.2g	10.5g	1.2g	12.2g	1.7g

1 Use short, puff or filo pastry. Bake blind.
2 Combine all the ingredients to make the filling.

32 Crab Malabar

Ingredient	4 portions	10 portions
Cooked crab meat, fresh or frozen and thawed	450g	1.2 kg
Vegetable oil	3 tbsp	7 tbsp
Onions, finely chopped	50g	125g
Cloves of garlic, finely chopped	3	7
Paprika	2 tsp	5 tsp
Thyme	½ tsp	1 tsp
Fennel seeds, crushed	½ tsp	1 tsp
Cayenne pepper	½ tsp	1 tsp
Fresh tomatoes, blanched, deseeded, diced	200g	500g
Salt, to taste		
Spring onions (both green and white parts), finely chopped	2	5
Lettuce leaves and chopped coriander to garnish		

Energy	Cal	Fat	Sat fat	Carb	Sugar	Protein	Fibre
1276 kJ	307 kcal	21.2g	2.6g	4.6g	2.5g	24.9g	1.4g

1 Pick over the crab meat and cut into 2 cm pieces.
2 Heat the oil in a large skillet with a lid over a moderate heat and cook the onions, stirring frequently, until golden but not brown. Add the garlic and cook for 1 minute. Add the paprika, thyme, fennel seeds, and cayenne pepper and cook for 2 more minutes. Add one-third of the tomatoes. Lower the heat and simmer covered for 15 minutes.

3 Remove from the heat and gently fold in the crab meat. Cover and refrigerate for 2 to 3 hours.
4 Immediately before serving, add salt to taste and fold in the remaining tomatoes and chopped spring onions.
5 Serve on a bed of lettuce, garnished with the chopped coriander.

33 Crab cakes with rocket and lemon dressing

Ingredient	4 portions	10 portions
Crab cakes		
Shallots, finely chopped	25g	60g
Spring onions, finely chopped	4	10
Fish/shellfish glaze	75ml	185ml
Crab meat	400g	1kg
Mayonnaise	75g	185g
Lemons, juice of	1	3
Plum tomatoes skinned, cut into concassé	2	5
Wholegrain mustard	1 tsp	3 tsp
Seasoning		
Fresh white breadcrumbs	200g	500g
Eggs, beaten with 100 ml of milk	2	5
Salad and lemon dressing		
Vegetable oil	170ml	425ml
White wine vinegar	25ml	60ml
Lemons, juice of	1	3
Seasoning		
Rocket, washed and picked	250g	625g
Reggiano Parmesan, shaved	100g	250g

Energy	Cal	Fat	Sat fat	Carb	Sugar	Protein	Fibre
3002 kJ	719 kcal	43.9g	10.5g	41.9g	4.4g	41.4g	4.1g

For the crab cakes

1 Mix the shallots, spring onions and the fish glaze with the hand-picked crab meat.

2 Add the mayonnaise, lemon juice, tomato concassé and mustard, check and adjust the seasoning.

3 Allow to rest for 30 minutes in the refrigerator.

4 Scale into 80–90g balls and shape into discs 1½ cm high, place in the freezer for 30 minutes to harden.

5 When firm to the touch, coat in breadcrumbs using the flour, egg and breadcrumbs.

6 Allow to rest for a further 30 minutes.

7 Heat a little oil in a non-stick pan, carefully place the cakes in and cook on each side until golden brown.

For the salad and dressing

1 Combine the oil, vinegar and lemon juice together, check the seasoning.

2 Place the rocket and Parmesan in a large bowl and add a little dressing, just to coat.

3 Place this in the centre of each plate, top with the crab cakes, garnish with radish and serve.

Professional tip

Any excess crab meat can be used up in this recipe – a quick, classic dish. The crab can be exchanged for salmon or most fresh fish trimmings.

34 Lobster Newburg

1 Gently reheat the lobster pieces in the butter.

2 Add Madeira and slowly reduce to a glaze.

3 Transfer to a different, cold pan (this reduces the risk of the egg splitting when the liaison is added). With the pan over gentle heat, pour in the liaison and allow to thicken by gentle continuous shaking; do not allow to boil. Correct seasoning, using a touch of cayenne if required.

4 Garnish with flat parsley. Serve with pilaff rice separately.

Lobster butter

A lobster butter made from the crushed soft lobster shells will improve the colour of the sauce.

1 Sweat the crushed lobster shells in 25–50g butter over a fierce heat, stirring well.

2 Moisten with stock or water, boil for 10 minutes, strain.

3 Clarify the butter by simmering to evaporate the liquid.

Ingredient		4 portions	10 portions
Cooked lobster meat cut into thickish pieces		400g	1¼ kg
Butter		50g	125g
Madeira		60ml	150ml
Cream	Liaison	120ml	250ml
Egg yolks		2	5

Energy	Cal	Fat	Sat fat	Carb	Sugar	Protein	Fibre
1556 kJ	375 kcal	30.7g	17.5g	0.6g	0.6g	24.1g	0.0g

Reduce the Madeira

During cooking, shake the pan to help thicken the sauce

35 Lobster tail gratin

Ingredient	4 portions	10 portions
Lobster		
Lobster tails (each from a 500–600g live lobster)	4	10
Butter	80g	200g
Dry sherry	20ml	50ml
Flour	20g	50g
Paprika	½ tsp	1 tsp
Cream	120ml	300ml
Seasoning		
Crumb topping		
Slices white bread	3	7
Butter	40g	100g
Chives, chopped	1 tbsp	2 tbsp
Seasoning		

Energy	Cal	Fat	Sat fat	Carb	Sugar	Protein	Fibre
1751 kJ	421 kcal	31.9g	19.5g	18.1g	1.6g	16.6g	0.6g

For the crumb topping

1 Remove crust from bread and place in food processor or grate finely.

2 Melt butter in a pan, add the breadcrumbs and cook until brown.

3 Add the chives and salt and pepper. When the lobster meat is returned to the shell, sprinkle over the meat.

For the lobster

1 Preheat the oven to 190°C.

2 Gently blanch the lobster tails until they are half done, then drain and cool.

3 Remove meat from shells and cut into small pieces, clean and save the shells.

4 Melt butter in a thick-bottomed pan. Stir in sherry and lobster, simmer for 2 minutes.

5 Stir in flour, paprika and cream until thickened, adjust the seasoning then return mixture to shells.

6 Place the filled and topped shells on a baking tray and bake in the oven for 10 minutes.

7 Serve immediately with a green salad or wilted greens.

> **Professional tip**
>
> Cornish or Scottish (native) lobsters are best for this recipe; their Canadian counterparts may be used but the native varieties will yield a better result.

36 Mussels

To prepare cooked, shelled mussels for use in other dishes, follow this recipe.

Ingredient	
Mussels	1 litre
Shallots or onions, chopped	25g

1 Scrape the shells to remove any barnacles, etc.

2 Wash well and drain in a colander.

3 Place the mussels and shallots in a thick-bottomed pan covered with a tight-fitting lid. Cook on a fierce heat for 4–5 minutes until the shells open completely.

4 Remove the mussels from their shells, checking carefully for sand, weed and beard.

5 Retain the liquid.

37 Moules marinières

Ingredient	4 portions	10 portions
Shallots, chopped	50g	125g
Parsley, chopped	1 tbsp	2 tbsp
White wine	60ml	150ml
Strong fish stock	200ml	500ml
Mussels	2kg	5kg
Butter	25g	60g
Flour	25g	60g
Seasoning		

Energy	Cal	Fat	Sat fat	Carb	Sugar	Protein	Fibre	Sodium
1900 kJ	452 kcal	14.3g	5.3g	18.1g	0.9g	61.4g	0.6g	1.5g

1 Take a thick-bottomed pan and add the shallots, parsley, wine, fish stock and the cleaned mussels.

2 Cover with a tight-fitting lid and cook over a high heat until the shells open.

3 Drain off all the cooking liquor in a colander set over a clean bowl to retain the cooking juices.

4 Carefully check the mussels and discard any that have not opened.

5 Place in a dish and cover to keep warm.

6 Make a roux from the flour and butter; pour over the cooking liquor, ensuring it is free from sand and stirring continuously to avoid lumps.

7 Correct the seasoning and garnish with more chopped parsley.

8 Pour over the mussels and serve.

38 Mussels with new potatoes

Ingredient	4 portions	10 portions
Mussels – cleaned	1 kg	2.5 kg
Small new potatoes or firm waxy potatoes, peeled and neatly diced	1 kg	2.5 kg
Fish stock	400ml	1 litre
Butter	25g	70g
Fennel	1 bulb	2½ bulbs
Shallots	2	5
Garlic	1 clove	2 clove
Lemongrass	1 stick	2½ sticks
Curry powder	2 teaspoons	5 teaspoons
Turmeric	½ teaspoon	1½ teaspoons
Double cream	200ml	500ml
Flat leaf parsley	2 tbsp	5 tbsp

Energy	Cal	Fat	Sat fat	Carb	Sugar	Protein	Fibre
2352 kJ	563 kcal	35.8g	20.9g	45.4g	5.2g	17.5g	5.4g

1　Clean and drain the mussels, finely shred the fennel, finely chop the lemongrass, garlic and shallots.

2　Place the mussels in a casserole and add 100ml boiling water. Put a lid on the pan and cook until the mussels open. Scoop them out of the pan and set aside. Strain the cooking liquid through a fine chinois and pour back into the (cleaned) casserole. Add the potatoes and the fish stock and simmer for about 15 minutes or until just tender.

3　Melt the butter in a small frying-pan or sauteuse and add the fennel, shallots and garlic. Cook these until soft, then add the lemongrass, curry powder and turmeric.

4　Add the contents of the frying-pan to the casserole containing the potatoes. Add the cream and bring to simmering point before adding the mussels.

5　Stir the mussels through the sauce, check seasoning then serve in bowls topped with chopped flat leaf parsley

39　Spanish-style clams

Ingredient	4 portions	10 portions
Cleaned clams	1 kg	2.5 kg
Tomato concassé	150g	450g
Dry white wine	150ml	450ml
Finely chopped onion	180g	450g
Paprika	1 tsp	2½ tsp
Small red chilli – finely chopped	½	1
Chopped coriander	1 tbsp	2 tbsp
Chopped flat leaf parsley	1 tbsp	2 tbsp
Olive oil	50ml	125ml
Garlic – crushed	1 clove	2 cloves
Salt and black pepper		

Energy	Cal	Fat	Sat fat	Carb	Sugar	Protein	Fibre
1521 kJ	364 kcal	14.4g	2.4g	10.7g	4.3g	41.3g	1.7g

1　Heat the olive oil in a large frying pan and sweat the onion until softened.

2　Stir in the paprika, garlic and chilli, cook for a further 30 seconds.

3　Add the concassé and wine and allow to simmer uncovered for five minutes, then add salt and pepper.

4　Add the cleaned and well drained clams, cover with a lid and shake in the sauce to coat them all well.

5　Cook for 3–5 minutes until all the clams are open.

6　Stir in the parsley and coriander, adjust seasoning and serve with crusty bread.

Variation
For a thicker sauce, add a little beurre manié before adding the clams.

40 Oyster fricassée

Ingredient	4 portions	10 portions
Oysters, shelled and juice retained	24	60
Cream	200ml	500ml
Butter	30g	75g
Wholegrain mustard	30g	75g
Parsley, finely chopped	1 tsp	3 tsp
Seasoning		
Pinch cayenne		

Energy	Cal	Fat	Sat fat	Carb	Sugar	Protein	Fibre
1389 kJ	337 kcal	34.3g	20.7g	2.2g	1.2g	5.2g	0.4g

1 Clean the oysters and retain on a clean tray in the refrigerator.

2 Heat the oyster liquor to boiling point, and strain through a double thickness of cheesecloth/muslin.

3 Add oysters to liquor and cook until plump, 1–2 minutes.

4 Remove oysters with a slotted/perforated spoon and place on a clean plate; cover with cling film.

5 Add the cream and butter to liquor and reduce to form a sauce consistency.

6 Add the wholegrain mustard (do not reboil), seasoning and parsley.

7 Return the oysters to the sauce, gently heat through, garnish with cayenne pepper and serve immediately.

41 Spiced coconut prawns

Energy	Cal	Fat	Sat fat	Carb	Sugar	Protein	Fibre
1192 kJ	285 kcal	20.9g	2.5g	9.9g	8.0g	18.5g	1.4g

Ingredient	4 portions	10 portions
Large raw prawns – heads removed	500g	1.5 kg
Fresh ginger	2cm piece	5cm piece
Green chillies	1	2
Garlic	1 clove	3 cloves
Coriander seeds	1 tbsp	2½ tbsp
Cumin seeds	1 tsp	2½ tsp
Ground turmeric	1 tsp	2½ tsp
Sesame seeds	1 tsp	2½ tsp
Ground almonds	50g	125g
Onion	1	2½
Spring onion	3	8
Fish stock	150ml	375ml
Sherry vinegar or white wine vinegar	1 tbsp	2½ tbsp
Nut oil or sunflower oil	2 tbsp	5 tbsp
Coconut milk	400ml	1 litre

Ingredient	4 portions	10 portions
To garnish		
Red chillies	½	1
Coriander leaves	2 tbsp	5 tbsp
Spring onion (optional)		

1 Peel the prawns but leave the tail piece intact and place the prawns in the vinegar with a little salt added.

2 Finely chop the ginger and the onion, cut the garlic into very thin slices, cut the chillies and spring onions into long, very thin shreds.

3 Place the seeds, ground turmeric and peppercorns into a grinder and grind to a powder.

4 Heat the oil in a frying pan or sauteuse and sweat the onion for five minutes. Add the ginger, green chilli and garlic and cook for a further minute.

5 Add the ground spices and cook gently for two minutes then add the ground almonds and spring onion.

6 Add the fish stock and coconut milk and cook for two minutes.

7 Add the prawns and cook just for 3–4 minutes so they don't overcook.

8 Adjust seasoning and scatter the red chillies and coriander, and perhaps strips of spring onion.

9 Serve with basmati rice.

42 Tempura scampi with mango salsa

Ingredient	4 portions	10 portions
Scampi, shell removed	20	50
For tempura batter		
Plain flour	50g	125g
Cornflour	50g	125g
Ice cold sparkling water	175ml	475ml
Seasoning		
For salsa		
Olive oil	20ml	55ml
Crushed garlic	1 clove	3 cloves
Finely chopped shallots	3	8
Long red pepper	1	3
Mango	1	3
Chopped coriander	2 tbsp	5 tbsp
Chopped flat leaf parsley	2 tbsp	5 tbsp
Basil leaves	2	5
Lemon juice	30ml	80ml
Lime juice	50ml	125ml
Red chilli – finely chopped	½	1½
Salt and pepper		

Energy	Cal	Fat	Sat fat	Carb	Sugar	Protein	Fibre
1056 kJ	250 kcal	6.6g	1.0g	29.6g	7.9g	20.0g	2.9g

1 Prepare the mango salsa by cutting the mango and red pepper into small, neat cubes and placing in a bowl with the shallots, garlic, oil, chilli, parsley and coriander. Add the lime and lemon juice and toss well.

2 Season with salt and pepper, and refrigerate until ready to serve.

3 Make the tempura batter by combining the flours and seasoning in a bowl. Gradually whisk in the ice cold sparkling water.

4 Dust the scampi lightly with flour then coat lightly in the batter and deep fry at 180°C.

5 Drain on kitchen paper and serve with the salsa, garnished with basil leaves.

Test yourself

1 (a) Name four flat fish and suggest a way that each of them could be cut/
       prepared into the required portion sizes.

   (b) Describe a suitable menu dish for a large flat fish stating the type of cut
       used.

2 What is an anadromous fish? Name one variety of anadromous fish and
   suggest three possible menu items using this fish.

3 (a) What causes the rapid spoilage of fish? How would you recommend to
       others the best way of storing fish to slow down spoilage?

   (b) What are the quality points you would look for in fish on delivery?

4 Describe how you would instruct a commis chef to:

   (a) Pin bone a fillet of cod.

   (b) Fillet a plaice.

   (c) Cut darnes from a salmon.

   (d) Gut a red mullet.

5 (a) Suggest five suitable methods for cooking fish and one type of fish suitable
       for each method.

   (b) What happens to the proteins in fish when it is cooked?

   (c) Why do many shellfish change from dark colours to bright pink or red when
       they are cooked?

6 (a) When catching sea fish what are the main methods used? Give an
       advantage and a disadvantage for each of them.

   (b) What is meant by 'aquaculture' and what are the advantages of
       aquaculture?

7 What are the differences between molluscs, crustaceans and cephalopods?
   Name two types of each and the preparation of one of them.

8 (a) Describe scallops.

   (b) How would you prepare them for a pan-fried recipe?

   (c) What are the advantages of using 'hand-dived scallops'?

9 (a) Most lobsters are purchased live and need to be killed when cooking them.
       What are the ways you could do this? Which method would you choose
       and why?

   (b) Suggest a lobster dish to put on the menu of a quality city centre
       restaurant.

10 When putting oysters on the menu for the first time, what instructions would
   you give to chefs about:

   (a) Checking the quality on delivery.

   (b) Proper storage.

   (c) How to prepare the oysters for service.

11 What months of the year would you have 'native' oysters on the menu?

11 Bread, dough and batter products

This chapter covers the following units:

NVQ:
→ Prepare, cook and finish complex bread and dough products.

VRQ:
→ Produce dough and batter products.

Introduction

Bread, as we would recognise it, first appeared in Ancient Egypt around 4000BC, and was the result of a happy accident between a brewery and a bakery. Froth from the fermenting beer blew over from the brewery and landed on some dough; the result was a lighter product. This raised (or leavened) bread was later helped by adding some of the previous day's dough, a method still used in the production of sourdough. It is generally accepted that the longest development time, with the minimum amount of yeast, will give bread the best flavour and character.

Although some methods have changed little over thousands of years, techniques are continually evolving and rules once written on tablets of stone are being broken as knowledge and understanding increase. For instance, it was taught that the liquid must be tepid, but in some modern recipes iced water is used (not just very cold water, but water with ice in it). A cold dough is easier to

shape. Also, improvers, used to speed up the production of factory bread, are now being used by artisan bakers, whose main interest is to improve the quality of the finished product.

Bread making involves few ingredients but is a complex science. Here we can only scratch the surface, but a thorough understanding of the reactions and effects of ingredients and processes is the difference between producing a good product and an excellent one.

Learning objectives

By the end of this chapter you should be able to:
→ Identify, prepare and finish a variety of fermented dough products to the recipe specifications, in line with current professional practice.
→ Explain techniques for the production of dough and batter products, including traditional, classical and modern skills, techniques and styles.
→ List appropriate flavour combinations.
→ Explain considerations when balancing ingredients in recipes for dough and batter products.
→ Explain the effects of preparation and cooking methods on the end product.
→ Describe how to control time, temperature and environment to achieve the desired outcome in dough and batter products.

Recipes included in this chapter

Food allergies and intolerance

It is estimated that up to 45 per cent of the population suffer from some sort of food intolerance which, while not life-threatening, can be very uncomfortable for sufferers. The symptoms vary from person to person and are numerous.

An allergy is an allergic reaction; it is much more serious and can be life-threatening.

An increasing number of people are intolerant to gluten (the protein found in wheat, barley and rye) and to a lesser extent, yeast. This is known as coeliac disease, a digestive condition which damages the lining of the small intestine. Sufferers of coeliac disease must avoid wheat- and flour-based products.

It is important that staff are well briefed on the ingredients used in fermented goods and that gluten-free and yeast-free products are kept separate to avoid contamination.

The main ingredients of fermented goods

Strong flour

This is the essential ingredient when making fermented products. Strong flour is a white wheat flour which has been processed to remove the outer skin (bran), the husk and germ. It has a high gluten or protein content. The gluten provides increased water absorption and elasticity, which is essential for allowing the dough to expand by trapping the gas produced by the yeast. Legislation requires that four specific nutrients in specific quantities are added to wheat flour; they are: iron, vitamin B1, nicotinic acid (niacin) and calcium carbonate.

Wholemeal and wheatmeal flour

As the names would suggest, these flours contain the whole wheat grain. Nothing is added or taken away, and they are considered to be a healthier alternative to white flour. Because the nutrients mentioned above are naturally present, these flours do not require any additions.

Rye flour

Rye is a type of grass which grows in harsh climates and is associated with Northern and Eastern Europe, where there is a tradition of rye breads. Rye was looked upon as being inferior to wheat. It has a low gluten content and will add flavour and texture when used with strong flour.

Spelt

A grain related to wheat but with less gluten and a nutty flavour.

Rice cones and rice flour

Coarsely ground rice can be sprinkled on the baking sheet before placing on the baking tray and on top of the bread, which adds texture and crunch to the products. Rice flour is ground much finer and is gluten-free.

Gram flour

This is flour made from chick peas.

Yeast

Yeast is a living organism which, when fed (on sugar or starch), watered and kept warm will multiply and produce carbon dioxide gas and ethyl alcohol. Yeast is essential to lighten or leaven fermented products. It comes compressed in a block (fresh) or dried, sometimes with the addition of ascorbic acid (vitamin C) which is an improver. Most recipes will state that yeast should be dissolved in the (usually warm) liquid, but more bakers are moving away from this and feel it is better to rub the yeast into the flour rather than 'drown' it in the liquid. Dried yeast is concentrated so remember to use half the quantity if using it in place of fresh yeast.

Using too much yeast can affect flavour and the products will stale more quickly. The best results are achieved by using the minimum quantity of yeast and allowing it to prove and develop over a longer time.

Improvers

Available in powder form, these usually have a vitamin C or ascorbic acid base, which speeds up the action of the yeast. This eliminates the need for bulk proving or BFT (bulk fermentation time) – see ADD method below. Fast-action dried yeast is a combination of dried yeast with an improver added.

Salt

In the last few years salt has had a bad press mainly because of its overuse in processed foods, but in truth we cannot live without salt in our diet, as a lack of it can lead to dehydration.

It is best to use sea or rock salt because unlike table salt, where most of the mineral content is destroyed by the high temperatures used in its production, sea and rock salts are relatively natural products.

As salt plays such a huge role in fermented goods, it must be measured carefully. It helps to stabilise the fermentation and strengthen the gluten, improves the crust texture, colour and flavour and lengthens the shelf life of products.

Professional tip

Salt must not come into direct contact with the yeast as it will slow or at worst kill the yeast and stop the fermentation.

Sugar

Sugar has also gained a bad reputation and is believed to make a significant contribution to obesity. Much of this is due to the food manufacturing industry using it in excessive quantities in their processed foods. Although yeast feeds on sugar, there is plenty naturally present in flour. The sugar added to a recipe is for flavour (it does, however, help speed up the fermentation).

> **Professional tip**
>
> Like salt, sugar should not come into direct contact with the yeast. You may come across recipes which recommend creaming the yeast and sugar together. Be aware that this is bad practice.

Liquid

Some bakers recommend using bottled water but this is by no means essential unless your tap water is heavily chlorinated. Milk, beer or buttermilk can also form part of the liquid content when making different products. The liquid is usually added at around body temperature, unless the air temperature is very hot or you want to delay the fermentation, in which case it is added cold. The quantity of liquid added may vary depending on the strength of the flour and the time of year (the flour tends to need more liquid during the winter).

Milk produces a softer dough. However, because milk contains an enzyme which inhibits yeast activity, they are not natural partners. Bringing the milk to the boil before adding it will neutralise this enzyme. Often, milk is added in the form of dried powder, as in several recipes featured in this chapter.

Fat

Butter is the most common fat in fermented goods recipes, but oils are also used. The advantage of using oil is that it does not have to be melted or rubbed in. Fat shortens the gluten strands, making the dough less elastic, so in most bread recipes only a little fat (or sometimes none at all) is used. In brioche, where the butter content is high (anything from 25–100 per cent of the flour weight), the texture is much softer and more like a cake.

Eggs

Eggs are used in enriched doughs. Good quality ingredients will result in good quality products (free-range hens are thought to produce eggs which contain a stronger albumen and have a deeper coloured yolk), and the fresher the egg the better. Most recipes will be based on a medium egg, but often the egg content is expressed as a liquid measurement which helps achieve greater accuracy. Pasteurised whole egg, yolks and whites are also available in litre cartons.

Classification of doughs

Fermented doughs can be divided into four categories: simple, enriched, laminated and batters.

Simple doughs

These include most types of bread. They consist of flour, salt, yeast, water and sometimes a small amount of fat.

Enriched doughs

Enriched doughs use the same ingredients as simple doughs, but are enriched with eggs and butter. Sometimes milk is used instead of water, and they may also include dried fruits, spices and a larger quantity of sugar.

Effects of adding ingredients which enrich a dough:
- increased nutritional content
- change in colour
- softer texture and a finer crumb
- may retard (slow down) the yeast
- increased shelf life.

Laminated doughs

These are either a simple dough which is layered with butter (such as croissant paste) or an enriched dough layered with butter (such as Danish pastry). The raising agent is a combination of yeast fermentation and lamination of the butter.

Batters

Batters, as the name would suggest, are much wetter than other doughs. They cannot be moulded or shaped and have to be cooked in moulds. Examples include savarin and blinis.

Methods

Straight dough

This is the simplest method and is often used when making simple bread doughs. All the ingredients are added at once and mixed together.

Sponge and dough

This method is often used for making enriched doughs (those which contain larger quantities of fat, sugar and eggs). As these additional ingredients slow down yeast activity, this method involves making a batter or 'sponge' with the yeast, liquid, and some of the flour to enable the yeast to start working before the other ingredients are added, giving the fermentation a 'kick start'.

Ferment and dough

This is very similar to the sponge and dough method and is used for the same reason. Here the yeast is mixed with the liquid and added to a well in the flour, a little of the flour is mixed in and more sprinkled over to cover. As this batter ferments, it 'erupts' through the flour crust, indicating it is ready to go to the next stage. This method is most often used when making very slack doughs or batters such as savarin and blinis.

Activated/accelerated dough development (ADD)

This process is primarily associated with the industrial manufacture of bread made by commercial methods. Such methods are concerned with saving time and reducing labour costs, which in turn saves money and maximises profit. Industrial bread produced by these methods is now often looked upon as contributing to an unhealthy diet – it is believed that high-speed milling reduces the nutritional content. Also, two or three times the usual amount of yeast is used, and the hydrogenated fats are often replaced with a fractionated variety. Breads produced using these methods have been linked to the increase in coeliac disease, gluten intolerance, yeast intolerance and irritable bowel syndrome.

The 'accelerator' referred to in this method is known as an 'improver' or conditioner, the main ingredient of which is more often than not ascorbic acid (vitamin C). The second feature of this process is the speed and length of mixing (very fast for a longer time). This generates heat in the dough, which speeds up the fermentation. This is also assisted by the improver, which develops the gluten much faster. The advantage of beating the dough in this way is that proving in bulk (see BFT below), as used in traditional methods, is not required, thus considerably reducing the production time. This method is now being adapted and developed by smaller bakeries, where the emphasis is on quality, rather than time and cost savings.

Retardation

By holding the dough at temperatures between 2°C and 4°C, yeast activity is stopped. This enables the dough to be made and held, then baked as needed. For example, a dough can be made, shaped and held overnight then baked in the morning.

This method is best undertaken using a piece of equipment known as a retarder–prover. A timer can be set allowing the dough to be proved slowly and be ready to bake at a specific time.

Such pieces of equipment go hand in hand with modern developments, and are contributing to quality improvements. For instance, using an improver with less yeast and a much longer final proving can result in a better flavour, character and shelf life – more like a sourdough where the proving is long and the fermentation natural.

Quality points

Although individual products will vary in terms of texture and crumb size, depending on the ingredients and methods used, they should all be:
- consistent in size and shape
- evenly coloured
- correctly finished according to type
- fresh.

The main equipment used

Planetary mixer

Every professional kitchen will have a planetary mixer. They are used when making anything other than the smallest quantities. The speed and length of mixing time are often stated in the recipes and affect the elasticity and development of the gluten.

Health and safety

Large mixers can be dangerous. All operatives must be over 18 and must be trained before using them. Some useful points to remember are:
- Do not use large equipment if working alone.
- If the mixer is table mounted always make sure it is secure.
- Use the correct attachment.
- All guards must be in place.
- Always start on the slow speed.
- Disconnect when not in use.

Prover

A prover provides the ideal conditions to encourage yeast activity. It provides warmth and moisture in a controlled environment. The optimum temperature is 24–26°C for most products. If no prover is available, then leave in a warm place covered with a large plastic bag.

Bakers' oven

Professional bakers' ovens are decked (separately controlled ovens stacked on top of each other) and often come with top/bottom heat controls and steam injection.

> **Professional tip**
>
> Most types of bread will benefit by being **baked with steam**. This helps to develop a crust, such as that found in a baguette. Baking in a dry heat will produce a softer finish.

> **Health and safety** !
>
> Care must be taken when removing baking sheets from the oven.
>
> - Always place the trays in the oven with the open end facing out.
> - Have a cooling oven rack next to the oven.
> - Always use two dry folded oven cloths to remove hot baking sheets.
> - Make others aware when removing hot baking sheets.

Digital scales

Accurate measurement is essential, particularly when the quantities are small. Salt, sugar and yeast all have a big effect on the speed and development of the fermentation.

Small equipment

It is expected that the kitchen will have a good supply of general items, such as:
- mixing bowls in various sizes
- sturdy baking sheets
- silicone mats
- cooling racks
- loaf/baking tins
- flour sieves
- flour brushes
- plastic scrapers
- measuring jugs
- rubber spatulas.

Definitions

Mixing and kneading

More often than not, unless very small quantities are required, doughs will be mixed mechanically. They should be mixed slowly on speed 1 to start with to make sure the flour is properly hydrated, and then mixed at speed 2 to develop the gluten. The dough should come away from the bowl, leaving the sides clean.

Kneading is a process done by hand when making smaller quantities. It consists of pulling and stretching the dough with the heel of the hand, changing the structure and creating a smooth, uniform, elastic mixture.

Biga/poolish/starter/levain

These are all terms for a soft ferment made with yeast, water, flour and sometimes other ingredients such as honey, yoghurt or raisins which all boost yeast activity and encourage the dough to ripen.

Proving

Most doughs are proved twice, first in bulk as explained below, and then a second time just before baking. During this process the yeast feeds on the sugars and ferments, producing carbon dioxide gas and ethyl alcohol. The gas is trapped by the gluten strands, which stretch and allow the dough to increase in size. This process develops character, texture and flavour.

Bulk fermentation time (BFT)

This is the period of time after the dough is made and before it is scaled and shaped. It is the time in which the dough is allowed to prove, developing the flavour and texture. The optimum time is usually one hour at 24–26°C.

Knocking back

After the bulk proving (BFT), the dough is 'knocked back' to expel the old gas and bring the yeast back into contact with the dough so it can prove a second time.

Scaling

Scaling is the process of dividing the dough by weight before shaping.

Baking

All fermented goods are baked in a hot to very hot oven which kills the yeast and stops the fermentation. Enriched doughs are baked at a lower temperature. Bread in particular will benefit from the addition of steam, which helps to develop the crust and colour. As stated earlier, modern bakers' ovens will often have a steam injection facility.

Eggwash

Many fermented products are eggwashed before baking. Be consistent – eggwash is best made to a recipe in a reasonable quantity, not just as you need it. For the best results use the yolks only, add 10 per cent water and a good pinch of salt. This will give a deep, rich glaze to the products.

Nutritional data

Nutritional data in this chapter is provided per loaf, per roll, etc.

Storage

If storing uncooked dough it should be flattened, placed inside a plastic bag and frozen. Some dough such as brioche can be left in the fridge overnight, but generally if doughs are stored in the fridge for too long the yeast will expire.

Cooked doughs can be kept at an ambient temperature for a short time but are best frozen if keeping for longer. Never store cooked dough products in the fridge as this will speed up staling.

When storing, all products should be placed in plastic bags or wrapped in cling film, labelled and dated. Cooked items should be kept separate from raw doughs and the stock rotated (using FIFO – first in, first out).

The Maillard reaction

The Maillard reaction was discovered by the French chemist Louis Maillard around the turn of the 20th century. It is the name given to the effect of baking – the reaction between carbohydrates (sugar) and amino acids (protein) when subjected to high temperatures. When heat is applied (baking), chemical changes take place on the surface, and the result is browning. This gives the products colour, aroma, flavour and texture. This process will not happen below 150°C.

Faults in fermented products

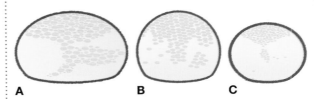

A   **B**   **C**

Under-ripe dough, sometimes referred to as a 'green' dough, means the dough is under-proved and will be small in volume and tough (C). When baked it will have a high crust colour, will possibly split at the side and have a very close texture.

Causes:
- not enough yeast and/or not enough time
- too much sugar, salt or spice.

Over-ripe dough has been over-proved, resulting in flat shapes where the products have risen and dropped back (A). When cooked they will have a loose, open texture and an anaemic colour.

Causes:
- too much yeast, too much proving
- lack of salt.

Test yourself

1. When making bread, what are the advantages of using the ADD method compared with traditional techniques using BFT?
2. Describe the production methods used to make the following products:
   (a) bread rolls
   (b) brioche à tête
   (c) Danish pastries.
3. Name and describe two products, other than croissants, made from croissant dough.
4. Describe the process used to 'retard' dough.
5. At what temperature does the Maillard reaction start?
6. Describe the process that leads to the Maillard reaction when baking products.
7. Describe three ways in which salt benefits fermented goods.

1 Seeded bread rolls

Ingredient	For 30 rolls
Strong flour	1 kg
Yeast	30g
Water at 37°C	600ml
Salt	20g
Caster sugar	10g
Milk powder	20g
Sunflower oil	50g
Eggwash	
Poppy seeds	
Sesame seeds	

Energy	Cal	Fat	Sat fat	Carb	Sugar	Protein	Fibre
550 kJ	130 kcal	1.7g	0.3g	26.0g	1.3g	4.6g	1.4g

Method: straight dough

1 Sieve the flour onto paper.

2 Dissolve the yeast in half the water.

3 Dissolve the salt, sugar and milk powder in the other half.

4 Add both liquids and the oil to the flour at once and mix on speed 1 for 5 minutes or knead by hand for 10 minutes.

5 Cover with cling film and leave to prove for 1 hour at 26°C.

6 'Knock back' the dough and scale into 50g pieces.

7 Shape and place in staggered rows on a silicone paper-covered baking sheet.

8 Prove until the rolls almost double in size.

9 Eggwash carefully and sprinkle with seeds.

10 Bake immediately at 230°C with steam for 10–12 minutes.

11 Break one open to test if cooked.

12 Allow to cool on a wire rack.

Professional tips

Instead of weighing out each 50g piece of dough, weigh out 100g pieces and then halve them.

Placing bread rolls in staggered rows means they are less likely to 'prove' into each other. The spacing allows them to cook more evenly and more will fit on the baking sheet.

Variations

Try using other types of seed such as sunflower, linseed or pumpkin.

For a beer glaze mix, together 150ml beer with 100g rye flour and brush on before baking.

2 Parmesan rolls

Use the bread recipe for seeded rolls (Recipe 1).

Ingredient	For 30 rolls
Grated Parmesan cheese	200g (approx.)

Energy	Cal	Fat	Sat fat	Carb	Sugar	Protein	Fibre
683 kJ	162 kcal	4.1g	1.6g	25.9g	1.2g	6.8g	1.4g

Method: straight dough

1 Follow method for seeded rolls (Recipe 1) up to Step 5.
2 Lightly flour work surface and roll the dough into a rectangle until 3cm thick.
3 Make sure the dough is not stuck to the surface.
4 Brush with water and cover with Parmesan.
5 Using a large knife, cut into squares 6 × 6cm.
6 Place on a silicone paper-covered baking sheet and leave to prove until almost double in size.
7 Bake at 230°C for 10–12 minutes with steam.
8 Cool on a wire rack.

Professional tips
● When making bread that requires rolling out (like Recipe 2) as opposed to being individually shaped (like Recipe 1), it is helpful to decrease the liquid content by 10 per cent so it will be easier to process.
● To ensure the squares are all the same size, mark a grid using the back of the knife before cutting.

3 Olive bread

Ingredient	Makes 4 loaves
Starter	
Yeast	40g
Water at 37°C	180ml
Strong flour	225g
Sugar	5g
Dough	
Strong flour	855g
Sugar	40g
Salt	20g
Water at 37°C	450ml
Olive oil	160ml
Green olives, cut into quarters	100g

Energy	Cal	Fat	Sat fat	Carb	Sugar	Protein	Fibre
5762 kJ	1367 kcal	46.7g	13.4g	215.5g	15.6g	34.8g	12.1g

Method: sponge and dough

1 For the starter, dissolve the yeast in the water, add the flour and sugar, mix well, cover and leave to ferment for 30 minutes.

2 For the dough, sieve the flour, sugar and salt into a mixing bowl, add the water followed by the starter and start mixing slowly.

3 Gradually add the oil and continue mixing to achieve a smooth dough.

4 Cover with cling film and prove for 1 hour or until double in size.

5 Knock back, add the olives and divide the dough into four.

6 Roll into long shapes and place on a baking sheet sprinkled with rice cones, return to the prover and leave until double in size.

7 Brush with olive oil and bake at 220°C for 20–25 minutes.

8 When cooked, the bread should sound hollow when tapped on the base.

9 Leave to cool on a wire rack.

Make up the starter

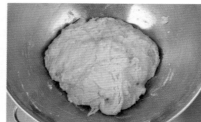
Starter ready for use after proving

Start mixing in the ingredients for the main dough, tearing up the starter

Continue mixing in the ingredients and working the dough

Shape the dough

Divide and roll into loaves

4 Bagels

Method: ferment and dough

1 Sieve the flour, place in a mixing bowl.
2 Make a well and add the yeast which has been dissolved in the water.
3 Mix a little of the flour into the yeast to form a batter, sprinkle over some of the flour from the sides and leave to ferment.
4 Mix together the salt, sugar, oil, egg yolk and milk.
5 When the batter has fermented add the rest of the ingredients and mix to achieve a smooth dough.
6 Cover and prove for 1 hour (BFT).
7 Knock back and scale at 50g pieces, shape into rolls and make a hole in the centre using a small rolling pin.
8 Place on a floured board and prove for 10 minutes.
9 Carefully drop into boiling water and simmer until they rise to the surface.
10 Lift out and place on a silicone-covered baking sheet, eggwash, sprinkle or dip in poppy seeds and bake at 210°C for 30 minutes.

Ingredient	Makes 10–12 bagels
Strong flour	450g
Yeast	15g
Warm water	150ml
Salt	10g
Caster sugar	25g
Oil	45ml
Egg yolk	20g
Milk	150ml
Poppy seeds	

Energy	Cal	Fat	Sat fat	Carb	Sugar	Protein	Fibre
942 kJ	223 kcal	7.1g	1.3g	37.3g	4.0g	6.8g	1.9g

Use a rolling pin to make a hole in the centre of each bagel

Poach the bagels in water

Eggwash the bagels and sprinkle with seeds before baking

5 Baguette

Ingredient	Makes 6 baguettes
Starter dough	
Yeast	5g
Water	135ml
Strong flour	100g
Rye flour	100g
Dough	
Cold water	680ml
Strong flour	1070g
Fine sea salt	15g
Yeast	22g

Energy	Cal	Fat	Sat fat	Carb	Sugar	Protein	Fibre
3100 kJ	728 kcal	3.1g	0.4g	159.7g	2.7g	25.4g	10.7g

Method: sponge and dough

1 For the starter dough, dissolve the yeast into the water.

2 Combine the two flours in a bowl, make a well, add the liquid and mix to a paste.

3 Cover the bowl and leave to ferment for 6 hours.

4 For the dough, add the cold water to the starter dough and mix well.

5 Place the flour in a mixing bowl, add the salt to one side and the yeast to the other, add the starter dough and mix slowly for 5 minutes.

6 Scrape down and continue to mix for 7 minutes on a medium speed until the dough is smooth and elastic. (A faster speed will generate heat and encourage fermentation.)

7 Prove until double in size.

8 Knock back, scale into 320g pieces and roll into long sticks.

9 Score by making 7 diagonal cuts with a sharp knife.

10 Bake at 250°C with steam for 20 minutes.

11 Cool on a wire rack.

Note: In France a baguette must weigh 320g and have 7 cuts along the top!

6 Ciabatta

Ingredient	Makes 4 loaves
Starter dough	
Yeast	10g
Water	180g
Strong flour	350g
Dough	
Strong flour	450g
Yeast	10g
Water	340g
Salt	20g
Olive oil	50g
Coarse semolina or rice cones	

Energy	Cal	Fat	Sat fat	Carb	Sugar	Protein	Fibre
3454 kJ	816 kcal	16.0g	4.9g	152.6g	4.2g	24.9g	8.3g

Method: sponge and dough

1 For the starter dough, dissolve the yeast in the water, add to the flour and mix to a dough. Place in a bowl, cover with cling film and leave for 24 hours.

2 For the dough, sieve the flour, rub in the yeast, break the starter dough into small pieces and add.

3 Add the water, salt and oil and mix on a slow speed for 5 minutes.

4 Place in an oiled bowl, cover and prove for 1 hour at 22°C.

5 Knock back the dough and divide into four pieces.

6 Roll into long cylinders and place on a baking sheet dusted with rice cones/semolina, brush with water and sprinkle over more rice cones/ semolina.

7 Prove until double in size and bake at 230°C for 18–20 minutes.

8 Cool on a wire rack.

Quality points

A good quality ciabatta should have:
- a flat shape (the name comes from the Italian word for slipper)
- a strong crust with small cracks
- a large, open-textured crumb.

7 Foccacia

Ingredient	Makes 8 × 15cm loaves
Starter dough	
Yeast	40g
Water at 37°C	180ml
Strong flour	225g
Sugar	15g
Dough	
Strong flour	855g
Sugar	30g
Salt	20g
Water at 37°C	480ml
Olive oil	180ml
Salamoia	
Water at 50°C	200g
Olive oil	200g
Salt	20g

Energy	Cal	Fat	Sat fat	Carb	Sugar	Protein	Fibre
3845 kJ	918 kcal	49.4g	7.1g	107.7g	7.8g	17.3g	5.6g

Possible garnishes
- Rosemary
- Chopped olives
- Thyme
- Sun-dried tomatoes
- Pesto.

1 For the starter dough, dissolve the yeast in the water, add the flour and sugar, mix well, cover and leave to ferment for 30 minutes.

2 For the dough, sieve the flour, sugar and salt into a mixing bowl, add the water followed by the starter dough and start mixing slowly.

3 Gradually add the oil and continue mixing to a smooth dough.

4 Cover with cling film and prove for 1 hour until double in size.

5 Brush individual round baking plates with olive oil.

6 Divide the dough into 250g pieces, roll into balls and then roll out on a lightly floured surface, keeping them round.

7 Place on baking plates, brush with olive oil and sprinkle over garnish.

8 Push your fingers into the dough to create dimples.

9 Prove for 30 minutes or until dough has risen slightly.

10 Bake at 230°C for 25–30 minutes.

11 Whisk together the salamoia ingredients to fully emulsify and brush over the cooked bread as soon as it comes out of the oven.

8 Wheatmeal loaf

Ingredient	Makes 2 loaves
Strong flour	125g
Wholemeal flour	625g
Yeast	25g
Water at 37°C	500ml
Sunflower oil	60g
Honey	40g
Salt	10g

Energy	Cal	Fat	Sat fat	Carb	Sugar	Protein	Fibre
6470 kJ	1531 kcal	37.9g	4.7g	262.5g	22.7g	51.4g	40.1g

Method: straight dough

1 Sieve the flours onto paper.
2 Dissolve the yeast in half the water.
3 Mix the honey, salt and oil with the rest of the water.
4 Add both liquids to the flours and mix well for 5 minutes or knead by hand for 10 minutes to achieve a soft and slightly sticky dough.
5 Place in a clean, oiled bowl, cover with cling film and prove at 26°C for 1 hour.
6 Knock back and divide in half, roll and shape, place into a prepared loaf tin and leave to prove until double.
7 Bake at 220°C for 35–40 minutes. When cooked, the bread should sound hollow when tapped on the base.
8 Leave to cool on a wire rack.

Variation
For walnut and sultana bread, add 50g walnuts and 50g sultanas to the recipe. When adding walnuts, first boil and then dry them before adding, or they will discolour the dough.

9 Sourdough

When it comes to bread, sourdough is the gold standard. It has a distinctive sour taste and a strong chewy crust.

The recipe looks long and complicated, but the different stages are necessary to get the process underway by making a 'sour' or starter dough. It is possible to have a 'sour dough' that is very old and has been passed down from one generation to the next. This will have a well developed and distinctive character that only comes with age.

A routine must be established to keep the dough alive. Even if you do not intend to bake, you must **refresh** the dough regularly by taking out what you would use and adding an equal quantity back. At least every two days, take out 500g dough. If you are not going to use it to bake, then discard it and add back in 500g flour with 250g water, mix, cover and place in the fridge.

It is important not to sanitise bowls and surfaces with something that destroys all known germs, as the wild airborne yeasts that you rely on for fermentation will not survive.

Day 1

Ingredient	Makes 2 loaves around 1 kg each
Sour	
Strong white flour	1 kg
Warm water	750g

Mix together in a bowl, cover with a plastic bag and secure with an elastic band and leave to ferment at room temperature.

Day 2

Ingredient	Quantity
Strong white flour	1 kg
Warm water	750g

Add the flour and water to the day 1 mixture, mix well, cover and leave as before. It is important to use the same bowl and the same plastic bag.

Day 3

Ingredient	Quantity
Strong white flour	500g
Warm water	250g

Repeat as day 2.

Days 4 and 5

Repeat as for day 3 and leave for a further two days before using.

The sourdough loaf

Ingredient	Quantity
Strong white flour	1 kg
Yeast	20
Salt	20g
Sour	500g
Olive oil	20g
Water	500g

1 Rub the yeast into the sieved flour and place in a mixing bowl.

2 Add the salt to one side and add the sour, oil and water.

3 Mix on the slow speed for 5 minutes, then turn up the speed and continue mixing for another 5 minutes.

4 Cover and rest for 4 hours.

5 Divide in half and shape into rounds.

6 Leave to prove until double in size.

7 Dust with flour and slash the top.

8 Bake at 230°C with steam for 30–35 minutes.

Sourdoughs do not usually contain any additional yeast but rely solely on natural airborne yeasts for fermentation. This recipe does include a small amount of additional yeast to boost fermentation and shorten the production time.

Quality points

A good quality sourdough will have:
- a slightly acidic taste with characteristic sour notes
- a very strong crunchy crust
- an open-textured dense crumb
- a dark colour with distinctive cuts
- excellent keeping qualities.

Professional tip

Allow the 'sour dough' to come up to room temperature before adding it to the other ingredients.

Variations
- Rye sourdough – use the above recipe but replace half the white flour with dark rye.
- Wholemeal sourdough – use the white starter dough but use 25% white flour to 75% wholemeal, and reduce the water by 10%.

10 Pain de campagne

Ingredient	Makes 2 large or 4 smaller loaves
Starter dough	
Strong flour	200g
Dark rye flour	50g
Yeast	5g
Water	175g
Salt	5g
Dough	
Strong flour	500g
Dark rye flour	100g
Yeast	5g
Water at 40°C	400ml
Salt	7g

Energy	Cal	Fat	Sat fat	Carb	Sugar	Protein	Fibre
6186 kJ	1453 kcal	6.5g	0.9g	320.7g	4.9g	48.2g	26.2g

Method: sponge and dough

1 For the starter dough, mix the two flours and rub in the yeast, add the water and salt and mix.

2 Place in an oiled bowl, cover with cling film and leave in the fridge overnight.

3 For the dough, place the flours in a mixing bowl and rub in the yeast, add the starter dough, water and salt and mix slowly for 5 minutes.

4 Cover and prove for 1 hour.

5 Knock back and divide; shape into round loaves.

6 Score the tops with a sharp knife, dust with flour and prove until double in size.

7 Bake at 220°C for 30–40 minutes.

11 Bun dough and varieties of bun

Ingredient	12 buns	24 buns
Strong flour	500g	1 kg
Yeast	25g	50g
Milk (scalded and cooled to 40°C)	250ml	500ml
Butter	60g	120g
Eggs	2	4
Salt	5g	10g
Sugar	60g	120g

Method: sponge and dough

1 Sieve the flour.

2 Dissolve the yeast in half the milk and add enough of the flour to make a thick batter, cover with cling film and place in the prover to ferment.

3 Rub the butter into the rest of the flour.

4 Beat the eggs and add the salt and sugar.

5 When the batter has fermented, add to the flour together with the liquid.

6 Mix slowly for 5 minutes to form a soft dough.

7 Place in a lightly oiled bowl, cover with cling film and prove for 1 hour at 26°C.

8 Knock back the dough and knead on the table, rest for 10 minutes before processing.

Bun wash

Milk	250ml
Caster sugar	250ml

Bring both ingredients to the boil and brush over liberally as soon the buns are removed from the oven. The heat from the buns will set the glaze and prevent it from soaking in, giving a characteristic sticky coat.

Sift the flour

Rub in the fat

Make a well in the flour, and pour in the beaten egg

Pour in the liquid

Fold the ingredients together

Knead the dough

Before and after proving: the same amount of dough is twice the size after it has been left to prove

▲ Chelsea buns, hot cross buns and Bath buns

Chelsea buns

Ingredient	12 buns	24 buns
Basic bun dough	1 kg	2 kg
Melted butter	60g	120g
Caster sugar	50g	100g
Currants	100g	200g
Sultanas	100g	200g
Mixed peel	30g	60g

1 Roll out the dough on a lightly floured surface into a rectangle 25cm deep.
2 Brush with melted butter and sprinkle over caster sugar, followed by the dried fruit and mixed peel.
3 Eggwash the far edge and roll up lengthways like a Swiss roll, pinch to seal.
4 Brush the outside with melted butter and cut into 3cm wide slices.
5 Line a deep-sided baking tray with silicone paper and lay in the slices so they are touching.
6 Allow to prove.
7 Bake at 220°C for 15–20 minutes.
8 Brush with bun wash as soon as they come out of the oven and break to separate.

Hot cross buns

Ingredient	12–14 buns	24 buns
Basic bun dough	1 kg	2 kg
Currants	75g	150g
Sultanas	75g	150g
Mixed spice	5g	10g
Crossing paste		
Strong flour	125g	250g
Water	250ml	500ml
Oil	25ml	50ml

1 Add the dried fruit and spice to the basic dough, mix well.
2 Scale into 60g pieces and roll.
3 Place on a baking sheet lined with silicone paper in neat rows opposite each other and eggwash.
4 Mix together the ingredients for the crossing paste. Pipe it in continuous lines across the buns.
5 Allow to prove.
6 Bake at 220°C for 15–20 minutes.
7 Brush with bun wash as soon as they come out of the oven.

Variation
To make **fruit** buns, proceed as for hot cross buns without the crosses.

Bath buns

Ingredient	Makes 12–14 buns
Basic bun dough	1 kg
Bun spice	20ml
Sultanas	200g
Sugar nibs	360g
Egg yolks	8

1 Mix the bun spice into the basic dough and knead.
2 Add the sultanas, two-thirds of the sugar nibs and all the egg yolks.
3 Using a plastic scraper, cut in the ingredients (it is usual for the ingredients not to be fully mixed in).
4 Scale into 60g pieces.
5 Place on a paper-lined baking sheet in rough shapes.
6 Sprinkle liberally with the rest of the nibbed sugar.
7 Allow to prove until double in size.
8 Bake at 200°C for 15–20 minutes.
9 Brush with bun wash as soon as they come out of the oven.

Swiss buns

Ingredient	Makes 12–14 buns
Basic bun dough	1 kg
Fondant	500g
Lemon oil	5ml

1 Scale the dough into 60g pieces.
2 Roll into balls then elongate to form oval shapes.
3 Place on a baking sheet lined with silicone paper, eggwash.
4 Allow to prove.
5 Bake at 220°C for 15–20 minutes.
6 Allow to cool then dip each bun in lemon-flavoured fondant. They can be decorated with confit of lemon.

Doughnuts

Ingredient	Makes 12 doughnuts
Basic bun dough	1 kg
Caster sugar	500g
Raspberry jam	250g

1 Scale the dough into 60g pieces.
2 For ring doughnuts, roll into balls and make a hole in the dough using a rolling pin. For jam doughnuts, shape as for bread rolls.
3 Prove on an oiled paper-lined tray.
4 When proved, carefully place in a deep fat fryer at 180°C.
5 Turn over when coloured on one side and fully cook.
6 Drain well on absorbent paper.
7 Toss in caster sugar.
8 For the jam doughnuts, make a small hole in one side and pipe in the jam.

Health and safety !

As a fryer is not a regular piece of equipment found in a patisserie, a portable fryer is often used to make doughnuts. Always make sure it is on a very secure surface in a suitable position. Never attempt to move it until it has completely cooled down. In addition, extreme care must be taken to avoid serious burns.
● Only use a deep fat fryer after proper training.
● Make sure the oil is clean and the fryer is filled to the correct level.
● Pre-heat before using but never leave unattended.
● Always carefully place the products into the fryer – never drop them in. Use a basket if appropriate.
● Never place wet products into the fryer.

Variation
The caster sugar can be mixed with ground cinnamon.

12 Savarin dough and derivatives

Ingredient	Makes 35 individual items
Basic dough	
Strong flour	450g
Yeast	15 g
Water at 40°C	125ml
Eggs	5
Caster sugar	60g
Salt	Pinch
Melted butter	150g

Energy	Cal	Fat	Sat fat	Carb	Sugar	Protein	Fibre
896 kJ	212 kcal	4.6g	2.5g	42.8g	33.2g	2.6g	0.8g

Method: ferment and dough

1 Sieve the flour and place in a bowl. Make a well.

2 Make a ferment by dissolving the yeast in the water and pour into the well.

3 Gradually mix the flour into the liquid, forming a thin batter. Sprinkle over a little of the flour to cover, then leave to ferment.

4 Whisk the eggs, sugar and salt.

5 When the ferment has erupted through the flour, add the eggs and mix to a smooth batter. Cover and leave to prove until double in size.

6 Add the melted butter and beat in.

7 Pipe the batter into prepared (buttered and floured) moulds one-third full.

8 Prove until the mixture reaches the top of the mould and bake at 220°C for 12–20 minutes, depending on the size of the mould.

9 Unmould and leave to cool.

Cream the yeast in milk to make a ferment

Add the dissolved yeast to the flour, and sprinkle a little flour over it

The mixture after fermentation

Add beaten eggs, sugar and salt

The dough after proving

Add the butter

After proving, pipe into moulds

After proving for the final time

Professional tip

Savarin and savarin-based products are never served without first soaking in a flavoured syrup. They are literally dry sponges and should be:

- golden brown in colour with an even surface
- smooth, with no cracks, breaks or tears
- evenly soaked without any hard or dry areas
- sealed by brushing with apricot glaze after soaking.

Cooked products can be stored in the fridge overnight, but if left for too long they will dry out and cracks will appear. They are best wrapped in cling film, labelled and stored in the freezer.

Savarin syrup

It is important that savarin syrup is at the correct density. It should measure 22° Baumé on the saccharometer (see page 477). If the syrup is too thin, the products are likely to disintegrate. If it is too dense it will not fully penetrate the product and leave a dry centre.

Ingredient	Makes 3 litres
Oranges	2
Lemons	2
Water	2 litres
Sugar	1kg
Bay leaf	2
Cloves	2
Cinnamon sticks	2

1 Peel the oranges and lemons and squeeze the juice.

2 Add all the ingredients into a large pan and bring to the boil, simmer for 2–3 minutes and pass through a conical strainer.

3 Allow to cool and measure the density (it should read 22° Baumé). Adjust if necessary. (More liquid will lower the density, more sugar will increase it.)

4 Reboil then dip the savarins into the hot syrup until they swell slightly. Check they are properly soaked before carefully removing and placing onto a wire rack with a tray underneath to drain.

5 When cooled, brush with boiling apricot glaze.

Savarin with fruit

A savarin is baked in a ring mould, either large or individual.

Before glazing, sprinkle with kirsch and fill the centre with prepared fruit. Serve with a crème anglaise or raspberry coulis.

Marignans Chantilly

A marignan is baked in an individual boat-shaped mould or barquette.

Split the glazed marignans and fill with Chantilly cream. Once split and before filling, they can be sprinkled with rum or Grand Marnier.

Decorate with cigarettes of white chocolate.

> **Professional tip**
>
> Savarin paste is notorious for sticking in the mould. Always butter moulds carefully and then flour them. After use do not wash the moulds but wipe clean with kitchen paper.

Blueberry baba

For a baba, currants are traditionally added to the basic savarin dough before baking.

As for marignans, split the glazed baba and fill with crème diplomate and blueberries or Chantilly cream. Decorate with blueberries and chocolate.

13 Brioche

▲ Brioche à tête (front) and brioche Nantaise

The quantity of butter added to a brioche dough can vary from 25 to 100 per cent of the flour weight according to type, as in the list below. The higher the butter content the more difficult it is to handle. (The standard is to use 50 per cent butter.)

- Brioche commune – uses 25 per cent butter.
- Brioche à tête – uses 50 per cent butter.
- Brioche mousseline – uses 75 per cent butter.
- Brioche surfine – uses 100 per cent butter.

Ingredient	Makes 2 × 500g loaves or 16 individual
Basic dough	
Strong flour	500g
Yeast	15g
Milk (scalded and cooled)	70ml
Eggs	4
Salt	15g
Sugar	30g
Butter, diced	250g

Energy	Cal	Fat	Sat fat	Carb	Sugar	Protein	Fibre
1696 kJ	406 kcal	23.8g	13.9g	41.4g	4.4g	9.2g	2.1g

Method: straight dough

1 Sieve the flour onto paper and place in a mixing bowl.
2 Dissolve the yeast in the milk.
3 Beat the eggs, sugar and salt.
4 Add both liquids to the flour and mix for 5 minutes on speed 1 with the dough hook, followed by 5 minutes on speed 2.
5 Replace the dough hook with a beater and gradually add the butter, and mix until smooth.
6 Rest in the fridge for 1 hour before processing.

Brioche à tête

This makes 16 brioches.

1 Scale the dough into 50g pieces.
2 First roll into balls and then, using the side of the hand, almost remove the top third, drop into buttered brioche moulds. Using a finger, push down so the head sits neatly in the centre.
3 Prove at 22°C until double in size.
4 Eggwash carefully and bake at 210°C for 15–20 minutes.
5 Unmould immediately and cool on a wire rack.

Brioche Nantaise

This makes three standard-sized loaves.

1 Scale the dough into 40g pieces
2 Roll each piece into a ball.
3 Place the balls side by side in a buttered loaf until full.

4 Prove at 22°C until mixture reaches the top of the tin.
5 Eggwash carefully and bake at 200°C for 25–30 minutes.
6 Unmould immediately and leave to cool on a wire rack.

Brioche mousseline

This makes four brioches.
The butter content can be increased to 75 per cent of the flour weight.

1 Divide the dough into four pieces and shape into balls.
2 Prepare the tins by buttering then lining with silicone paper. (Washed A2½ tins can be used in place of a specialist mould – just make sure the paper comes about 4cm above the top of the tin.)
3 Elongate the brioche and carefully drop into the tin, snip the top with a pair of scissors to form a cross.
4 Prove at 22°C until the mixture reaches the top of the tin.
5 Bake at 210°C for 20–25 minutes.
6 Unmould immediately and cool on a wire rack.

Variation: Croûte Bostock

Slice the brioche mousseline into 2cm rounds, spread each with almond cream or frangipane, slightly domed in the centre, and sprinkle with flaked almonds. Place on a paper-lined baking sheet and bake for 12–15 minutes at 210°C. When baked, dust with icing sugar and glaze under the salamander. Serve with crème anglaise.

Brioche en tresse

This makes two loaves.

1 Divide the dough into half.
2 Divide each piece into three or five.
3 Roll each piece into long strands and plait.
4 Prove at 22°C until double in size.
5 Eggwash and bake at 210°C for 20–25 minutes.
6 Cool on a wire rack.

Brioche couronne

This makes two loaves.

1 Divide the dough into two pieces and shape into round balls.
2 Using a rolling pin, make a hole in the centre and, moving the rolling pin in circles, make the hole bigger until the brioche is around 20cm in diameter.
3 Place on a silicone paper-covered baking sheet. Carefully eggwash and snip around the top edge with scissors.
4 Prove slowly at 22°C until almost double in size. Carefully re-eggwash around the base and bake at 205°C for 20–25 minutes.
5 Cool on a wire rack.

Quality points

A good-quality brioche will have:
● a soft crust with a deep golden colour
● a fine crumb and a short cake-like texture.

---

14 Croissants

Ingredient	Makes 35 pieces
The détrempe or basic paste	
Strong flour	960g
Yeast	25g
Cold water	600ml
Salt	12g
Sugar	30g
Milk powder	30g
Pastry butter (beurre sec)	500g

Energy	Cal	Fat	Sat fat	Carb	Sugar	Protein	Fibre
867 kJ	207 kcal	12.1g	7.5g	22.1g	1.8g	3.8g	1.1g

1 Sieve the flour onto paper and place in a mixing bowl.
2 Dissolve the yeast in half the water.
3 Dissolve the salt and sugar in the rest of the water and whisk in the milk powder. Add both liquids to the flour and mix slowly for 5 minutes.
4 Take a large tray (approximately 30cm × 60cm) and cover with cling film.
5 Roll the dough into a large rectangle the same size as the tray, lay on the tray, cover with cling film and rest for 25 minutes.
6 Process the butter by rolling out to two-thirds the size of the dough inside a plastic bag.
7 Place the butter on the dough, leaving one third uncovered, seal the edges.
8 Fold over the uncovered dough and then fold over the other third, giving three layers of dough and two layers of butter.
9 Roll out into a rectangle the same size as before and shake before folding into three.
10 Rest in a cool place for 30 minutes.
11 Repeat steps 9 and 10 twice more, giving three single turns in total.

12 Roll out into a rectangle 20cm deep and 3mm thick, shake well.

13 Cut into isosceles triangles. Make a small cut in the short edge of each triangle and roll away from yourself as tightly as possible, stretching the dough at the same time. Seal the tip with eggwash.

14 Lay on a silicone paper-covered baking sheet and bend the ends around to meet in the middle to form the crescent shape. Arrange in staggered rows, allowing the croissants room to expand.

15 Prove at 22°C for 35–40 minutes until double in size. Carefully eggwash and bake at 230°C for 15 minutes. Cool on a wire rack.

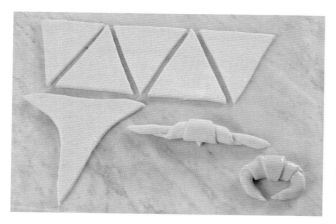

▲ Folding croissants

Variations

Before rolling and shaping into crescents, they can be filled with Parma ham or gruyère cheese, or for a sweet variation use almond cream and sprinkle the top with a few flaked almonds after eggwashing.

Pain aux raisins

Ingredient	Makes 35 pieces
Crème pâtissière	500g
Raisins, pre-soaked in rum	50g

1 Follow the instructions for croissants up to Step 12.

2 Spread the dough with the softened crème pâtissière and sprinkle with the raisins.

3 Roll up as for a Swiss roll and seal the edge with eggwash.

4 Cut into 1–2cm slices and lay flat on a silicone paper-covered baking sheet, spacing as for croissants.

5 Prove at 22°C for 35–40 minutes until almost double in size. Carefully eggwash and bake at 220°C for 20 minutes. Cool on a wire rack.

6 Brush with boiled apricot glaze.

Pain au chocolat

Ingredient	Makes 30 pieces
Chocolate batons	50 (approx.)

1 Follow the instructions for croissants up to Step 12.

2 Cut into rectangles 5 × 8cm.

3 Lay on two batons (or one if preferred) of chocolate. Eggwash and fold over the dough.

4 Space as before on a papered baking sheet and prove at 22°C until double in size.

5 Carefully eggwash and bake at 230°C for 15 minutes. Cool on a wire rack.

▲ Filling and folding pain au chocolat

15 Danish pastries

Method: straight dough and lamination

1 Sieve the flour onto paper and place in a mixing bowl.

2 Dissolve the yeast in the milk.

3 Whisk together the eggs, oil, salt and sugar.

4 Add both liquids to the flour and mix to a soft dough.

5 Take a large tray (approximately 30cm × 60cm) and cover with cling film.

6 Roll the dough into a large rectangle the same size as the tray. (A word of caution – Danish pastry is a soft dough and needs careful handling; the work surface should be lightly dusted with flour at regular intervals.) Lay on the tray, cover with cling film and rest.

7 Process the butter by rolling out to two-thirds the size of the dough inside a plastic bag.

8 Place the butter on the dough, leaving one-third uncovered, and seal the edges.

9 Fold over the uncovered dough and then fold over the other third, giving three layers of dough and two layers of butter.

10 Roll out into a rectangle the same size as before and shake before folding into three.

11 Rest in a cool place for 30 minutes.

12 Repeat Steps 10 and 11 twice more, giving three single turns in total.

13 Process according to the variety/varieties required (see recipes and methods below).

Ingredient	Makes 20 pieces
The détrempe or basic dough	
Strong flour	450g
Yeast	20g
Cold milk (previously scalded)	200ml
Eggs	2
Sunflower oil	50ml
Salt	5g
Sugar	70g
Pastry butter (beurre sec)	250g

Place the butter on the dough, leaving one-third uncovered

Fold over the uncovered dough

Many different shapes and fillings can be used to make Danish pastries. Four examples are shown here.

Sultana roulade

Ingredient	Makes 20 pieces
Almond cream	500g
Sultanas	100g

1 Carefully roll out the dough into a rectangle 18cm wide and 3mm thick.

2 Spread the almond cream over the paste, leaving 2cm of the long edge uncovered. Eggwash before rolling up as for a Swiss roll.

3 Pinch the edge to seal and cut into 1–2cm slices.

4 Lay on a papered baking sheet in staggered rows and prove at 22°C until nearly double in size.

5 Carefully eggwash and bake at 210°C for 25 minutes.

6 Allow to cool before brushing with boiled apricot glaze.

7 Brush over a thin coat of water icing. Serve at room temperature.

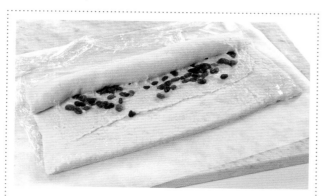

Roll up the paste

Shape the roulade

Apple envelopes

Ingredient	Makes 20 pieces
Almond cream	500g
Large apples, sliced	2
Cinnamon sugar	100g

1 Roll out the dough into a large square 3mm thick.

2 Cut into smaller 8cm squares.

3 Pipe a line of almond cream diagonally on the paste.

4 Lay on two slices of apple which have been dipped in cinnamon sugar.

5 Fold over the corners diagonally and seal with eggwash.

6 Lay on a papered baking sheet in staggered rows and prove at 22°C until nearly double in size.

7 Carefully eggwash and bake at 210°C for 25 minutes.

8 Allow to cool before brushing with boiled apricot glaze.

9 Brush over a thin coat of water icing. Serve at room temperature.

▲ Filling and folding the apple envelopes

Windmills

Ingredient	Makes 20 pieces
Pastry cream	500g
Apricot halves or plum halves	20

1 Roll out the dough into a large square 3mm thick.
2 Cut into smaller 8cm squares.
3 Make a cut on each corner and pipe a bulb of pastry cream in the centre.
4 Place the fruit on the pastry cream.
5 Fold alternate corners to meet in the centre, sealing with eggwash.
6 Lay on a papered baking sheet in staggered rows and prove at 22°C until nearly double in size.
7 Carefully eggwash and bake at 210°C for 25 minutes.
8 Allow to cool before brushing with boiled apricot glaze.
9 Brush over a thin coat of water icing. Serve at room temperature.

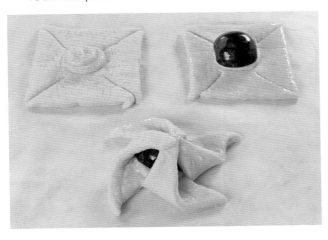

▲ Filling and folding the windmills

Variations
All the fillings and combinations of fillings used above are interchangeable.
● Diced dried apricots or dried cranberries can be used in place of sultanas, or pears in place of apple.
● The pastry cream can be flavoured with chocolate or praline paste.
● Lemon curd and jams can also be used as fillings or flavourings.

Cockscombs

Ingredient	Makes 20 pieces
Pastry cream	500g
Cherries, stoned	250g

1 Roll out the dough into a long strip 10cm wide and 3mm thick.
2 Pipe the pastry cream down the length of the paste using a large plain piping tube.
3 Cover with halves of cherry, eggwash one edge and fold the other over to enclose the filling.
4 Make a series of 1cm cuts down the length where the dough meets.
5 Cut into 8cm lengths and bed outwards, causing the cuts to open up.
6 Lay on a papered baking sheet in staggered rows and prove at 22°C until nearly double in size.
7 Carefully eggwash and bake at 210°C for 25 minutes.
8 Allow to cool before brushing with boiled apricot glaze.
9 Brush over a thin coat of water icing. Serve at room temperature.

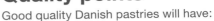

▲ Shaping the cockscombs

Quality points

Good quality Danish pastries will have:
● uniform shapes and sizes
● soft light texture with no dryness
● uniform colour and even glaze

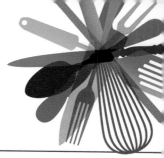

12 Petits fours

This chapter covers the following **VRQ** unit:
→ Produce petits fours.

Introduction

Petits fours are small confections that are usually eaten in one or two bites and are served at the end of a large dinner, or as an accompaniment to tea or coffee with afternoon tea. Originally a French term, the name literally translates as 'small oven' due to the fact that traditionally cakes in France were baked in coal-fuelled ovens which were hotter than normal ovens and quite difficult to control the heat to the correct temperature. After the initial baking of large cakes, the oven was then cooled and it was at this stage the small petits fours confections were baked.

All petits fours contain a very high percentage of sugar and are hygroscopic. This means that they will absorb moisture from the atmosphere quickly and should always be kept in airtight containers to prolong their shelf life. Fruits au caramel can be prepared in advance and then dipped into a boiled sugar solution, which is boiled to hard-crack temperature of 156°C, just before service. This is essential otherwise the surface will become sticky and start to dissolve.

There are many alternative names that can be used on menus for petits fours. These include:
→ friandes
→ gourmandises
→ sweetmeats
→ frivolities
→ mignardises.

Learning objectives

By the end of this chapter you should be able to:
→ Explain techniques for producing petits fours, including traditional, classical and modern skills, techniques and styles, and the importance of consistency.
→ List appropriate flavour combinations.
→ Explain considerations when balancing ingredients in recipes for petits fours.
→ Explain the effects of preparation and cooking methods on the end product.
→ Describe how to control time, temperature and environment to achieve the desired outcome.
→ Produce and finish petits fours.

Recipes included in this chapter

Types of petits fours

Generally petits fours can be divided into three categories: glacé, sec and confiserie variée.

Glacé

Petits fours glacés have a glazed finish on the outer surface which can be achieved by using fondant, boiled sugar, chocolate, etc. A particular type of fruit-based petit four glacé is referred to as a 'déguisé' which translates as a 'disguise' fruit. This is achieved by neatly filling small fruits with coloured marzipan flavoured with kirsch, pistachio compound, praline, coffee extract and other suitable products, then dipping them in boiled sugar cooked to hard-crack temperature. The finished confection will now have a crunchy coating of sugar on the outside with a soft textured marzipan fruit inside.

Sec

Petits fours secs are 'dry' confections (unlike petits fours glacés) which require further embellishment once baked to give them a decorative finish. Examples of petits fours secs are cats' tongues, macaroons and Dutch biscuits.

Confiserie variée

The term confiserie variée covers all the other petits fours confections that do not come under the glacé or sec categories. Petits fours in this category include truffles, fudges, caramel mou and pralines.

Food allergies

Health and safety

It is very important to be aware that almost any substance can trigger an allergic reaction in someone. Although peanuts are the most common food to trigger such a reaction, other common potential 'problem foods' can have the same effect.

Petits fours are normally served on a tray with several different varieties. There are many different ingredients in their make-up and some could cause an allergic reaction. The pastry chef should train staff in product knowledge – they should know exactly what ingredients are used in the dishes and the ones that may cause an allergic reaction, and communicate this information to the service staff so that they are aware of the make-up of each petits fours dish.

In any catering establishment, should a customer enquire if a substance or food to which they are allergic is included in the recipe's ingredients, it is vital that staff give an accurate and clear answer (for example, a person who is allergic to nuts would need to know if there were walnuts in the chocolate fudge, marzipan filling in the fruit déguisés, baton almonds in the chocolate rochers, flaked almonds in the Florentine squares or ground almonds in the macaroons). Failure to provide the correct information could have dire consequences.

Commodities that may trigger an allergic reaction in petits fours:
- Nuts: cashew, pecan, walnut, ground almonds, nib almonds, brazil nuts, pistachio nuts, ground hazelnuts, praline, gianduia, caramelised nuts.
- Sesame seeds, caraway seeds, poppy seeds.
- Milk, eggs, cream, yoghurt, crème fraiche.
- Chocolate, coffee, oranges, red fruits.
- Yeast, wheat, soya, sugar.

Presentation

When petits fours are received by a customer at the end of the meal they need to appeal to their senses of sight and smell, even before taste. For this reason, you need to consider presentation early in the design, include different types of petits fours and ensure that there is a wide variety of colours, textures, shapes and flavours in the presentation. Remember, petits fours are the last memory of a dining experience.

Storage

Because petits fours are generally made from large concentrations of sugar, they are hygroscopic and tend to absorb moisture from the atmosphere, so they need to be stored carefully. To keep biscuits crisp and prevent a sticky surface from forming on boiled sugar products, store them in airtight containers. Chocolate petits fours, however, need to be stored in cool surroundings at approximately 15°C, away from any humidity, which could result in sugar bloom, and away from foods like onions and garlic from which they could absorb strong smells. Once removed from moulds, place the chocolates in silicone-lined gastronorm trays with lids. If the chocolates are exposed to heat or direct sunlight this could result in fat bloom.

Handmade chocolates

Chocolate for petits fours often needs to be tempered. Refer to Chapter 16 for details.

Using convenience shells

Convenience shells come in various forms – spheres, squares, rectangles, rounds.

1 Fill the shell with ganache, using a closing template.

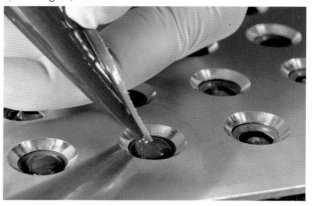

2 Seal the top with tempered couverture. Leave to set.

3 Coat each chocolate in tempered couverture.

4 To finish, dip in cocoa powder using a dipping fork.

Moulding pralines

Polish the mould with cotton wool before use. Never use an abrasive as any marks made on the surface of the mould will be picked up on the outer surface of the chocolate.

1 Embellish the mould with coloured, tempered cocoa butter and plain tempered couverture, or just leave plain.

2 Fill to the top with tempered couverture, then tap the mould to release any air bubbles.

3 Invert the mould and tap the side; most of the chocolate will drain out, leaving a thin coating in the cavity.

4 Using a chocolate scraper, scrape away and reserve any excess chocolate from the top of the mould. Place the mould, inverted, on a flat surface covered with silicone paper and leave the chocolate to set.

5 Fill the cavities with ganache or any other suitable filling to just below the top of the mould. Allow the ganache to harden.

6 Spread or pipe tempered chocolate over the top.

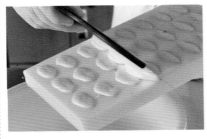

7 Scrape off any excess, leaving a flat, smooth surface enclosing the ganache. Leave the chocolate to set.

8 Turn the mould upside down and tap the chocolates out.

Always wear cotton gloves when handling chocolates to maintain the shine on the tempered surface.

Shelf life

The shelf life of chocolates generally depends on the water activity in the ganache being used as a filling. The less water (cream is classified as water) in the recipe, the longer the shelf life. Always use UHT cream in the making of ganache as this is the most sterile cream to use and should give a longer shelf life. (Once UHT cream is opened, store and use as fresh cream.)

Fondant

Fondant is made by boiling sugar, water and glucose to the soft ball stage (115°C). Once it reaches this temperature, it is then poured onto a lightly oiled marble slab and worked with a palette knife, entrapping air into the mixture until it becomes a white, opaque and solid mass.

Commercial fondant is purchased ready-made in a white opaque block.

To process, fondant is broken down into chucks and placed over a bain marie of simmering water. Stock syrup is added to thin down the fondant to the desired consistency and the temperature is taken to 37°C. If the fondant is taken above this temperature, then the sugar particles within the fondant break down, resulting in them not reflecting the light from the surface, and thus creating a dull appearance.

Fondant can be flavoured with chocolate, raspberry, vanilla or mint, piped into starch moulds (these are made by compacting cornflour in a container and then making indentations), allowed to set for a few days, then enrobed in chocolate and served as petits fours (for example, peppermint creams).

Marzipan (almond paste)

Marzipan is widely used in the production of petits fours (for example, fruits déguisés, fondant glacé and cut-out pieces of marzipan lightly grilled to give a caramelised surface).

For more information about marzipan, and a recipe, see Chapter 16.

Regional recipe – Northney fudge

Contributed by Andrew Jessup, South Downs College

Ingredient	70 pieces
Northney cream or whipping cream	600 ml
Sugar	700 g
Salted butter	200 g
Dark chocolate	300 g

1 Place all ingredients into a pan on a low heat until dissolved.
2 Increase the heat and boil to just above soft ball stage (118°C).
3 Stir until the mixture is cool and has 'grained'.
4 Place in a greased tin.
5 Cut and place in bags.

This recipe is made using cream from Northney Farm, Hayling Island, Hampshire – a local dairy farm. It is sold in the South Downs College shop.

1 Poppy seed tuiles

Ingredient	Makes 80 tuiles
Glucose	50g
Granulated sugar	500g
Water	200g
Soft flour	150g
Butter, melted	250g
Almonds, nibbed	250g
Poppy seeds	250g

1 Add the glucose and sugar to the water. Bring to the boil.
2 Remove from the heat. Stir in the flour and allow to stand for a few minutes.
3 Add the rest of the ingredients and mix well.

4 Place ½ teaspoon portions of the mixture onto a silicone mat, with space between them. Batten out with a fork.
5 Bake at 180°C for 3–4 minutes. Remove from the oven and shape over a rolling pin for crescents, a dariole mould for baskets or a round wooden spoon handle for cigars.

Variation
Sesame seeds can be used instead of poppy seeds.

▲ Tuiles (left to right: banana, poppy seed, sesame seed, chocolate, coconut)

2 Coconut tuiles

Ingredient	Makes 50 tuiles
Icing sugar	300g
Flour	100g
Desiccated coconut	100g
Egg whites	250g
Butter, melted, not hot	200g
Vanilla essence	

1 Mix the dry ingredients together well in a food processor.
2 Mixing on a slow speed, gradually add the egg whites, then the melted butter and vanilla essence.
3 Pipe onto silicone mats, in bulbs about 2½ cm in size.
4 Tap the tray to flatten the bulbs.
5 Bake at 205–220°C.
6 Shape as in Recipe 1.

3 Chocolate tuiles

Ingredient	Makes approx. 60 tuiles
Cocoa powder	30g
Sugar	360g
Water	250g
Pectin	10g
Sugar	90g
Chocolate paste, made with 100% cocoa chocolate	150g
Unsalted butter	150g

1 Place the cocoa powder and 360g of sugar in the water. Bring to the boil.
2 Whisk in the pectin and the remaining sugar.
3 Simmer on a medium heat for 10 minutes.
4 Remove from the heat and add the chocolate paste and butter. Mix together.
5 Place ¼ teaspoon portions of the mixture onto silicone mats, well spaced apart.
6 Bake at 200°C for 9 to 10 minutes, until they start to bubble. To check that they are done, place one onto marble and test that it snaps when broken.
7 Shape as in Recipe 1.

4 Brandy snaps

Ingredient	Makes approx. 20
Strong flour	225g
Ground ginger	10g
Golden syrup	225g
Butter	250g
Caster sugar	450g

1 Combine the flour and ginger in a bowl on the scales. Make a well.

2 Pour in golden syrup until the correct weight is reached.

3 Cut the butter into small pieces. Add the butter and sugar.

4 Mix together at a slow speed.

5 Divide into 4 even pieces. Roll into sausage shapes, wrap each in cling film and chill, preferably overnight.

6 Slice each roll into small rounds. Place on a baking tray, spaced well apart.

7 Flatten each round using a fork dipped in cold water, keeping a round shape.

9 Bake in a pre-heated oven at 200°C until evenly coloured and bubbly.

10 Remove from oven. Allow to cool slightly, then lift off and shape over a dariole mould for baskets, a rolling pin for crescents or a round wooden spoon handle for the traditional cigar shape.

11 Stack the snaps, no more than 4 together, on a stainless steel tray and store.

5 Cigarette paste biscuits

Ingredient	Makes approx. 30
Icing sugar	125g
Butter, melted	100g
Vanilla essence	3–4 drops
Egg whites	3–4
Soft flour	100g

1 Proceed as for Steps 1–3 of Recipe 8 (cats' tongues).

2 Using a plain tube, pipe out the mixture onto a lightly greased baking sheet into bulbs, spaced well apart. Place a template over each bulb and spread it with a palette knife.

3 Bake at 150°C, until evenly coloured.

4 Remove the tray from the oven. Turn the cornets over but keep them on the hot tray.

5 Work quickly while the cornets are hot and twist them into a cornet shape using the point of a cream horn mould. (For a tight cornet shape it is best to set the pieces tightly inside the cream horn moulds and leave them until set.) If the cornets set hard before you have shaped them all, warm them in the oven until they become flexible.

The same paste may also be used for cigarettes russes (tight rolls), coupeaux (spirals) and other shapes.

6 Madeleines

Ingredient	Makes 45 pieces
Caster sugar	125g
Eggs	3
Vanilla pod, seeds from	1
Flour	150g
Baking powder	1 tsp
Beurre noisette	125g

1 Whisk the sugar, eggs and vanilla seeds to a hot sabayon.

2 Fold in the flour and the baking powder.

3 Fold in the beurre noisette and chill for up to 2 hours.

4 Pipe into well-buttered madeleine moulds and bake in a moderate oven (180°C) for 8–10 minutes.

5 Turn out and allow to cool.

7 Macaroons

Ingredient	Makes approx. 40
Ground almonds	195g
Icing sugar	195g
Egg white	75g
Italian meringue	
Caster sugar	190g
Water	50ml
Egg white	75g
Cream of tartar	Pinch

1 Sieve the icing sugar and ground almonds together.

2 Beat in the egg white until a smooth paste is formed with no lumps.

3 Make an Italian meringue cooked to 121°C (see page 440).

4 Beat the Italian meringue into the almond mixture until a shiny surface is achieved.

5 Place into a piping bag and pipe 1cm small rounds onto a non-stick mat.

6 Tap the tray to even out the mixture.

7 Leave to stand in a warm place for 30 minutes to form a crust on the surface.

8 Bake at 150°C for 10 minutes. Allow to cool, then sandwich with a filling.

The biscuit may be flavoured or coloured delicately.

Fillings
Possible fillings include:
- ganache
- lemon cream
- flavoured crème mousseline
- buttercream.

8 Cats' tongues

Ingredient	Makes approx. 40
Icing sugar	125g
Butter	100g
Vanilla essence	3–4 drops
Egg whites	3–4
Soft flour	100g

1 Lightly cream the sugar and butter, add the vanilla essence.

2 Add the egg whites one by one, continually mixing and being careful not to allow the mixture to curdle.

3 Gently fold in the sifted flour and mix lightly.

4 Pipe on to a lightly greased baking sheet using a 3mm plain tube, 2½ cm apart.

5 Bake at 230–250°C, for 3–4 minutes.

6 The outside edges should be light brown and the centres yellow.

7 When cooked, remove on to a cooling rack using a palette knife.

9 Almond biscuits

Ingredient	Makes 30
Butter	200g
Icing sugar	80g
Soft flour	200g
Salt	Pinch
Whole egg	25g
Ground almonds	100g
Granulated sugar	100g

1 Rub together the butter, icing sugar, flour and salt until the mixture resembles fine breadcrumbs.

2 Add the egg and work to a smooth paste.

3 Roll into a cylinder 2cm in diameter.

4 Brush with egg white and roll in a mixture of equal quantities of granulated sugar and ground almonds.

5 Refrigerate to firm up the paste.

6 Cut into discs and bake at 180°C until lightly golden.

Variation
Almond biscuits can also be flavoured with lemon zest or chopped glacé cherries.

10 Japonaise

1. Sieve the ground almonds and flour together onto a sheet of greaseproof paper.
2. Whisk the egg white with the cream of tartar to full foam, adding the sugar gradually to form a firm meringue.
3. Fold the meringue through the dry ingredients.
4. Spoon into a piping bag with a plain nozzle.

Batons au chocolat

Pipe in 3cm lengths onto a greased and floured baking tray, sprinkle with nib almonds and bake at 180°C. When cold, spin with chocolate or dip on an angle into melted chocolate and sandwich together with a ganache.

Boules de neige

Using a plain nozzle, pipe out small domed bulbs of the mixture on to a greased and floured baking tray, sprinkle with nibbed almonds and bake at 180°C. When cold, sandwich together with ganache and dust the surface with icing sugar to represent snow balls.

Base mixture

Ingredient	Makes approx. 20
Ground almonds	90g
Soft flour	15g
Egg white	90g
Cream of tartar	Pinch
Caster sugar	120g

11 Rothschilds

Ingredient	Makes approx. 25
Caster sugar	125g
Ground almonds	125g
Strong flour	25g
Egg whites	5
Caster sugar	25g
Cream of tartar	Pinch

1. Sieve the 125g sugar, almonds and strong flour together onto a sheet of greaseproof paper.
2. Whisk the egg whites, the 25g of sugar and cream of tartar to a firm meringue.
3. Gradually mix the meringue into the dry ingredients.
4. Deposit into a piping bag, pipe out into 3cm shells and sprinkle with nib almonds.
5. Bake at 180°C for 20 minutes.
6. When cool, spin with melted chocolate.

12 Chocolate Viennese biscuits

Ingredient	Makes approx. 20
Butter	125g
Icing sugar	50g
Egg white	20g
Soft flour	130g
Bitter cocoa powder	25g
Salt	Pinch
Vanilla essence	Drop
Walnut or pecan pieces	50g

1 Cream the butter and sugar together until white.
2 Gradually beat in the egg white.
3 Fold in the flour and cocoa powder. Add the salt and vanilla essence.
4 Pipe into 3cm lengths onto a greased and floured tray.
5 Bake at 180°C.
6 Dust with icing sugar or spin with chocolate. Place a small piece of pecan or walnut on each.

13 Dutch biscuits

Ingredient	Makes approx. 50
White paste	
Soft flour	300g
Icing sugar	100g
Vanilla	4 drops
Butter	200g
Egg yolk	1
Chocolate paste	
Soft flour	285g
Icing sugar	100g
Bitter cocoa power	20g
Butter	200g
Egg yolk	1
Chocolate colour to darken if required	

For the white paste

1 Sieve the flour and icing sugar. Add the vanilla.

2 Rub in the butter until a breadcrumb texture is achieved.

3 Bind with the egg yolk and refrigerate until the mixture is firm.

For the chocolate paste

1 Sieve the flour, icing sugar and cocoa powder together.

2 Rub in the butter until a breadcrumb texture is achieved.

3 Bind with egg yolk and refrigerate until the mixture is firm.

Using the white and dark Dutch biscuit mixture, shape into various designs. Cut into petits fours size and bake at 180°C for 10 minutes.

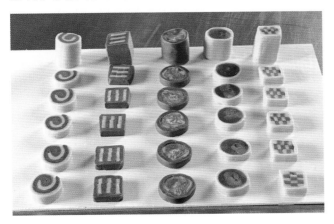

▲ The patterns are formed (top) and then the biscuits are sliced up before baking

Design 1

1 Roll the white paste into a cylindrical shape 1.5cm diameter.

2 Brush the surface of the white paste with water.

3 Place onto a square sheet of thinly rolled out chocolate paste the same width as the white cylinder.

4 Roll the white cylinder in the chocolate paste, forming a thin coating of chocolate paste around the cylinder.

5 Refrigerate until firm and then cut into thin discs ready for baking.

Design 2

1 Mix equal amounts of white and chocolate paste together to form a marbled effect.

2 Roll into a cylindrical shape 1.5cm diameter.

3 Brush the surface with water.

4 Place onto a square sheet of thinly rolled out white or chocolate paste the same width as the cylinder.

5 Roll the marbled cylinder in the white or chocolate paste, forming a thin coating around the cylinder.

6 Refrigerate until firm and then cut into thin discs ready for baking.

Design 3 (Battenburg design)

1 Roll out two sheets of paste, one white and one chocolate, 5mm thick.

2 Brush the surface of one paste with water and then attach the second.

3 Cut the paste in half, placing one on top of the other to give four alternating coloured pastes.

4 Refrigerate to set the pastes.

5 Cut one slice 5mm thick. Place on the working surface, and brush the surface with water.

6 Cut a second slice, rotate this before placing on the first slice to create two layers of alternating lines of white and dark paste.

7 Repeat this process until you have four layers (all alternating).

8 Brush the surface.

9 Place onto a square sheet of thinly rolled out white or chocolate paste.

10 Roll in the desired coloured paste and refrigerate.

11 Cut into thin squares revealing a Battenburg design ready for baking.

14 Petit Genoese fondants (fondant dips)

Either pain de Gêne or heavy Genoise can be used for this recipe.

Ingredient	Makes 50 pieces
Heavy sponge (pain de Gêne or heavy Genoise – see below)	500g
Marzipan	150g
Apricot glaze	50g
Fondant	500g
Stock syrup	
Pain de Gêne (2 × 46cm × 30cm trays)	
Butter	750g
Sugar	750g
Ground almonds	750g
Soft flour	150g
Cornflour	150g
Whole eggs	750g
OR heavy Genoise (2 × 46cm × 30cm trays)	
Soft flour	455g
Caster sugar	340g
Milk powder	42g
Salt	6g
Margarine	340g
Egg	425g
Glycerine	35g
Glucose	115g
Water	175g
Soft flour	70g
Baking powder	15g

To make pain de Gêne

1 Cream the butter and sugar until aerated and white in colour.
2 Sieve the ground almonds, soft flour and cornflour together onto a sheet of greaseproof paper.
3 Gradually add the whole eggs to the creamed butter and sugar mixture.
4 Carefully fold in the sieved dry ingredients.
5 Deposit into trays lined with silicone paper and bake at 180°C for 30 minutes.

To make heavy Genoise sponge

1 Sieve the caster sugar, milk powder, salt and 455g of flour together.
2 Place into a mixing bowl with the margarine and mix on speed 1 to a crumbly consistency. (It is very important that the mixture does not form into a paste.)
3 Add the egg and glycerine gradually, cream for 5–7 minutes on speed 2. Scrape down as required.
4 Dissolve the glucose and water, add half to the above ingredients on speed 1.
5 Blend with a few turns of the beater. Do not over-mix as this will toughen the gluten in the flour.
6 Add the remaining 70g of sieved flour and baking powder and blend in.
7 Add the remaining liquid and mix to a smooth batter.
8 Deposit into a tray lined with silicone paper and bake at 195°C for approximately 30 minutes.

To prepare the heavy sponge for petit Genoese fondants

1 Remove the top crust from the sponge.
2 Clear surplus crumb.
3 Cut the sponge, sandwich with buttercream and brush with kirsch-flavoured stock syrup.

To prepare and apply the marzipan

1 Soften the marzipan in readiness for rolling.
2 Roll the marzipan out thinly using icing sugar to prevent sticking.
3 Spread apricot glaze on the sponge or on the correct size of rolled out marzipan.
4 Apply the marzipan to the sponge.

To cut the heavy sponge into pieces

1 Using metal bars or rulers, cut the sponge into pieces of the required shape and size with the marzipan face down on the work bench. (Choose shapes that limit the amount of waste.)
2 Turn the shapes over so that the marzipan is facing upwards and lightly glaze the surface with apricot glaze – this forms a key for the fondant to stick to.

To prepare the fondant

Prepare the fondant by adding stock syrup and warming gently over a bain marie of hot water to bring the temperature of the fondant to 37°C with the correct fluidity.

The dipping procedure

1 Prepare the work surface with the fondant pan in the centre, the sponge pieces to one side and the dipping wire on the other side with a drip tray underneath.

2 Dip each piece, marzipan surface down, into the fondant using a dipping fork, remove from the fondant, allow surplus to run off, and then place on wire to drain. (Place each piece side by side on the wire and do not allow any fondant drips to pass over previously dipped pieces.)

To decorate

Any fondant piped lines used in the decoration should be very thin, especially when stronger colours are used. A good piping mix to use is equal quantities of melted chocolate and piping jelly, mixed to the required consistency with stock syrup.

To serve

1 Allow fondants ample time to set.
2 Place in suitable paper cases and display neatly.
3 Any surplus fondant dripped through the wires and free of crumb may be returned to the pan.

Professional tip

When preparing fondant, it is important not to take it above 37°C, otherwise the finished product will have a dull appearance.

You can make 20 petit Genoese fondants in four colours, starting from one container of white fondant, by following this sequence:
● white: dip five pieces in the white fondant
● pink: colour the fondant with pink colouring, then dip five more pieces
● orange: add yellow colouring to the pink fondant, then dip five more pieces
● brown: add chocolate colouring to the orange fondant and dip the last five pieces.

15 Fruits au caramel

Faults

A thick base, commonly known as 'feet on dipped fruits', is caused by dipping the fruit in boiled sugar which has thickened as it has cooled down. This results in a thick outer coating on the fruit which runs to the base once the fruit is placed onto the non-stick surface.

Another cause is not wiping the dipped fruits against the side of the pan to remove excess sugar syrup prior to placing on a non-stick surface.

Ingredient	Makes 50 pieces
Granulated sugar	400g
Glucose	60g
Water to saturate the sugar	
Fruits suitable for dipping (grapes, Cape gooseberries, cherries, strawberries, satsuma segments)	50 pieces

1 Place the sugar in a small saucepan and just saturate with water. Add the glucose and gradually bring to the boil following the principles of sugar boiling (see Chapter 16).

2 When the sugar reaches 155°C hard crack, dip the pan into a container of cold water to stop the sugar from cooking any further.

3 Remove the pan of sugar from the cold water and, with the aid of tweezers, dip the fruits in the sugar solution ensuring that the fruits have an even coating.

4 Stand the fruits upright on a silicone mat.

16 Nougat Montelimar

Ingredient	Makes 50–60 pieces
Granulated sugar	350g
Water	100g
Honey	100g
Glucose	100g
Egg white	35g
Glacé cherries, cut into quarters	50g
Pistachio nuts, chopped	50g
Nibbed almonds	25g
Flaked almonds or flaked hazelnuts	25g

1 Place the sugar and water into a suitable pan, bring to the boil and cook to 107°C.

2 When the temperature has been reached, add the honey and glucose and cook to 137°C.

3 Meanwhile, whisk the egg whites to full peak in a machine, then add the syrup at 137°C slowly, while whisking on full speed.

4 Reduce speed, add the glacé cherries, chopped pistachio nuts, and the nibbed and flaked almonds.

5 Turn out onto a lightly oiled tray or rice paper and mark into pieces while still warm.

6 When cold, cut into pieces and place into paper cases to serve.

Professional tip

Instead of using rice paper, the tray may be dusted with neige-décor.

17 Marshmallows

Ingredient	Makes approx. 50 pieces
Granulated or cube sugar	600g
Egg whites	3
Leaf gelatine, soaked in cold water	35g
To finish	
Cornflour	2 tbsp
Icing sugar	5 tbsp

This is a foam set with a gum.

1 Place sugar into a suitable saucepan with 125ml water and boil to soft ball stage, 118°C.

2 When the sugar is nearly ready, whisk the egg whites to a firm peak.

3 Pour in the boiling sugar and continue to whisk.

4 Squeeze the water from the gelatine and add the gelatine to the mixture.

5 Add colour and flavour if desired.

6 Turn out onto a tray dusted with cornflour and dust with more cornflour.

7 Cut into sections and roll in a mixture of icing sugar and cornflour.

18 Turkish delight

Ingredient	Makes approx. 50 pieces
Water	500ml
Caster sugar	450g
Icing sugar	100g
Lemon juice	1 tbsp
Cornflour	150g
Cream of tartar	1 tsp
Rose water	1 tbsp
Grenadine	1 tbsp
To finish	
Cornflour	2 tbsp
Icing sugar	5 tbsp

1 Dust a baking tray measuring 20 × 25cm with cornflour. Pour half of the water into a heavy-based saucepan and add the sugars and lemon juice. Heat until the sugar has dissolved and then bring to the boil.

2 Reduce the heat and simmer until the mixture reaches 115°C on a sugar thermometer (soft ball stage). Remove from the heat.

3 In a separate saucepan, mix the cornflour and cream of tartar together with the remaining water until the mixture is smooth. Cook over a medium heat until the mixture thickens.

4 Gradually pour the hot sugar syrup into the cornflour paste, stirring continuously. Return the mixture to the heat and simmer for about one hour, until the mixture is pale and feels stringy when a little of the cold mixture is pulled between the fingers. Stir in the rose water and Grenadine.

5 Pour the mixture into the prepared baking tin and leave to set overnight.

6 Cut into squares. Mix the cornflour and icing sugar together and toss the Turkish delight in it. Store in an airtight container, between layers of greaseproof paper.

19 Chocolate truffles (general purpose ganache)

Ingredient	Makes approx. 60 truffles
Whipping cream	175g
Milk	75g
Butter	75g
Inverted sugar (see page 480; e.g. Trimoline)	75g
Powder sorbitol (optional, gives a better shelf life to the ganache)	12g
Milk couverture	165g
Plain couverture	333g
Alcohol concentrate	10g

1 Boil the cream, milk, butter, inverted sugar and sorbitol to 80°C.

2 Partially melt both chocolates and gradually add the boiled liquid, working from the centre of the chocolate, forming a good elastic emulsion.

3 Finish with a stem blender to achieve a smooth and shiny ganache. Add the alcohol concentrate.

4 Pour into a bowl and leave to cool and crystallise. Periodically stir the outside mixture to the centre.

5 When the ganache is firm, pipe out into the desired shapes.

20 Caramel truffles

Ingredient	Makes approx. 120 truffles
Caster sugar	325g
Inverted sugar syrup (Trimoline)	100g
Whipping or double cream	500g
Plain chocolate	575g
Milk chocolate	75g
Butter	10g
Chocolate spheres	120

1 Place the sugar and Trimoline together in a pan and take to a caramel, being mindful that it will turn from caramel to burnt quickly.

2 Remove from the heat and slowly add the boiled cream. Return to the heat to dissolve the set caramel.

3 Once dissolved, add the chocolate and emulsify using a stem blender.

4 Add the butter.

5 Remove and allow to chill naturally.

6 Once the mixture has reached room temperature, place it in a disposable piping bag.

7 Snip off the end of the bag and carefully pipe the mixture into the chocolate spheres.

8 Allow to set and carefully close the top of each sphere with melted chocolate.

9 Once set, roll in the desired chocolate, allow to set and then serve.

Professional tip

Chocolate spheres can be purchased from good provision distributors in milk, dark and white; all the major confectioner's suppliers make the product.

21 Cut pralines

Ingredient	Makes 80–100
General purpose ganache (Recipe 19)	2 kg
Couverture	

1 Place a sheet of acetate in the bottom of a praline frame.

2 Prepare general-purpose ganache and pour into the praline frame. Tap to level.

3 Leave the ganache for 12 hours to crystallise and firm up.

4 Remove the ganache from the praline frame and coat one side in tempered couverture.

5 Place the couverture side down onto the guitar base and cut through the ganache. Alternatively, use a sharp knife.

6 Lift the slab of ganache off the guitar base using a metal take-off sheet. Raise the wire and wipe clean.

7 Place the slab of ganache back on the guitar in the opposite direction and cut through once again, giving uniform cut squares.

8 Using a dipping fork, place the cut praline, chocolate-coated side down, into tempered couverture.

9 Lift out of the couverture using a dipping fork and tap on the side of the container to remove excess chocolate.

10 Carefully deposit the praline on a sheet of silicone paper.

11 The top of the pralines can be decorated by placing transfer sheets or textured sheets on top or by giving a ripple effect by using the dipping fork on the surface.

Allow the ganache to firm up in the frame, then coat the surface with tempered couverture using a paint roller

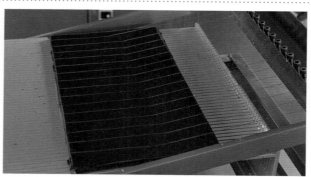

Place the couverture side down onto the guitar base and cut through the ganache

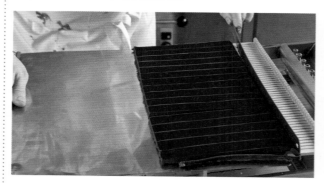

Lift the slab of ganache off the guitar base using a metal take-off shee

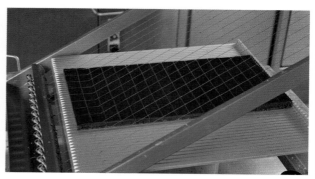

Wipe the guitar clean, then carefully place the slab of ganache back on the guitar in the opposite direction and cut through once again, creating uniform squares

Professional tip

A guitar (see photos) is used to achieve neat, even cuts.

Layered pralines can be made by pouring general-purpose ganache into a frame, allowing to set, then topping with a layer of pâte de fruits. Once completely set, process in the normal manner, giving a two-layered praline.

22 Chocolate rochers

First you need to prepare the caramelised almonds (used as a base for the rochers) using one of the following methods.

Caramelised almonds (method one)

1 Pass baton almonds through egg yolk, then caster sugar, and spread out onto silicone paper.

2 Bake in the oven at 160°C until golden and caramelised.

Caramelised almonds (method two)

1 Boil 400g sugar and 160g water to 115°C.

2 Add 1200g baton almonds or nib almonds and mix.

3 Reheat over an open flame in a round-bottomed bowl until the mixture starts to caramelise, keeping the mixture moving around the bowl at all times.

4 Add 50g butter and mix in to separate the almonds.

5 Store in an airtight container until required.

Orange rochers

Ingredient	Makes approx. 30 pieces
Orange couverture, tempered	500g
Cocoa butter	50g
Caramelised almonds	600g
Orange confit	200g

Milk chocolate rochers

Ingredient	Makes approx. 30 pieces
Milk couverture, tempered	500g
Cocoa butter	50g
Caramelised almonds	600g
Puffed rice cereal	100g

White rochers

Ingredient	Makes approx. 30 pieces
White couverture, tempered	500g
Cocoa butter	50g
Pine kernels	400g
Pistachio nuts, chopped	100g
Orange confit	200g
Puffed rice cereal	50g

1 For all three rochers, mix all the ingredients together and spoon onto silicone paper.

2 Remove from the paper once set.

23 Pâte de fruits

Ingredient	Makes approx. 60 pieces
Blackcurrant pulp	400g
Caster sugar	360g
Granulated sugar	100g
Pectin mixture	
Pectin	30g
Caster sugar	40g

1 Heat the blackcurrant pulp to 50°C.
2 Add the caster sugar and then bring to the boil and skim.
3 Add the pectin mixture and cook to 105°C.
4 Remove from the heat and leave to settle for 10 seconds.

5 Pour the mixture into a prepared tray. Wrap in cling film.
6 Leave to set for 4 hours.
7 Just before serving, cut into 2cm cubes. Roll the pieces in the granulated sugar and place into petit four cases.

Professional tip
These jellies should not be stored in the fridge.

Variations
Use other varieties of fruit purée (pulp) for different flavours of jelly.

24 White and dark chocolate fudge

1 Line a shallow tray with silicone paper.
2 Boil the cream, glucose and sugar to 120°C.
3 Pour onto the melted chocolate and add the appropriate nuts and fruit for the type of chocolate used. Add the butter.
4 Pour into the prepared tray and allow to set.
5 Once set, remove from the tray. Spread the surface of the white fudge with over-tempered plain couverture and the dark fudge with over-tempered white couverture. (Over-tempering takes the couverture above the recommended tempering temperature so that the chocolate does not become too brittle and splinter when being cut.)
6 Just before both couvertures set, comb the surface using a toothed scraper.
7 Once the chocolate décor has set, cut into neat squares.

Ingredient	Makes approx. 50 pieces
Glucose	275g
Caster sugar	800g
Whipping cream	250g
Couverture (white or plain)	400g
Butter	60g
Walnuts (chopped) for the dark fudge or pistachios for the white	150g
Dried cranberries for the white fudge	30g

Professional tip
It is important that the sugar boils to the correct temperature or the fudge will be too hard.

25 Chocolate lollipops

1 Line the base of a magnetic lollipop mould with a transfer sheet.
2 Place a lollipop frame on top of the transfer sheet.

3 Fill with tempered couverture and place lollipop sticks in the indentations.
4 Decorate the top of the chocolate with crisp pearls, dried raspberries or Bres (caramelised nuts).
5 When the couverture has set, remove the lollipops from the mould.

Variation: using transfer sheets
- Cut strips of transfer sheets 5cm wide.
- Pipe round discs of tempered couverture onto the transfer sheet.
- Place a lollipop stick in the centre of the disc of chocolate.
- Decorate the top of the chocolate, as in the main recipe.
- When the couverture has set, remove the lollipops from the transfer sheet.

26 Frozen chocolate lollipops

1 Pipe fruit water ice (see Chapter 14) into praline spheres.
2 Place lollipop sticks into the water ice.
3 Return to the freezer and freeze until solid.
4 Once frozen, seal the lollipop stick at the opening of the sphere with melted couverture.
5 Once set, dip the sphere in over-tempered couverture.
6 Serve frozen as a petit four.

Professional tip

For presentation purposes, serve in an iced socle on a bed of dry ice. (A socle is a moulded ice base used for the presentation of food.) Just before serving, add warm water to the dry ice to form a mist.

27 Lemon feuilletée

Ingredient	Makes approx. 80
Pasteurised egg yolks	600g
Pasteurised egg whites	720g
Lemon juice	1 litre
Lemon zest	36g
Sugar	1.5 kg
Butter, melted	600g
Gelatine	20 leaves
Chocolate glaze	
Double cream	320ml
Plain couverture	300g
Butter, soft	100g

1 Whisk the egg yolks, egg whites, lemon juice, lemon zest and sugar over a bain marie of simmering water until the mixture thickens and reaches a temperature of 80°C.

2 Add the butter and pre-soaked gelatine.

3 Pour into a sided mould lined with silicone paper and *biscuit jaconde*.

4 Once set, remove from the mould and coat the surface with chocolate glaze.

5 Cut into petit-four size pieces with a warm knife.

To make the chocolate glaze

1 Boil the cream, add in the chocolate and mix well.

2 Blend in the butter, ensuring that there are no bubbles.

28 Florentine squares

Ingredient	Makes approx. 80
Sweet paste, cooked sheet	
Preparation A	
Cream	500ml
Honey	500g
Butter	500g
Granulated sugar	1 kg
Preparation B	
Flaked almonds	750g
Ground almonds	250g
Strong flour	250g
Glacé fruit (cherries/angelica), chopped	1 kg

1 Mix together and heat the ingredients for preparation A until it is at soft ball stage (116°C). Add all the preparation B ingredients and mix in well.

2 Turn out the mixture onto a sheet of cooked sweet paste in a sided container.

3 Place in an oven until the mixture bubbles at 180°C.

4 When cold, cut into squares and dip in chocolate if desired.

Test yourself

1 Name the three classifications of petits fours, giving two examples of each.
2 Briefly describe why it is important to form a good emulsion when making a ganache.
3 Name four convenience products that may be used in the production of petits fours.
4 List the steps you would carry out to produce a tray of moulded pralines.
5 Translate the term 'petits fours'.
6 Why is timing important when dipping fruits in caramel?
7 Describe the process of making fondant.

This chapter covers the following units:

NVQ:
➜ Prepare and cook and finish complex pastry products.

VRQ:
➜ Produce paste products.

It is very important to note that many of the basic items prepared in this chapter are also used as foundations in other areas, and are therefore referred to in other chapters. For example, a filling such as crème pâtissière is commonly used in paste products but is also utilised in the preparation of a variety of desserts (Chapter 14), petits fours (Chapter 12) and dough products (Chapter 11). Another example is ganache, which is also used in a variety of ways.

Introduction

This chapter is focussed on the various pastes, fillings, creams and preparations used to produce a wide range of pastry products, and then takes the use of these base products further to produce a variety of pastry goods.

The range of pastes includes simple pastes such as shortcrust and sweet pastes, as well as pastes produced in other ways such as choux paste and puff paste.

Learning objectives

By the end of this chapter you should be able to:
➜ Identify and prepare a range of fillings, creams and sauces that are used to finish paste products.
➜ Explain techniques for the production of paste products, including construction and traditional, classical and modern skills, techniques and styles.
➜ List appropriate flavour combinations.
➜ Explain considerations when balancing ingredients in recipes for paste products.
➜ Explain the effects of preparation and cooking methods on the end product.
➜ Describe how to control time, temperature and environment to achieve the desired outcome in paste products.
➜ Produce, finish and present a variety of pastry products to the recipe specifications, in line with current professional practice.

Recipes included in this chapter

Key points

The specialist area of paste production requires close attention to the following key points:

- Check all weighing scales for accuracy.
- Follow recipes carefully.
- Check all storage temperatures are correct.
- Always work in a clean, tidy and organised way; clean all equipment after use.
- Always store ingredients correctly: eggs should be stored in a refrigerator, flour in a bin with a tight-fitting lid, sugar and other dry ingredients in closed storage containers.
- Keep equipment clean and dry.

Hygienic working practices are essential in all areas of the pastry kitchen and it is paramount that food safety practices are followed throughout all stages of preparation and cooking. Items must be stored appropriately using the following guidelines:

- chilled items in a refrigerator between 1 and 4°C
- frozen items in a freezer between −18 and −22°C.

> **Professional tip**
>
> Service of frozen items can involve removal of the product from the freezer before serving. This will allow the product to acclimatise and reach its ideal service temperature of between −8 and −10°C.

When storing items, they should be:

- clearly labelled and dated
- covered or wrapped
- positioned on an appropriate shelf
- rotated with other stock items.

Health and safety

Patisserie items are often viewed as an indulgence or a treat and people allow themselves to enjoy items that are often high in sugar and fat and therefore calories. However, there are opportunities to produce pastry products with low fat and low sugar options and also the opportunity to use wholemeal flour in place of white flour (e.g. in short pastry).

In terms of the health and welfare of customers, it is important to identify the ingredients that make up patisserie products. Flour-based products, of which there are countless types, would not be suitable for a customer with a gluten intolerance or coeliac disease. Another regularly used ingredient in the production of pastry products is sugar. Products of this type would be unsuitable for a diabetic customer.

Techniques in pastry work

Adding fat to flour

Fats act as a shortening agent in short and sweet pastes. The fat has the effect of shortening the gluten strands in flour, which are easily broken when eaten, making the texture of the product more crumbly. Short and sweet pastes are used to produce the lining for tarts and flans, filled with creams, purées and custards, such as crème pâtissière.

In contrast to the production of such pastes, the development of gluten in puff pastry is very important as it is needed to support the expanding steam during the baking process – this is what makes the paste rise. Therefore, depending on the type of paste being produced, the type of flour to be used will differ according to the way in which the gluten will act in the preparation and cooking processes. Pastes such as short and sweet use soft flour, low in gluten, whereas choux and puff pastes require strong, high gluten flour. As a general rule, any pastry that rises during baking is made from strong flour.

Fat can be added to pastes in a number of ways depending on the type of paste being produced. Fat can be rubbed into flour, melted in water or layered between an existing paste, to produce a range of paste with different textures and uses. Generally, the more fat in proportion to flour, the richer the paste and, in the case of short and sweet pastes, the more shortening properties (light and crumbly) it will possess. However, as fat softens as temperature rises, pastes with a high ratio of fat to flour will become increasingly difficult to handle. Some chefs will therefore choose to use a paste with a lower ratio of fat to flour in warmer conditions.

The various methods of adding fat to flour and examples of their use are described below.

Rubbing in

Rubbing in (for example, for short pastry) can be carried out by machine or by hand.

With this method it is better to work with the fat if it is cold – it will be much easier to produce a fine crumb.

Creaming

In the production of sweet pastry, the fat can be creamed with sugar, followed by egg before the flour is added. The creaming can be achieved by machine or by hand.

When creaming, it is better to work with fat that is 'plastic' (at room temperature). This will make it easier to cream.

Remember to always cream the fat and sugar well, before adding the liquid.

> **Professional tip**
>
> When the liquid in a paste is egg, rather than water, it is much less likely to activate the gluten within the flour, and will therefore produce a short, crumbly and light pastry product.

Lamination

Puff pastry is produced using the lamination method, by making a series of alternating layers of a flour-based paste and a fat of the same texture. This is done using a series of either single or double turns (see diagram) during the preparation of the paste.

Puff pastry is very versatile in both the patisserie and the savoury kitchen. Its texture is light and crisp and it is buttery and crunchy to the palate. It can be combined with all types of food in sweet and savoury dishes. One of the differences in making puff pastry is in the fat used. The taste and texture of a puff pastry made with butter is considered as the finest in comparison to those made with other fats.

With this method, the flour paste and the fat are laid in successive folds and rolled between each turn, rather than kneaded, so the two elements do not bind completely. The fat forms a separating layer which, when cooked, retains the steam generated by the water in the dough and produces the layer-separation effect. The flour paste, which includes part of the fat, becomes crunchy and takes on a pleasant golden tone rather than becoming hard and dry.

Single turn

1.

2.

3.

At step 3, roll out and repeat five more times to give six turns in total.

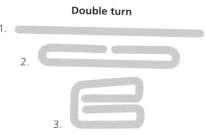

Double turn

1.

2.

3.

At step 3, roll out and repeat three more times to give four turns in total.

Boiling

Choux pastry is produced by placing butter in water and melting it by bringing it to the boil. Once boiled and melted, flour is mixed into the liquid and cooked until a smooth paste (panada) is produced. This is then cooled before beaten eggs are mixed in to produce a paste of piping consistency.

During the baking process, the moisture within the paste produces steam and the eggs and starch in the flour form a coating or case in which the steam is captured. As the paste cooks, it naturally aerates to form a hollow centre in which fillings such as crème Chantilly or crème pâtissière can be piped once the paste is cooked. The most well known choux pastry products are profiteroles and chocolate éclairs.

Handling pastry

Techniques used to work pastry include:

- Folding: folding the initial paste when making puff pastry to create its layers, as in a vol-au-vent or gâteau pithiviers.
- Kneading: using your hands to work dough or puff pastry in the first stage of making.
- Relaxing: keeping pastry covered with a damp cloth, cling film or plastic to prevent a skin forming on the surface and to help prevent the pastry from shrinking during the baking process.

- Shaping: when producing flans, tartlets, barquettes and other goods with short, sweet or lining paste; this also refers to the crimping with the back of a small knife when using the finger and thumb technique.
- Docking: this is the piercing of raw pastry with small holes to prevent it from rising during baking, as when cooking tartlets blind (without a filling).

Rolling

- Roll the pastry on a lightly floured surface; turn the pastry regularly but delicately to prevent it sticking. Keep the rolling pin lightly floured and free from the pastry.
- Always roll with care, handling the pastry lightly – never apply too much pressure.
- Always apply even pressure when using a rolling pin.
- Handle as lightly and quickly as possible.

In a commercial environment, a pastry break is often used, due to the quantity of product. This is a machine that rolls pastry to the desired thickness.

Cutting

- Always cut with a sharp, damp knife.
- When using cutters, always flour them before use by dipping in flour. This will give a sharp, neat cut.
- Only use a lattice cutter on firm pastry; if the pastry is too soft, you will have difficulty lifting the lattice.

Glazing

A glaze is something that gives a product a smooth, shiny surface. Glazes used for pastry dishes include:

- Hot clear gel, which is produced from a pectin source obtainable commercially for finishing flans and tartlets; always use this while it is still hot. Cold gel is exactly the same except that it is used cold. Both gels give a sheen to the products and keep out oxygen, which might otherwise cause discoloration.
- Apricot glaze, produced from apricot jam, acts in the same way as a hot gel.
- Eggwash, applied prior to baking, produces a rich glaze during the cooking process.
- Icing sugar dusted on the surface of the product caramelises in the oven or under the grill.
- Fondant gives a rich sugar glaze, which may be flavoured and/or coloured.
- Water icing gives a transparent glaze, which may also be flavoured and/or coloured.

Lining a flan case

1 Grease the flan ring and baking sheet, or use a non-stick liner such as a silicon baking mat (for example, Silpat).
2 Roll out the pastry so that the circumference is 2cm larger than the flan ring. The pastry may be rolled between cling film, greaseproof or silicone paper.
3 Place the flan ring on the baking sheet.

4 Carefully place the pastry on the flan ring, by rolling it loosely over the rolling pin, picking it up and unrolling it over the flan ring.

6 Press the pastry into shape without stretching it, being careful to exclude any air.
7 Allow a 0.5cm ridge of pastry on top of the flan ring.
8 Cut off the surplus paste by rolling the rolling pin firmly across the top of the flan ring. The rim can be left straight or moulded using the thumb and forefinger, pressing the pastry neatly to form a corrugated pattern. The back of a small paring knife can also be used to press a rim between the forefinger and thumb.

9 Bake blind, lined with cling film, silicone or greaseproof paper, filled with beans, if the recipe requires.

Finishing and presentation

It is essential that all products are finished according to the recipe requirements. Finishing and presentation are key stages in the process, as failure at this point can affect sales. The way products are presented is an important part of the sales technique. Each product of the same type must be of the same shape, size, colour and finish. The decoration should be attractive, delicate and in keeping with the product range. All piping should be neat, clean and tidy.

Some methods of finishing and presentation are as follows:
- dusting: a light sprinkling of icing sugar on a product using a fine sugar dredger or sieve, or muslin cloth
- piping: using fresh cream, chocolate or fondant
- filling: with fruit, cream, pastry cream, etc. (Be careful never to overfill as this will often give the product a clumsy appearance and may be problematic for the customer to eat.)

Piping fresh cream

- The piping of fresh cream is a skill and, like all other skills, it takes practice to become proficient. Finished items should look attractive, simple, clean and tidy, with neat piping.
- Modern practice is to use a disposable piping bag. If using a washable piping bag, it should be sterilised and dried after each use.
- All the equipment for piping must be hygienically cleaned before and after use to avoid cross-contamination.

Other considerations

- Ensure all cooked products are cooled before finishing.
- Always plan your time carefully.
- Understand why pastry products react in different ways according to the production process. Understand why pastry items must be rested or relaxed and docked. This will prevent excessive shrinkage in the oven, and docking will allow the air to escape through the product, preventing any unevenness.
- Use silicone paper or specialist silicone mats for baking in preference to greaseproof paper.

Sauces, creams and fillings used in patisserie

Sauces, creams and fillings used in pastry work have changed dramatically over the past 15 years and have gone from the classic anglaise, Chantilly and coulis preparations to what we see in modern restaurants today, such as foams, oils, cold creams stabilised with gelatine, syrups and convenience fruit purées.

Cold sauces, creams and fillings include:
- **Coulis** – made from various soft fruit such as strawberries, raspberries, blackberries.
- **Crème anglaise** – also used as a base for desserts such as bavarois, ice cream and oeufs à la neige.
- **Crème pâtissière** – which can be transformed into crème chiboust, crème diplomat or crème mousseline and used for filling pastry products.
- **Cold set creams** – made from cream which is sweetened and flavoured, set with gelatine and, when cold and set, is whisked and spooned/dragged on the dessert plate.
- **Crème chantilly** – sweetened cream lightly whipped to piping consistency and flavoured with vanilla.
- **Crème fouettée** – this is lightly whipped cream with no flavourings or sweetener, and is used to enrich and aerate certain pastry products such as mousses, parfaits and cold soufflés.
- **Crémeux** – literally means creamy/smooth and is made from a crème anglaise base emulsified onto chocolate, which is whipped when cold to a smooth, shiny chocolate cream used for filling pastry products.
- **Pâte à bombe** – this is made by whipping egg yolks until aerated, slowly pouring on sugar cooked to 121°C and whisking until cold: it is used in the base of chocolate mousses, parfait glacé, bombe glacée and fruit gratins.
- **Lemon cream/curd** – this is made by whisking egg yolks, lemon juice/zest and sugar over a simmering bain marie of water until the mixture thickens and reaches a temperature of 80°C. Once thickened, melted butter is added and the mixture is then chilled down and piped into pastry cases or used as a filling for petits fours.
- **Ganache** – this is made by heating two parts double cream to 80°C and emulsifying onto one part couverture using a stem blender (ratios may vary depending on the use of the ganache).
- **Sabayon** – a mixture of whole egg or egg yolk with caster sugar, whisked over a pan of simmering water to form a cooked, aerated mass; alcohol may be added, e.g. Marsala for a zabaglione.

> **Professional tips**
> - Ultratex is a relatively new product used in patisserie work. It is a modified pre-gelatinised starch derived from tapioca. It comes in white powder form and has the property to thicken cold solutions such as fruit purées, thus maintaining the freshness of the fruit which could be affected by heat.
> - A key factor in both modern and classic approaches is that each must work with the principle ingredient of the product. A sauce and a cream are integral parts of the dish, whether it be sweet or savoury, and must be treated as such – not just added to the dish for aesthetic reasons. Essentially, a sauce and cream must complement the dish.

1 Short paste (pâte à foncer)

Short pastry is used in fruit pies, Cornish pasties, etc.

Ingredient	Makes 400g	Makes 850g
Flour (soft)	250g	500g
Salt	Pinch	Large pinch
Butter or block/cake margarine	125g	250g
Water	40–50ml	80–100ml

1 Sieve the flour and salt.
2 Rub in the fat to achieve a sandy texture.
3 Make a well in the centre.
4 Add sufficient water to make a fairly firm paste.
5 Handle as little and as lightly as possible. Refrigerate until firm before rolling.

▲ From left to right: short paste, rough puff paste (Recipe 7) and sweet paste (Recipe 2)

Professional tips

The amount of water used varies according to:
● the type of flour (a very fine soft flour is more absorbent)
● the degree of heat (for example, prolonged contact with hot hands, and warm weather conditions).

Different fats have different shortening properties. For example, paste made with a high ratio of butter to other fat will be harder to handle.

Variations
● For wholemeal short pastry, use wholemeal flour in place of half to three-quarters of the white flour.
● Short pastry for sweet dishes such as baked jam roll may be made with self-raising flour.
● Lard can be used in place of some or all of the fat (the butter or cake margarine). Lard has excellent shortening properties and would lend itself, in terms of flavour, to savoury products, particularly meat-based ones. However, many people view lard as an unhealthy product, as it is very high in saturated fat. It is also unsuitable for anyone following a vegan or vegetarian diet as it is an animal product.

Faults

Possible reasons for faults in short pastry are detailed below.

Hard:
● too much water
● too little fat
● fat rubbed in insufficiently
● too much handling and rolling
● over-baking.

Soft–crumbly:
● too little water
● too much fat.

Blistered:
● too little water
● water added unevenly
● fat not rubbed in evenly.

Soggy:
● too much water
● too cool an oven
● baked for insufficient time.

Shrunken:
● too much handling and rolling
● pastry stretched while handling.

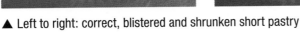

▲ Left to right: correct, blistered and shrunken short pastry

2 Sweet paste (sugar paste, pâte à sucre)

Sugar pastry is used for products such as flans, fruit tarts and tartlets.

Ingredient	Makes 400g	Makes 1 kg
Sugar	50g	125g
Butter or block/cake margarine	125g	300g
Eggs	1	2–3
Flour (soft)	200g	500g
Salt	Pinch	Large pinch

Method 1 – sweet lining paste (rubbing in)

1 Sieve the flour and salt. Lightly rub in the margarine or butter to achieve a sandy texture.
2 Mix the sugar and egg until dissolved.
3 Make a well in the centre of the flour. Add the sugar and beaten egg.
4 Gradually incorporate the flour and margarine or butter, and lightly mix to a smooth paste. Allow to rest before using.

Method 2 – traditional French sugar paste (creaming)

1 Taking care not to over-soften, cream the butter and sugar.
2 Add the beaten egg gradually, and mix for a few seconds.
3 Gradually incorporate the sieved flour and salt. Mix lightly until smooth.
4 Allow to rest in a cool place before using.

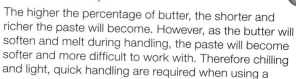

Professional tips

The higher the percentage of butter, the shorter and richer the paste will become. However, as the butter will soften and melt during handling, the paste will become softer and more difficult to work with. Therefore chilling and light, quick handling are required when using a sweet paste with a high butter content.

This also applies to the working environment. For example, in a particularly warm kitchen, it will be more difficult to work with a paste of this structure than in a cooler kitchen.

The butter in this recipe could be reduced from 125g to 100g to make handling easier.

Measure out the sugar and cut the butter into small chunks

Cream the butter and sugar together

Add the beaten egg in stages, thoroughly mixing each time

Incorporate the flour and salt

Press into a tray and leave to chill

The paste will need to be rolled out before use in any recipe

3 Lining paste

Lining paste may be used in place of sweet paste. Lining paste is not as rich and sweet as sweet (sugar) paste.

Ingredient	Makes 450g/ 3 × 15cm tarts
Soft flour	250g
Caster sugar	10g
Salt	5g
Butter	125g
Water	40ml
Egg	1

1 Combine the dry ingredients together; rub in the butter.
2 Add the water and egg to gently form a dough.
3 Wrap in cling film and leave to rest in the refrigerator for several hours before using.

4 Sablé paste

Sablé paste may be used for petits fours, pastries and as a base or platform for other desserts. 'Sablé' means a sandy texture.

Ingredient	Makes 500g
Egg	1
Caster sugar	75g
Butter or block/cake margarine	150g
Soft flour	200g
Salt	Pinch
Ground almonds	75g

1 Lightly cream the egg and sugar without over-softening.
2 Lightly mix in the butter – do not over-soften.
3 Incorporate the sieved flour, salt and ground almonds.
4 Mix lightly to a smooth paste.
5 Chill in the refrigerator before use.

Variation
Sablé paste may also be made using a creaming method. An example of this is the paste used in the recipe for Gâteau MacMahon (see Chapter 14).

5 Choux paste

Choux paste is used to make products such as éclairs, profiteroles and gâteau Paris-Brest.

Ingredient	Makes 750g	Makes 1.5 kg
Water	250ml	500ml
Sugar	Pinch	Large pinch
Salt	Pinch	Large pinch
Butter or block/cake margarine	100g	200g
Flour (strong)	150g	300g
Eggs	4–5	8–10

1 Bring the water, sugar, salt and fat to the boil in a saucepan. Remove from the heat.

2 Add the sieved flour and mix in with a wooden spoon.

3 Return to a moderate heat and stir continuously until the mixture leaves the sides of the pan. (This is known as a panada.)

4 Remove from the heat and allow to cool.

5 Gradually add the beaten eggs, beating well. Do not add all the eggs at once – check the consistency as you go. The mixture should just flow back when moved in one direction (it may not take all the egg).

Faults

Greasy and heavy paste:
● the basic mixture was over-cooked.

Soft paste, not aerated:
● flour insufficiently cooked
● eggs insufficiently beaten in the mixture
● oven too cool
● under-baked.

Split or separated mixture:
● egg added too quickly.

▲ The choux buns on the left are light and well risen; those on the right are poorly aerated.

Variation
50 per cent, 70 per cent or 100 per cent wholemeal flour may be used to make choux paste.

Cut the butter into cubes and then melt them in the water

Add the flour

When the panada is ready, it will start to come away from the sides

Add egg until the mixture is the right consistency – it should drop from a spoon under its own weight

6 Puff paste (French method)

Ingredient	Makes 1.5 kg
Flour (strong)	560g
Salt	12g
Pastry butter or pastry margarine	60g
Water, ice-cold	325ml
Pastry butter or pastry margarine	500g
Lemon juice, ascorbic or tartaric acid, or white vinegar	A few drops

1 Sieve the flour and salt.
2 Rub in the 60g of pastry butter/pastry margarine.
3 Make a well in the centre.
4 Add the water and lemon juice or acid (to make the gluten more elastic), and knead well into a smooth dough in the shape of a ball.
5 Relax the dough in a cool place for 30 minutes.
6 Cut a cross halfway through the dough and pull out the corners to form a star shape.
7 Roll out the points of the star square, leaving the centre thick.

8 Knead the remaining butter/margarine to the same texture as the dough. This is most important – if the fat is too soft it will melt and ooze out, if too hard it will break through the paste when being rolled.
9 Place the butter or margarine on the centre square, which is four times thicker than the flaps.
10 Fold over the flaps.
11 Roll out to 30cm × 15cm, cover with a cloth or plastic and rest for 5–10 minutes in a cool place.
12 Roll out to 60cm × 20cm, fold both the ends to the centre, fold in half again to form a square. This is one double turn.
13 Allow to rest in a cool place for 20 minutes.
14 Half-turn the paste to the right or the left.
15 Give one more double turn; allow to rest for 20 minutes.
16 Give two more double turns, allowing to rest between each.
17 Allow to rest before using.

Rub the first batch of butter into the flour

Mix in the water and lemon juice

Knead into a smooth dough

Roll the dough out into a cross shape

Knead the remaining butter in a plastic bag, then place it on the centre of the dough

Fold over each flap

Roll into a neat rectangle, then take each end and fold to meet in the centre

Fold again from the top end of the paste to the bottom

These photos have shown one double turn. When resting the turned and folded paste, leave an indented finger mark on the surface to show the number of turns completed

Professional tips

- Care must be taken when rolling out the paste to keep the ends and sides square.
- When rolling between each turn, always roll with the folded edge to the left.
- The addition of lemon juice (or other acid) helps to strengthen the gluten in the flour, thus helping to make a stronger dough so that there is less likelihood of the fat oozing out.
- The rise is caused by the fat separating layers of paste and air during rolling. When heat is applied by the oven, steam is produced, causing the layers to rise and give the characteristic flaky formation. This is aeration by lamination.

Faults

The pastry on the left is unevenly laminated. Possible reasons for this:

- The paste was not folded equally.
- The paste was rolled too thinly.
- Re-used scraps of paste were used, instead of making up a virgin paste.

7 Rough puff paste (Scottish method)

Ingredient	Makes 475g	Makes 1.2 kg
Flour (strong)	200g	500g
Salt	2g (large pinch)	4g (2 large pinches)
Butter or block/cake margarine (lightly chilled)	150g	375g
Water, ice-cold	125ml	300ml
Lemon juice, ascorbic or tartaric acid	10ml	25ml

1 Sieve the flour and salt.

2 Cut the fat into small pieces and lightly mix them into the flour without rubbing in.

3 Make a well in the centre.

4 Add the liquid and mix to a dough. The dough should be fairly tight at this stage.

5 Turn on to a floured table and roll into an oblong strip, about 30 × 10cm, keeping the sides square.

6 Give one double turn (as for puff pastry).

7 Allow to rest in a cool place, covered with cloth or plastic for 30 minutes.

8 Give three more double turns, resting between each. (Alternatively, give six single turns.) Allow to rest before using.

Professional tip

Each time you leave the paste to rest, gently make finger indentations, one for each turn you have given the paste. This will help you to keep track.

Make a well in the centre of the flour and butter, and add the liquid

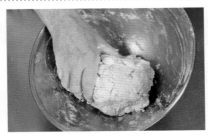

Mix to a fairly stiff dough

Roll out and fold the ends to the middle

Keep rolling, folding and turning

The finished paste, ready to rest and then use

8 Strudel paste

Ingredient	Makes 1.25 kg
Strong flour	680g
Eggs, whole	3
Egg yolks	3
Oil	3 tbsp
Salt	7g
Water, cold	To make up to 575ml

1 Sift the flour and place into a mixer.

2 Place the eggs, egg yolks, oil and salt into a measuring jug. Add cold water to the 575ml mark.

3 Add the liquid to the flour and mix with a hook attachment to make a smooth dough. If the dough is very sticky, add more flour.

4 Divide the dough into 4 equal pieces. Wrap each piece in cling film and leave to rest in a cool area between oiled plates.

5 Cover a free-standing table, away from the wall, with a large, clean cloth. Dust the cloth with flour.

6 Roll out the paste as far as possible across the table with a rolling pin.

7 Stretch and pull the paste out by hand.

There is a recipe for apple strudel in Chapter 14.

Professional tip

Make sure you have all the ingredients and equipment needed for the strudel, including the filling, before you start rolling out the paste. As strudel paste is so fine and delicate, it is important that the process of making the strudel is completed quickly to prevent the paste either drying out (if left uncovered for too long) or becoming soggy (if in contact with a filling and not baked immediately).

Half this recipe will be enough for ten portions, but when making strudel it is advisable to make more than you need in case of mistakes being made at the stretching stage.

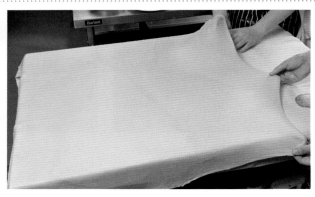

Pulling the dough

Fully stretched paste

9 Hot water paste

Hot water paste is used to produce savoury pies and the lining for terrines.

Ingredient	Makes 500g	1.2 kg
Flour (strong)	250g	625g
Salt		
Lard, butter or block/cake margarine	125g	300g
Water	125ml	312ml

Note: Not all pies are cooked in moulds. Instead they are hand-raised using a hot water paste. A well known example is a pork pie (see Chapter 4).

1 Sift the flour and salt into a basin.
2 Make a well in the centre.
3 Boil the fat with the water and pour immediately into the centre of the flour.
4 Mix with a wooden spoon until cool.
5 Mix to a smooth paste and use while still warm.

Professional tip

You can use four parts lard to one part butter or block/cake margarine.

10 Suet paste

Suet paste is used for steamed fruit puddings, steamed jam rolls, steamed meat puddings and dumplings.

Ingredient	Makes 400g	Makes 1 kg
Flour (soft) or self-raising flour	200g	500g
Baking powder	10g	25g
Salt	Pinch	Large pinch
Prepared beef or vegetarian suet	100g	250g
Water	125ml	300ml

1 Sieve the flour, baking powder and salt.
2 Mix in the suet. Make a well. Add the water.
3 Mix lightly to a fairly stiff paste.

Professional tip

Self-raising flour already contains baking powder so this element could be reduced in the recipe if using self-raising flour.

Vegetarian suet is also available to enable products to be meat-free.

Faults

Possible reasons for faults in suet paste:
● Paste is heavy and soggy – it may be that the cooking temperature was too low.
● Paste is tough – it may have been handled too much or over-cooked.

11 Chantilly cream

Ingredient	Makes 500ml
Whipping cream	500ml
Caster sugar	100g
Vanilla arome/fresh vanilla pod	A few drops to taste/seeds from 1 vanilla pod

1 Place all ingredients in a bowl. Whisk over ice until the mixture forms soft peaks. If using a mechanical mixer, stand and watch until the mixture is ready – do not leave it unattended as the mix will over-whip quickly, curdling the cream.

2 Cover and place in the fridge immediately.

12 Pastry cream (crème pâtissière)

▲ Left to right: pastry cream, crème diplomat and crème chiboust

Ingredient	Makes approx. 750ml
Milk	500ml
Vanilla pod	1
Egg yolks	4
Caster sugar	125g
Soft flour	75g
Custard powder	10g

1 Heat the milk with the cut vanilla pod and leave to infuse.

2 Beat the sugar and egg yolks together until creamy white. Add the flour and custard powder.

3 Strain the hot milk, gradually blending it into the egg mixture.

4 Strain into a clean pan and bring back to the boil, stirring constantly.

5 When the mixture has boiled and thickened, pour into a plastic bowl, sprinkle with caster sugar and cover with cling film.

6 Chill over ice and refrigerate as soon as possible. Ideally, blast chill.

7 When required, knock back on a mixing machine with a little kirsch.

Professional tip

At step 4, the microwave may be used effectively. Pour the mixture into a plastic bowl and cook in the microwave for 30-second periods, stirring in between, until the mixture boils and thickens.

The recipes below are based on using the recipe for crème pâtissière, with additional ingredients.

Crème mousseline

Beat in 100g of soft butter (a pomade).
The butter content is usually about 20 per cent of the volume but this can be raised to 50 per cent depending on its intended use.

Crème diplomate

When the pastry cream is chilled, fold in an equal quantity of whipped double cream.

Crème chiboust

When the pastry cream mixture has cooled slightly, fold in an equal quantity of Italian meringue (Recipe 16).

Variations

Additional flavourings can also be added to crème pâtissière, crème diplomat or crème chiboust.

13 Frangipane (almond cream)

Ingredient	Makes 300g
Butter	100g
Caster sugar	100g
Eggs	2
Ground almonds	100g
Flour	10g

1 Cream the butter and sugar until aerated.
2 Gradually beat in the eggs.
3 Mix in the almonds and flour (mix lightly).
4 Use as required.

Cut the butter into small pieces and add to the sugar

Cream the butter and sugar together

Beat in the eggs (before adding the flour)

Variation

Try adding lemon zest or vanilla seeds to the recipe.

14 Ganache

Ingredient	Makes 750g
Version 1 (for decoration)	
Double cream	300ml
Couverture, cut into small pieces	350g
Unsalted butter	85g
Spirit or liqueur	20ml
Ingredient	**Makes 1 kg**
Version 2 (for a filling)	
Double cream	300ml
Vanilla pod	½
Couverture, cut into small pieces	600g
Unsalted butter	120g

1 Boil the cream (and the vanilla for Version 2) in a heavy saucepan.

2 Pour the cream over the couverture. Whisk with a fine whisk until the chocolate has melted.

3 Whisk in the butter (and the liqueur for Version 1).

4 Stir over ice until the mixture has the required consistency.

15 Pâte à bombe

This is used as a base for parfaits or soufflés.

Ingredient	Makes 700ml
Caster sugar	300g
Water	200ml
Glucose	20g
Egg yolks, large	10

1 Boil the sugar, water and glucose together in a heavy-based pan. Continue to gently boil. Wash down the sides of the pan to prevent crystallisation.

2 Meanwhile put the eggs yolks in a food mixer and start mixing, while watching the sugar.

3 When the sugar reaches 121°C, pour onto the eggs in a steady stream.

4 Carry on whisking until the mixture has increased in volume and is cold.

16 Italian meringue

Ingredient	Makes 250g	Makes 625g
Granulated or cube sugar	200g	500g
Water	60ml	140g
Cream of tartar	Pinch	Large pinch
Egg whites	4	10

Professional tip

To ensure the sugar is not heated beyond the hard-ball stage, it is advisable to remove from the heat at 115°C as the sugar will continue to rise in temperature, and this will provide a little time to ensure the egg whites are whipped to the correct point.

1 Boil the sugar, water and cream of tartar to hard-ball stage of 121°C.

2 While the sugar is cooking, beat the egg whites to full peak and, while stiff, beating slowly, pour on the boiling sugar.

3 The mixture will stand up in stiff peaks when it is ready. Use as required

17 Apricot glaze

Ingredient	Makes 150ml
Apricot jam	100g
Stock syrup or water	50ml

1 Boil the apricot jam with a little syrup or water.

2 Pass through a strainer. The glaze should be used hot.

Professional tip

A flan jelly (commercial pectin glaze) may be used as an alternative to apricot glaze. This is usually a clear glaze to which food colour may be added.

18 Treacle tart

1 Lightly grease an appropriately sized flan ring, or barquette or tartlet moulds if making individual portions.
2 Line with pastry.
3 Warm the golden syrup, water and lemon juice; add the crumbs.
4 Place into the pastry ring and bake at 170°C for about 20 minutes.
5 Dust with icing sugar.

Variations

This tart can also be made in a shallow flan ring. Any pastry debris can be rolled and cut into 0.5cm strips and used to decorate the top of the tart before baking.

Try sprinkling with vanilla salt as a garnish.

Ingredient	Makes 4 portions	Makes 10 portions
Short paste	125g	300g
Golden syrup	100g	250g
Water	1 tbsp	2½ tbsp
Lemon juice	3–4 drops	8–10 drops
Fresh white bread or cake crumbs	15g	50g

Energy	Cal	Fat	Sat fat	Carb	Sugar	Protein	Fibre
1818 kJ	433 kcal	21.2g	13.1g	58.7g	17.7g	5.5g	2.2g

19 Plum and soured cream tart

Ingredient	Makes 1 × 20cm flan	Makes 3 × 20cm flans
Sweet paste	200g	600g
Filling		
Eggs, beaten	1	4
Plums	6	18
Unsalted butter	45g	130g
Caster sugar	60g	180g
Grated nutmeg	Small pinch	Pinch
Soured cream	140ml	400ml
Semolina	15g	50g
Zest and juice of lemon	½	1
Topping		
Soft flour	75g	200g
Butter	50g	150g
Demerara sugar	25g	70g

Energy	Cal	Fat	Sat fat	Carb	Sugar	Protein	Fibre
1459 kJ	349 kcal	21.6g	13.0g	36.8g	19.1g	4.6g	1.8g

1 Line 20cm flan rings with the pastry and chill in the fridge.

2 Eggwash the tart cases and bake blind for 15 minutes until golden brown: start at 190°C for 2 minutes, then lower the oven to 150°C for the remaining time.

3 Cut the plums in half, remove the stones, and arrange them cut side up over the base of the tart. Cream the butter and the sugar together until light and fluffy.

4 Gradually beat in the remaining egg, and then stir in the nutmeg, soured cream, semolina, lemon zest and juice. Pour the mixture over the plums and bake for 25 minutes until lightly set.

5 Meanwhile, for the topping, rub the flour and butter together until it resembles fine breadcrumbs, and stir in the demerara sugar.

6 After the 25 minutes has elapsed, sprinkle the crumble mixture over the top of the tart and bake for another 25 minutes.

7 Decorate with sliced plums.

20 Pear and almond tart

Ingredient	Makes 1 × 20cm ring (8 portions)
Sweet paste	200g
Apricot jam	25g
Almond cream	350g
Poached pears	4
Apricot glaze	
Flaked almonds	
Icing sugar	

Energy	Cal	Fat	Sat fat	Carb	Sugar	Protein	Fibre
2775 kJ	668 kcal	53.0g	27.3g	42.2g	31.3g	8.3g	4.4g

1 Line a buttered 20cm flan ring with sweet paste. Trim and dock.

2 Using the back of a spoon, spread a little apricot jam over the base.

3 Pipe in almond cream until the flan case is two-thirds full.

4 Dry the poached pears. Cut them in half and remove the cores and string.

5 Score across the pears and arrange on top of the flan.

6 Bake in the oven at 200°C for 25–30 minutes.

7 Allow to cool, then brush with apricot glaze.

8 Sprinkle flaked almonds around the edge and dust with icing sugar.

21 Egg custard tart

Ingredient	Makes 8 portions (1 × 20cm flan)
Sweet paste	250g
Egg yolks	9
Caster sugar	75g
Whipping cream, gently warmed and infused with 2 sticks of cinnamon	500ml
Nutmeg, freshly grated	

Energy	Cal	Fat	Sat fat	Carb	Sugar	Protein	Fibre
2023 kJ	488 kcal	39.9g	22.7g	27.5g	15.9g	6.6g	0.7g

1 Roll out the pastry on a lightly floured surface, to 2mm thickness. Use it to line a flan ring, placed on a baking sheet.

2 Line the pastry with food-safe cling film or greaseproof paper and fill with baking beans. Bake blind in a preheated oven at 190°C, for about 10 minutes or until the pastry is turning golden brown. Remove the paper and beans, and allow to cool. Turn the oven down to 130°C.

3 To make the custard filling, whisk together the egg yolks and sugar. Add the cream and mix well.

4 Pass the mixture through a fine sieve into a saucepan. Heat to 37°C.

5 Fill the pastry case with the custard to 5 mm below the top. Place it carefully into the middle of the oven and bake for 30–40 minutes or until the custard appears to be set but not too firm.

6 Remove from the oven and cover liberally with grated nutmeg (and/or icing sugar). Allow to cool to room temperature.

22 Fruit tart, tartlets and barquettes

Ingredient	Makes 1 × 20cm flan
Sweet paste	250g
Fruit (e.g. strawberries, raspberries, grapes, blueberries)	500g
Pastry cream	500ml
Glaze	5 tbsp

Energy	Cal	Fat	Sat fat	Carb	Sugar	Protein	Fibre
1197 kJ	285 kcal	12.8g	6.9g	38.4g	21.4g	6.3g	2.4g

1 Line an appropriately sized flan ring with paste and cook blind at 190°C. Allow to cool.

2 Pick and wash the fruit, then drain well. Wash and slice/segment, etc. any larger fruit being used.

3 Pipe pastry cream into the flan case, filling it to the rim. Dress the fruit neatly over the top.

4 Coat with the glaze. Use a glaze suitable for the fruit chosen; for example, with a strawberry tart, use a red glaze.

Professional tip

Brush the inside of the pastry case with melted couverture before filling. This forms a barrier between the pastry and the moisture in the filling.

Faults

Although this strawberry tart may appear to be fine at first glance, the husks of the strawberries are visible. It would be better to present the strawberries with their tops pointing upwards, or sliced and overlapping.

There is also quite a wide gap between the rows of strawberries, showing the crème pâtissière underneath. This should also be avoided.

The second photo shows the importance of ensuring that fillings are prepared and/or cooked properly. In this case, the crème pâtissière has not been cooked sufficiently or prepared accurately as the filling is not structured sufficiently to support the fruit once the tart has been cut.

For tartlets

1 Roll out pastry 3mm thick.

2 Cut out rounds with a fluted cutter and place them neatly in greased tartlet moulds. If soft fruit (such as strawberries or raspberries) is being used, the pastry should be cooked blind first.

3 After baking and filling (or filling and baking) with pastry cream, dress neatly with fruit and glaze the top.

Professional tip

Certain fruits (such as strawberries and raspberries) are sometimes served in boat-shaped moulds (barquettes). The preparation is the same as for tartlets.

Tartlets and barquettes should be served allowing one large or two small per portion.

23 French apple flan (flan aux pommes)

Ingredient	4 portions	10 portions
Sweet paste	100g	250g
Pastry cream	250ml	625ml
Dessert apples, large	3	8
Sugar	50g	125g
Apricot glaze	2 tbsp	6 tbsp

Energy	Cal	Fat	Sat fat	Carb	Sugar	Protein	Fibre
1849 kJ	440 kcal	19.4g	8.4g	60.6g	45.9g	9.6g	2.9g

1 Line an appropriately sized flan ring with the sweet paste. Pierce the bottom several times with a fork.

2 Pipe a layer of pastry cream into the bottom of the flan.

3 Peel, core, halve and wash the apples.

4 Cut into thin slices and lay carefully on the pastry cream, in overlapping slices starting at the outside and working towards the centre. Ensure that each slice points to the centre of the flan, then no difficulty should be encountered in joining the pattern up neatly.

6 Sprinkle a little sugar on the apple slices and bake the flan at 200–220°C for 30–40 minutes.

7 When the flan is almost cooked, remove the flan ring carefully, and return to the oven to complete the cooking. Mask with hot apricot glaze or flan gel.

Pipe the filling neatly into the flan case

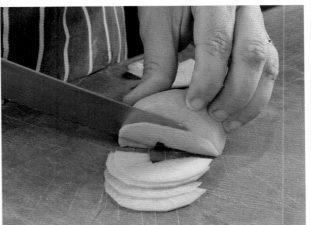

Slice the apple very thinly

Arrange the apple slices on top of the flan

Complete the arrangement of apple slices

24 Lemon tart (tarte au citron)

Ingredient	8 portions
Sweet paste	200g
Lemons	Juice of 3, zest from 4
Eggs	8
Caster sugar	300g
Double cream	250ml

Energy	Cal	Fat	Sat fat	Carb	Sugar	Protein	Fibre
2094 kJ	501 kcal	29.7g	16.3g	52.9g	43.6g	8.8g	0.5g

1 Prepare 200g of sweet paste, adding the zest of one lemon to the sugar.

2 Line a 20cm flan ring with the paste.

3 Bake blind at 190°C for approximately 15 minutes.

4 Prepare the filling: mix the eggs and sugar together until smooth, add the cream, lemon juice and zest. Whisk well.

5 Seal the pastry, so that the filling will not leak out. Pour the filling into the flan case, and bake for 30–40 minutes at 150°C until just set. (Take care when almost cooked as overcooking will cause the filling to rise and possibly crack.)

6 Remove from the oven and allow to cool.

7 Dust with icing sugar and glaze under the grill or with a blowtorch. Portion and serve.

Note: The mixture will fill one 16 × 4cm or two 16 × 2cm flan rings. If using two flan rings, double the amount of pastry and reduce the baking time when the filling is added.

Professional tip

If possible, make the filling one day in advance. The flavour will develop as the mixture matures.

Variation
Limes may be used in place of lemons. If so, use the zest and juice of 5 limes or use a mixture of lemons and limes.

25 White chocolate and citrus meringue tart

The following two recipes are examples of tarts with inserts of different flavours and textures. In addition to the eating qualities, the inserts also add an extra element to the presentation of the tarts, providing a layered visual effect.

Ingredient	Makes 1 × 20cm flan
20cm flan case, baked blind using sablé paste	1
White chocolate lemon cream	
Lemon juice	250g
Lemon zest	1
Caster sugar	80g
Whole eggs	5
Gelatine leaves	2
White couverture	200g
Cocoa butter	10g
Lime meringue (insert/filler)	
Gelatine (approximately 4 leaves)	10g
Caster sugar	115g
Water	135g
Lime juice	125g

Energy	Cal	Fat	Sat fat	Carb	Sugar	Protein	Fibre
1964 kJ	469 kcal	23.2g	8.6g	58.7g	44.8g	9.8g	0.8g

1 Wash and zest the lemon.
2 Mix the lemon juice with the sugar, zest and eggs.
3 Cook slowly over a low heat until the mixture thickens.
4 Remove from the heat at 85°C. Add the softened gelatine.
5 Pour the lemon cream over the melted chocolate and cocoa butter.
6 Emulsify using a stem blender.

For the lime insert

1 Soak the gelatine in cold water.
2 Boil the water, sugar and lime juice and add the gelatine.
3 Allow to set in the fridge until solid.
4 Partially melt in the microwave and whip up at full speed to aerate like a meringue.
5 Pour into a tray and freeze.
6 Once frozen, cut out a disc slightly smaller than the baked pastry case.

Assembly

1 Half fill the pastry case with warm lemon cream.
2 When almost set, push in the frozen lime meringue disc.
3 Top up to the rim of the pastry case with the remaining cream and refrigerate.
4 Cut out a wedge and decorate the top with raspberry and passion fruit coulis thickened with Ultratex, green pistachio nuts and squares of lime meringue.

26 Chocolate and ginger tart with insert of mango jelly and praline

Ingredient	Makes 1 × 20cm flan
20cm flan case, baked blind	1
Disc of mango pâte de fruits (see Chapter 12 Recipe 23)	1
Disc of praline (see below)	1
Chocolate and ginger ganache	
Whipping cream	200g
Fresh ginger	22g
Preserved ginger	22g
Plain couverture	250g
Unsalted butter	50g

Energy	Cal	Fat	Sat fat	Carb	Sugar	Protein	Fibre
3112 kJ	745 kcal	47.6g	21.3g	76.4g	60.6g	7.5g	3.3g

1 Boil the cream with the finely grated fresh ginger and allow to infuse for 10 minutes.

2 Add the chopped preserved ginger. Strain, and pour onto the melted chocolate and butter. Emulsify with a stem blender.

The praline disc

Ingredient	1 disc
Plain couverture, melted, tempered	100g
Praline paste (commercial product)	100g
Pâte à feuilletine	100g

1 Warm the praline paste in a microwave.

2 Mix with the melted couverture.

3 Add the pâte à feuilletine.

4 Roll out thinly between sheets of silicone paper.

5 Refrigerate until solid.

6 Cut out a disc using a flan ring narrower than the baked pastry case.

Assembly

1 Half fill the tart case with the ganache and allow to partially set. Place the discs of mango jelly and praline feuilletine, cut slightly smaller than the flan case, on top.

2 Top with the remaining ganache to fill the pastry case and refrigerate.

3 Once set, cut a wedge using a warm knife and, just before service, run the flame of a blow torch over the top of the ganache to give a shine. Decorate with gold leaf.

4. Serve with pear water ice decorated with white chocolate.

27 Puff pastry cases (bouchées, vol-au-vents)

Ingredient	12 bouchées/ 6 vol-au-vents
Puff pastry	200g

1 Roll out the pastry approximately 5mm thick.
2 Cut out with a round, fluted cutter.
3 Place on a greased, dampened baking sheet; eggwash.

4 Dip a plain cutter of a slightly smaller diameter into hot fat or oil and make an incision 3mm deep in the centre of each.
5 Allow to rest in a cool place.
6 Bake at 220°C for about 20 minutes.
7 When cool, remove the caps or lids carefully and remove all the raw pastry from inside the cases.

Professional tip

Bouchées are filled with a variety of savoury fillings and are served hot or cold. They may also be filled with creams or curds as a pastry.

Larger versions are known as vol-au-vents. They may be produced in one-, two-, four- or six-portion sizes; a single-sized vol-au-vent would be approximately twice the size of a bouchée. When preparing one- and two-portion sized vol-au-vents, the method for bouchées may be followed. When preparing larger vol-au-vents, it is advisable to have two layers of puff pastry, each 5mm thick, sealed together with eggwash. One layer should be a plain round, and the other of the same diameter with a circle cut out of the centre.

28 Cheese straws (paillettes au fromage)

Ingredient	8–10 portions	16–20 portions
Puff paste or rough puff paste	100g	250g
Cheese, grated	50g	125g
Cayenne pepper		

Energy	Cal	Fat	Sat fat	Carb	Sugar	Protein	Fibre
248 kJ	60 kcal	4.2g	2.2g	3.9g	0.1g	1.9g	0.0g

1 Roll out the pastry to 60 × 15cm, 3mm thick.
2 Sprinkle with the cheese and cayenne pepper.
3 Roll out lightly to embed the cheese.
4 Cut the paste into thin strips by length.
5 Twist each strip to form rolls in the strip.
6 Place on a silicone mat.
7 Bake in a hot oven at 230–250°C for 10 minutes or until a golden brown. Cut into lengths as required.

29 Gâteau pithiviers

Ingredient	Makes 1 × 22cm gâteaux	Makes 2 × 22cm gâteaux
Puff pastry	500g	1 kg
Pastry cream	30g	60g
Frangipane	300g	600g
Eggwash (yolks only)		
Granulated sugar		

Energy	Cal	Fat	Sat fat	Carb	Sugar	Protein	Fibre
1755 kJ	420 kcal	24.8g	8.0g	44.1g	20.8g	7.8g	0.0g

1 Divide the paste into four equal pieces. Roll out each piece in a circle with a 22cm diameter, 4mm thick.

2 Rest in the fridge between sheets of cling film, preferably overnight.

3 Lightly butter two baking trays and splash with water. Lay one circle of paste onto each tray and dock them.

4 Mark a 16cm diameter circle in the centre of each.

5 Beat the pastry cream, if desired, and mix it with the frangipane.

6 Using a plain nozzle, pipe the cream over the inner circles, making them slightly domed. (The paste may be brushed with apricot glaze first, if desired.)

7 Eggwash the outer edges of the paste. Lay one of the remaining pieces over the top of each one, smooth over and press down hard.

8 Mark the edges with a round cutter. Cut out a scallop pattern with a knife, or use a cut piping nozzle.

9 Eggwash twice. Mark the top of both with a spiral pattern.

10 Bake at 220°C for 10 minutes. Remove from the oven and sprinkle with granulated sugar. Turn the oven down to 190°C and bake for a further 20 to 25 minutes.

11 Glaze under a salamander.

Adding the filling to the rolled base

Trimming the edge

Marking the top

30 Mille-feuilles (puff pastry slices)

7 Place the second strip on top and pipe with pastry cream.

8 Place the last strip on top, flat side up. Press gently. Brush with boiling apricot glaze to form a key.

To decorate by feather-icing

1 Warm the fondant to 37°C (warm to the touch) and correct the consistency with sugar syrup if necessary.

3 Pour the fondant over the mille-feuilles in an even coat.

4 Immediately pipe the melted chocolate over the fondant in strips, ½ cm apart.

6 With the back of a small knife, wiping after each stroke, mark down the slice at 2cm intervals.

7 Quickly turn the slice around and repeat in the same direction with strokes in between the previous ones.

8 Allow to set and trim the edges neatly.

9 Cut into even portions with a sharp, thin-bladed knife; dip into hot water and wipe clean after each cut.

For a traditional finish, crush the pastry trimmings and use them to coat the sides.

Ingredient	10 portions
Puff pastry trimmings	600g
Pastry cream	400ml
Apricot glaze	
Fondant	350g
Chocolate	100g

Energy	Cal	Fat	Sat fat	Carb	Sugar	Protein	Fibre
1959 kJ	466 kcal	19.3g	9.5g	71.1g	46.0g	6.4g	0.2g

1 Roll out the pastry 2mm thick into an even-sided square.

2 Roll up carefully on a rolling pin and unroll onto a greased, dampened baking sheet.

3 Dock well.

4 Bake in a hot oven at 220°C for 15–20 minutes; turn over after 10 minutes. Allow to cool.

5 Using a large knife, cut into three even-sized rectangles.

6 Keeping the best strip for the top, pipe the pastry cream on one strip.

Variations
- Whipped fresh cream may be used as an alternative to pastry cream.
- The pastry cream or whipped cream may also be flavoured with a liqueur if so desired, such as curaçao, Grand Marnier or Cointreau.

Pipe cream between layers of pastry

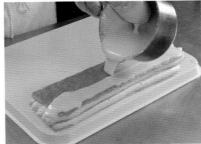

Ice the top with fondant

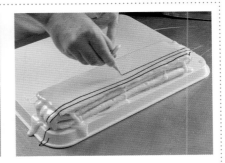

Decorate with chocolate

31 Mille-feuilles Napoleon

Ingredient	10 portions
Puff paste	600g
Crème diplomat (see below)	400ml
Strawberries, small or halves	225g
Icing sugar	20g
Crème diplomat	
Double cream, whipped	300ml
Pastry cream	300g

Energy	Cal	Fat	Sat fat	Carb	Sugar	Protein	Fibre
1843 kJ	443 kcal	32.2g	17.6g	34.9g	10.8g	5.6g	0.4g

1 Roll out the puff paste in a very thin sheet. Dock well.

2 Rest the pastry with a clean baking sheet on top to prevent it from rising. Bake at 220°C.

3 Prepare the crème diplomat by folding the whipped double cream into the chilled pastry cream.

4 Cut the paste neatly into rectangles, approximately 9 × 5cm. Allow 3 pieces per portion.

5 Take one rectangle of paste and pipe on bulbs of crème diplomat, placing strawberry halves between the bulbs.

6 Place a second rectangle on top and layer with crème diplomat and strawberries as before.

7 Dust the third rectangle of paste with icing sugar and sear in diagonal lines with a hot poker. (This technique is referred to as 'quadrillage'.) Place gently on top of the dessert.

8 Serve with strawberry coulis or crème anglaise.

32 Strawberry soleil

Ingredient	Makes 1 (8 portions)
Puff pastry	250g
Pastry cream	400g
Slice of sponge	1
Grand Marnier or rum stock syrup	20ml
Strawberries, hulled	2 punnets
Strawberry glaze	200ml

Energy	Cal	Fat	Sat fat	Carb	Sugar	Protein	Fibre
1120 kJ	267 kcal	12.2g	5.2g	35.0g	17.9g	5.1g	1.2g

1 Roll the pastry. Rest it, then cut and fold it into a soleil (sun) shape (see box). Dock the central area, eggwash the points and press down.

2 Bake blind at 205°C for about 15 minutes. Dock the base again at least once during cooking.

3 Beat the pastry cream until smooth. Pipe it in the centre of the pastry to within 2cm of the inside edge. Pipe in a cone shape so that it is domed towards the centre.

4 Cut a slice of sponge to fit and lay it on top of the cream. Brush with the Grand Marnier stock syrup.

5 Spread a little more pastry cream on top of the sponge.

6 Arrange whole strawberries all over the sponge, pointing outwards. Brush with boiling strawberry glaze, making sure any gaps are filled in.

Professional tip: the soleil shape

- Cut out a 30cm circle of pastry.
- Mark, but do not cut, an inner circle with a 22cm diameter.
- Starting from the inner circle, make 16 cuts, regularly spaced.
- Fold over each piece so that edge a aligns with edge b (see diagram), forming points.

33 Pear jalousie

1 Roll out two-thirds of the pastry 3mm thick into a strip 25 × 10cm and place on a greased, dampened baking sheet.
2 Pierce with a docker. Moisten the edges.
3 Pipe on the frangipane, leaving 2cm free all the way round. Place the pears on top.
4 Roll out the remaining one-third of the pastry to the same size. Chill before cutting.
5 Cut the dough with a trellis cutter to make a lattice.
6 Carefully open out this strip and neatly place onto the first strip.
7 Trim off any excess. Neaten and decorate the edge. Brush with eggwash.
8 Bake at 220°C for 25–30 minutes.
9 Glaze with apricot glaze. Dust the edges with icing sugar and return to a very hot oven to glaze further.

Ingredient	8–10 portions
Puff pastry	200g
Frangipane	200g
Pears, poached or tinned (cored and cut in half lengthways)	5

Energy	Cal	Fat	Sat fat	Carb	Sugar	Protein	Fibre
922 kJ	221 kcal	14.8g	6.2g	19.6g	11.8g	3.7g	1.3g

34 Profiteroles and chocolate sauce

1 Spoon the choux paste into a piping bag with a plain nozzle (approx. 1.5 cm diameter).
2 Pipe walnut-sized balls of paste onto the greased baking sheet, spaced well apart. Level the peaked tops with the tip of a wet finger.
3 Bake for 18–20 minutes at 200°C, until well risen and golden brown. Remove from the oven, transfer to a wire rack and allow to cool completely.
4 Make a hole in each and fill with Chantilly cream.
5 Dredge with icing sugar and serve with a sauceboat of cold chocolate sauce, or coat the profiteroles with the sauce.

Ingredient	10 portions
Choux paste	200ml
Chocolate sauce (see Chapter 14)	250ml
Chantilly cream	250ml
Icing sugar, to serve	

Variations
Alternatively, coffee sauce may be served and the profiteroles filled with non-dairy cream. Profiteroles may also be filled with chocolate-, coffee- or rum-flavoured pastry cream.

Energy	Cal	Fat	Sat fat	Carb	Sugar	Protein	Fibre
1327 kJ	320 kcal	28.6g	16.6g	13.8g	10.0g	2.8g	0.3g

35 Chocolate éclairs (éclairs au chocolat)

Ingredient	12 éclairs
Choux paste	200ml
Whipped cream/Chantilly cream	250ml
Fondant	100g
Chocolate couverture	25g

Energy	Cal	Fat	Sat fat	Carb	Sugar	Protein	Fibre
741 kJ	179 kcal	14.0g	7.8g	12.4g	9.1g	1.5g	0.3g

Video: éclairs
http://bit.ly/1iv6s9T

1 Place the choux paste into a piping bag with a 1cm plain tube.

2 Pipe into 8cm lengths onto a lightly greased baking sheet.

3 Bake at 200–220°C for about 30 minutes.

4 Allow to cool. Make two small holes in the base of each.

5 Fill with Chantilly cream (or whipped cream), using a piping bag and small tube.

6 Warm the fondant to 37°C, add the finely cut chocolate and allow to melt slowly, adjusting the consistency with a little sugar and water syrup if necessary. Do not overheat or the fondant will lose its shine.

7 Glaze the éclairs by dipping them in the fondant; remove the surplus with a finger. Allow to set.

Note: Traditionally, chocolate éclairs were filled with chocolate pastry cream.

Variations

The continental fashion is to fill with pastry cream.

For coffee éclairs (éclairs au café), add a few drops of coffee extract to the fondant instead of chocolate; coffee éclairs may also be filled with pastry cream flavoured with coffee.

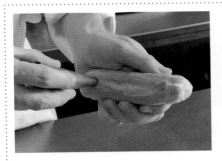

Pierce the éclair with a cream horn mould or similar

Pipe in the filling

Dip the éclair in fondant; wipe the edges to give a neat finish

36 Gâteau Paris-Brest

Ingredient	1 large or 8 individual
Choux paste	200ml
Crème diplomat	400ml
Praline	
Flaked almonds, hazelnuts and pecans (any combination)	200g
Granulated sugar	250g

Energy	Cal	Fat	Sat fat	Carb	Sugar	Protein	Fibre
2089 kJ	501 kcal	32.1g	10.8g	47.5g	39.8g	8.2g	0.4g

To make the praline

1 Place the nuts on a baking sheet and toast until evenly coloured.

2 Place some of the sugar in a large heavy stainless steel saucepan. Set the pan over a low heat and allow the sugar to caramelise. Do not over-stir, but do not allow the sugar to burn. Gradually feed in the rest of the sugar.

3 When the sugar is evenly coloured, remove from the heat and stir in the warm nuts.

4 Immediately deposit the mixture on a Silpat mat. Place another mat over the top and roll as thinly as possible.

5 Allow to go completely cold. Break up and store in an airtight container. A food processor such as a Robot-Coupe can be used to break up the praline. This will produce quite a fine mix, particularly if ground multiple times.

For the Paris-Brest

1 Pipe choux paste into rings (approximately 8cm), sprinkle with flaked almonds (optional) and bake at 200°C for 30 minutes. Pierce two holes in the bottom of each to let steam escape.

2 Once cooled, slice each ring in half and fill by piping with a mixture of crème diplomat and praline.

3 Dust with icing sugar.

Pipe the paste into the shape required

Sprinkle the raw paste with flaked almonds before baking

37 Croquembouche

The quantities required depend on the size of the croquembouche.

Ingredient		
Water	400ml	
Granulated sugar	1 kg	
Glucose	200g	
Profiteroles, piped and baked		
Crème diplomate		
Nougatine		

1 Boil the water, sugar and glucose to make a caramel.
2 Dip each profiterole in caramel and allow to cool.
3 When cool, fill with crème diplomat.
4 Using caramel as glue, stick the profiteroles together around a conical croquembouche mould.
5 Assemble the nougatine in shapes to form a base.
6 Once the caramel has set, turn the profiteroles out of the mould onto the nougatine base.
7 Decorate with cut-out nougatine shapes and pulled sugar.

Note: This dish is traditionally served at weddings in France.

The embellishment techniques used to create croquembouche (such as poured sugar, etc.) can be found in Chapter 16.

38 Gâteau St Honoré

1 Pipe the choux paste in a ring around the edge of the puff paste disc. Also pipe a set of profiteroles onto a baking sheet.

2 Bake at 220°C for 10 minutes, then at 165°C for 15 to 20 minutes.

3 Beat the pastry cream, flavour it with kirsch and fold in the whipped double cream.

4 Dip the profiteroles in caramel. Fill them with the cream.

5 Stick the profiteroles to the ring of choux paste on the base.

6 Fill the centre with quenelles of the cream using a St Honoré piping nozzle.

Ingredient	1 (10 portions)
Choux paste	500g
Puff paste disc	16cm
Pastry cream	300ml
Kirsch	30ml
Double cream, whipped	300ml
Caramel	1 kg

Professional tip

Traditionally, this gateau is filled with crème chiboust (crème St Honoré), which is pastry cream with leaf gelatine added (6g per 250ml) while hot, with Italian meringue folded through.

Energy	Cal	Fat	Sat fat	Carb	Sugar	Protein	Fibre
3255 kJ	781 kcal	51.8g	28.9g	75.8g	60.1g	5.8g	0.6g

Test yourself

1 Describe the difference between crème chiboust and crème mousseline, providing an example of a use for each.

2 What is Ultratex used for in the pastry kitchen?

3 Describe the difference between a single turn and a double turn when laminating pastry.

4 Name two products that can be made using off-cuts of (or secondary rolled) puff pastry.

5 Which paste(s) is used to make the following products?

(a) Steamed fruit puddings

(b) Gateau St Honoré

(c) Pork pies

(d) Gâteau pithiviers

6 What is the name of the pastry product that is often served at weddings in France?

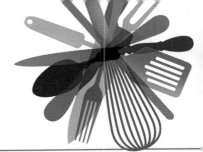

14 Hot, cold and frozen desserts

This chapter covers the following units:

NVQ:
→ Prepare, cook and finish complex hot desserts.
→ Prepare, cook and finish complex cold desserts.

VRQ:
→ Produce hot, cold and frozen desserts.

Introduction

The list of recipes here is by no means exhaustive – there are many more desserts than those included. There is, however, an example from each category, and variations and examples of alternatives have been given. This chapter includes traditional puddings and a variety of desserts that are perhaps more in tune with a modern lifestyle.

Learning objectives

By the end of this chapter you should be able to:
→ Prepare and finish hot desserts and puddings to the recipe specifications, in line with current professional practice.
→ Work safely and hygienically to produce cold desserts, using the correct equipment.
→ Apply quality points and evaluate the finished cold dessert.

Recipes included in this chapter

Ingredients used in desserts

Milk

Full-cream, skimmed or semi-skimmed milk can be used for the cold and frozen desserts in this chapter.

Milk is a basic and fundamental element of Western diets. It is composed of water, sugar and fat (with a minimum fat content of 3.5 per cent). It is essential in an infinite number of products, from creams, ice creams, yeast doughs, mousses and custards to certain ganaches, cookies, tuiles and muffins.

Milk has a slightly sweet taste and little odour. Two distinct processes are used to conserve it:

- **Pasteurisation** – the milk is heated to 72°C for 15 seconds, then cooled quickly to 4°C.
- **Ultra-heat treatment (UHT)** – the milk is heated to between 140°C and 150°C for 2 seconds, then cooled quickly.

Milk is homogenised to disperse the fat evenly, since the fat has a tendency to rise to the surface (see 'Cream', below).

Here are some useful facts about milk:

- Pasteurised milk has a better taste and aroma than UHT milk.
- Milk is a useful for developing flavour in sauces and creams, due to its lactic fermentation.
- There are other types of milk, such as sheep's milk, that are very interesting to use in many restaurant desserts.
- Milk is much more fragile than cream. In recipes, adding it in certain proportions is advisable for a much more subtle and delicate final product.

Cream

Cream is used in many recipes because of its great versatility and capabilities.

Cream is the concentrated milk fat that is skimmed off the top of the milk when it has been left to sit. A film forms on the surface because of the difference in density between the fat and liquid. This process is speeded up mechanically in large industries by heating and using centrifuges.

Cream should contain at least 18 per cent butterfat. Cream for whipping must contain more than 35 per cent butterfat. Commercially frozen cream is available in 2 kg and 10 kg slabs. Types, packaging, storage and uses of cream are listed in Table 14.1.

Whipping and double cream may be whipped to make them lighter and to increase volume. Cream will whip more easily if it is kept at refrigeration temperature. Indeed, all cream products must be kept in the refrigerator for food safety reasons. They should be handled with care and, as they will absorb odour, they should never be stored near onions or other strong-smelling foods.

As with milk, there are two main methods for conserving cream:

- **Pasteurisation** – the cream is heated to 72°C for 15 seconds and then cooled quickly; this cream retains all its flavour properties.
- **UHT** – this consists of heating the cream to between 140°C and 150°C for 2 seconds; cream treated this way loses some of its flavour properties, but it keeps for longer.

Always use pasteurised cream whenever possible; for example, in the restaurant when specialities are made for immediate consumption, such as 'ephemeral' patisserie or desserts with a short life (for example, a chocolate bonbon or a soufflé).

Here are some useful facts about cream:

- Cream whips with the addition of air, thanks to its fat content. This retains air bubbles formed during beating.
- Understand how to use fresh cream; remember that it is easily over-whipped.
- Cream adds texture and enriches.
- To whip cream well, it must be cold (around 4°C).
- Cream can be infused with other flavours when it is hot or cold. The flavour of cream will develop if left to infuse prior to preparation.
- Always ensure that the bowl being used to whip the cream is cold.
- Once cream is boiled and mixed or infused with other ingredients to add flavour, it will whip again if first left to cool; when preparing a chocolate Chantilly, for example.
- Stabilisers may be added to whipped cream which helps to prevent water leakage and gives a firmer cream for piping.

Table 14.1 Types of cream

Type of cream	Legal minimum fat content (%)	Processing and packaging	Storage	Characteristics and uses
Half cream	12	Homogenised; may be pasteurised or ultra-heat treated	2–3 days	Does not whip; used for pouring; suitable for low-fat diets
Cream (single cream)	18	Homogenised; pasteurised by heating to 72°C for 15 seconds, then cooling to 4.5°C; packaged in bottles and cartons, sealed with foil caps; may be available in bulk	3–4 days under refrigeration; observe use-by dates and storage guidance	A pouring cream suitable for coffee, cereals, soup or fruit; added to cooked dishes and sauces; does not whip
Whipping cream	35	Not homogenised; pasteurised and packaged like single cream		Ideal for whipping; suitable for piping, cake and dessert decoration; used in ice cream, cake and pastry fillings
Double cream	48	Slightly homogenised; pasteurised and packaged like single cream		A rich pouring cream; will whip; floats on coffee or soup
'Thick' double cream	48	Heavily homogenised; pasteurised and packaged like single cream; usually only sold in domestic quantities		A rich, spoonable cream; cannot be poured; if stirred, returns to original consistency
Clotted cream	55	Heated to 82°C then cooled for about 4½ hours; the cream crust is then skimmed off; packed in cartons, usually by hand; may be available in bulk		Very thick; has its own special flavour and colour; used with scones, fruit and fruit pies
Ultra-heat treated (UHT) cream	12 (half), 18 (single) or 35 (whipping)	Homogenised; heated to 132°C for 1 second, then cooled immediately; aseptically packaged in polythene and foil-lined containers; available in catering-size packs	6 weeks if unopened; does not need refrigeration; usually date stamped	A pouring cream; 35% fat UHT will whip

Key term

Homogenisation – forcing milk or cream through very fine mesh under high pressure. This breaks up fat globules and distributes them evenly, which prevents the milk or cream from separating with a thicker cream layer at the top.

Eggs

Eggs are one of the principal ingredients in cooking and essential for many desserts. Their great versatility and extraordinary properties as a thickener, emulsifier and stabiliser make their presence important in various creations in patisserie: sauces, creams, sponge cakes, custards and ice creams. Although it is not often the main ingredient, the egg plays specific and determining roles in terms of texture, taste and aroma. It is fundamental in products such as brioches, crème anglaise, sponge cakes and crème pâtissière. The extent to which eggs are used (or not) makes an enormous difference to the quality of the product.

A good custard cannot be made without eggs, as they cause the required coagulation and give it the desired consistency and finesse.

Eggs are also an important ingredient in ice cream, where their yolks act as an emulsifier, due to the lecithin they contain, which aids the emulsion of fats.

Eggs are used for several reasons:

- They act as a texture agent in, for example, crème pâtissière and ice creams.
- They enhance flavours.
- They give volume to whisked sponges and batters.
- They act as a thickening agent, e.g. in crème anglaise.
- They act as an emulsifier in products such as mayonnaise and ice cream.
- They aerate and give lightness to pastry products, for example, mousses and parfaits.

Quality and nutritional value

Important facts about eggs:

- A fresh egg (in shell) should have a small, shallow air pocket inside it.
- The yolk of a fresh egg should be bulbous, firm and bright.
- The fresher the egg, the more viscous (thick and not runny) the egg white.
- Eggs should be stored away from strong odours as their shells are porous and smells are easily absorbed.
- In a whole 60g egg, the yolk weighs about 20g, the white 30g and the shell 10g.

Egg yolk is high in saturated fat. The yolk is a good source of protein and also contains vitamins and iron. The egg white is made up of protein (albumen) and water. The egg yolk also contains lecithin, which acts as an emulsifier in dishes and commodities such as mayonnaise and chocolate – it helps to keep the ingredients mixed, so that the oils and water do not separate.

Working with egg whites

- To avoid the danger of salmonella, if the egg white is not going to be cooked or will not reach a temperature of 70°C, use pasteurised egg whites. Egg white is available chilled, frozen or dried.
- Equipment must be thoroughly clean and free from any traces of fat, as this prevents the whites from whipping; fat or grease prevents the albumen strands from bonding and trapping the air bubbles.
- Take care that there are no traces of yolk in the white, as yolk contains fat.
- A little acid (cream of tartar or lemon juice) strengthens the egg white, extends the foam and produces more meringue. The acid also has the effect of stabilising the meringue.
- If the foam is over-whipped, the albumen strands, which hold the water molecules with the sugar suspended on the outside of the bubble, are overstretched. The water and sugar make contact and the sugar dissolves, making the meringue heavy and wet. This can sometimes be rescued by further whisking until it foams up, but very often you will find that you may have to discard the mixture and start again.

Beaten egg white forms a foam that is used for aerating sweets and many other desserts, including meringues (see Recipe 12).

Fruit

Fruit is used as an ingredient in many desserts.

Quality and purchasing

Fresh fruit should be:

- whole and fresh looking (for maximum flavour the fruit must be ripe but not overripe)
- firm, according to type and variety
- clean, and free from traces of pesticides and fungicides
- free from external moisture
- free from any unpleasant foreign smell or taste
- free from pests or disease
- sufficiently mature; it must be capable of being handled and travelling without being damaged
- free from any defects
- characteristic of the variety in shape, size and colour
- free of bruising and any other damage.

Soft fruits deteriorate quickly, especially if they are not sound. Take care to see that they are not damaged or overripe when purchased. Soft fruits should look fresh; there should be no signs of wilting, shrinking or mould. The colour of certain soft fruits is an indication of their ripeness (for example, strawberries or dessert gooseberries).

Food value

Fruit is rich in antioxidant minerals and vitamins. Antioxidants protect cells from damage by oxygen, which may lead to heart disease and cancer. The current recommendation is to eat five portions of fruit and vegetables each day.

Storage

Hard fruits, such as apples, should be left in boxes and kept in a cool store. Soft fruits, such as raspberries and strawberries, should be left in their punnets or baskets in a cold room. Stone fruits, such as apricots and plums, are best placed in trays so that any damaged fruit can be seen and discarded. Peaches and citrus fruits are left in their delivery trays or boxes. Bananas should not be stored in too cold a place because their skins will turn black.

Healthy eating and desserts

Desserts and puddings remain popular with the consumer, but there is now a demand for products with reduced fat and sugar content, as many people are keen to eat healthily.

Chefs will continue to respond to this demand by modifying recipes to reduce the fat and sugar content; they may also use alternative ingredients, such as low-calorie sweeteners where possible and unsaturated fats. Although salt is an essential part of our diet, too much of it can be unhealthy, and this is something else that chefs should take into consideration.

Egg custard-based desserts

The essential ingredients for an egg custard are whole egg and milk.

Cream is often added to egg custard desserts to enrich them and to improve the feel in the mouth (mouth-feel) of the final product; sugar is also added to sweeten and vanilla to flavour.

Egg custard mixture provides the chef with a versatile basic set of ingredients that covers a wide range of sweets. Often the mixture is referred to as crème anglaise or crème renversée. Although the ingredients are very similar, crème renversée is cooked in the oven, whereas crème anglaise is cooked on top of the stove.

Some examples of sweets produced using this mixture are:
- crème caramel
- bread and butter pudding
- diplomat pudding
- cabinet pudding
- queen of puddings
- baked egg custard.

Savoury egg custard is used to make:
- quiches
- tartlets
- flans.

Two other products based on an egg custard are pastry cream (also known as confectioner's custard or crème pâtissière; see Chapter 13) and sauce à l'anglaise (see Recipe 39; it is also used as a base for some ice creams).

When a starch such as flour is added to the ingredients for an egg custard mix, this changes the characteristic of the end product, as in crème pâtissière, for example.

Basic egg custard sets by coagulation of the egg protein. Egg white coagulates at approximately 60°C, egg yolk at 70°C. Whites and yolks mixed together will coagulate at 66°C. If the egg protein is overheated or overcooked, it will shrink and water will be lost from the mixture, causing undesirable bubbles in the custard. This loss of water is called syneresis, commonly referred to as scrambling or curdling. This will occur at temperatures higher than 85°C. Therefore, sauce à l'anglaise should be ideally cooked between 70°C and 85°C.

The sauce will become thicker as it comes closer to 85°C but is at risk of curdling beyond this temperature.

Modern methods of cooking crème anglaise use a water bath to cook 'sous vide' at 85°C for the required length of time, and the temperature is checked using a special probe designed for sous vide cooking. The phrase 'sous vide' literally translates to 'under vacuum'.

Hot desserts

Fresh ingredients

The principal fresh ingredient when making hot desserts is egg. When using eggs, obviously, they should be as fresh as possible, and if you use free range you will gain a deeper coloured yolk and a stronger white – this is particularly important when making a soufflé or desserts which contain meringue.

All dry ingredients must be stored correctly at around 20°C in a well-ventilated environment protected from contamination by pests or moisture.

Cooking methods

Boiling, simmering, baking and deep-fat frying are all covered elsewhere in this book, but the following three methods deserve a special mention in relation to this chapter.

Steaming

This involves cooking in steam produced by boiling water. It is a gentle method of cookery that suits dishes such as suet puddings, allowing the hard fat to melt more easily over a longer cooking time. Because of the low temperature, steaming does not give any colour to the food, hence the old-fashioned name for a steamed suet roll is 'dead man's leg'. When steaming, make sure the products being steamed are properly sealed so they don't become waterlogged.

Combination ovens can provide a facility which allows baking (dry heat) with steam or steaming at a temperature higher than 100°C. This, unlike steam on its own, can add some colour to the product.

Health and safety

Be careful if using a high-pressure commercial steamer – stand aside and open the door carefully to avoid being burnt by the escaping steam.

Bain marie

This is often used as a cookery method when baking desserts which contain custard. It consists of placing the products in a tray, half filling with hot water and baking in the oven. This allows the products to cook without boiling.

Health and safety

Always place the bain marie tray in the oven before adding the hot water, and be extremely careful when removing it after the cooking is complete. Have somewhere cleared to put the tray down and do not attempt to carry it across the kitchen.

Stewing, poaching and roasting

Fruit can be stewed in syrup or fruit juice, although nowadays this is rarely seen on a menu. More often than not, fruit is either poached or roasted. For a stew, the fruit would usually be cut into pieces and the liquid thickened. Poached fruit is often cooked whole, with the liquid un-thickened and served cold. The term 'roasted', more commonly associated with meat, is now often used to describe vegetables and fruit which are cooked in a dry heat with the aid of fat.

Making soufflés

Hot soufflés have a largely undeserved reputation for being problematic, but if the following points are observed they should pose no problems for a competent chef.

Mould preparation is very important. The rule is to butter the sides twice and the bottom once. Always use soft, not melted, butter (then it stays where you put it).

1 Give the moulds one coat all over to start with, then place in the fridge to set.
2 Give the sides only a second coat (giving the base two coats results in a puddle of butter left in the bottom).
3 After the second coat of butter, the moulds are usually coated with sugar.

Video: soufflé
http://bit.ly/1qslkvg

When beating the whites, adding sugar, lemon juice and cornflour helps to strengthen the gluten and stabilise the mixture. When whisked, the whites should be firm but creamy. Over-whisking ruins them. Scald the bowl first to remove any trace of fat, which will prevent the whites from reaching their full potential. Recipe 6 uses more egg white than is needed – always take from the centre and leave behind the egg white from the edge, which is never good. The quantity of whites added is approximately the same in volume as the base, but whatever the method used, the key points are always the same.

Be organised with your timings. Make sure you have the moulds fully prepared and ready, and make sure the oven is up to heat well before the whites are whisked.

Once mixed, the soufflé must be cooked (and served) immediately, although some recipes (including Recipe 6) may be held before baking for up to 2 hours.

Quality points

The following quality points identify a good soufflé.
- The point of a soufflé is that it will rise out of the dish in which it is cooked – it should rise evenly, at least 3–4 cm.
- The term used to describe the centre of a cooked soufflé is 'baveuse', which means slightly undercooked.
- A soufflé should have flavour, which can sometimes almost get overlooked with the emphasis on getting an impressive 'rise'. Too much egg white can dilute the flavour.
- Finally, a soufflé should look like it tastes – if the flavour is raspberry then that should be obvious from the colour.

Making a soufflé

1 Prepare the mould.

2 Knock one third of the meringue into the panada.

3 Fold in the remaining meringue.

4 Thumb the edge.

Food safety and hygienic practice

It is recommended where a product or dish is cooked at a low temperature (bread and butter pudding, for example) or where the dish is required to have an undercooked centre (soufflé), that pasteurised eggs are used.

Allergies

It is estimated that up to one in three people will at some time in their lives suffer an allergy. Food is one of the most common allergens. Nuts, dairy produce and wheat products can all cause an allergic reaction. Anaphylaxis, usually associated with nuts, is one of the most serious allergic reactions, and if not treated quickly may cause death.

It is therefore vital that dishes are correctly labelled and staff are thoroughly briefed regarding the ingredients they contain. If a dish is sold as suitable for a particular dietary requirement, then care should be taken to ensure it is not contaminated at any time during or after its production.

1 Apple Charlotte

Ingredient	8 individual or 2 small Charlotte moulds
Dessert apples (Cox's)	1 kg
Butter	30g
Caster sugar	100g
Lemon zest	1
Breadcrumbs	
Large, thin-sliced loaf of bread	1
Clarified butter	250g

Energy	Cal	Fat	Sat fat	Carb	Sugar	Protein	Fibre
1554 kJ	374 kcal	29.0g	18.2g	29.0g	27.7g	1.1g	3.4g

1 Peel, core and cut the apples into thick slices.

2 Melt the butter in a pan, add the sugar and finely grated lemon zest.

3 Add the apples and simmer until barely cooked, stir in some breadcrumbs to absorb any liquid.

4 Cut out circles of bread for the top and base of the moulds.

5 Cut the crusts and the rest of the bread into fingers 2–3 cm wide, depending on whether you are making individual or larger Charlottes.

6 Butter the moulds, dip half the circles in the clarified butter and place them butter-side down in the base of each mould.

7 Next, dip the bread fingers and line them around the outside of the moulds, slightly overlapping.

8 Fill the centre with the apple filling, pressing in carefully.

9 Dip the rest of the circles in the butter and place on top, pressing firmly.

10 Bake at 230°C for 30–40 minutes until the bread is coloured and crisp.

11 Allow to cool slightly before unmoulding, serve with hot apricot sauce or crème anglaise.

Hard margarine was used instead of butter for the nutritional analysis.

Faults

A common fault is that the Charlotte collapses when unmoulded (the larger versions are more prone to this). The main reasons for this are:

● The filling is too wet. To avoid this, ensure that cooking apples are never used, and do not overcook.
● The bread is not baked crisp enough to withstand the pressure of supporting the filling.

If making individual Charlottes, be careful the ratio of bread to filling is not compromised by using bread that is cut too thick or overlapping too much.

Professional tip

As this dessert will most likely be plated and sent from the kitchen, it will more often than not be made in individual portions. Larger versions (such as that in the photograph) would be suitable for family or silver service, but could not be cut and made to look presentable if served from the kitchen.

Variation
Replace the apples with pears or use a mixture of both.

2 Tarte Tatin

The tarte Tatin was created in 1888 by French sisters Caroline and Stéphanie Tatin, who owned and ran the Hotel Tatin in the Loire valley.

Ingredient	10 portions
Soft butter	120g
Caster sugar	120g
Dessert apples	5
Caramel, crushed	120g
Puff paste	150g
For the caramel	
Granulated sugar	500g

1 Make a dry caramel by placing the granulated sugar in a hot heavy-based saucepan. When the sugar reaches a deep amber colour, pour it out onto a silicone mat and leave to cool completely.

2 When cold, crush the caramel into small pieces.

3 Take a medium-sized sauteuse (or small individual moulds) and spread the butter thickly around the base and sides. Sprinkle over the caster sugar and then sprinkle over the caramel. (Any spare caramel can be stored in an airtight container for later use.)

4 Peel, core and halve the apples (if large, cut into quarters) and pack into the sauteuse core-side up and with the cores running horizontally, not facing outside.

5 Roll out the pastry 2mm thick and leave to rest.

6 Place the pan on a medium heat for 10–12 minutes to allow the caramel to melt and infuse.

7 Quickly lay over the pastry and trim, tucking the edges down the side of the pan.

8 Bake at 220°C for 15 minutes until the pastry is crisp and the apples cooked through.

9 Invert onto a hot plate – please be aware this procedure can be tricky and requires two dry cloths and very careful handling.

10 Serve with cream, crème fraiche or ice cream.

Energy	Cal	Fat	Sat fat	Carb	Sugar	Protein	Fibre
1914 kJ	454 kcal	15.2g	9.0g	83.2g	77.9g	1.3g	1.0g

Cook the melted butter, sugar and caramel in a pan of the desired size

Lay in the apple pieces

Lay the pastry over the top once the apples are half-cooked

Tuck in the edges

Turn out the tart carefully

The finished tart should have neat layers of apple and caramel

Professional tip

Not all apples are suitable for this dish. Never use cooking apples, as if they contain a high water content they are likely to collapse and turn to purée.

Variations

● This dish can be made using pears instead of apples.
● Tarte Tatin can also be made in individual moulds (as shown in the first photo) if preferred.

Health and safety

Extreme care should be taken when preparing the caramel as the sugar reaches very high temperatures and can cause serious burns.

3 Apple crumble tartlets

Ingredient	4 tartlets	10 tartlets
Sweet paste	200g	500g
Dessert apples	2	5
Filling		
Soured cream	200ml	500ml
Caster sugar	25g	70g
Plain flour	30g	75g
Egg	½	1
Vanilla extract		
Crumble		
Plain flour	35g	80g
Walnuts, chopped	25g	60g
Brown sugar	25g	65g
Ground cinnamon	Pinch	Pinch
Salt	Pinch	Pinch
Unsalted butter, melted	20g	65g
Icing sugar, to garnish		

Energy	Cal	Fat	Sat fat	Carb	Sugar	Protein	Fibre
2334 kJ	557 kcal	27.7g	14.5g	72.3g	30.4g	9.3g	3.2g

1 Line individual tartlet moulds with the sweet paste.

2 Peel, core and finely slice the apples, and divide between the tartlets.

3 Whisk together the soured cream, sugar, flour, egg and a few drops of vanilla, and pass through a conical strainer.

4 Pour over the apples and bake at 190°C for 10 minutes.

5 Combine the dry crumble ingredients and mix with the melted butter.

6 Divide the crumble mixture between the tartlets and bake for a further 10 minutes.

7 Allow to cool slightly before unmoulding. Dust with icing sugar and serve with sauce à l'anglaise.

Variation
This dish could be made with pears or plums instead of apples.

4 Chocolate fondant

Ingredient	4 portions	10 portions
Unsalted butter	110g	260g
Dark couverture	110g	260g
Eggs, pasteurised	50g	120g
Egg yolks, pasteurised	16g	40g
Caster sugar	60g	150g
Instant coffee	2g	5g
Plain flour	50g	110g
Baking powder	2g	5 g
Cocoa powder	30g	75g
Salt	Pinch	Pinch

Energy	Cal	Fat	Sat fat	Carb	Sugar	Protein	Fibre
1996 kJ	479 kcal	33.0g	19.6g	42.0g	32.4g	6.1g	2.1g

1 Melt the butter and couverture together.

2 Warm the eggs, egg yolks, sugar and coffee and whisk to the ribbon stage.

3 Sieve all the dry ingredients twice.

4 Fold the chocolate and butter into the eggs.

5 Fold in the dry ingredients.

6 Pipe into individual stainless steel rings lined with silicone paper and placed on a silicone paper-lined baking sheet.

7 Bake at 190°C for 5 minutes.

8 Carefully slide off the rings and serve with vanilla ice cream dusted with cocoa powder.

Melt the chocolate and butter in small pieces

Fold the melted chocolate into the egg mixture

Add the dry ingredients

To make a contrasting centre, add white chocolate pieces on a base of the chocolate mixture

Pipe in more of the chocolate mixture until the mould is full

Professional tips

- These fondants can be kept in the refrigerator and cooked to order. If they are chilled, then extend the cooking time by 2 minutes.
- Chocolate fondant should have a liquid centre with a rich, buttery, chocolate taste. Because of the liquid centre, they are very delicate; if piped inside a ring they are much easier and quicker to serve, rather than trying to turn them out of a mould.
- Precise timing is essential or the centre of the fondant will not be liquid.
- Like most recipes, the quality of the finished product relies on the quality of the ingredients. Always use good quality chocolate (couverture) which contains a high percentage of cocoa butter and solids.

Variations

- Try adding salted caramel to the centre by making and freezing it in ice cube trays.
- Alternatively, serve with malt ice cream (just add malt powder instead of vanilla and mix in some crushed chocolates) or replace the cream with crème fraiche to give a less rich ice cream.
- Prepare fondants in moulds lined with melted butter and roasted sesame seeds.

5 Baked Alaska (omelette soufflé surprise)

Ingredient	10 individual portions
Vanilla ice cream or parfait	10 × 5cm diameter rings
Roulade sponge (Chapter 15, Recipe 3)	1 sheet
Stock syrup flavoured with rum or kirsch	50ml
Italian meringue	500g

Energy	Cal	Fat	Sat fat	Carb	Sugar	Protein	Fibre
2237 kJ	531 kcal	18.0g	9.9g	86.1g	79.7g	10.1g	0.3g

1 Sit the ice cream or parfait on a base of sponge.

2 Cut more sponge to fit and completely cover (as in the photo).

3 Brush all over with the syrup.

4 Set on squares of silicone paper, coat with the meringue and decorate by piping on a design with a small plain tube.

5 Dust with icing sugar and place in a very hot oven at 230°C for 2–3 minutes until the meringue is coloured.

6 Serve immediately with crème anglaise or a fruit coulis. Garnish with caramelised banana.

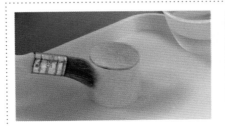
Brush the sponge with syrup

Pipe meringue to cover

Pipe swirls of meringue to decorate

Professional tips

● Baked Alaska is best made in advance and held in the freezer, then flashed through the oven just before serving. It is now common practice to colour these with a blowtorch, but the meringue will have a much better texture and more even colouring if it is finished in the oven.

● If making individual baked Alaskas, as in the photographs, take care not to upset the balance between filling and meringue – when scaled down it is easy to pipe on too much meringue.

Variations

Classic variations are omelette soufflé milady, which contains poached sliced peaches with vanilla or raspberry ice cream, and omelette soufflé milord, which contains poached sliced pear with vanilla ice cream.

Health and safety

Under no circumstances should this dessert be refrozen once it has been removed from the freezer and baked. Ice cream is highly susceptible to contamination by bacteria which can cause food poisoning.

6 Vanilla soufflé

Ingredient	10 individual soufflés
Milk	500ml
Vanilla pods	2
Butter	75g
Strong flour	60g
Egg yolks	10
Caster sugar	50g

For every 400g of the above mixture, use the following quantity of egg whites, sugar, cornflour and lemon juice.

Egg whites	150g
Caster sugar	60g
Cornflour	12g
Lemon juice	2–3 drops

Energy	Cal	Fat	Sat fat	Carb	Sugar	Protein	Fibre
932 kJ	223 kcal	13.7g	6.7g	19.5g	13.9g	6.6g	0.2g

1 Rinse out a heavy saucepan with cold water and add the milk, split the vanilla pod, scrape out the seeds and add both to the milk. Put on the heat to boil.

2 Melt the butter in another heavy pan. Add the flour and cook out to form a white roux. Gradually start adding the boiling milk, mixing in each addition before adding the next.

3 When all the milk has been added, allow to simmer for a few minutes.

4 Whisk the egg yolks and sugar. Add to the mixture in the saucepan and keep stirring over the heat until the mixture starts to bubble around the edges. This forms the panada.

5 Pour onto a clean tray and cover with cling film to prevent a skin forming. Allow to cool. (This can be kept in the fridge until needed, as soufflés must be cooked to order.)

6 Take 400g of the base and beat in a clean bowl until smooth.

7 Whisk the whites, sugar, cornflour and lemon juice to form firm peaks.

8 Add one-third of the whites to the base and mix in, then very carefully fold in the remaining whites.

9 Carefully fill prepared individual china ramekins (see page 464). Level the top and run your thumb around the edge, moving the mixture away from the lip of the mould.

10 Space well apart on a solid baking sheet (if they are close together they will not rise evenly and will bake stuck together).

11 Place immediately in the oven at 215°C for 12–14 minutes.

12 The soufflés should rise out of the moulds by around 5–6cm and have a flat top with no cracks.

13 Dust with icing sugar and serve immediately with fruit coulis and/or ice cream (chocolate or vanilla).

Faults

Soufflé does not rise:
- under- or overbeaten whites
- wrong proportion of whites to base
- mixture left to stand before cooking
- moulds not buttered.

Soufflé does not rise evenly:
- moulds not prepared correctly (mixture has stuck to the mould on one side)
- uneven heat in the oven.

Soufflé rises but drops back:
- too much egg white used.

Soufflé has a cracked top:
- too much egg white used
- egg white is overbeaten.

Variations

The vanilla recipe is basic and can easily be adapted. For example, a chocolate soufflé can be made by adding melted chocolate and cocoa powder to the base. This will firm up the base mixture, so the whites mixture will need to be increased by 25–30 per cent to compensate.

Soufflés can be made in many different flavours and combinations. Recipes can vary considerably. For example, fruit soufflés can be made using a sabayon or a boiled sugar base.

Liqueur soufflés can be made using a béchamel (like the one above) or a crème pâtissière base. For liqueur soufflés, sprinkle the liquor over a small dice of sponge fingers and fold into the mixture just before the whites. This will trap the liquor in small pockets, concentrating the flavour, and it will not evaporate during cooking.

As an alternative to coating the moulds with sugar, if compatible, dust them with cocoa powder, grated chocolate or try adding some cinnamon to the sugar.

7 Lemon curd flourless soufflé

As this dessert is gluten-free it would be suitable for coeliacs or those with a wheat intolerance.

Ingredient	10 individual soufflés
Lemon curd	
Eggs	2
Caster sugar	100g
Lemons (juice of)	5
Unsalted butter	60g
Cornflour	15g
Soufflé	
Eggs	9
Caster sugar	190g
Lemons (zest and juice)	5
Cream of tartar	Pinch
Egg-white powder	Pinch

Energy	Cal	Fat	Sat fat	Carb	Sugar	Protein	Fibre
1179 kJ	280 kcal	13.3g	5.7g	31.7g	30.4g	10.1g	0.1g

For the lemon curd

1 Prepare by whisking all the ingredients over a pan of simmering water until the mixture thickens.

2 After preparing the soufflé moulds as described on page 464, divide the lemon curd mixture between the prepared ramekins.

For the soufflé

1 Separate the eggs. Place together the yolks, sugar, lemon zest and juice and whisk together well.

2 In a separate bowl, whisk the whites with the cream of tartar and egg-white powder until soft but in strong peaks.

3 Carefully fold the two mixtures together.

4 Carefully fill the prepared moulds, level the top and run your thumb around the edge, moving the mixture away from the lip of the mould.

5 Place well apart on a heavy baking sheet and bake at 200°C for 16–18 minutes.

6 Dust with icing sugar and serve immediately.

8 Crêpe soufflé

This recipe makes 10 portions.

1 Prepare crêpes – they should be undercooked.

2 Prepare the soufflé mixture from Recipe 6, appropriately flavoured.

3 Fold the pancakes into four and fill the two pockets with soufflé mixture. This is best done using a piping bag and a large plain tube, or simply fold in half and fill.

4 Place well apart on a silicone paper-covered baking sheet and bake immediately at 210°C for 6–7 minutes.

5 Dust with icing sugar and serve with a fruit coulis and/or ice cream, or a flavoured syrup.

Variation

An alternative to this dish uses two pancakes per portion. Lay one pancake flat and place some soufflé mixture in the centre. Eggwash the edge, place the second pancake on top and seal. After baking as normal, the result should be a pancake ball.

9 Apple strudel

Energy	Cal	Fat	Sat fat	Carb	Sugar	Protein	Fibre
1250 kJ	298 kcal	12.9g	6.9g	41.2g	8.5g	6.8g	2.4g

Ingredient	10 portions
Strudel paste (Chapter 13, Recipe 8)	500g
Melted butter	120g
Fresh breadcrumbs	Approx. 100g
Filling	
Lemons	1
Large dessert apples	5
Caster sugar	75g
Ground cinnamon	1 tsp
Nibbed almonds	50g
Sultanas	50g

Make the paste as specified in Chapter 13, Recipe 8 and leave to rest between two oiled plates for an hour.

For the filling

1 Grate the zest and juice the lemon; place in a large bowl.

2 Peel and core the apples before cutting into 2mm-wide batons. Add to the lemon juice and zest.

3 Add the other ingredients to taste. (The quantity of sugar will be influenced by the sweetness of the apples.)

4 Mix well with a spoon, cover and leave to stand. About half an hour before rolling out the paste, tip the filling into a colander to drain.

Processing the paste

This must be done on a large cloth (an old tablecloth or similar, but without holes).

1 Lay the cloth over a table you can walk all the way around (i.e. not one against a wall).

2 Lightly flour the cloth and ease the paste off the plate using a plastic scraper.

3 Do not knead or work at all – just lightly flour the top and start rolling carefully, lifting the paste up and flouring underneath until it reaches a size where it is too big to roll.

4 Reach underneath the paste and, using the backs of your hands, carefully stretch, until the paste is gossamer thin and covers the table. It should be thin enough to read a newspaper through, literally.

5 You must now work quickly – brush the paste with melted better, and sprinkle a good layer of crumbs over half the paste only.

6 Place a layer of apples on the crumbs only, and start to roll up using the cloth, as shown in the final photo.

7 Twist the ends, lift onto a silicone paper-covered baking sheet, and brush the outside with melted butter.

8 Bake at 190°C for 30–35 minutes.

9 Allow to settle, dust with icing sugar, cut diagonally with a serrated sponge knife and serve with soured cream.

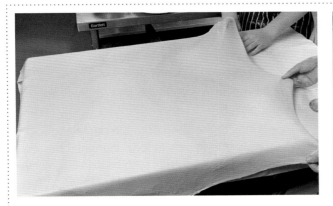

Fully stretched pastry

After dusting, add the filling

Fold the side in

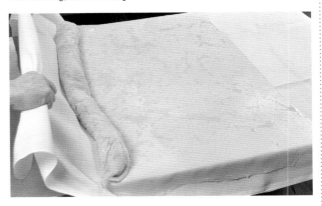

Roll up in a cloth

Faults

The paste develops holes during stretching:
- Rough handling, or taking too long to process the paste so it starts to dry out. This is a difficult and an awkward process for one person – ask a friend and get them to work opposite you; this will make the work easier and speed up the process.

The strudel splits during cooking:
- the filling is too wet
- not enough breadcrumbs used – these act as insulation and are there to absorb moisture
- the apples have been cut too large and/or in irregular shapes so they puncture the paste.

Tastes like cardboard with a hint of apple:
- Paste too thick, under-filled and overbaked.

Professional tip

As with so many processes in the pastry kitchen, it is vital to have all ingredients prepared and ready before the paste is processed.

Health and safety

When mixing fruit always use a spoon, never your hands, which can cause the fruit to ferment.

Variations

- Cake crumbs, ground almonds, brioche crumbs or any combination of these can be used in place of breadcrumbs.
- Instead of breadcrumbs, make a chapelure by shallow-frying breadcrumbs to a golden brown in butter.
- Replace the apples with pears or stoned cherries soaked in kirsch and replace the almonds with pine nuts.
- Use dried cranberries or dried apricots in place of sultanas, and try adding diced stem ginger if using pears.
- The almonds can be replaced by hazelnuts, and mixed spice used instead of cinnamon.
- It is traditional to serve strudel with soured or acidulated cream but crème fraiche, sauce à l'anglaise and/or ice cream will also compliment the dish well.

10 Almond and apricot samosas

1 Open the spring roll paste and keep covered with a damp cloth so it does not dry out.

2 Cut the paste in half, which should give two rectangles 15 × 10cm.

3 Pipe a little of the filling onto the paste and fold diagonally three times to form triangular parcels. Brush the exposed edge with beaten egg white and seal.

4 Deep-fry at 180°C for 5 minutes until coloured and an even golden brown. Drain on absorbent paper.

5 Allow two per portion. Dust with icing sugar and serve with caramel ice cream (beurre de Paris).

Variation

- Fill with a mixture of diced poached pears, stem ginger and mascarpone cheese.
- Fill with semi-poached and dried rhubarb with stem ginger, and serve with a white chocolate sauce.
- Instead of making into triangles, the pastry could be made into spring rolls.

All these recipes would be suitable to serve as a pre-dessert if they were made smaller.

Ingredient	10 portions
Spring roll paste	1 packet
Beaten egg white or flour paste to seal	
Filling	
Almond cream	250g
Dried apricots, diced, soaked in rum	50g

Energy	Cal	Fat	Sat fat	Carb	Sugar	Protein	Fibre
938 kJ	225 kcal	15.5g	5.9g	17.2g	11.1g	4.0g	0.7g

Hot desserts

1 Describe the correct procedure when scaling up (increasing) or scaling down (decreasing) a recipe.
2 List the differences between a soufflé and a pudding soufflé.
3 What is the difference between the processes of pasteurising and sterilising milk?
4 List four egg custard-based desserts.

Cold and frozen desserts

Iced confections

Traditional ice cream is made from a basic egg custard sauce (sauce à l'anglaise). The sauce is cooled and mixed with fresh cream. It is then frozen by a rotating machine where the water content forms ice crystals and the mixture is aerated.

Ice cream should be served at around −13°C; this is the correct eating temperature. It is too hard if it is any colder and too soft if it is any warmer. Long-term storage should be between −18°C and −20°C.

The traditional method of making ice cream uses only egg yolks, sugar and milk/cream in the form of a sauce à l'anglaise base. Modern approaches to making ice cream use stabilisers and different sugars as well as egg whites. This can help to reduce the sometimes high wastage of egg whites that might otherwise occur in the pastry section.

Classification

Water ices, sorbets, and granita

Water ices are a simple syrup flavoured with a fruit purée and lemon juice. They contain no milk, eggs or cream, so they are not subject to the ice cream regulations. They are frozen in the same manner as normal ice-cream.

Also included in this type of ice is sorbet. This is a very light, slightly less sweet water ice, usually lemon, to which Italian meringue is added during churning, after the base mixture has partially frozen. It is sometimes flavoured with champagne, liqueurs or wine.

Granita is similar to a water ice, less sweet than a sorbet and contains no meringue. It has a slightly granular texture normally achieved by pouring into a shallow tray, freezing, then scraping the surface into frozen granules and serving in frozen glasses.

For many pastry dishes (for example, sorbets), sugar syrups of a specific density are required. The density is measured by a hydrometer known as a saccharometer – a device that measures the thickness of stock syrups. This is measured in degrees Baumé. The instrument is a hollow glass tube sealed at each end. One end is weighted with lead shot so that when it is placed in the solution it floats upright. The scale marked on the side of the saccharometer is calibrated in the Baumé scale from 0 to 45, where 0 represents the density of water and 45 represents the density of a heavy-saturated sugar solution. The instrument thus measures the amount of sugar in the solution. To increase the density, add more sugar to the warm syrup; to decrease it, add water.

Table 14.2 Baumé measurements for different water ices

Water ice	17° Baumé
Sorbet	15° Baumé
Granita	14° Baumé
Sorbet syrup	700g sugar to 1 litre boiling water (17° Baumé)

Another instrument that measures the amount of sugar in a syrup is a refractometer. This piece of equipment measures the percentage size of the sugar crystal and is calibrated according to the Brix scale. 17° Baumé equates to 30 per cent Brix.

Table 14.3 Conversion chart for Baumé to Brix scale and common uses in patisserie

Baumé	Brix	Product
14°	26%	Granita
15°	28%	Sorbet
17°	31%	Water ice
20°	37%	Compote of fruits
22°	40%	Savarin syrup
28°	52%	Syrup for pâte à bombe
34°	63%	Confiture of fruits
17°	31%	Sorbet syrup

Cream ice

This is a mixture of milk, sugar, eggs, cream and flavouring. The base is made from a crème anglaise using these ingredients.

The soft creamy texture is achieved by churning and aerating as the mixture is freezing; this can only be achieved by freezing the mixture in an ice cream machine known as a sorbetière. Modern methods of processing cream ices use Pacojet machines. These machines give a very smooth texture and allow the ice cream to be churned prior to each service, ensuring consistency in texture.

Biscuit glacé

Biscuit glacé is a light, solid-type ice cream that can be produced without the aid of a sorbetière. The aeration is achieved from a sabayon made from egg yolks and stock syrup whisked over a bain marie of hot water.

Once cold, whipped cream, Italian meringue and flavourings are folded through. Traditionally the mixture is then deposited into a 'biscuit glacé mould' and then frozen.

The glacé may be made in two or more flavours arranged in layers and could also contain various garnishes such as nuts and crystallised fruits. The moulds used are rectangular in shape and for service the biscuit is sliced into portions.

Bombe glacée

These are also frozen in moulds. Bombe moulds are dome-shaped and are fitted with a screw plug in the bottom to facilitate removal. The moulds are first lined 'chemise' with cream or water ice and then the centre is filled with a pâte à bombe mixture, producing two or more different flavours, colours and textures. They are decorated and served whole. Bombes are made of various combinations of ices depending on the title of the bombe.

Table 14.4 Iced confection identification chart

Name of mixture	Egg yolk	Water	Sugar	Syrup	Cream	Fruit purée	Egg white	Milk	Other
Vanilla ice cream	✓		✓		✓			✓	Vanilla, sometimes inverted sugar, ice cream stabiliser, glucose
Pâte à bombe	✓	✓	✓		✓				
Pâte à biscuit glacé	✓	✓	✓		✓		Italian meringue		Flavour
Cream-based soufflé glacé	✓	✓	✓		✓		Italian meringue		Flavour
Fruit-based soufflé glacé					✓	✓	Italian meringue		
Parfait glacé	✓	✓	✓		✓				Flavour
Sorbet		✓		15° Baumé			Italian meringue		Flavour
Water ice				17° Baumé					Flavour
Granita				14° Baumé					Flavour
Spoons				20° Baumé			Same as sorbets with double quantity of Italian meringue		Flavour, wine
Marquise				17° Baumé	✓		Italian meringue		Diced strawberries and pineapple macerated in kirsch
Cassata					✓		Italian meringue		Glacé fruits, mould is lined with three flavours of ice cream

Parfait glacé

Parfait gets its name from the mould it is shaped in. Modern interpretations make the parfait and deposit it into ring moulds lined with jaconde sponge, then freeze it, remove it from the moulds and plate with a suitable decoration.

Soufflé glacé

There are two different types of soufflé glacé – cream based and fruit purée based. Cream-based iced soufflé is made by preparing a pâte à bombe and folding through whipped cream, Italian meringue and flavourings. Purée-based iced soufflé is prepared with fruit purée and folded through with whipped cream and Italian meringue.

Both types are frozen in a soufflé mould with a paper collar. Once frozen, the paper collar is removed and the exposed sides can be decorated with chopped pistachio nuts, Bres (caramelised nuts), toasted coconut, chocolate shavings, etc.

Ice cream regulations

The Dairy Products (Hygiene) Regulations 1995 apply to the handling of milk-based ice cream and the Ice Cream Heat Treatment Regulations 1959 and 1963 apply to non-milk-based ice cream in any catering business or shop premises. The production process must also take into consideration the Food Hygiene Regulations of 2006.

The regulations state that:
- Ice cream must be obtained from a mixture which has been heated to one of the temperatures in Table 14.5 for the time specified.
- The mixture must be reduced to a temperature of not more than 7.2°C within 1½ hours. This temperature must not be exceeded until freezing begins.
- If the ice cream becomes warm (above −2.2°C) it cannot be sold/used until it has been heated again as described above.
- A complete cold mix which is reconstituted with water does not need to be pasteurised first to comply with these regulations.
- A complete cold mix reconstituted with water must be kept below −2.2°C once it has been frozen.

Table 14.5 Ice cream mixture regulation temperatures

Temperature	Time (not less than)
65.5°C	30 minutes
71.1°C	10 minutes
79.4°C	15 seconds

Ice cream needs this treatment to kill harmful bacteria. Freezing without the correct heat treatment does not kill bacteria; it allows them to remain dormant. The storage temperature for ice cream should not exceed −20°C ideally, although standard freezers operate between −18°C and −22°C.

The rules for sterilised ice cream are the same except that:
- The temperature for the heat treatment must not be less than 149.9°C for at least 2 seconds.
- If the sterilised mix is kept in unopened, sterile and air-tight containers, there is no requirement to refrigerate the mixture before it is frozen.
- In the case of non-milk based products, the temperature of opened containers must not exceed 7.2°C, except where food mixtures are added that have a pH of 4.5 or less to make water ice or similar products and the combined product is frozen within 1 hour of combination.

Any ice cream sold must comply with the following compositional standards:
- It must contain not less than 5 per cent fat and not less than 2.5 per cent milk protein (not necessarily in natural proportions).
- It must conform to the Dairy Product Regulations 1995.

For further information contact the Ice Cream Alliance (see www.ice-cream.org).

The ice-cream making process

1 Weighing: ingredients should be weighed precisely in order to ensure the best results and, what is more difficult, regularity and consistency.
2 Pasteurisation: this is a vital stage in making ice cream. Its primary function is to minimise bacterial contamination by heating the mixture of ingredients to 85°C, then quickly cooling it to 4°C.
3 Homogenisation: high pressure is applied to cause the explosion of fats, which makes ice cream more homogenous, creamier, smoother and much lighter. This is not usually done for home-made ice cream.
4 Ripening: this basic but optional stage refines flavour, further develops aromas and improves texture. This occurs during a rest period (4–24 hours), which gives the stabilisers and proteins time to act, improving the overall structure of the ice cream. This has the same effect on a crème anglaise, which is much better the day after it is made than it is on the same day.
5 Churning: the mixture is frozen while at the same time air is incorporated. The ice cream is removed from the machine at about −10°C.

The main components of ice cream

- **Sucrose** (common sugar) not only sweetens ice cream, but also gives it body. An ice cream that contains only sucrose (not recommended) has a higher freezing point. The optimum sugar percentage of ice cream is between 15 and 20 per cent.

  As much as 50 per cent of the sucrose can be substituted with other sweeteners, but the recommended amount is 25 per cent.

- Ice cream that contains **dextrose** (another type of sugar) has a lower freezing point, and better taste and texture. The quantity of dextrose used should be between 6 and 25 per cent of the substituted sucrose (by weight).

- **Glucose** (another type of sugar) improves smoothness and prevents the crystallisation of sucrose.

  The quantity of glucose used should be between 25 and 30 per cent of the sucrose by weight.

- **Atomised glucose** (glucose powder) is more absorbent of water, so helps to reduce the formation of ice crystals.

- **Inverted sugar** is a paste or liquid obtained from heating sucrose with water and an acid (such as lemon juice). Using inverted sugar in ice cream lowers the freezing point.

  Inverted sugar also improves the texture of ice cream and delays crystallisation.

  The quantity of inverted sugar used should be a maximum of 33 per cent of the sucrose by weight. It is very efficient at sweetening and gives the ice cream a low freezing point.

- **Honey** has very similar properties to inverted sugar.

- The purpose of **cream** in ice cream is to improve creaminess and taste.

- **Egg yolks** act as stabilisers for ice cream due to the lecithin they contain – they help to prevent the fats and water in the ice cream from separating.

  Egg yolks improve the texture and viscosity of ice cream.

- The purpose of other **stabilisers** (e.g. gum Arabic, gelatine, pectin) is to prevent crystal formation by absorbing the water contained in ice cream and making a stable gel.

  The quantity of stabilisers in ice cream should be between 3g and 5g per kilo of mix, with a maximum of 10g.

  Stabilisers promote air absorption, making products lighter to eat and also less costly to produce, as air makes the product go further.

Stabilisers

Gelling substances, thickeners and emulsifiers are all stabilisers. They are products we use regularly, each with its own specific function, but their main purpose is to retain water to make a gel. The case of ice cream is the most obvious, as they are used to prevent ice crystal formation. They are also used to stabilise the emulsion, increase the viscosity of the mix and give a smoother product that is more resistant to melting.

There are many stabilising substances, both natural and artificial.

Edible gelatine

Edible gelatine is extracted from animals' bones (for example, pork and veal) and, more recently, fish skin. It is not suitable for vegetarians. Sold in sheets of 2g, it is easy to precisely control the amount used and to manipulate it. Gelatine sheets must always be washed thoroughly with lots of cold water to remove impurities and any remaining odours. They must then be drained before use.

Gelatine sheets melt at 40°C and should be melted in a little of the liquid from the recipe before adding it to the base.

Pectin

Pectin is another commonly used gelling substance because of its great absorption capacity. It comes from citrus peel (orange, lemon, etc.), though all fruits contain some pectin in their peel.

It is a good idea to mix pectin with sugar before adding it to the rest of the ingredients.

Agar agar

Agar agar is a gelatinous marine algae found in Asia. It is sold in whole or powdered form and has a great absorption capacity. It dissolves very easily and, in addition to gelling, adds elasticity and resists heat (this is classified as a non-reversible gel).

Other stabilisers

Carob gum, which comes from the seeds of the carob tree, makes sorbets creamier and improves heat resistance.

Guar gum and **carrageenan** are, like agar agar, extracted from marine algae. They are some of the many other gelling substances available, but they are used less than agar agar.

Professional tips

What you need to know about ice cream and water ices:
- Hygienic conditions are essential while making ice cream – personal hygiene and high levels of cleanliness in the equipment and the kitchen environment must be maintained.
- An excess of stabilisers in ice cream will make it sticky.
- Stabilisers should always be mixed with sugar before adding, to avoid lumps.
- Stabilisers should be added at 45°C, which is when they begin to act.
- Cold stabilisers have no effect on the mixture, so the temperature must be raised to 85°C.
- Ice cream should be allowed to 'ripen' for 4–24 hours. This is a vital step that helps improve its properties. The syrup should be ripened before the fruit is added, because its acidity would damage the stabiliser.
- Ice cream should be cooled quickly to 4°C, because micro-organisms reproduce rapidly, particularly between 20°C and 55°C.
- Water ices and sorbet are generally more refreshing and easier to digest than ice cream.
- Fruit for water ices and sorbets must always be of a high quality and perfectly ripe.
- The percentage of fruit used in water ices and sorbets varies according to the type of fruit, its acidity and the properties desired.
- The percentage of sugar will depend on the type of fruit used.
- The minimum sugar content in water ice and sorbets is about 13 per cent.
- Stabiliser is added to water ices in the same way as for ice cream.

Portion control with cold and frozen desserts

Portion control is extremely important to obtain the required portions from a set recipe and to calculate its food cost and selling price. It is also essential to calculate the number of portions for a set number of customers, especially for large functions. There are a number of ways of achieving good portion control:
- use measured ladles
- use individual moulds, for example, ramekins, ring moulds
- use a torten divider which will divide the top of a mousse into uniform portions
- use ice-cream scoops
- use individual silicone moulds (these come in various individual shapes with 10, 15 or 20 units per mat)
- piped crème Chantilly can be used to indicate a portion from a multi-portion dessert (for example, on a cheesecake or trifle)
- following the dish specification
- use cases made from meringue (nests), pastry (tartlets) or chocolate.

Faults in cold and frozen desserts

Water ice or sorbet once frozen is hard and difficult to scoop:
- There is not enough sugar in the recipe – test prior to freezing with a saccharometer or refractometer and add more sugar if required.

Water ice or sorbet not setting firm enough:
- There is too much sugar in the recipe – test prior to freezing with a saccharometer or refractometer and dilute sugar content by adding water until the correct Baumé/Brix scale is achieved.

Curdled appearance to baked egg custard dessert such as crème caramel:
- It has been baked at too high a temperature, causing the protein in the egg custard to over-coagulate.

Regional recipe – Hazelnut and apricot parfait with nougatine and chocolate sorbet

Contributed by Fabrice Teston, Leeds City College

Parfait

1 Cook the sugar at 115°C, add to the egg yolks and make a sabayon. Whisk until cold.

2 Whip the cream. Make meringue. Fold the hazelnut praline into the sabayon then gently fold into the semi-whipped cream.

3 Gradually add the crushed hazelnut and the chopped apricots, fold in the meringue, pour into metal rings.

4 Freeze for at least 2 hours.

Nougatine

1 Make a golden caramel with the glucose and the fondant. Warm up the almonds, and then add them to the caramel.

2 Pour onto an oiled tray, or Silpat mat. Leave to cool down. Grind into powder in the Robot-Coupe. Sprinkle into metal rings. Bake at 180°C for 2 or 3 minutes.

Apricot coulis

1 Boil together the purée and the stock syrup, and then pass through a chinois.

2 Add a few drops of lemon juice.

Chocolate sorbet

1 Boil the water, sugar, stabiliser, cocoa powder and glucose together.

2 Pour onto the crushed chocolate, cool down, then churn.

Presentation

1 De-ring the parfait, place between 2 discs of nougatine.

2 Serve with a quenelle of sorbet on the top and chocolate spaghetti, surrounded by apricot coulis.

Ingredient	4 portions	10 portions
Hazelnut and apricot parfait		
Egg yolks	4	10
Egg whites	4	10
Sugar	125g	312g
Double cream	250ml	625ml
Hazelnut praline	50g	125g
Hazelnuts, roasted	70g	175g
Dried apricots	80g	220g
Nougatine		
Glucose syrup	62.5g	156g
Fondant	62.5g	156g
Nibbed almonds	50g	125g
Apricot coulis		
Apricot purée	62.5g	156g
Stock syrup	25g	62g
Chocolate sorbet		
Water	750g	152g
Sugar	150g	375g
Glucose	62g	155g
Sorbet stabiliser	2g	5g
Cocoa powder	35g	87g
Valrhona 64% chocolate	195g	340g

Energy	Cal	Fat	Sat fat	Carb	Sugar	Protein	Fibre
5874 kJ	1403 kcal	77.7g	33.5g	168.0g	151.7g	18.3g	8.9g

11 Vanilla panna cotta

1 Boil the milk and cream. Add the aniseeds, and infuse with the vanilla pod, removing it after infusion.
2 Heat again and add the soaked gelatine and caster sugar. Strain through a fine strainer.
3 Place in a bowl set over ice and stir until it thickens slightly; this will allow the vanilla seeds to suspend throughout the mix instead of sinking to the bottom.
4 Fill individual dariole moulds.
5 Turn out the panna cotta, garnish with poached rhubarb and dried, split vanilla pod.

The nutritional analysis includes a fruit accompaniment.

Ingredient	6 portions
Milk	125ml
Double cream	375ml
Aniseeds	2
Vanilla pod	½
Gelatine, soaked	2 leaves
Caster sugar	50g

Energy	Cal	Fat	Sat fat	Carb	Sugar	Protein	Fibre
1565 kJ	378 kcal	34.0g	21.1g	16.1g	16.1g	2.9g	1.5g

12 Meringue

▲ Unfilled meringues and vacherins

Ingredient	4 portions	10 portions
Lemon juice or cream of tartar		
Egg whites, pasteurised	4	10
Caster sugar	200g	500g

1 Whip the egg whites stiffly with a squeeze of lemon juice or cream of tartar.
2 Sprinkle on the sugar and carefully mix in.
3 Place in a piping bag with a large plain tube and pipe onto silicone paper on a baking sheet.
4 Bake in the slowest oven possible or in a hot plate (110°C). The aim is to dry out the meringues without any colour whatsoever.

Optionally, add flaked almonds before baking. For a chocolate version, see Recipe 13.

Whipping egg whites

The reason egg whites increase in volume when whipped is because they contain so much protein (11 per cent). The protein forms tiny filaments, which stretch on beating, incorporate air in minute bubbles then set to form a fairly stable puffed-up structure expanding to seven times its bulk. To gain maximum efficiency when whipping egg whites, the following points should be observed.

- Because of possible weakness in the egg-white protein, it is advisable to strengthen it by adding a pinch of cream of tartar and a pinch of dried egg-white powder. If all dried egg-white powder is used no additions are necessary.
- Eggs should be fresh.
- When separating yolks from whites no speck of egg yolk must be allowed to remain in the white; egg yolk contains fat, the presence of which can prevent the white being correctly whipped.
- The bowl and whisk must be scrupulously clean, dry and free from any grease.
- When egg whites are whipped, the addition of a little sugar (15g to 4 egg whites) will assist efficient beating and reduce the chances of over-beating.

13 Black Forest vacherin

Ingredient	12 portions
Pasteurised egg whites	250ml
Caster sugar	500g
Lemon juice	5ml
Vanilla essence	Drop
Cornflour, sieved	30g
Cocoa powder, sieved	50g
Small discs of chocolate sponge	12
Kirsch syrup	100ml
Cherries (fresh, tinned or griottines)	60–72
Pastry cream	200g
Kirsch	20ml
Gelatine, soaked	2 leaves
Couverture, melted	200g (approx.)
Double cream, whipped	400ml
Chocolate shavings	
Icing sugar	
Cocoa powder (to dust)	

Energy	Cal	Fat	Sat fat	Carb	Sugar	Protein	Fibre
2751 kJ	657 kcal	34.2g	18.3g	79.5g	71.4g	7.3g	2. g

1 Whisk the egg white and one-quarter of the sugar until firm. Continue to whisk while streaming in half of the sugar.

2 Add the lemon juice and vanilla. Fold in the cornflour and cocoa powder, and the remaining quarter of the sugar.

3 Pipe this vacherin mixture into 12 rounds, 80mm in diameter. Bake at 150°C for approximately 1 hour.

4 Place a sponge disc on each vacherin. Moisten the sponge with Kirsch syrup and place 5 or 6 cherries on top.

5 Beat the pastry cream. Dissolve the gelatine in the warm kirsch and then beat it into the pastry cream.

6 Beat in the melted couverture to taste. Fold in the cream.

7 Pipe the chocolate mixture onto the prepared bases in a spiral.

8 Cover with chocolate shavings. Dust with icing sugar first, then cocoa powder. Serve with crème anglaise.

14 Baked blueberry cheesecake

Ingredient	1 cheesecake (8–12 portions)
Base	
Digestive biscuits	150g
Butter, melted	50g
Filling	
Full-fat cream cheese	350g
Caster sugar	150g
Eggs	4
Lemon, zest and juice of	1
Vanilla essence	5ml
Blueberries	125
Soured cream	350ml

Energy	Cal	Fat	Sat fat	Carb	Sugar	Protein	Fibre	Sodium
1791 kJ	431 kcal	33.5g	19.5g	28.0g	19.7g	6.4g	0.4g	0.3g

1 Blitz the biscuits in a food processor. Stir in the melted butter. Press the mixture into the bottom of a lightly greased cake tin with a removable collar.

2 Whisk together the cheese, sugar, eggs, vanilla and lemon zest and juice, until smooth.

3 Stir in the blueberries, then pour the mixture over the biscuit base.

4 Bake at 160°C for approximately 30 minutes.

5 Remove from the oven and leave to cool slightly for 10–15 minutes.

6 Spread soured cream over the top and return to the oven for 10 minutes.

7 Remove and allow to cool and set. Chill. Decorate with blueberries and orange zest.

15 Baked apple cheesecake

Ingredient	16 portions
Base	
Biscuit crumbs	225g
Butter, melted	110g
Caster sugar	30g
Filling	
Apples, cooked, halved	Approx. 8
Cream cheese, full fat	800g
Caster sugar	230g
Cornflour	75g
Eggs	2 (120g)
Vanilla arome or essence	4 drops
Double cream	290ml

Energy	Cal	Fat	Sat fat	Carb	Sugar	Protein	Fibre
2251 kJ	542 kcal	43.3g	25.4g	36.3g	26.3g	4.0g	1.5g

1 Combine the ingredients for the base. Press the mixture into the bottom of two lined cake tins.

2 Place the cooked apple halves into the tins.

3 Cream the cheese and sugar together. Stir in the cornflour, eggs, vanilla and double cream.

4 Divide the filling between the two tins.

5 Bake at 160°C for approximately 40 minutes.

6 Allow to cool slightly, then remove from the mould and dust with icing sugar.

7 Decorate with dried apple slices.

16 Rich vanilla ice cream

Ingredient	10 portions
Whipping cream	250ml
Milk	250ml
Pasteurised egg yolk	100g
Caster sugar	150g
Vanilla pod	1
Unsalted butter	65g

Energy	Cal	Fat	Sat fat	Carb	Sugar	Protein	Fibre
1052 kJ	253 kcal	19.4g	11.2g	17.6g	17.6g	3.0g	0.0g

1 Make a crème anglaise in the normal manner using the cream, milk, egg yolks, sugar and vanilla pod.

2 Stir in the unsalted butter and churn down in the sorbetière.

Professional tip

The ice cream will have a better flavour if the custard is chilled down quickly and matured in a refrigerator at 3°C for 12 hours prior to freezing down.

17 Caramel, lemon curd and peach ice creams

▲ Caramel, lemon curd and peach ice creams

Caramel ice cream

Ingredient	8 portions
Crème anglaise	
Milk	500ml
Egg yolks	5
Caster sugar	25g
Whipping cream	100ml
Inverted sugar (Trimoline)	25g

Ingredient	8 portions
Caramel	
Glucose	20g
Caster sugar	100g
Butter	10g
Water, boiling	40ml

Energy	Cal	Fat	Sat fat	Carb	Sugar	Protein	Fibre
862 kJ	206 kcal	12.0g	6.4g	21.7g	20.6g	4.2g	3.4g

1 Make the crème anglaise in the normal manner, then add the inverted sugar.

2 To make the caramel, melt the glucose in a thick-bottomed pan.

3 Add half the sugar to the melted glucose and heat until a caramel colour starts to appear.

4 Gradually add the remaining sugar and continue to cook until a golden caramel is obtained.

5 Add the butter and the boiling water to arrest the cooking of the sugar and dilute the caramel.

6 Add the caramel to the crème anglaise and freeze down in the sorbetière.

Health and safety

Remember that caramel is extremely hot; be very careful when pouring onto the crème anglaise.

Professional tip

Always have a frozen metal container in the deep freezer to transfer the ice cream into. This will prevent the base of the ice cream melting.

Lemon curd ice cream

Ingredient	6–8 portions
Lemon curd	250g
Crème fraiche	125g
Greek yoghurt	250g

Energy	Cal	Fat	Sat fat	Carb	Sugar	Protein	Fibre	Sodium
774 kJ	185 kcal	8.9g	5.2g	24.8g	16.7g	2.8g	0.0g	0.1g

1 Mix all ingredients together.

2 Churn in the ice-cream machine.

Peach ice cream

Ingredient	8 portions
Milk	250ml
Caster sugar	175g
Orange rind	1
Lemon rind	1
Stabiliser (Trimoline)	25g
Single cream	250ml
Peach purée	250ml
Lemon juice	10ml

Energy	Cal	Fat	Sat fat	Carb	Sugar	Protein	Fibre
755 kJ	179 kcal	7.3g	4.6g	27.8g	27.8g	2.5g	4.1g

1 Slowly bring the milk, sugar, rinds and stabiliser to the boil.

2 Remove from the heat and leave to cool slightly.

3 Add the cream and leave to cool.

4 When cold, add the peach purée and lemon juice. Leave overnight to mature.

5 Pass, then churn in the ice-cream machine.

6 Place into a frozen container. Store in the freezer.

18 Apple sorbet

▲ Fruits of the forest, apple and chocolate sorbets

Ingredient	8–10 portions
Granny Smith apples, washed and cored	4
Lemon, juice of	1
Water	400ml
Sugar	200g
Glucose	50g

Energy	Cal	Fat	Sat fat	Carb	Sugar	Protein	Fibre
673 kJ	158 kcal	0.1 g	0.0 g	41.5 g	38.7 g	0.4 g	2.4 g

1 Cut the apples into 1cm pieces and place into lemon juice.

2 Bring the water, sugar and glucose to the boil, then allow to cool.

3 Pour the water over the apples. Freeze overnight. Blitz in a food processor.

4 Pass through a conical strainer, then churn in an ice-cream machine.

Professional tip

For best results, after freezing, process in a Pacojet.

Variation

Fruits of the forest sorbet: use a mixture of forest fruits instead of apples.

19 Chocolate sorbet

Ingredient	8 portions
Water	400ml
Skimmed milk	100ml
Sugar	150g
Ice-cream stabiliser	40g
Cocoa powder	30g
Dark couverture	60g

1 Combine the water, milk, sugar, stabiliser and cocoa powder. Bring to the boil slowly. Simmer for 5 minutes.
2 Add the couverture and allow to cool.
3 Pass and churn in an ice-cream machine or sorbetière.
4 Place in a frozen container. Freeze until required.

Energy	Cal	Fat	Sat fat	Carb	Sugar	Protein	Fibre
544 kJ	129 kcal	3.0g	1.8g	25.4g	24.9g	1.6g	6.3g

20 Blackberry sorbet

Ingredient	12 portions
Sugar	150g
MSK standard sorbet stabiliser	50g
Water	500g
Blackberry purée	500g
MSK malic acid	4g

1 Combine the sugar with the stabiliser and add to the water. Heat this mixture until the sugar is dissolved.
2 Add the blackberry purée and malic acid and allow to cool.
3 Freeze in Pacojet containers.
4 Churn in the Pacojet machine and serve.

Energy	Cal	Fat	Sat fat	Carb	Sugar	Protein	Fibre
261 kJ	61 kcal	0.1g	0.0g	15.8g	15.8g	0.3g	5.7g

21 Grapefruit water ice

Ingredient	8–10 portions
Water	250ml
Sugar	100g
Grapefruit juice	250ml
Orange juice	100ml
White wine	5ml
Lemon (juice of)	1

1 Bring the water and sugar to the boil.
2 Add the rest of the ingredients, mix well and cool.
3 Churn in a sorbetière and place into a frozen container.
4 Freeze until required.

Energy	Cal	Fat	Sat fat	Carb	Sugar	Protein	Fibre
221 kJ	52 kcal	0.0g	0.0g	13.5g	13.5g	0.2g	0.0g

22 Champagne water ice

Base syrup	
Water	500ml
Sugar	500g
Lemon, zest of	1
Orange, zest of	1
Vanilla pod	½

1 Boil all the ingredients of the syrup and infuse for 12 hours, then strain.

2 For every 750ml of Champagne add 650ml of the syrup and 10g of lemon juice.

3 Check the density and adjust if necessary until it is 17° Baumé.

4 Churn in a sorbetière.

23 Pistachio and chocolate crackle ice cream

Ingredient	12 portions
Milk	500ml
Cream	500ml
Egg yolks	10
Caster sugar	200g
Ice-cream stabiliser	50g
Pistachio compound	40g
Chocolate crackle crystal	50g

Energy	Cal	Fat	Sat fat	Carb	Sugar	Protein	Fibre
1104 kJ	265 kcal	17.0g	8.1g	23.3g	22.8g	6.0g	4.8g

1 Bring the milk and cream to the boil in a saucepan.

2 Combine the egg yolks, sugar and stabiliser in a bowl. Whisk until the mixture is very pale and leaves a trail when the beaters are lifted. Gradually whisk in the milk mixture.

3 Return the mixture to the pan and cook it over a very low heat, or cook it in the top of a double boiler, stirring constantly until the custard is thick enough to coat the back of a wooden spoon.

4 Remove the custard from the heat to cool, stirring it from time to time to prevent a skin forming. Once the mixture has cooled, stir in the pistachio compound, pour into the Pacojet containers and freeze down.

5 Once frozen, place the container in the Pacojet machine and process.

6 When the ice cream is creamed down, fold in the chocolate crackle crystal and serve.

Professional tip
Never fill the Pacojet containers to the top, as the mixture will spill out during processing.

24 Peach Melba

1 Poach the peaches. Allow to cool, then peel, halve and remove the stones.

2 Dress the fruit on a ball of the ice cream in an ice-cream coupe, or in a tuile basket.

3 Finish with the sauce. The traditional presentation is to coat the peach in Melba sauce or coulis, and decorate with whipped cream. In this picture, the peach is garnished with crushed fresh pistachios and covered with a caramel cage; the basket is then placed carefully onto a base of coulis.

Professional tip

If using fresh peaches, dip them in boiling water for a few seconds, cool them by placing into cold water, then peel and halve.

Ingredient	4 portions	10 portions
Peaches	2	5
Vanilla ice cream	125ml	300ml
Melba sauce or raspberry coulis	125ml	300ml

Energy	Cal	Fat	Sat fat	Carb	Sugar	Protein	Fibre
607 kJ	145 kcal	2.6g	1.3g	30.5g	30.2g	1.6g	1.3g

Variation

Fruit Melba can also be made using pear or banana instead of peach. Fresh pears should be peeled, halved and poached. Bananas should be peeled at the last moment.

25 Pear belle Hélène

1 Serve a cooked pear on a ball of vanilla ice cream in a coupe.

2 Decorate with whipped cream. Serve with a sauceboat of hot chocolate sauce.

Alternatively, present the ingredients on a plate as shown here.

Energy	Cal	Fat	Sat fat	Carb	Sugar	Protein	Fibre
1826 kJ	440 kcal	35.1g	20.8g	28.1g	28.1g	4.6g	0.2g

26 Bombe glacée

Pâte à bombe (large scale)

Ingredient	15 portions
Caster sugar	250g
Water	62ml
Pasteurised egg yolks	250g
Cream, lightly whipped	500g

1 Boil the water and sugar to hard ball stage.
2 Whisk the egg yolks until aerated and gradually pour on the sugar solution.
3 Whisk until cold, then fold in the whipped cream and flavouring.

Creating the bombe

1 Stand the bombe mould in a bowl packed with ice.
2 Fill the bombe mould to the top with either ice cream or water ice.
3 Scoop out the centre with a warm spoon.
4 Fill with pâte à bombe and any flavouring.
5 Skim the top of the bombe with ice cream or sorbet and freeze.
6 To remove the bombe mould, dip into warm water, turn upside down and release the screw in the base, and turn out onto a round of sponge. Glaze and decorate as required.

Pâte à bombe (small scale)

Ingredient	10 portions
Pasteurised egg yolk	200g
Sugar syrup at 32° Baumé (250g sugar + 250ml water = 32° Baumé)	250g
Cream, lightly whipped	250g

1 Blend the egg yolks with the syrup. Whisk over a heated bain marie (as for a Genoise sponge) and continue to whisk over ice until the mixture is completely cold.
2 Add the whipped cream and any flavourings.

Variations
- As shown in the photo: line with chocolate ice cream and fill with tutti frutti (pâte à bombe with chopped glacé fruits).
- **Bombe Nesselrode:** lined with chestnut ice cream, interior filled with vanilla pâte à bombe.
- **Bombe Sultane:** lined with chocolate ice cream, interior filled with pâte à bombe praline.
- **Bombe Othello:** lined with praline ice cream, interior filled with vanilla pâte à bombe with diced peaches.
- **Bombe diplomate:** lined with vanilla ice cream, interior flavoured with maraschino and glacé fruits.

Fill the mould to the top, then scoop out the centre

Fill the centre with pâte à bombe and then cover with some of the ice cream removed earlier

To remove from the mould after freezing, release the screw in the base

27 Raspberry parfait

1 Make up the Italian meringue at 120°C (see Chapter 13, Recipe 16).

2 Combine the egg yolks and caster sugar over a pan of simmering water, to make a sabayon.

3 Drain the gelatine and dissolve it in the liqueur and lemon juice.

4 Fold the gelatine mixture into the sabayon, then fold in the raspberry purée.

5 Fold in half the Italian meringue, then fold in the whipped cream.

6 Place into prepared moulds lined in the base with sponge and freeze.

7 Remove from mould onto a dessert plate. Serve with raspberries and a raspberry coulis.

Ingredient	6–8 portions
Egg yolks, pasteurised	80g
Caster sugar	60g
Gelatine, soaked	1½ leaves
Raspberry liqueur	10ml
Lemon juice	10ml
Raspberry purée	120g
Whipped cream	150ml
Italian meringue	
Caster sugar	150g
Glucose	20g
Water	80ml
Egg whites	200g

Energy	Cal	Fat	Sat fat	Carb	Sugar	Protein	Fibre
981 kJ	234 kcal	10.7g	5.6g	31.3g	30.2g	4.7g	0.5g

Variation: Kirsch cream parfait

Water	70g
Caster sugar	300g
Pasteurised whole egg	300g
Pasteurised egg yolk	100g
Cream, lightly whipped	1 litre
Kirsch	50ml

1 Boil the water and sugar to soft ball stage.

2 Whip the whole eggs and egg yolks to a sabayon.

3 Slowly add the boiled sugar solution, and whip until cold.

4 Lightly fold in the whipped cream and kirsch and deposit into a mould lined in the base with sponge.

5 Freeze.

28 Orange and Cointreau iced parfait/soufflé

Ingredient	8 portions
Oranges (juice and zest)	3
Oranges (juice only)	8 (400ml)
Whipping cream, half whipped	800ml
Egg yolks	100g
Caster sugar	75g
Gelatine, soaked in cold water	2 leaves
Cointreau	20ml
Italian meringue	
Caster sugar	100g
Glucose	10g
Water	80ml
Egg whites	100g

Energy	Cal	Fat	Sat fat	Carb	Sugar	Protein	Fibre
2344 kJ	564 kcal	44.2g	26.3g	35.9g	35.3g	6.6g	1.4g

To prepare a chocolate wrap

1 Cut a piece of patterned acetate to fit the mould being used precisely, making sure that the edges of the acetate meet securely without overlapping.

2 Once measured, lay the acetate flat and spread tempered white chocolate thinly to cover it.

3 Carefully place the acetate into the mould, with the chocolate facing inwards, and leave to set.

To make the parfait

1 Use a Microplane grater or similar to zest three oranges. Bring a small pan of water to the boil, add the zest and simmer for 5 minutes. Refresh the zest in cold water and reserve.

2 Juice all the oranges (11 in total). Measure the juice; pass and reduce until you have 200ml. Leave to cool.

3 Whip the cream to soft peaks and chill.

4 Make a sabayon by whisking the egg yolks and sugar over a bain marie until the mixture reaches 75°C. Mix the sabayon in a food processor until it goes cold.

5 Make the Italian meringue by carefully boiling the sugar, glucose and water in a clean stainless steel pan until it reaches 110°C. Whisk the egg whites until they form soft peaks. Pour the sugar mixture onto the whites gradually and keep mixing until the meringue is tepid.

6 Drain and melt the gelatine. Heat it in the Cointreau with the orange zest and reduced juice until it becomes a liquid. Pass.

7 Fold the gelatine into the sabayon.

8 Fold the Italian meringue into the sabayon mixture, then fold in the cream.

9 Fill the mould with the parfait mix and freeze until set, preferably overnight.

10 To serve, remove the parfait from the mould and carefully peel away the acetate. The pattern should have transferred from the acetate onto the lining chocolate.

Variations

To serve as an iced soufflé (without the chocolate wrap):

1 Line a soufflé mould (ramekin) with silicone paper or acetate to form a collar 2cm higher than the mould. Fill the lined mould with the parfait. The collar will secure the mixture, allowing it to be poured above the level of the mould to create the soufflé effect when served.

2 Freeze until set, preferably overnight.

3 To serve, remove the paper collar and garnish, e.g. with crème Chantilly and orange confit.

This recipe is for a cream-based soufflé. A fruit-based iced soufflé can be made from fruit purée, whipped cream and Italian meringue.

29 Tiramisu torte

Ingredient	2 tortes
Biscuit or sponge bases	4
Pasteurised egg yolks	60g
Sugar	150g
Gelatine	3 leaves
Mascarpone cheese	600g
Double cream	200g
Coffee syrup	100ml
Rum	40ml
Cocoa powder for dusting	
Meringue bulbs, cooked, to decorate (optional)	

Energy	Cal	Fat	Sat fat	Carb	Sugar	Protein	Fibre
1691 kJ	408 kcal	32.6g	17.1g	24.7g	18.2g	4.1g	0.4g

1 Cut the biscuit or sponge bases into shape: cut two to the size of the torte ring, and two to the same shape, but slightly smaller.

2 Mix the egg yolks and sugar. Cook over a bain marie to 75°C, to form a sabayon.

3 Soak the gelatine in iced water. Drain and add it to the sabayon.

4 Beat the cheese well. Add the sabayon.

5 Lightly whip the cream and fold it into the mixture.

6 Place a large biscuit or sponge base into each torte ring, on a board. Soak the base with a mixture of coffee syrup and rum.

7 Half fill each ring with the cheese mixture.

8 Place the smaller circles of biscuit or sponge on top of the filling. Again, soak with syrup and rum.

9 Fill the rest of the ring with the cheese mixture, to a level top.

10 Chill in the fridge overnight.

11 Decorate with inverted cooked meringue blobs and dust with cocoa powder.

30 Chocolate mousse

1 Boil the syrup.

2 Place the yolks into the bowl of a food mixer. Pour over the boiling syrup and whisk until thick. Remove from the mixer.

3 Add all the melted couverture at once and fold it in quickly.

4 Drain the gelatine, melt it in the microwave and fold it into the chocolate sabayon mixture.

5 Add all the whipped cream at once and fold it in carefully.

6 Place the mixture into prepared moulds, which may be lined in the base with sponge. Refrigerate or freeze immediately.

7 Remove from the mould to serve. Glaze or decorate if desired.

Ingredient	8 portions	16 portions
Stock syrup at 30° Baumé (equal quantities of sugar and water will give 30° Baumé)	125ml	250ml
Pasteurised egg yolks	80ml	160ml
Bitter couverture, melted	250g	500g
Gelatine	2 leaves	4 leaves
Whipping cream, whipped	500ml	1 litre

Energy	Cal	Fat	Sat fat	Carb	Sugar	Protein	Fibre
1928 kJ	46 kcal	37.0g	21.9g	29.7g	29.5g	5.0g	1.0g

Faults

Possible causes of a heavy texture in chocolate mousse:
- The pâte à bombe is under-aerated.
- The cream is insufficiently whipped.
- The mix has been over-worked when folding in the cream and Italian meringue.

31 Orange mousse with biscuit jaconde

Ingredient	6 portions
Patterned biscuit jaconde (Chapter 15, Recipe 5)	1 sheet
Orange mousse	
Orange juice	200g
Pasteurised egg whites	30g
Caster sugar	30g
Gelatine, soaked in ice water	2 leaves
Lemon juice	Few drops
Whipping cream	225g
Orange segments, to garnish	
Glaze (jelly)	
Stock syrup	50ml
Gelatine, pre-soaked and drained	1 leaf
Orange juice	100ml

Energy	Cal	Fat	Sat fat	Carb	Sugar	Protein	Fibre
1949 kJ	467 kcal	28.9g	14.1g	41.0g	35.4g	13.1g	0.6g

1 Reduce the orange juice by half.

2 Whisk the egg whites and half of the sugar over a bain marie until firm peaks form (this is Swiss meringue). Remove from the heat and whisk in the rest of the sugar.

3 Drain the gelatine and dissolve it in the reduced orange juice. Add a few drops of lemon juice.

4 Fold the meringue into the juice, then fold in the cream.

5 Line individual rings with the sponge to ¾ height. Place a disc of sponge in the base of each ring. Fill the lined moulds to the top and level off.

6 Chill to set.

7 To make the glaze, heat the stock syrup. Add the gelatine and stir until it dissolves. Add the orange juice and pass through muslin.

8 When the mousses are cold, spoon the glaze over the top. Return to the fridge to set.

9 Remove from the moulds and decorate with orange segments.

Faults

A white granular texture in fruit mousse has two possible causes:
- The egg white has been over-whipped.
- The cream has been over-whipped.

32 Strawberry Charlotte

1 To make the mousse, drain the gelatine. Warm it with the liqueur and one quarter of the fruit purée, until the gelatine has dissolved. Add the rest of the purée, then fold this mixture into the Italian meringue. Fold in the whipped cream.

1 Line the inside of two torte rings with a strip of biscuit à la cuillère to ¾ of the height of the ring.

2 Place a disc of sponge inside each hoop to form a base. Moisten the sponge with kirsch syrup.

3 Fold the diced strawberries into the strawberry mousse.

4 Half fill each ring with mousse. Place a second disc of sponge on top.

5 Fill to the top of each ring with mousse and level off.

6 Allow to set in the fridge for at least 12 hours.

7 Spread the top with cream.

8 Remove the torte ring with the aid of a blow torch.

9 Decorate with crème Chantilly and cut-out chocolate shapes and shavings.

Ingredient	2 Charlottes
Biscuit à la cuillère	2 strips
Sponge (vanilla or jaconde)	2 discs
Kirsch syrup (½ kirsch, ½ syrup)	200ml
Strawberries, diced	100g
Whipping cream, half whipped	200ml
Strawberry mousse	
Gelatine, soaked in ice water	10 leaves
Cassis liqueur	66g
Strawberry purée	666g
Italian meringue	340g
Whipping cream, ¾ whipped	666g

Energy	Cal	Fat	Sat fat	Carb	Sugar	Protein	Fibre
1633 kJ	392 kcal	23.9g	14.2g	34.9g	31.9g	5.6g	0.1g

33 Floating islands (îles flottantes or oeufs à la neige)

Ingredient	12 portions
Compote	
Strawberries	500g
Caster sugar	50g
Champagne	125ml
Anglaise	
Vanilla pods, split	3
Double cream	750ml
Egg yolks	240ml
Caster sugar	160g
Poached meringue	
Egg whites	250ml
Caster sugar	500g
Lemon juice	2 drops

Energy	Cal	Fat	Sat fat	Carb	Sugar	Protein	Fibre
2662 kJ	638 kcal	39.7g	22.6g	65.8g	65.8g	6.5g	0.6g

To make the compote

1. Cut any large strawberries in half.
2. Place all the fruit in a clean pan and sprinkle with caster sugar.
3. Stir over heat until hot and starting to produce liquid.
4. Douse with Champagne and chill over ice.

To make the anglaise

1. Add the split vanilla pods to the cream. Bring to the boil slowly.
2. Whisk the egg yolks and sugar together.
3. Pour half the boiling cream over the egg yolk mixture.
4. Return this to the pan of cream. Cook to 84°C.
5. Pass and chill over ice.

To make the poached meringue

1. Whisk the egg whites, lemon juice and 125g of the sugar at a medium speed until soft peaks form.
2. Increase the speed and rain in 250g of the sugar.
3. Fold in the remaining sugar by hand.
4. Pipe or spoon into prepared spherical moulds. Poach in a bain marie, or steam, until firm. Alternatively, shape into quenelles and poach in a water syrup, stock syrup or milk.

To serve

1. Fill glasses with a layer of compote followed by a layer of anglaise.
2. Top each glass with a meringue.
3. Garnish with bubble sugar.

Variation

For a classical presentation without fruit compote, dress the meringues with a coating of cold crème anglaise and crushed praline.

▲ Classical presentation

34 Crème brûlée

1 Warm the milk, cream and vanilla essence in a pan.

2 Mix the eggs, egg yolk and caster sugar in a basin and add the warm milk. Stir well and pass through a fine strainer.

3 Pour the cream into individual dishes and place them into a tray half-filled with warm water.

4 Place in the oven at approximately 160°C for about 30–40 minutes, until set.

5 Sprinkle the tops with demerara sugar and glaze under the salamander or by blowtorch to a golden brown.

6 Clean the dishes and serve with a tuile.

Variations

Sliced strawberries, raspberries, peaches, apricots or other fruits may be placed in the bottom of the dish before adding the cream mixture, or placed on top after the tops are caramelised.

A bain marie is not always used. As an alternative, use a shallow mould and cook at 100°C.

Ingredient	4 portions	10 portions
Milk	125ml	300ml
Double cream	125ml	300ml
Natural vanilla essence or pod	3–4 drops	7–10 drops
Eggs	2	5
Egg yolk	1	2–3
Caster sugar	25g	60g
Demerara sugar		

Energy	Cal	Fat	Sat fat	Carb	Sugar	Protein	Fibre
1633 kJ	392 kcal	23.9 g	14.2 g	34.9 g	31.9 g	5.6 g	0.1 g

35 Unmoulded crème brûlée with spiced fruit compote

Ingredient	16 portions
Crème brûlée	
Whipping cream	750g
Milk	250g
Vanilla pods	2
Egg yolks, fresh	300g
Sugar	150g
Demerara sugar	
Spiced fruit compote	
Plums	500g
Peaches	500g
Star anise	2
Vanilla pods	2
Cinnamon sticks	2
Cloves	2
Caster sugar	100g
Red wine	150ml
Stock syrup	100ml

Energy	Cal	Fat	Sat fat	Carb	Sugar	Protein	Fibre
1450 kJ	349 kcal	25.3g	13.9g	25.3g	25.3g	5.0g	1.4g

To make the crème brûlée

1 Prepare individual stainless steel rings: cover the bases with two layers of cling film and bake them to seal.

1 Boil the cream, milk and vanilla pods.

2 Whisk the egg yolks and sugar together.

3 Pour the boiling liquid onto the egg mixture. Pass, then stir well.

4 Using a dropper, pour into prepared individual rings. Bake in a fan oven at 100°C for 30 minutes.

5 Remove from the moulds onto plates.

6 Sprinkle demerara sugar on top and caramelise with a blow torch. Place slices of dried strawberry below the caramelised sugar, to decorate.

To make the compote

1 Cut up the fruit and break up the spices.

2 Spread the fruit and spices over a roasting tray. Sprinkle with caster sugar and three-quarters of the wine.

3 Roast in a moderate oven until the fruit starts to soften.

4 Carefully remove the fruit. Deglaze the pan with more wine and a little stock syrup.

5 Pass the sauce back onto the fruit. Chill.

36 Gâteau MacMahon

Ingredient	10 portions
Almond sweet paste (sablé paste)	
Butter	300g
Icing sugar	150g
Caster sugar	62g
Ground almonds	62g
Vanilla essence	
Soft flour	125g
Eggs	2
Soft flour	375g
Strawberry bavarois	
Strawberry purée	500ml
Gelatine	7 leaves
Lemon juice	
Cream, lightly whipped	500ml
Caster sugar	180g

Energy	Cal	Fat	Sat fat	Carb	Sugar	Protein	Fibre
3677 kJ	881 kcal	56.7g	33.0g	83.4g	45.9g	14.3g	2.3g

1. For the sablé paste, cream together the butter, sugars, almonds and vanilla essence using a paddle beater.
2. Add 125g of soft flour and the eggs.
3. Take off the machine and add 375g of soft flour; do not over-work.
4. Chill the pastry in the refrigerator.
5. Roll out two discs the size of a torte ring 25cm diameter and 5cm high. Using the torte ring, cut out into two circles on the baking tray.
6. Mark one disc into eight sections and lightly bake both.
7. Once baked, cut the marked disc into sections and place the second inside the torte hoop ring.
8. Place the torte hoop ring in the deep freeze.
9. Prepare the strawberry bavarois (see below). Fill the torte hoop to the top with bavarois and level off, giving a flat and smooth surface. Place in the refrigerator.
10. Dust four of the cut sablé shapes with icing sugar and coat the other four with red glazing jelly.
11. Remove the torte ring with the aid of a blow torch.
12. Neatly arrange the sablé pastry on the top of the bavarois alternating white (dusted with icing sugar) and red (coated with jelly) pieces. May be decorated with whipped cream and chocolate shapes.

For the bavarois

1. Heat the fruit purée, but do not boil.
2. Add the pre-soaked gelatine and lemon juice.
3. Partially set over ice.
4. Lightly fold in the whipped cream.

> **Professional tip**
>
> When pouring gelatine mousses and bavarois into moulds lined on the bottom with sponge, it is good working practice to place them in the freezer on the tray. This helps to prevent any leaks.

37 Crème beau rivage

For the crème renversée

1 Whisk together the whole egg, sugar and vanilla essence.
2 Boil milk and pour onto the eggs. Chill.
3 Pour into a savarin mould lightly greased with melted butter and coated with crushed praline.
4 Place in a bain marie of hot water and bake at 150°C until the surface of the custard feels firm.
5 Chill down and refrigerate for 12 hours.
6 Dip the mould into hot water and turn out onto a round serving dish.
7 Decorate with cornets made from cigarette paste (see Chapter 12, Recipe 5) filled with praline cream and decorated with crystallised violets (optional).

For the praline cream

1 Lightly whip the cream.
2 Fold through the sieved crushed praline.

Ingredient	10 portions
Crème renversée	
Whole egg	250g
Caster sugar	50g
Vanilla essence or pod	4 drops or 1 pod
Milk	500ml
Praline cream	
Whipping cream	200ml
Crushed praline	80g

Energy	Cal	Fat	Sat fat	Carb	Sugar	Protein	Fibre
876 kJ	211 kcal	16.0g	7.3g	11.5g	11.4g	5.9g	0.4g

Professional tip

Crushed praline is very hygroscopic (it attracts moisture from the atmosphere) so it needs to be stored in airtight containers to prevent it sticking together.

Test yourself

Cold and frozen desserts

5 Explain the difference between a water ice, a granita and a sorbet.
6 Name two pieces of equipment that can be used to measure the sugar concentration in syrups.
7 Name four gelatine-based desserts, giving a brief description of each.
8 Name three iced confections that can be made without using a sorbetière.
9 Name the legislation that governs ice cream manufacture. What are the key points of hygiene control in this legislation?

Sauces, espumas and foams

The word espuma directly translates from Spanish into 'foam' or 'bubbles'. An espuma is created using a thermo-whip (classic cream-whipper), which is a stainless steel vessel fitted with a screw top and a non-return valve which you charge with nitrogen dioxide (which constitutes 78 per cent of the air we breathe). This has minimum water solubility, therefore it will not affect the product that is being charged.

The principle role of the gas is to force the liquid out of the canister under pressure through two nozzles, making the cream more voluminous (increasing its volume) due to the mechanical disturbance of the fats. (Although this statement may seem quite complicated it is necessary to explain the mechanics and function of this equipment.)

In simple terms, the canister, once charged, will whip cream the same way as a whisk. The key factors that are essential to a successful preparation are detailed below.

Cold fat-based foams

For a litre canister, 750g is the maximum amount of product to be placed inside. Depending on the viscosity required, one or two charges can be used – for low viscosity (thin mixtures) use two charges; for high viscosity (thicker mixtures) use one charge.

Once the product has been charged, it will need to be treated like any fat-based product that has been aerated, and not stored at room temperature as the aeration will reduce dramatically.

Warm fat-based foams

For a litre canister, 600g is the maximum amount of product to be placed inside.

Warm products tend to need two charges to ensure good aeration.

50–55°C is the optimum temperature to have the canister charged and ready for use. Any hotter and the expansion in the canister will be too great and uncontrollable when the trigger is pressed. If the canister is too cold, the fat molecules will tend to coat the tongue and not give optimum flavour.

Gelatine-based foams

For a litre canister, 750g is the maximum amount of product to be placed inside.

The product will be liquid when it is poured into the canister. It will need to be charged immediately, placed in the fridge and shaken every 10–15 minutes to prevent total setting.

This preparation will give you a purer flavour as there is little or no fat involved. Fat coats the tongue, therefore the absence of fat in this preparation will increase flavour.

Why use espumas?

The boundaries of gastronomy have changed dramatically over the last 20 to 30 years and will no doubt continue to do so for the next 20 to 30, but the current approach is 'less volume, more flavour'.

By offering more flavours, the dining experience will be heightened; by reducing the volume that is taken, more flavour combinations can be offered – espumas are excellent vehicles to achieve such a result.

However, this is a technique that should be used in moderation as too much on one menu will become repetitive to the palate, and what was initially a motivation for using them will become the norm.

Test yourself

Sauces, espumas and foams

10 Name three classical sauces that are often served with desserts.

11 Which gas is used to aerate foams when using a thermo-whip?

38 Sabayon sauce (sauce sabayon)

Ingredient	450–500 ml
Egg yolks, pasteurised	6
Caster or unrefined sugar	100g
Dry white wine	250ml

1 Whisk the egg yolks and sugar in a 1-litre pan or basin until white.
2 Dilute with the wine.
3 Place the pan or basin in a bain marie of boiling water.
4 Whisk the mixture continuously until it increases to four times its bulk and is firm and frothy.

Note: Sauce sabayon may be offered as an accompaniment to any suitable hot sweet (e.g. soufflés or pudding soufflé).

Variation

A sauce sabayon may also be made using milk in place of wine, which can be flavoured according to taste (for example, vanilla, nutmeg, cinnamon).

39 Fresh egg custard sauce (sauce à l'anglaise)

Ingredient	300ml	700ml
Egg yolks, pasteurised	40ml	100ml
Caster or unrefined sugar	25g	60g
Vanilla extract or vanilla pod (seeds)	2–3 drops/½ pod	5–7 drops/1 pod
Milk, whole or skimmed, boiled	250ml	625ml

1 Mix the yolks, sugar and vanilla in a bowl.

2 Whisk in the boiled milk and return to a thick-bottomed pan.

3 Place on a low heat and stir with a wooden spoon until it coats the back of the spoon. Do not allow the mix to boil or the egg will scramble. A probe can be used to ensure the temperature does not go any higher than 85°C.

4 Put through a fine sieve into a clean bowl. Set on ice to arrest the cooking process and to chill rapidly.

Variation

Other flavours may be used in place of vanilla, for example:
- coffee
- curaçao
- chocolate
- Cointreau
- rum
- Tia Maria
- brandy
- whisky
- star anise
- cardamom seeds
- kirsch
- orange flower water.

40 Fruit coulis

Ingredient	1.4 litres
Fruit purée	1 litre
Caster sugar	500g
Lemon juice	10g

1 Warm the purée.

2 Boil the sugar with a little water to soft-ball stage (121°C).

3 Pour the soft-ball sugar into the warm fruit purée while whisking vigorously. Add the lemon juice. Bring back to the boil.

4 This will then be ready to store.

Professional tips

The reason the soft-ball stage needs to be achieved when the sugar is mixed with the purée is that this stabilises the fruit and prevents separation once the coulis is presented on the plate.

Adding lemon juice brings out the flavour of the fruit.

41 Chocolate sauce (sauce au chocolat)

Method 1

Ingredient	300ml	750ml	
Double cream	150ml	375ml	
Butter	25g	60g	
Milk or plain couverture callets	180g	420g	

1 Place the cream and butter in a saucepan and gently bring to a simmer.
2 Add the chocolate and stir well until the chocolate has melted and the sauce is smooth.

Method 2

Ingredient	300ml	750ml
Caster sugar	40g	100g
Water	120ml	300ml
Dark chocolate couverture (75 per cent cocoa solids)	160g	400g
Unsalted butter	25g	65g
Single cream	80ml	200ml

1 Dissolve the sugar in the water over a low heat.
2 Remove from the heat. Stir in the chocolate and butter.
3 When everything has melted, stir in the cream and gently bring to the boil.

42 Butterscotch sauce

Ingredient	300ml	750ml	
Double cream	250ml	625ml	
Butter	62g	155g	
Demerara sugar	100g	250g	

1 Boil the cream, then whisk in the butter and sugar.
2 Simmer for 3 minutes.

43 Caramel sauce

Ingredient	1 litre
Granulated sugar	400g
Oranges	4
Lemons	4
Apricot purée or jam	170g
Crème de cacao	100ml

1 Place the sugar in a large, heavy saucepan on a very low gas to melt and caramelise. Do not stir until the sugar has melted.

2 While the sugar melts, grate the orange and lemon zests and squeeze the juice. Place the zest, juice and apricot purée in a saucepan and heat slowly.

3 When the sugar has caramelised, add the fruit mixture and stir carefully until dissolved.

4 Pass through a chinois into a suitable container. Once cold, stir in the crème de cacao liqueur.

5 Chill thoroughly before using.

44 Hot chocolate espuma

Ingredient	Makes 650g
Milk chocolate couverture	300g
Dark chocolate couverture	50g
White chocolate couverture	100g
Hot water	200g

1 Melt the three types of chocolate over a bain marie until the temperature reaches 45°C.

2 Add the hot water and whisk until smooth.

3 Pour mix into an espuma gun and charge with two cartridges.

4 Place in a bain marie to keep warm.

45 Pernod foam

Ingredient	Makes 800ml
Skimmed milk	450ml
Sugar	170g
Leaf gelatine, pre-soaked	3 leaves
Pernod	160g

1 Boil the milk and sugar. Add the pre-soaked gelatine leaves and allow them to dissolve.
2 Allow the mixture to cool.
3 Add the Pernod.
4 Pour into an espuma siphon and charge with two cartridges. Refrigerate.
5 Once chilled, shake well before using.

46 Raspberry foam

Ingredient	300ml
Still mineral water	100ml
Raspberry purée	200ml
Sugar	25g
Gelatine leaves	2

1 Mix together the mineral water and the raspberry purée.
2 Soak the gelatine in cold water.
3 Warm a quarter of the raspberry purée mixture. Squeeze the excess water from the gelatine and dissolve in the warmed raspberry mixture.
4 Strain this into the remaining raspberry mixture.
5 Pour into an espuma siphon, charged with one cartridge, and refrigerate for 30 minutes.
6 Once chilled, shake well before using.

15 Biscuit, cake and sponge products

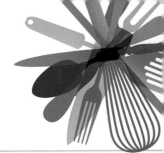

This chapter covers the following units:

NVQ:
→ Prepare, cook and finish complex cakes, sponges, biscuits and scones.

VRQ:
→ Produce biscuits, cakes and sponges.

Introduction

Cakes are often associated with special occasions such as a birthday, wedding, Christmas or a Sunday tea time. They are not desserts to be eaten after a main course but more often with a cup of tea mid-afternoon.

The word 'biscuit' means twice cooked (bis-cuit) and dates back to a time when the idea was to make them last as long as possible – for example, ship's biscuits, which had to stay edible over many weeks or months. These biscuits were made by baking, slicing and baking again so as to remove all moisture and prevent the growth of bacteria – an early, but usually unsuccessful attempt at convenience food. Ship's biscuit or hardtack was notorious for being so hard and inedible that without the worms with which they were infested making lots of small holes, they would have been impossible to break. It is said most sailors preferred to eat them at night so they couldn't see what they were eating.

Learning objectives

By the end of this chapter you should be able to:
→ Explain techniques for the production of biscuits, cakes and sponges, including construction and traditional, classical and modern skills, techniques and styles.
→ List appropriate flavour combinations.
→ Explain considerations when balancing ingredients in recipes for biscuits, cakes and sponges.
→ Explain the effects of preparation and cooking methods on the end product.
→ Describe how to control time, temperature and environment to achieve the desired outcome in biscuits, cakes and sponges.
→ Prepare and cook biscuit, cake and sponge products to the recipe specifications, in line with current professional practice.

Recipes included in this chapter

Cakes and sponges

Virtually all types of cake and sponge are made using butter, eggs, sugar and soft (low-gluten) flour – the difference between them is in the proportions of ingredients used. For example, a pound cake is so named because it uses equal amounts of the main ingredients (1 pound of butter, 1 pound of eggs, etc.) – this will produce a denser, firmer texture than found in a sponge which contains more eggs and far less butter and flour, giving a much lighter product. Air is an essential ingredient in both, and this is achieved by whisking eggs and sugar (sponges) or beating fat with sugar (cakes).

Some cakes may need additional help incorporating air in the form of baking powder or bicarbonate of soda (depending on what the recipe states). Both of these will produce carbon dioxide gas when brought into contact with heat and moisture (see 'Baking powders' below).

Successfully making a sponge is all about damage limitation. Warmed eggs whisked with sugar will create a stable foam, but if the flour is not added carefully (see 'Folding in' below), the air will be knocked out and the volume lost, resulting in a heavy, close texture. Sponges are often moistened before filling and not intended to be eaten on their own.

Techniques

Whisking

When whisking, the idea is to physically add air into eggs (for some recipes these are separated) and sugar. The first rule to remember is that cold eggs will not aerate very well, so it is necessary to either heat the sugar in the oven or warm the eggs and sugar over hot water (not boiling). Second, it is best to whisk at a medium speed – vigorous whisking will create larger but more unstable air bubbles which will burst more easily, thus providing less volume and a heavier result. When a stable foam has been achieved this is called the 'ribbon stage' (the test is to lift up the whisk and trail a figure of eight on top). After this the dry ingredients are carefully added (for example, folded in).

Beating or creaming

This is sometimes referred to as the sugar batter method.

This technique also involves physically adding air, this time into fat and sugar. Like eggs, cold fat will not aerate successfully. As the fat will more than likely come from the fridge, cut it into small, even pieces, mix with the sugar, soften over hot water, and then beat the mixture. The finished mixture should increase in volume by approximately 40 per cent and will be lighter in colour. The eggs (which should be at room temperature) are then beaten into the aerated fat and sugar in two or three lots, and finally the dry ingredients are folded in.

Folding in

The final stage for both of the techniques above is to add the dry ingredients by 'folding in'. The aim is to mix in the dry ingredients while losing as little volume as possible. This is achieved by 'cutting in' with a rubber spatula (large mixes are best done using one hand), 'folding' and turning the bowl at the same time.

The all-in-one method

As the name suggests, the all-in-one method consists of adding all the ingredients together and mixing them. Sometimes, the fat used in these recipes is oil, and they depend on chemical raising agents to give them lift as they are not physically aerated. Products made using this method tend to have less volume and a denser texture, but should be moist in the centre.

Flour-batter method

The flour-batter method is a combination of whisking and creaming. It involves whisking the eggs and sugar, creaming the fat and half the flour, and then folding everything together.

Chemical aeration

If cakes are made using equal quantities of ingredients, as illustrated by the pound cake, as long as the ratios are correct and properly aerated, the ingredients will not need any extra help. However, this can be an expensive way of producing cakes, so to reduce costs they can be bulked out by adding more of the less expensive ingredients (flour and sugar). In these cases, because the flour has been increased they will need more moisture (usually milk) and extra help to rise. This is where baking powder and bicarbonate of soda come in. Heat from the oven and moisture in the cake will allow these raising agents to produce carbon dioxide gas, thus helping the cake to rise.

The above techniques are essential for producing a good result, but it should be remembered that while these cover the main principles, individual recipes will vary and may not be exactly consistent with the techniques above.

Making a Genoise sponge

1 Whisk the eggs and sugar together over hot water.

2 Carry on whisking as the mixture warms up.

3 When the mixture is ready, it will form ribbons and you can draw a figure eight with it (at this stage it is known as a sabayon).

4 Fold in the flour.

5 Add part of the flour mixture to the butter and mix

6 Add this to the rest of the flour mixture and fold in

7 Place the mixture into prepared tins and level off.

8 After baking, turn upside down to remove from tins.

Making sponge fingers

1 Whisk the egg yolks and sugar to a sabayon.

2 Whip the egg whites until they are firm.

3 Add some egg whites to the sugar mixture.

4 Mix lightly and pour back into the egg white container.

5 Fold in the sieved flour.

6 Pipe out in finger shapes.

7 Dust with icing sugar and bake immediately.

The main ingredients

Eggs

All recipes are based on a medium egg, which should of course be fresh. Free-range eggs are thought to contain a stronger albumen and a deeper coloured yolk. Whole eggs, yolks and whites that have been pasteurised by passing them through an infra-red beam are now available in cartons and are measured by weight or volume. For recipes made using the creaming method, the eggs should be at room temperature to avoid splitting the mixture.

Sugar

Sugar obviously sweetens, but it also plays a vital role in the aeration of cakes and sponges. Sugar enables fat and eggs to retain more air when beaten or whisked.

For most recipes, caster sugar is used, occasionally icing sugar, but not granulated sugar because it can give a gritty texture. Icing sugar is used to finish several products (dusting). Brown sugars like muscovado or demerara will add caramel flavours and colour to products and are used in some cake recipes, but they do not aerate well and should not be used in whisked egg sponges.

Fat

There are special cake fats available which are neutral in flavour. These will aerate better than traditional fats, but are more commonly used for commercial large volume production (high-ratio cakes). Oil is used in some recipes, but is not a substitute for solid fat. Where a solid fat is required, the recipes in this chapter use butter because it will give a better flavour.

Flour

The types of flour used in this chapter are soft (low-gluten), self-raising (medium strength plus the addition of a raising agent) and plain (medium strength). Weaker (low-gluten) flours are always used in sponges and give the products a lighter, shorter texture.

Flour should always be sieved to remove any foreign bodies and to aerate, helping it to disperse and mix more easily with other ingredients. It is important to store flour in a cool, dry environment, protected from contamination.

Baking powders

Bicarbonate of soda or baking soda can be used as a raising agent only if the other ingredients include an acid, such as those found in citrus juices or buttermilk. Without an acid, there would be no chemical reaction and consequently no lift. The acid also helps to counteract any unpleasant aftertaste.

Baking powder, of which there are several different kinds, is a complete raising agent that already contains an acid. It consists of one part alkali (bicarbonate of soda) to two parts acid (usually cream of tartar). When hydrated and heated, baking powder will release CO_2, causing the product to expand and rise.

Always add baking powder to the flour and then sieve twice onto paper so it is evenly distributed.

Do not buy in large quantities unless using a lot. Baking powder becomes less effective if it is left on the shelf for too long.

Chocolate

Products made with chocolate will only be as good as the quality of the chocolate used. Never use a compound or baker's chocolate – they will contain a low percentage of cocoa solids, and too much sugar and hydrogenated vegetable fat. The chocolate (couverture) recommended is one where the majority of the fat is cocoa butter with a minimum of 60 per cent cocoa mass.

Combinations

Flavours and textures should be compatible, complimentary and provide contrast. Opera is a good example – chocolate, coffee and rum all work well together, as do the spices with red wine in a Christmas cake, or strawberries with kirsch or an orange liqueur. An effective combination of textures is illustrated by the Himmel cheese torte in Recipe 12, which contrasts a light shortcake with a soft, rich filling, offset by the acidity of the raspberries.

Essential equipment

Equipment used for biscuit, cake and sponge products includes:

- planetary mixers (three-speed electric mixers)
- digital scales
- food processors
- fine-mesh sieves
- fine-wire balloon whisks

- a range of Teflon®-coated baking tins (loaf, Genoise, cake, etc.)
- plain and fluted pastry cutters
- heavy-duty baking sheets (small, medium and large)
- silicone baking mats
- disposable piping bags and a range of plain and star piping tubes
- cooling racks
- a range of small equipment such as stainless steel bowls, rubber and plastic spatulas, palette knives, scissors etc.

Professional tips

- As with all pastry work, always weigh and measure accurately.
- All equipment must be prepared and ready before the final mixing. Likewise, make sure an oven is set at the correct temperature and available. In particular, sponges need to be baked immediately and will lose volume if left standing.
- The fat used to grease the tins or moulds should be soft (never melted), so it does not run down the sides of the cake tin and collect in a puddle on the bottom.
- The shape, volume and density of the products will affect the baking time and temperature.
- Allow to cool slightly before removing from moulds, but do not leave for too long in the tins or they will 'sweat'.
- All ovens behave differently. What is the correct temperature for one may be higher or lower for another; get to know your oven and make a note of the correct temperature and top/bottom heat settings (if these are a feature).

Regional recipe – Chocolate orange jaconde

Contributed by Paul Robinson, Grimsby Institute of Further and Higher Education

Ingredient	8 portions
Jaconde paste	
Unsalted softened butter plus 15g melted butter	250g
Icing sugar	250g
Pasteurised egg whites	250g
Plain flour	250g
Food colouring (optional)	
Jaconde sponge	
Pasteurised egg whites	200g
Caster sugar	35g
Ground almonds	250g
Icing sugar sifted to remove lumps	250g
Eggs	6
Plain flour	50g
Cocoa powder	50g
Melted butter (clarified)	100g
Chocolate and orange mousse	
Zest of large orange (with juice)	1
Powdered gelatine	2 tsp
Good quality plain chocolate, broken into pieces	200g
Eggs, separated	2
Double cream, whisked to soft peaks	350ml

Energy	Cal	Fat	Sat fat	Carb	Sugar	Protein	Fibre
5782 kJ	1386 kcal	92.9g	39.7g	119.2g	88.2g	25.9g	3.4g

To make the jaconde paste

1 Preheat the oven to 220°C then line two 45cm × 30cm baking trays with silicone paper, lightly brushed with melted butter.

2 Cream the butter and sugar until light and fluffy then gradually add the egg whites, beating continuously until smooth. Fold in the flour. If you have chosen to use food colouring, mix this in to suit.

3 Spoon the mixture into a piping bag fitted with a small plain nozzle and pipe the mixture onto the baking trays in a criss-cross or swirl pattern, then place in a freezer.

To make the sponge

1 Whisk the egg whites until soft peaks then add the sugar and continue whisking until stiff peaks are formed.

2 Cover with cling film then separately beat the almonds, icing sugar and eggs in a bowl until the mixture is light and fluffy.

3 Mix in the flour and cocoa powder into the mix then gently fold in the meringue mixture using a large spatula.

4 Mix the clarified butter into the sponge batter, then remove the baking trays with the decorated sheets of silicone paper from the freezer.

5 Divide the batter evenly between the two baking trays, spreading it smoothly over the decorations and ensuring it is level.

6 Bake for 5–7 minutes or until the sponges are golden brown around the edges.

7 Once out of the oven, cover each sponge with a sheet of either silicone or greaseproof paper, then upturn the baking trays onto the work surface and peel off the paper to reveal the pattern.

8 Cut the sponge into strips to line the sides of the desired cake tin, ensuring the pattern is facing outwards, then cut a circle of sponge to line the base. Cut a second circle to make the top of the cake, unless trimming short to show the mousse once filled.

For the mousse

1 Sieve the orange juice into a small bowl and sprinkle over the gelatine. Set the mixture aside for five minutes then place the bowl over a pan of simmering water, ensuring the bottom of the bowl does not touch the water, and stir gently until the gelatine has dissolved.

2 Place the chocolate in a bowl then melt it over a bain marie. Once melted, mix the orange zest and egg yolks into the chocolate, then stir in the gelatine mixture and fold in the whipped cream.

3 Whisk the egg whites until they form stiff peaks then gently fold into the chocolate mixture.

4 Pour the mousse into the lined cake tins then chill in the fridge until the mousse has set, then decorate.

1 Spiced hazelnut shortbread

1 Cream together the butter and icing sugar until light and soft.

2 Whisk the eggs and vanilla and beat into the butter and sugar.

3 Sieve the flour and clove powder and fold in, with the hazelnuts.

4 Line a plastic or stainless steel tray with silicone paper and press the mixture in. Wrap in cling film and place over a second tray. Chill under weights.

5 Remove from the tray, peel off the paper and cut into batons.

6 Lay flat on a baking sheet lined with silicone paper.

7 Bake at 170°C for approximately 15 minutes.

8 Transfer to a cooling rack.

9 While still hot dust with the spiced sugar.

Professional tips

- Be careful when adding the clove powder, as the flavour is very strong.
- To cut, make sure the paste is chilled and use a large, sharp chopping knife to avoid dislodging the whole nuts.

Variation

These shortbreads can be cut smaller and used as petits fours secs.

Ingredient	Approx. 100 pieces
Butter	500g
Icing sugar	375g
Eggs	2
Vanilla seeds	1 pod
Soft flour	750g
Clove powder	5g
Hazelnuts, whole	250g
Hazelnuts, chopped	250g
For dusting	
Caster sugar	25g
Clove powder	Pinch
Ground cinnamon	5g

2 Plain Genoise sponge (Genoise nature)

▲ A thinly sliced Genoise filled with cream and dusted with icing sugar

Ingredient	1 × 16cm sponge	3 × 16cm sponges
Eggs	4	12
Caster sugar	100g	300g
Soft flour	100g	300g
Butter, melted (not hot)	25g	75g

1. Make sure the mixing bowl is clean, dry and has no traces of grease.

2. Prepare Genoise tins by brushing with soft butter and flouring (bang out the excess). Set the oven at 185°C.

3. Place the eggs and sugar in the mixing bowl, and stir over hot water until warm.

4. Whisk to the 'ribbon' stage.

5. Sieve the flour and any other dry ingredients onto paper and melt the butter.

6. Carefully fold in the flour.

7. Finally take out a small amount of mixture, add it to the butter and mix. Add this back to the rest of the mixture and fold in.

8. Deposit equal amounts of mixture into the tins, level the tops and place in the oven immediately. Bake for approximately 25 minutes.

9. When cooked, the sponge should spring back when gently pressed.

10. Cool slightly then turn upside down to remove from tins.

11. When cold, wrap in cling film, label, date and store in the freezer until needed.

Variations
- Chocolate Genoise – replace 50 per cent of the flour with 25 per cent cocoa powder and 25 per cent cornflour.
- Coffee Genoise – add some strong coffee essence to the foam just before folding in the flour.

Professional tips
- Taking out some of the mixture to add to the butter will prevent it from 'dropping' through, allowing it to be mixed in more easily and minimising volume loss.
- A fatless sponge can be made exactly as above but leaving out the butter. The shelf life will be shorter.
- Genoise is best made the day before it is needed, as it can be cut much more easily.

Faults

▲ The smaller cake has a low volume and a heavy texture because it was over mixed

There are three reasons why a Genoise sponge might have a low volume and a heavy texture:
- Weak foam – not whisked enough; ingredients not warmed first; traces of fat in the bowl.
- Air knocked out when adding dry ingredients; dry ingredients not folded in correctly.
- Left standing before baking – lack of preparation; moulds not prepared; oven not set at the correct temperature.

Note: Genoise sponges should be light and springy, golden brown with an even texture and a fine crumb.

3 Roulade sponge

▲ You should be able to bend a roulade sponge

Ingredient	2 sheets	4 sheets
Eggs	8 (450ml)	16 (900ml)
Egg yolk	2 (40ml)	4 (80ml)
Caster sugar	260g	520g
Soft flour	170g	340g

1 Make sure the mixing bowl is clean, dry and free from grease.

2 Line the baking sheets with silicone paper cut to fit.

3 Set the oven at 230°C.

4 Place the eggs and sugar in a mixing bowl, and stir over hot water until warm.

5 Whisk to the 'ribbon' stage and sieve the flour onto greaseproof paper.

6 Carefully fold in the flour.

7 Divide equally between the baking sheets and spread evenly with a drop blade palette knife. Place immediately in the oven for between 5 and 7 minutes.

8 As soon as the sponge is cooked, turn it out onto sugared paper, place a damp, clean cloth over it and lay the hot baking sheet back on top, then leave to cool. (This will help keep the sponge moist and flexible as it cools.)

Professional tip

Each sheet should be left on the paper on which it is cooked, individually wrapped, labelled, kept flat and stored in the freezer to stop it from drying out and losing flexibility.

Faults

There are two reasons why a roulade sponge might become hard and crisp, instead of being pliable:

● Baked at too low a temperature for too long.
● Mixture spread too thin.

Variations

For a lighter sponge, replace 20 per cent of the flour with cornflour.

A chocolate version can be made by substituting 30–40 per cent of the flour for cocoa powder. For a coffee sponge, add coffee extract to the eggs after whisking.

A roulade sponge may also be made using a 'split-egg' method, separating the eggs.

4 Flourless chocolate sponge

Ingredient	10 portions
Egg yolks	230g
Sugar	180g
Inverted sugar	30g
Cocoa powder	100g
Cornflour	50g
Egg whites	300g
Cream of tartar	½ tsp
Sugar	150g

1 Mix together the egg yolks, inverted sugar and 180g of sugar.

2 Whisk until well aerated.

3 Sieve the cocoa powder and cornflour.

4 Whisk the egg whites with cream of tartar, using a little sugar at the beginning of the process and the rest at the end.

5 Transfer the egg yolk sabayon to a mixing bowl. Using a whisk, beat in half of the meringue, then fold in the cocoa powder and cornflour. Fold in the other half of the meringue.

6 Pipe out in round discs and bake at 230°C for 10–15 minutes.

5 Patterned biscuit jaconde

This sponge is used as a lining for desserts such as Charlottes and petit gâteaux. The term 'biscuit' is used whenever egg whites and yolks are whisked separately or, as in this case, additional egg whites are whisked and added.

Ingredient	4 medium baking sheets
Decorating paste (coloured)	
Butter, melted	100g
Icing sugar	100g
Egg whites	50g
Soft flour, sieved	50g
Colour	
Or chocolate decorating paste	
Butter, melted	80g
Icing sugar	80g
Egg whites	80g
Soft flour	60g
Cocoa powder, sieved	30g
Biscuit jaconde	
Soft flour	100g
Ground almonds	250g
Butter, melted	75g
Eggs	500ml
Icing sugar	375g
Egg whites	375ml
Caster sugar	50g

For the decorating paste

1 Place all ingredients in a bowl and mix to a paste.
2 Spread thinly onto a silicone baking mat and, using a comb scraper, make a pattern.
3 Set in the freezer.

For the sponge

1 Set the oven to 230°C.
2 Sieve the flour and ground almonds onto paper and melt the butter.
3 Warm the eggs and icing sugar, then whisk them to the ribbon stage.
4 Separately whisk the egg whites and caster sugar to a meringue.

5 Fold the meringue into the eggs and sugar, then fold in the dry ingredients.
6 Finally add a small amount of mixture to the butter and mix; add this back to the rest and fold in.
7 Divide between the patterned trays and spread evenly.
8 Place in the oven for approximately 6 minutes.
9 As soon as the sponge is cooked, turn out onto sugared paper, place a damp, clean cloth over it and lay the hot baking sheet back on top.
10 When cooled, carefully peel back the mat.
11 If not using immediately, wrap individual sheets in cling film, label and freeze, making sure they are laid flat.

Spread the decorating paste

Make the pattern, then freeze

Spread a layer of sponge over the frozen pattern

Professional tips

- It is usual to pattern the paste with a comb scraper to create stripes or waves, but an effective alternative can be achieved by simply making circular movements with your finger or the handle of a spoon.
- Jaconde sponges should be light and flexible, they should not be too thick and have a clearly defined pattern.

6 Sponge fingers (biscuits à la cuillère)

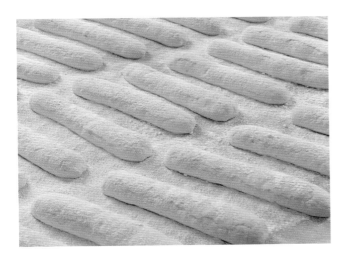

The literal translation of this is 'spoon biscuits', which comes from a time when the mixture would have been shaped between two spoons instead of being piped. They are traditionally used to line the mould for a Charlotte Russe, although the 'spooned' version would not lend itself to that.

Ingredient	Approx. 60 × 8cm fingers
Egg yolks	180g
Caster sugar	125g
Vanilla essence	Few drops
Soft flour	125g
Cornflour	125g
Egg whites	270g
Caster sugar	125g

1 Prepare a baking sheet by lining with silicone paper cut to fit, and set the oven at 160°C. Have ready a piping bag fitted with a medium plain tube. Scald two mixing bowls to ensure they are clean and free of grease.

2 Whisk the yolks, sugar and vanilla over a bain marie until warm, then continue whisking off the heat until a thick, sabayon-like consistency is reached.

3 Sieve the flours onto paper.

4 In a second mixing bowl, whisk the whites with the sugar to a soft meringue.

5 Add the whisked yolks to the meringue and start folding in. Add the flour in 2 or 3 portions, working quickly but taking care not to overwork the mixture.

6 Using a plain piping tube, immediately pipe onto the prepared baking sheet in neat rows.

7 Dust evenly with icing sugar and place straight in the oven for approximately 25 minutes.

8 When cooked, slide the paper (and biscuits) onto a cooling rack.

9 When cool, remove from the paper and store in an airtight container at room temperature, or leave on the paper and store in a dry cabinet.

Professional tips

- It is easier to pipe the fingers all the same length if a template marked with parallel lines is placed under the silicone paper.
- A common problem with this recipe is over-mixing and/or not working fast enough or being disorganised, which results in biscuits that collapse.
- Sponge fingers should be pale in colour, very light in texture, be dusted with icing sugar and have a rounded shape. They should also be identical in length and width.

Variations

For a chocolate version, instead of 125g each of soft flour and cornflour, use 120g soft flour, 60g cornflour and 70g cocoa powder.

To make **Othellos**, use the above recipe to make small, domed sponges. Hollow them out and fill them with crème mousseline. Sandwich pairs together and coat with coloured fondant.

7 Strawberry gâteau

This is based on the classic French gateau, 'le fraisier'.

Ingredient	1 gateau (12 portions)
Plain Genoise sponge (Recipe 2)	1 × 16cm
Stock syrup flavoured with Grand Marnier or kirsch	50ml
Strawberry jam	50g
Fresh strawberries	125g
Crème pâtissière	400g
Gelatine	5 leaves
Double cream	800ml
Grand Marnier or kirsch	80ml
Pink colouring	
Marzipan	80g

Energy	Cal	Fat	Sat fat	Carb	Sugar	Protein	Fibre
2430 kJ	585 kcal	43.1g	25.1g	37.3g	25.0g	9.8g	0.6g

1 Split the Genoise equally. Place the base on a cake board and moisten with syrup, then spread with a thin layer of strawberry jam.

2 Reserve 3 or 4 strawberries for decoration, hull the rest and cut them in half.

3 Place a deep stainless steel ring over the sponge and stand the halved strawberries all the way around, cut sides against the ring.

4 To make the filling, beat the crème pâtissière until smooth, soak the gelatine and whip the cream.

5 Squeeze the gelatine, add to the liquor and heat to dissolve. Pass onto the crème pâtissière and beat vigorously to blend in. Finally, fold in the cream.

6 Place the filling in a piping bag and pipe into the ring almost to the top, then place on the other half of the sponge and moisten with syrup. Press down evenly. Set in the fridge.

7 Colour the marzipan and roll out 2–3mm thick, cut to the same size as the gateau and lay on the top. (It is a good idea to spread a little jam on top of the sponge so the marzipan sticks.)

8 Using melted chocolate or royal icing, pipe on a decorative border.

9 Decorate with the reserved strawberries which have been dipped in crack sugar.

Professional tips

● The marzipan is traditionally piped with the words 'le fraisier' and coloured green.
● When soaking gelatine, always use iced water to prevent the sheets from breaking up.
● The finish on the marzipan in the photograph has been achieved using a patterned rolling pin, but alternatively it could be marked using the back of a knife, or the edge marked with marzipan crimpers.

Quality points

A good-quality strawberry gâteau should have a moist sponge and a good balance between sponge and filling. The filling should be smooth and lump-free, the strawberries evenly sized and the decoration (piping) restrained and tasteful.

Variation

If preferred, this dish could be filled with Chantilly cream or crème mousseline.

8   Coffee gâteau

Ingredient	1 × 16cm gateau
Plain Genoise sponge (Recipe 2)	1 × 16cm
Stock syrup flavoured with rum	50ml
Coffee buttercream	750g
Coffee marzipan	100g
Fondant	500g
Crystallised violets	
Chocolate squares	

Energy	Cal	Fat	Sat fat	Carb	Sugar	Protein	Fibre
2667 kJ	634 kcal	25.6g	14.3g	100.9g	92.5g	4.6g	0.4g

1 Carefully split the sponge into three, and line up the three pieces

2 Place the sponge base on a cake card and moisten with rum syrup.

3 Pipe on an even layer of buttercream, no thicker than that of the sponge.

4 Place on the next layer of sponge, moisten with the syrup and repeat to give three layers of sponge and two of buttercream. Moisten the top with syrup.

5 Put in the fridge for 1–2 hours to firm up.

6 Work some coffee essence into the marzipan, roll out to 2mm thick and lay over the gateau, working the sides to prevent any creases.

7 Warm the fondant to blood heat, flavour with coffee essence and the adjust consistency with syrup.

8 Place the gâteau on a wire rack with a tray underneath to catch the fondant.

9 Starting in the centre and moving outwards, pour over the fondant to completely cover. Draw a palette knife across the top to remove the excess.

10 Add some melted chocolate to some of the fondant, adjust the consistency and squeeze through muslin. Decorate the gâteau by piping on a fine line design.

11 Finish the sides with squares of chocolate and the top with crystallised violets.

Professional tips

- Mark the sponge by cutting a 'v' on the side before splitting horizontally; when reassembling, line up the marks so it goes back together exactly as it came apart.
- Turn the sponge upside down before splitting so the base becomes the top; this is the flattest surface and will give the best finish.
- It is best practice to use a Genoise that was made the day before – fresh sponges do not cut well and are susceptible to falling apart.
- Fondant should never be heated above 30°C, as the shine will be lost.

Quality points

A good-quality coffee gâteau should have a moist sponge and a good balance between sponge and filling (as a guide, the thickness of the sponge and the depth of the buttercream should be equal). The coffee flavour should not be in question, and the decoration should reflect and complement the coffee theme. (It is sometimes easy to get carried away, so it is good to remember when decorating, 'less is definitely more'.)

Variations

- Instead of enrobing with fondant, the top and sides can be covered with buttercream, the sides can be either comb-scraped or masked with toasted nibbed/ flaked almonds or grated chocolate. The top can be piped with buttercream and/or decorated with coffee marzipan cut-out shapes.
- To add another texture, place a disc of meringue or dacquoise on the bottom layer. Dacquoise is an Italian meringue with the addition of toasted ground hazelnuts, spread or piped onto a silicone mat and baked at 180°C for 15–20 minutes. Cut out the desired shape half way through cooking.

9 Dobos torte

▲ Traditional Dobos torte

This is a Hungarian speciality which consists of seven layers of biscuit. Six are sandwiched with chocolate buttercream. The seventh is coated with caramel, cut into triangles and placed on top.

Ingredient	1 × 20cm torte
Biscuit discs	
Butter	120g
Icing sugar	120g
Soft flour	120g
Eggs	2
Vanilla extract	½ tsp
Chocolate buttercream	500g
Caramel	
Water	100ml
Granulated sugar	250g
Glucose	25g

Energy	Cal	Fat	Sat fat	Carb	Sugar	Protein	Fibre
2893 kJ	689 kcal	32.3g	19.9g	102.3g	89.2g	3.1g	0.6g

1 Cream the butter and sugar until soft and light.

2 Sieve the flour.

3 Mix together the eggs and vanilla and beat in to the butter and sugar mixture.

4 Mix in the flour.

5 Spread the paste thinly onto a silicone mat and bake at 180°C until evenly coloured.

6 Turn the paste over onto the hot baking sheet, remove the mat and quickly cut out the seven discs.

7 Reserve a disc for the top.

8 Build the torte by piping on thin layers of chocolate buttercream.

9 Coat the top and sides with buttercream, comb-scrape the sides and level the top.

10 Make the caramel.

11 When the caramel reaches a pale amber colour, place the final disc of biscuit on a lightly oiled marble slab and carefully pour over the caramel. Trim around the edge and quickly cut into segments using a large, oiled knife.

12 Finish by piping small bulbs of buttercream on top and set the caramel biscuit pieces at an angle.

13 The bottom edge can be finished with grated chocolate or roast chopped almonds.

Professional tips
- The biscuits are very fragile so it is advisable to make extra.
- When cutting out the biscuits and the caramel top it is essential to work quickly.
- Do not cut out the biscuits on the baking mats.

Quality points

A good quality Dobos torte will have:
- six even layers
- a good ratio of biscuit to buttercream
- a neat, light caramel top, in even-sized segments without any breaks or chips
- light, chocolate-flavoured buttercream.

Variation

For a more modern interpretation, the layers of biscuit and buttercream can be replaced with a layer each of chocolate and orange mousse on a base of almond (jaconde) sponge. The outside and top are coated with Chantilly cream.

▲ Modern Dobos torte

10 Sachertorte

This is a classic chocolate cake from Vienna. The long-fought legal battle over who had the original recipe, between the Hotel Sacher and the patisserie Demel, has helped to make this cake famous. The hotel finally won in 1950, but both establishments make their own versions.

Ingredient	1 × 20cm torte
Eggs	6
Unsalted butter, soft	125g
Caster sugar	125g
Dark chocolate, melted	125g
Plain flour	125g
Apricot jam	200g
Chocolate glaze	200ml

Energy	Cal	Fat	Sat fat	Carb	Sugar	Protein	Fibre
2404 kJ	574 kcal	31.1g	17.9g	69.8g	57.6g	8.0g	1.7g

1 Set the oven at 170°C and butter a 20cm Genoise tin. Line the base with a disc of paper, re-butter and dust with flour.

2 Separate the eggs.

3 Cream the butter with 100g of the sugar until soft and light.

4 Add the egg yolks gradually.

5 Add the chocolate gradually.

6 Sieve the flour and fold in.

7 In a separate bowl, whisk the egg whites with the remaining 25g sugar to soft peaks.

8 Fold the meringue into the cake mixture, but do not over-mix.

9 Deposit into the prepared cake tin and bake for approximately 45 minutes.

10 Test with a metal skewer.

11 Leave to cool slightly before turning out onto a cooling rack.

12 Rest overnight in an airtight container.

13 Cut in half, fill with apricot jam and assemble.

14 Boil the remaining jam, strain it and brush over the top and sides.

15 Place on a wire rack with a tray underneath, pour over the warmed chocolate glaze, and remove excess with a palette knife. Allow to set.

16 Finish by piping the word 'Sacher' on top using some of the remaining glaze.

Quality points

A good quality Sachertorte will have:

- a smooth, dark, shiny glaze evenly coating the cake
- the word 'Sacher' written on the top
- rich moist chocolate cake with one layer of apricot jam.

11 Opera

Ingredient	1 cake (12 × 15cm)
Plain biscuit jaconde (Recipe 5)	3 sheets
Rum and coffee stock syrup	150ml
Coffee buttercream	800g
Ganache	800g
Chocolate glaze	500ml

Energy	Cal	Fat	Sat fat	Carb	Sugar	Protein	Fibre
2532 kJ	606 kcal	37.0g	17.0g	57.8g	51.0g	11.7g	1.3g

1. Build the opera on a flat baking sheet covered with silicone paper.
2. Place on the first sheet of sponge, and brush with syrup.
3. Pipe on an even layer of coffee buttercream and level.
4. Place on a second layer of sponge and repeat, this time using ganache.
5. Place on a final layer of sponge, cover with paper, place on another baking sheet and gently press.
6. Chill in the fridge to set.
7. Pour over the chocolate glaze and quickly smooth with a palette knife. Leave to set.
8. If presenting as individual slices, using a large knife dipped in hot water, or a guitar (see page 418), carefully cut into 8 × 3cm rectangles.
9. Traditionally, the full-size gâteau is finished by piping on the word 'Opera'.

Variations
- Decorate with a few flecks of gold leaf.
- If you do not feel confident enough to write the word 'Opera', you can decorate with piped dots decreasing in size, spun with chocolate, or tastefully decorate with moulded chocolate decorations.

▲ Individual slices

12 Himmel cheese torte

1 Cream the butter and sugar until soft and light.

2 Sieve the flour.

3 Mix together the egg yolks and vanilla, and beat into the butter and sugar mixture.

4 Mix in the flour.

5 Divide the mixture between six buttered 20cm flan rings set on silicone mats. Level and bake at 180°C for approximately 20 minutes.

6 Transfer to cooling racks.

7 For the filling, mix together the cream cheese, sugar and vanilla (but do not overwork). Set aside.

8 Whip the cream to firm peaks, add to the cream cheese and fold in until evenly mixed.

9 Set a biscuit base on a cake card and spread with jam, then place on the second biscuit.

10 Pipe the filling in concentric circles, and fill the gaps with the fruit.

11 Invert the final biscuit layer on top and gently press down.

12 Smooth the sides and dust the top with icing sugar. Serve with a fruit coulis.

Ingredient	2 × 20cm tortes
Himmel bases	
Butter	675g
Caster sugar	340g
Soft flour	675g
Vanilla extract	½ tsp
Egg yolks	9
Filling	
Cream cheese	900g
Caster sugar	225g
Vanilla extract	½ tsp
Double cream	570ml
Raspberry jam	100g
Raspberries	400g
Icing sugar to dust	

Energy	Cal	Fat	Sat fat	Carb	Sugar	Protein	Fibre
4476 kJ	1077 kcal	84.2g	51.6g	76.3g	44.9g	8.3g	2.6g

Professional tips

● Add a couple of gelatine leaves to the filling for a slightly firmer set.

● To achieve an even spread, the biscuit base can be piped inside the flan rings.

● Transfer the biscuit bases onto a cooling rack by sliding a cake card underneath – take care, they are very fragile!

● The top can be pre-cut before placing it on; this will make it much easier to serve.

Variation

Any red berries can be used in this recipe.

13 Fruit scones

Ingredient	8 scones	20 scones
Self-raising flour	200g	500g
Baking powder	5g	12g
Salt	Pinch	Large pinch
Butter or margarine	50g	125g
Caster sugar	50g	125g
Milk or water	95ml	250ml
Sultanas, washed and dried	50g	125g

Energy	Cal	Fat	Sat fat	Carb	Sugar	Protein	Fibre
678 kJ	162 kcal	5.8 g	2.5 g	26.3 g	7.5 g	2.7 g	1.0 g

Professional tips

To produce a light scone, it is essential to mix the ingredients quickly into a soft dough, handling it quickly and lightly.

For precisely formed scones, roll out the dough to approx. 2 cm thick and cut scones with a 4–5 cm cutter.

Variations
- 50 per cent wholemeal flour may be used.
- Add other ingredients instead of sultanas: try coconut or dried cranberries.

1 Sieve the flour, baking powder and salt. Rub in the fat to achieve a sandy texture.

2 Dissolve the sugar in the liquid.

3 Make a well in the centre of the flour. Gradually incorporate the liquid into the flour, mixing lightly.

4 Mix in the sultanas.

5 Roll out into rounds, 1cm thick. Place on a greased baking sheet. Cut a cross halfway through each round with a large knife. Milkwash.

6 Bake at 200°C for 15–20 minutes.

Video: scones
http://bit.ly/1g4txwJ

14 Butter icing

Ingredient	Makes 350g
Icing sugar	150g
Butter	200g

1 Sieve the icing sugar.
2 Cream the butter and icing sugar until light and creamy.
3 Flavour and colour as required.

Variations
● Rum buttercream – add rum to flavour and blend in.
● Chocolate buttercream – add melted chocolate, sweetened or unsweetened according to taste.

15 Royal icing

Ingredient	Makes 400g
Icing sugar	400g
Pasteurised egg whites	3
Lemon, juice of	1
Glycerine	2 tsp

Professional tip
An alternative is to use commercial egg white substitute or dried egg whites. Always follow the manufacturer's instructions for quantities.

1 Sift the icing sugar. Mix it well with the egg whites in a basin, using a wooden spoon.
2 Add a few drops of lemon juice and glycerine and beat until stiff.

16 Chocolate glaze

An example of a product finished with this glaze is Opera (Recipe 11).

Ingredient	Makes 600ml
Double cream	188g
Water	175g
Caster sugar	225g
Cocoa powder	75g
Gelatine, soaked in cold water	4½ leaves (10g)

1 Bring the cream, water, sugar and cocoa to the boil slowly in a heavy-bottomed saucepan.
2 Simmer for 2 to 3 minutes, then remove from the heat.
3 Drain the gelatine and add it to the mixture. Stir until dissolved.
4 Pass, then cool by stirring over ice.
5 Store in a plastic container with cling film pressed directly onto the surface.
6 If using to glaze a frozen product, warm the glaze in the microwave before use.

17 Crémeux

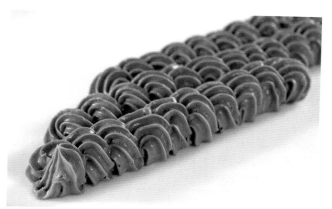

Ingredient	15 portions
Base anglaise	
Whipping cream	200ml
Milk	200ml
Egg yolk	80g
Sugar	40g
Crémeux	
Base anglaise (see above)	500g
Plain couverture	220g

1 Boil the milk and cream and add the mixture of sugar and yolks. Cook to 84°C. Strain and weigh 500g.

2 Part melt the couverture and slowly pour onto the warm anglaise.

3 Emulsify with a stick blender.

4 Refrigerate.

5 When firm, beat to a piping consistency.

Test yourself

1 List the ingredients used to make each of the following:
   (a) dacquoise
   (b) biscuit jaconde.

2 What are the differences between mechanical and chemical aeration? Give examples of products which use each technique.

3 Describe the split-egg method of making sponges. Name two products made in this way.

4 As well as sweetness, describe the other contribution sugar makes to the development of cakes and sponges.

5 List the components that make up 'baking powder' and give their ratios.

6 Give four examples of why the pre-preparation of commodities and equipment is so important when producing cakes and sponges.

7 List three reasons why a Genoise sponge might be low in volume with a heavy texture.

16 Decorative items

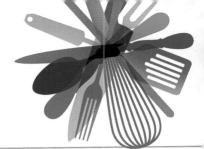

This chapter covers the **VRQ** unit:
→ Produce display pieces and decorative items.

Introduction

Decorative items help to provide the excitement and 'wow' factor that impress diners and other guests and visitors. Just as a chef in the kitchen garnishes dishes to make them attractive, a pastry chef can use a range of commodities and techniques to make their dishes and products visually striking.

This chapter covers decorative items and products using the four mediums of chocolate, sugar, pastillage and marzipan. Chocolate, along with certain products made with sugar, are the most frequently used in the industry today. A skilled pastry chef should be comfortable tempering couverture and producing items such as chocolate motifs and transfers, as well as lining moulds and making garnishes such as chocolate cigarettes or swirls using acetate. Equally, it is important for a pastry chef to work confidently with sugar, utilising the various properties and products that can be achieved from sugar cooked to varying temperatures.

Display pieces, for ornamental purposes rather than consumption, are not as frequently seen in current times as they were in the past. The modern business has to operate very efficiently to compete and survive, and display pieces, although highly impressive, are expensive to produce in terms of labour and materials. Furthermore, display pieces do not make any direct return through sales as they are not produced to be eaten. Therefore, businesses tend not to invest their resources (staff and materials) in the production of such creations. However, the skills used in the production of display pieces are highly transferable to other work in the pastry kitchen and it is important that they are retained for the development of tomorrow's pastry chefs.

Learning objectives

By the end of this chapter you should be able to:
→ Produce and finish display pieces and decorative items to design specifications, using traditional, classical and modern construction techniques, skills and styles.
→ Carry out quality checks during production, and correct pieces that do not meet quality requirements.
→ Explain techniques for the production of display pieces and decorative items.
→ Design display pieces and decorative items.
→ Explain key design considerations.
→ Describe how to control time, temperature and environment to achieve the desired outcome in display pieces and decorative items.

Recipes included in this chapter

Chocolate

Translated, the French word 'couverture' means 'covering' or 'coating' in English. Coverture has a very high percentage of cocoa butter (at least 30 per cent) and is used to flavour patisserie products such as ice creams, mousses, ganaches, soufflés, etc. For flavouring purposes, couverture only requires melting and adding to the desired product. For moulding and setting purposes, the couverture needs to be 'tempered'. This is a process whereby the chocolate is taken through different temperatures to stabilise one particular chocolate crystal known as the 'beta crystal'. This crystal has all the characteristics of good tempered chocolate (snap, shine and retraction), enabling the prepared couverture to literally fall out of the mould it has been set in, giving a solid, shiny piece of chocolate with a perfect snap.

Types of couverture

Couverture comes in dark chocolate, milk chocolate, and white chocolate varieties. You can also purchase coloured and flavoured couverture. The purchasing unit is either callets or a solid block. For tempering purposes, callets are preferred as they melt uniformly, making tempering of the chocolate more effective.

White couverture is chocolate which does not contain the dark-coloured cocoa solids derived from cocoa beans. It only contains 30 per cent cocoa butter (the fatty substance derived from cocoa beans), milk and sugar. White chocolate is sweet, with a slight vanilla taste, and has a light flavour which is not too heavy or intense.

Milk couverture has added dried milk powder, along with cocoa butter, 40 per cent sweeteners and flavourings; it contains a minimum of 10 per cent chocolate liquor and 12 per cent milk solids.

Plain couverture has a higher content of cocoa butter (60–70 per cent), which gives the chocolate more viscosity, and cocoa solids, which give the chocolate its colour. It is more fluid than the white and milk varieties and is used for decorations, moulding, enrobing and flavouring.

All three chocolates are factory-tempered before being packaged. (For small quantities it is possible to melt the chocolate while keeping the tempering qualities in the chocolate. This will be discussed in more detail under tempering of chocolate.)

Compound chocolate is used in food manufacturing. It gives a crisp, hard coat. It may also contain vegetable oil, hydrogenated fats, coconut and/or palm oil, and sometimes artificial chocolate flavouring. This type of chocolate does not need to be tempered in order to set. It is inferior to couverture in taste and quality, but less expensive.

Cocoa

Cocoa bean

This was once called a 'cocoa almond' or 'cocoa grain'. It is the seed that is found in the pods of cacao trees. After being treated, it is packed and sent to be sold on the international market. It is from this bean that cocoa butter, chocolate liquor, cocoa powder and cocoa nibs are extracted.

Cocoa nibs

These are roasted, shelled cocoa beans broken into small pieces. This is a very interesting product with an intense flavour – 100 per cent cocoa. It gives aroma, flavour and texture to many preparations, like sponge cakes, chocolate pralines, muffins, ice creams, cookies and cake decorations. Care should be taken not to use excess quantities so that the balance with the other ingredients is not upset.

Chocolate liquor

This is a smooth, liquid paste. In addition to being the base for other cocoa derivatives, such as cocoa butter or cocoa powder, it can be used in all types of desserts and cakes (toffee, for example). One of its main characteristics is that it contains no sugar, which gives it a slightly bitter flavour in its pure state.

Cocoa butter

Chocolate liquor is pressed to extract the fat (cocoa butter) and separate it from the dry extract. Cocoa butter is the 'spine' of chocolate, since its proper crystallisation determines whether chocolates (couvertures) have adequate densities and melting points. We would recommend melting cocoa butter at 55°C (it begins melting at 35°C) to achieve proper decrystallisation.

Cocoa butter is used to coat with a spray gun (mixed with chocolate in greater or lesser quantity), for chocolate praline moulds, desserts, cakes and artistic pieces, or in pure form for moulds and marzipan figurines. Cocoa butter is available in various colours for decoration purposes.

Mycryo cocoa butter is a commercial product for use in tempering chocolate. It can save time. To use, follow the manufacturer's instructions.

Cocoa powder

Two products are extracted from pressed chocolate liquor – cocoa butter in liquid form, and dry matter, which is ground and refined to make cocoa powder. The quality of cocoa powder is a function of its finesse, its fat content, the quantity of impurities it contains, its colour and its flavour. It is very important to store it in a dry place and in an airtight container.

Roasting
After being cleaned, the cocoa beans are roasted which develops the distinctive flavour of the cocoa bean.

↓

Winnowing
After roasting, the beans are put through a winnowing machine which removes the outer husks or shells, leaving behind the roasted beans, now called nibs.

↓

Milling – making cocoa liquor
The nibs are then ground into a thick liquid called chocolate liquor (this is cocoa solids suspended in cocoa butter). Despite its name, chocolate liquor contains no alcohol and has a strong unsweetened taste.

↓

Pressing for the production of cocoa powder and cocoa butter
The next stage is to press the cocoa liquor and extract the cocoa butter. This leaves behind a solid mass which is then processed into cocoa powder.

↓

Making the chocolate
The following ingredients are mixed together to make the three different types of chocolate:

White chocolate: made from the same ingredients as milk chocolate (cocoa butter, milk, sugar) but without the chocolate liquor. White chocolate must contain at least 20% cocoa butter and 14% total milk ingredients.

Milk chocolate: a combination of chocolate liquor, cocoa butter, sugar and milk or cream.

Plain chocolate: a combination of chocolate liquor, cocoa butter and sugar. Must contain at least 35% chocolate liquor.

↓

Refining
The next step is to pass the chocolate through heavy rollers to form a fine flake. Additional cocoa butter is added at this stage and an emulsifying agent called lecithin. The mixture is now mixed to a paste.

↓

Conching
The process of conching kneads the chocolate through heavy rollers which develops the flavour.

↓

Tempering
The chocolate is now tempered ready for use which gives it shine, snap and retraction.

↓

Moulding
The liquid tempered chocolate is now deposited into solid block moulds or shaped into callets ready for use.

↓

Use in the patisserie kitchen
For small quantities, melt the couverture in the microwave following the technique for microwave tempering, or fully melt to the stated temperature for the type of chocolate and then follow the tempering procedure (see below).

▲ The process of manufacturing chocolate

Tempering chocolate

As already mentioned, cocoa butter is a vital component of chocolate, since the final result depends on its crystallisation. It determines good hardness, balance, texture and shine, and it prevents excessive hardening, whitening and the formation of beads of oil on the surface.

When we melt chocolate, the cocoa butter melts and its particles separate. To achieve a perfect result, we must re-bond them by cooling the chocolate (recrystallising the cocoa butter).

Tempering allows us to manipulate chocolate and combine it with other ingredients or make artistic pieces that, when recrystallised, regain the texture and consistency of the chocolate before it was melted.

Tempering is necessary because of the high proportion of cocoa butter and other fats in the chocolate. It stabilises the fats in the chocolate to give a crisp, glossy finish when dry.

▲ This couverture was tempered and spread over textured sheets to give the desired finish

It is essential that a thermometer is used for the tempering process.

The melting and working temperatures are given in Table 16.1 as a guideline. Some brands of chocolate may vary. Always check the tempering instructions on the packaging.

Table 16.1 Temperatures for tempering the three types of couverture

Initial melted temperature	Finished working temperature
Plain couverture 45°C	31°C–32°C
Milk couverture 40°C	30°C–31°C
White couverture 40°C	28°C–30°C

Tempering: Table-top method

1 Melt carefully to the specific temperature for the type of couverture, avoiding steam, moisture and over-heating. (The use of a chocolate melting tank is ideal for this.)
2 Once the couverture has reached the melting temperature, remove from the heat source.
3 Pour 70 per cent of it onto a very clean and dry marble surface/slab. Work continuously by spreading outwards and pulling back to the centre with a step palette knife until the couverture starts to thicken and good beta crystals form.

4 Quickly add the couverture back to the remaining 30 per cent, stirring continuously, dispersing and seeding the beta crystals into the liquid chocolate until it reaches its finished working temperature (see Table 16.1).
5 Check the finished temperature with a digital probe. If the chocolate is still too warm, pour a small amount once more onto the marble and repeat the steps above until the chocolate reaches the desired temperature.

Tempering: Injection (or seeding) method

1 Take 30 per cent of the couverture being tempered.
2 Melt the remaining 70 per cent of the couverture following Steps 1 and 2 of the table-top method above.
3 Remove the container of melted chocolate and stand it on the table top with a folded cloth underneath (this is to prevent the chocolate from setting on the base).

4 Gradually add the remaining 30 per cent of the chocolate callets to the melted couverture, stirring continuously until the finished working temperature is achieved (see Table 16.1).

For large-scale tempering using the injection method, a wheel tempering tank is used.

Tempering: Microwave method (for small quantities)

As previously discussed, couverture is packaged already tempered.
1 Place 500g of couverture into a plastic bowl.
2 Warm the chocolate in short intervals in a microwave oven set at 50 per cent, until the chocolate partially melts but there are still signs of solid chocolate.

3 Remove from the microwave and continue to stir the chocolate until the solid chocolate pieces melt down. Any solid pieces are still tempered and the gentle mixing and melting will seed the newly melted chocolate.

Professional tips

- If the working temperature is exceeded by more than 3°C, the process will have to be repeated as the couverture will not be correctly tempered and faults will occur.
- If the temperature of the chocolate drops during processing, it can be gently reheated to the working temperature with a heat gun, without any detrimental effects to the characteristics of the chocolate.
- Always keep water away from melted couverture and never store in humid conditions.
- The ideal room temperature for working with chocolate is 18°C with 60 per cent humidity.
- Chocolate products should be stored in a dry place at 15–16°C and at 20 per cent humidity.
- Chocolate absorbs all odours and should therefore be stored well covered.

- The higher its fat content, the faster chocolate melts in your mouth.
- In tempering, it is essential to check the temperature with a thermometer and to perform the 'paper test'. This is done by dipping a piece of paper in the tempered chocolate. The tempering is optimal if, in about 2 minutes, it has crystallised with a flawless, uniform shine and without stains or fat drops on the surface.
- A glossy surface is a sign of good tempering.
- Two important points are to be followed to achieve good tempered couverture – correct temperature and continuous movement of the chocolate. Movement develops beta crystals in the chocolate.

Faults

The two most common faults in prepared couverture are fat bloom and sugar bloom.

The first picture shows fat bloom. This is caused by:
- poor tempering of the chocolate
- incorrect cooling methods
- covering a confectionery product that is too cold
- warm storage conditions.

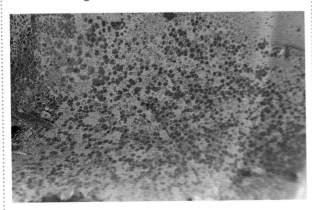

The second picture shows sugar bloom. This is caused by:
- storage of chocolate in damp conditions
- working in humid conditions
- using hygroscopic ingredients
- products which have high liquid content packaged and stored in a warm area. (Vapour is given off which is entrapped in the packaging, creating a layer of moisture on the surface of the chocolate.)

Sugar work

Boiled sugar

Sugar is boiled for a number of purposes – pastry work, bakery and sweet-making.

Soaked sugar (approximately 125ml water per 250g sugar) is boiled steadily without being stirred. Any impurities on the surface should be carefully removed (otherwise the sugar is liable to crystallise). Once the water has evaporated, the sugar begins to cook – you will notice that the bubbling in the pan will get slower. It is then necessary to keep the sides of the pan free from crystallised sugar – this can be done with a wet pastry brush. The brush should be dipped in ice water or cold water, rubbed round the inside of the pan and then quickly dipped back into the water.

Health and safety

Care should always be taken when boiling sugar due to the very high temperature of the boiling liquid.

The cooking of the sugar then passes through several stages, which may be tested with a sugar thermometer or by the hand-testing method (dip the fingers into ice water, then into the sugar and quickly back into the ice water).

Degrees of cooking sugar

- **Thread (104°C)** – when a drop of sugar held between thumb and forefinger forms small threads when the finger and thumb are drawn apart.
- **Pearl (110°C)** – when testing as for thread, the threads are more numerous and stronger. Used for crystallising fruits, making fruit liqueur and some icings.
- **Soufflé/blown (113°C)** – if a metal loop is inserted into the sugar and blown, a small, thin, aerated ball can be formed.
- **Feather (115°C)** – if the skimmer (used for cleaning the sugar) is dipped into the sugar solution and then given a sudden jerk as if to throw the sugar away from you, long, thin, fine strings are formed. Used for fruit confit.
- **Soft ball (118°C)** – when testing as for thread, the sugar rolls into a soft ball. Used in the production of fondant, fudge, pralines, pâte à bombe and peppermint creams.
- **Hard ball (121°C)** – as for soft ball, but the sugar rolls into a firmer ball. Used for products such as Italian meringue, boiled buttercream, nougat and marshmallows.
- **Soft crack (140°C)** – the sugar lying on the finger peels off in the form of a thin, pliable film, which sticks to the teeth when chewed.
- **Hard crack (155–160°C)** – the sugar taken from the end of the fingers, when chewed, breaks cleanly between the teeth, like glass. Used for dipping fruits and the production of poured, pulled and spun sugar.
- **Caramel (176°C)** – cooking is continued until the sugar is a golden-brown colour. Used for crème caramel, caramel sauce, nougatine, croquant and praline.

Inversion or 'cutting the grain'

Monosaccharides are single sugars, such as fructose and glucose. They are more stable and there is less chance of premature crystallisation during the boiling process.

Disaccharides are double sugars, such as granulated sugar (sucrose), which are made from two single sugars (fructose and glucose) which crystallise easily during sugar boiling.

As granulated sugar is a key ingredient when sugar boiling and is likely to crystallise, acid is sometimes added in the form of lemon juice or tartaric acid. This breaks down double sugars into single sugars, making the solution more stable and less likely to crystallise. This process is known as 'inversion' or 'cutting the grain'.

Professional tips

- The purpose of glucose in sugar boiling is to add single sugars, making the solution more stable.
- The addition of acid will make the sugar elastic for pulling.

Faults

The sugar in the jar has crystallised

Causes of premature crystallisation:
- working with dirty equipment
- working with dirty sugar
- intermittent boiling of the sugar solution
- stirring the sugar solution once boiling starts
- not removing sugar crystals from the sides of the pan; they will crystallise and cause the main sugar solution to do the same
- incorrect dipping procedure for items such as fruits déguisés.

Sugar preparations

Table 16.2 shows a list of sugar preparations and the equivalent term in French.

Table 16.2 Sugar preparations

Poured sugar	Sucre coulé
Spun sugar	Sucre filé
Pulled sugar	Sucre tiré
Blown sugar	Sucre soufflé
Straw sugar	Sucre paille
Rock sugar	Sucre roche

There is a range of commercial products on the market that greatly assist the pastry chef in the production of specialised sugar work – isomalt, for example. This product is not as hygroscopic as normal sugar solutions, thus enabling finished goods to be stored relatively easily. It can be used several times over and has a long shelf life. Such commercial products are simple to use, quick and labour-saving.

Pulling sugar

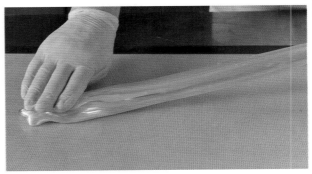

1 Pull sugar under a lamp to keep it at the right consistency.

2 Reheat poured sugar slowly, until pliable, and then begin to pull it.

3 Pull and fold repeatedly until the sugar is shiny, lighter in colour and less opaque.

4 Form into shape as required.

Professional tips

- Never attempt to cook sugar in a damp atmosphere, when the humidity is high. The sugar will absorb water from the air and this will render it impossible to handle.
- Work in clean conditions, as any dirt or grease can adversely affect the sugar.
- The choice of equipment is important – copper sugar boilers are ideal as these conduct heat rapidly. Induction hobs are used as they cook the sugar very fast with no naked flames (which might crystallise the sugar on the sides of the pan).
- Never use wooden implements for working with or stirring sugar. Wood absorbs grease, which can ruin the sugar.
- Make sure the sugar is cooked to temperature according to the specific purpose of the product being made.
- If you are colouring sugar, it is advisable to use powdered food colourings as they tend to be brighter than liquid ones. Before use, dilute the powder with a few drops of 90 per cent-proof alcohol or water. Add the colourings to the boiling sugar when it reaches 140°C and then continue to cook to the desired temperature. For poured sugar with a transparent effect, add the colour while the sugar is cooking.

- Once the sugar is poured onto a silicone mat and it becomes pliable, it should be transferred to a special, very thick and heat-resistant plastic sheet.
- To keep the sugar pliable, it should be kept under infra-red or radiant heat lamps.
- For a good result with poured sugar, use a small gas jet to eliminate any air bubbles while you pour it.
- Ten per cent calcium carbonate (chalk) may be added to sugar before pouring to give an opaque effect and to improve its shelf life. This should be added at a pre-mixed ratio of 2 parts water to 1 part calcium carbonate and added to the boiled sugar solution at 140°C. The sugar is then further boiled to 160°C.
- To store completed sugar work, place in airtight containers, the bottom of which should be lined with a dehydrating compound, such as silica gel, carbide or quicklime. Pulled sugar pieces can be vacuum-sealed for storage.
- If you are using a weak acid, such as lemon juice or cream of tartar, to prevent crystal formation, it is advisable to add a small amount of acid towards the end of the cooking. Too much acid will over-invert the sugar, producing a sticky, unworkable product.
- The best sugar to use is granulated sugar, straight from a 1 kilo bag.

Marzipan (almond paste)

There are two methods of producing marzipan: boiling or cooking (see Recipe 19), and raw (uncooked). For cooked marzipan, sugar and water are boiled to 116°C and mixed with ground almonds and egg yolk. The mixture is then worked on a table top lightly dusted with icing sugar until it is smooth. Raw marzipan is made by mixing together ground almonds and icing sugar and using egg to bind the ingredients together. It is worked on a table top lightly dusted with icing sugar to form a smooth paste.

Marzipan can be purchased ready-made, as it is a consistent product. There are various types on the market, such as that suitable for diabetics, organic and specialist modelling marzipan.

Almonds are the key ingredient in marzipan and are grown principally in California and Spain, with small quantities coming from other countries around the Mediterranean. Mediterranean almonds are generally considered to be superior to Californian because they are cultivated in more natural and wilder surroundings with more favourable climatic conditions.

High-grade marzipan generally has a lower sugar content (around 35 per cent), whereas marzipans used for figures or decorations will have a higher sugar content (up to 70 per cent) to make them more suitable for use as a modelling mass.

> **Professional tip**
>
> Due to the high sugar content in marzipan, it dries very quickly when exposed to the air and should be kept covered at all times, during processing and when being stored.

Marzipan is used extensively in pastry work. It can be rolled out like pastry (but using icing sugar instead of flour) to cover cakes (for example, simnel cake, which is traditionally eaten at Easter). It can be left smooth or textured using a marzipan roller (as in the photo of a strawberry gâteau in Chapter 15). Other uses include the filling of fruits déguisés, or in fondant dips (Chapter 12, Recipe 14). Marzipan with a high almond percentage can be used as a base for almond tuiles, lightly toasted shapes used as petit fours and for modelling.

Pastillage (gum paste)

Pastillage is a white paste that is made from icing sugar, cornflour, lemon juice and gelatine or gum tragacanth (see Recipe 18), which makes the pastillage pliable. The paste is rolled out very thinly, dusting the working surface using cornflour in a muslin bag, and then cut into various shapes using templates of set designs. It is then left to dry until solid on flat boards, turning over periodically. Once fully dry, the pastillage is stuck together using royal icing into the shape of the template – square boxes, heart-shaped caskets, etc. Centrepieces can be made by combining pastillage cut-out shapes with pulled sugar. As the pastillage is set firm, it can only be used for decorative purposes.

Test yourself

1 Give four reasons why a sugar solution may crystallise.
2 Briefly describe three methods of tempering chocolate.
3 What is meant by the phrase 'sugar is hygroscopic'?
4 From the principles that have been covered in this chapter, design a chocolate centrepiece, drawing it to scale and incorporating five different chocolate techniques.
5 Briefly explain the term 'inversion', as applied to sugar solutions.
6 State four points to consider before preparing a pastillage centrepiece.

1 Bubble sugar

Ingredient	Makes 750g
Water	100ml
Fondant	450g
Glucose	300g
Unsalted butter	20g
Colouring	

1 Heat the water, fondant and glucose to 150°C.
2 Remove the thermometer. Add the butter and colouring. Swirl to mix the ingredients.
4 Pour onto a silicone mat and leave to set.
5 Break into pieces. The pieces can then be stored in an airtight container until needed.
6 When needed, blitz the pieces of sugar in a mixer.
7 Sieve them over a silicone mat in a slightly uneven layer.
8 Melt in the oven at 165°C.
9 Allow to set. The sugar sets very thin, with bubbles throughout.
10 Store in an airtight container and keep dry.

Use bubble sugar to decorate sweet dishes.

2 Dried fruits

1 Boil the water, sugar, lemon juice and glucose together to make a syrup.
2 Prepare the fruit. Leave the skin on. Remove the cores from apples and pears. Slice the fruit very thinly using a meat slicer. Brush slices of apple or pear with lemon juice to prevent browning.
3 Pass each slice through the hot syrup. Lay them on silicone mats.
4 Leave to dry in the oven overnight, or use a commercial dehydrating cabinet.

Water	500ml
Granulated sugar	300ml
Lemon juice	25ml
Glucose	50g
Fruit (lemons, limes, oranges, apples, pears)	

3 Spun sugar

Granulated sugar	500g
Water to saturate	
Liquid glucose	60g

1 Place the sugar into a sugar boiler.

2 Add water until the sugar is just saturated. Stir with a metal spoon to distribute the water, ensuring that all the sugar is moistened.

3 Gently dissolve the sugar, removing any scum as it rises to the surface.

4 Add the liquid glucose and boil to 160°C.

5 Once the temperature is reached, arrest the cooking by placing the pan into a bowl of cold water.

6 Remove the sugar pan from the cold water and leave until the boiled sugar syrup slightly thickens.

7 Using a sawn-off whisk, spin the sugar over a lightly oiled steel, forming thin strands of brittle sugar.

Professional tip

The fondant used in Recipe 4 can also be used to make spun sugar.

4 Nougatine sheets

Nougatine can be used to decorate plated desserts and tortes.

Ingredient	Makes 1.65 kg
Caster sugar	500g
Fondant	500g
Glucose	500g
Flaked almonds	150g

1 Cook the fondant, sugar and glucose to a blond caramel colour.

2 Roast the flaked almonds on a baking tray in the oven and mix into the sugar when cooked.

3 Pour onto a non-stick mat and cool down until cold.

4 Crush in a food processor until it becomes a sandy texture, pass through a sieve and keep in a sealed container until required.

5 When required, place a plastic cut-out template onto a non-stick mat.

6 Dust over the template with the nut powder, remove the template.

7 Bake at 180°C for approximately 10 minutes.

8 When cold and firm, remove from the mat.

5 Piped sugar spirals

Ingredient	Makes approx. 50
Water	100ml
Fondant	500g

1 Place the water in a copper sugar boiler. Add the fondant.
2 Cook until a pale caramel forms.
3 Allow to stand for 5 minutes.
4 Pipe onto a silicone mat in spirals.
5 Allow to set.
6 Warm the spirals and pull them up to the desired height.

6 Transfer-sheet chocolate swirls

1 Thinly spread tempered couverture onto the reverse side of a transfer sheet.
2 Allow to partially set, then mark with a thin-bladed knife, avoiding cutting through the plastic sheet. (A ruler or pastry cutter may be used to create various shapes.)
3 Place a sheet of silicone paper on top of the chocolate.
4 Roll up and allow to set.
5 Unfold the plastic sheet to release the chocolate swirls.

Spread tempered couverture over the acetate transfer sheet

When partially set, mark with a knife, but do not cut through the acetate

Cover with silicone paper, roll up and leave to set

7 Chocolate cigarettes

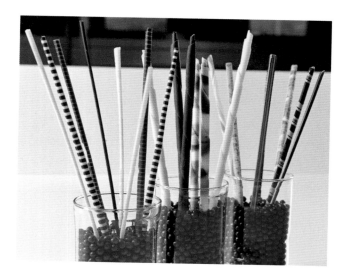

White or plain cigarettes

1 Spread either white or plain tempered couverture thinly onto a marble slab.
2 When almost set, scrape up into cigarettes.

Two-tone coloured cigarettes

1 Spread tempered white couverture onto a marble slab and comb using a grooved scraper, or the plastic-toothed edge from a cling film box.
2 When the chocolate has almost set, cover with a thin coat of tempered plain couverture.
3 Once the plain couverture has set, scrape up into thin two-tone cigarettes.

Multi-coloured cigarettes

1 Using crinkled cling film, dab different coloured tempered cocoa butter onto a marble slab.
2 Thinly spread with tempered white couverture.
3 Once set, scrape up into white cigarettes with a multi-coloured outer surface.

Professional tips
- It is advisable to use a special scraper for this task.
- Never attempt to make these cigarettes in a warm kitchen.

Spread tempered white couverture on marble, and comb through it

When the white couverture has almost set, pour tempered plain couverture over it

Spread the plain couverture thinly over the white

Clear away excess couverture from the edges

Working at the edge of the chocolate, scrape up thin cigarettes

8 Chocolate for spraying

Plain couverture	200g
Cocoa butter (melted red cocoa butter may be added to give a richer chocolate effect)	200g

1 Melt the couverture and cocoa butter together.

2 Allow to cool to 35°C.

3 Use a spray gun to spray chocolate onto a centrepiece or dessert. To achieve a smooth, matt finish, spray at room temperature. For a velvet finish, the centrepiece must be cold: refrigerate for half an hour before spraying.

Professional tips

- Always ensure that the spray gun is warm before putting the chocolate mixture in, otherwise the cocoa butter will set inside the chamber of the gun as it is being sprayed.
- Never clean the spray gun with water; simply spray vegetable oil through the gun until the oil becomes clear.

▲ Spraying chocolate to decorate a frozen dessert, giving a velvet finish

9 Chocolate teardrops

1 Place acetate strips onto a flat surface.

2 Embellish the strips using coloured cocoa butter, or tempered white, milk or plain couverture. Allow to set.

3 Thinly spread the acetate strip with tempered couverture.

4 When the couverture is almost set, join both ends of the acetate together and secure with a paper clip.

5 Stand inside a metal ring and shape into a teardrop.

6 Once set, remove the acetate and fill the teardrop shell with mousses.

Variation

To make tubes, follow the same process as the teardrop but this time join the acetate to make a tube, and stand inside a metal ring until set.

10 Sugar boiling

Methods 1 to 3 can be used for pulled sugar, blown sugar, ribbons, etc.

Method 1

Water	350ml
Granulated sugar	1 kg
Cream of tartar	1–2g
Glucose	50g

1 Place the water, sugar and cream of tartar in a pan and stir on a low heat. Do not boil. Before the mixture reaches boiling point, skim using a tea strainer dipped into water, and wash the sides of the pan. (This process will take approximately 20 minutes.)

2 Once the sugar is clean and still simmering, warm the glucose in the microwave and add to the sugar solution. Bring to the boil.

3 Boil for one minute, then pour into a kiln jar almost to the top. Store until required (for up to three months).

4 Cook 1 kg at a time to 165°C. Skim and colour at any stage. (Mix the colour with water and warm in the microwave to bring together.)

5 Pour onto silicone paper, cool and break up. Store with a de-humidifying product.

6 When required, soften either in a microwave or under a lamp and then pull to obtain a glossy shine.

7 Place under a heated lamp and process as required.

Method 2

Granulated sugar	1 kg
Water	500g
Glucose	200g
Tartaric acid, prepared (see below)	

1 Bring the water and sugar to a simmer. Clean as in the previous method.

2 Add the glucose.

3 Boil to 165°C, then add 12–24 drops of tartaric acid, depending on the elasticity you need, using a pipette.

4 Pour onto a silicone mat and pull until a glossy sheen is achieved.

5 Place under a heated lamp and process.

To prepare tartaric acid for this recipe, mix together equal quantities of tartaric acid and water. Heat in the microwave until the mixture becomes clear. Keep in a pipette bottle.

Method 3

Glucose	400g
Fondant	600g

1 Warm the glucose in a sugar boiler.

2 Add the fondant, cut into chunks.

3 Heat until the liquid clears.

4 Boil to 165°C.

5 Finish as in previous methods.

Sugar syrups for soaking sponges

Base syrup	
Water	1 kg
Sugar	1.4 kg

To make a liqueur-flavoured syrup, for every 1 kg of this base syrup, add 400g of the liqueur (for example, rum, kirsch, Grand Marnier or Curaçao).

Variation: Vanilla syrup

Water	1.25 kg
Sugar	800g
Vanilla pods, seeds from	5

Faults in sugar preparations

Poured sugar not setting:
- sugar insufficiently boiled
- sugar thermometer giving inaccurate reading.

Uneven texture of spun sugar:
- not allowing the sugar solution to thicken slightly before spinning.

Milky-coloured pulled sugar:
- too much acid used, over-inverted
- pulling the sugar too soon.

Sugar cracking when poured sugar pieces are assembled using a blowtorch:
- sugar bases are too cold.

11 Nougatine (croquant)

Ingredient	Makes 1 kg
Granulated sugar	500g
Water	200ml
Glucose	100g
Flaked almonds	375g

1 Boil the sugar, water and glucose to a light caramel.
2 Remove from the heat and immediately stir in the almonds.
3 Roll out between silicone mats on a baking sheet.
4 Cut out into shapes as required.

12 Praline

Ingredient	Makes 1 kg
Granulated sugar	500g
Water	200ml
Glucose	100g
Whole almonds	150g
Hazelnuts	150g

1 Boil the sugar, water and glucose to a light caramel.
2 Remove from the heat and immediately stir in the nuts.
3 Pour the mixture on a non-stick mat and cool until solid.
4 Break down to a fine powder using a food processor.

13 Blown sugar swan

1 Warm pulled sugar using either isomalt or traditional sugar solution.

2 Shape the sugar into a smooth ball and make an indentation using your finger.

3 Insert a warm copper tube into the indentation and press the outer sugar to seal the pipe inside the ball.

4 Apply pressure with a sugar pump, shaping the sugar into a swan.

5 Cool using a hair drier on a cold setting until the sugar has set.

6 To remove the copper tube once the sugar has set into the shape of a swan, gently warm the tube 6cm away from the sugar figure and gently twist to release.

7 Attach the beak made from pulled yellow and black sugar.

8 Make the wings by pulling white sugar and shaping in a leaf mould. Warm on the blow torch and attach to the body of the swan.

9 Stand on a poured sugar tin foil base.

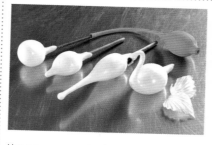

Use a sugar pump to shape a ball of warm pulled sugar into the swan. This picture shows the shapes that form during this process. Make the wings over a leaf mould.

Release the swan from the pump and trim to shape using a hot knife

Warm each part with the blow torch so that it can be attached. Start by attaching the swan's body to a base.

Attach the wings

Attach the tail

14 Meringue sticks

1 Make a cold meringue with the egg white, sugar and cream of tartar.

2 Using a plain nozzle, pipe out in straight lines 4cm in diameter onto a non-stick mat.

3 Sprinkle with cocoa nibs.

4 Place in a dehydrator or an oven at 80°C for 2 hours until fully dry and crisp.

Professional tip

The meringue can also be spread out thinly onto a non-stick mat and dried as above. For decoration, break into irregular shapes.

Variation

For raspberry meringue, add 50g raspberry purée and red colour at the full meringue stage. Decorate with crushed dried raspberries before cooking.

Ingredient	Makes 12–16
Egg white	100g
Caster sugar	200g
Cream of tartar	Pinch
Cocoa nibs	

15 Poured clear sugar centrepiece

Water	500ml
Granulated cane sugar	1 kg
Glucose	200g
Soluble strong powdered colours, pre-mixed (if required)	Few drops

1 Clean out a copper sugar boiler with salt and lemon. Rinse but do not wipe dry.

2 Place the water in the pan, then the granulated sugar, then the glucose.

3 Bring to the boil slowly. Stir carefully – do not scrape the bottom or sides.

4 Skim off impurities.

5 Cook on a fast boil. Place a thermometer in the pan.

7 If coloured sugar is required, when the temperature of the sugar reaches 150°C, add a few drops of colour solution.

8 Cook until the temperature reaches between 155°C and 160°C. Remove from the heat and arrest cooking by placing the pan in a bowl of cold water. Allow to stand.

9 Pour the sugar out into moulds. Always pour into the centre of the mould in a steady stream.

10 Allow to set completely before trying to move or touch the sugar.

11 To assemble a shape from moulded pieces, heat the edge of each piece with a spirit burner to melt the sugar slightly. Hold it in place and it will harden, welding the two pieces together.

12 To finish, blast with cold air from a hairdryer to quickly set the joins.

Pour the sugar into the moulds

Once set, dip the edge of each piece in hot sugar solution (or heat it with a burner) and then stick it into place

16 Pulled sugar using isomalt

1 Bring the isomalt and water to the boil. Whisk to make sure that the isomalt dissolves completely.

2 Cook until the temperature reaches 165°C (not more than 20 minutes, or it will get too hot).

3 Plunge the pan into cold water to arrest the cooking.

4 Pour out and allow to set.

5 Place the sugar on a board under a lamp and allow it to reheat slowly. When it starts to run, turn it. When the sugar becomes pliable, it is ready to be pulled.

6 Pull and fold the sugar evenly, 20 to 30 times, until it is completely opaque. Use the lamp to keep it at the right consistency.

7 After this point, if the sugar gets cold and hard, it can be brought back by microwaving on a low heat setting in 8- to 10-second bursts.

8 Form into shape as required.

| Isomalt | 1 kg |
| Water | 100ml |

Note

Isomalt is sugar rearranged with hydrogen. It is less susceptible to moisture than sugar. It was developed as a sweetener for diabetics, but unfortunately it acts as a laxative.

17 Pulled sugar ribbon

1 Align two or three strips of malleable pulled sugar in different colours, under a heated lamp.

2 Gently pull and stretch until double in length.

3 Cut the strip in half and join side by side (thus giving six strips if you started with three).

4 Stretch again, keeping the sugar the same thickness along the strip.

5 Continue to stretch until you have a thin, shiny, multi-coloured ribbon.

6 Cut into lengths with a hot knife, and heat to bend into shape.

18 Pastillage

1 Separate the gelatine leaves and soak them in iced water.

2 Sieve the icing sugar onto paper twice.

3 Drain the gelatine and squeeze out the water. Add the gelatine to the lemon juice and warm to dissolve.

4 Place approximately half the icing sugar in a clean bowl, make a well in the centre and add the dissolved gelatine and lemon juice. Mix to a smooth paste.

5 Gradually add the rest of the icing sugar and work/knead to obtain a smooth, firm paste.

6 Place the paste in a clean plastic bag and cover with a damp cloth.

7 Allow to rest for 20 minutes before using.

8 Roll out as thinly as possible on a smooth surface.

9 Lay on a dusted board. Cut out around a template, using a thin, sharp-bladed knife.

10 Leave the pieces to dry overnight, turning them over after the first 2–3 hours.

11 Assemble the pieces into a centrepiece using royal icing as the adhesive.

Ingredient	Makes 500g
Gelatine	2½ leaves
Icing sugar	450g
Lemon juice	25g

Professional tips

● Keep the paste covered as much as possible during processing, to prevent it drying out.

● Use cornflour tied in a muslin bag to dust the work surface.

● The centrepiece may be embellished by spraying it with liquid colours.

● Alternatively, sprinkle 25g of gum tragacanth into 1 kg of royal icing. (The icing must not contain glycerine.) Mix in a machine on low speed, then knead to a smooth, dough-like consistency on a slab.

Add the dissolved gelatine to the sugar

Mixing

Kneading

Roll out thinly

Cut out each piece around a template (all templates must be exactly the right size

19 Marzipan

1 Place the water and sugar in a pan and boil. Skim as necessary.

2 When the sugar reaches 116°C, draw aside and mix in the ground almonds, then add the egg yolks and essence and mix in quickly to avoid scrambling.

3 Knead well until smooth.

▲ Marzipan may be shaped into fruits, figures, etc.

Ingredient	Makes 400g
Water	250ml
Caster sugar	1 kg
Ground almonds	400g
Egg yolks	3
Almond essence	2–3 drops

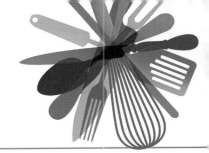

17 Food product development

This chapter covers the following units:

NVQ:
→ Contribute to the development of recipes and menus.

VRQ:
→ Food product development.

Introduction
The aim of this chapter is to enable you to develop the necessary advanced skills, knowledge and understanding of the principles in the research and development of dishes, using a variety of specialist pieces of equipment. The emphasis in this chapter is to develop more refined and advanced techniques in cooking and finishing dishes to meet individual needs or a specific brief.

Learning outcomes
By the end of this chapter you should be able to:
→ Identify suitable commodities and recipes to meet requirements.
→ Explain how to select the correct type, quality and quantity of products and ingredients to meet recipe, dish and customer requirements.
→ Design products and dishes to meet a brief
→ Evaluate and finalise products or dishes through development.
→ Demonstrate a range of techniques using the correct tools and equipment appropriate to the different cooking methods.
→ Explain the considerations when developing a dish or food product
→ Explain the benefits of new technology and modern equipment for dish and food product development.

Recipes included in this chapter

No	Recipe	Page
1	Water-bath smoked goose breast	558
2	Slow-cooked boiled eggs	558
3	Sous vide minted baby heirloom carrots	559

Developing recipes and menus

Why produce new recipes and new menus? There are certainly plenty available through the internet, magazines, TV shows and so on, but this is something that has been happening in kitchens the world over as technology develops, the availability of ingredients spreads, and customer expectations change in terms of dishes and flavour combinations.

While the creator of new recipes may enjoy the experience, it must be clear that the exercise is twofold: to satisfy customer demand and to meet management requirements. Therefore the prime consideration is the cost of the development and whether it going to be cost-effective in terms of selling price. These are considerations for a simple dish development for a restaurant menu, or for a mainline product for a food distributor.

Whatever the reason for the development, it is essential to evaluate the following:
● The cost of the development.
● The effect the change or changes will have on existing products or dishes.
● The ability of the staff to cope with the development.
● Availability of equipment and supplies for ingredients etc.
● The presentation of the dishes or products to market.
● The format of the menu.

Competitions to develop new products or menus are a good professional development opportunity and can boost staff morale and develop team spirit.

Essential considerations in menu planning

Prior to compiling menus, there are a number of essential considerations:
● **Location of an establishment** – there should be easy access to both customers and suppliers as and when required. A difficult journey can be off-putting, no matter how good the quality of food on offer, and can affect repeat business and profitability. If the establishment is in an area noted for regional speciality foods or dishes, the inclusion of a selection of these on the menu can give extra menu appeal.

- **Competition** – it is important to be aware of what is offered by competitors, including their prices and particularly their quality. Knowing this information enables an establishment to make decisions about how to compete effectively.
- **Suitability of a particular establishment to a particular area** – a self-service restaurant situated in an affluent residential district, or a very expensive seafood restaurant in a rundown inner city area may not be very successful. Anticipating and analysing the nature of demand that the operation is planning to appeal to will contribute to ensuring that the menu is appropriate.
- **Spending power of the customer** – a most important consideration is how much the potential customer is able and willing to pay.
- **Customer requirements** – analysis of dish popularity is necessary and those dishes that are not popular should not stay on the menu. Customer demand must be considered and traditional dishes and modern trends in food fashions need to be taken into account.
- **Number of items and price range of menus** – it is essential to determine the range of dishes and whether table d'hôte or à la carte types of menu are to be offered. A table d'hôte menu may be considered with an extra charge or supplement for more expensive dishes, or several table d'hôte menus of different prices may be more suitable.
- **Throughput** – if space is limited, or there are many customers (and control of the time the customer occupies the seat is needed), then the menu can be adjusted to increase turnover, e.g. more self-service items or quick preparation items, or separate service for beverages.
- **Space and equipment.**
- **Amount, availability and capability of labour.**
- **Supplies and storage** – menu planning is dependent on availability of supplies, that is, frequency of deliveries of the required amounts. Storage space and seasonal availability of foods need to be taken into account when planning menus.
- **Cost factor** – when an establishment is run for profit, the menu is a crucial consideration, but even when working to a budget, the menu is no less crucial. Costing is the crux of the success of compiling any menu.
- **Nutritional information** – there are various initiatives to encourage people to be aware of the relationship between health and diet and also to address the problems associated with obesity. These initiatives include providing the nutritional content information on menus.

Developing a menu policy

When compiling the menu for an operation, it is necessary to consider the creation of a menu policy that will govern the approach to the composition of the menu. This policy will determine the methods the operation will take to:

- Establish the essential and social needs of the customer.
- Accurately predict what the customer is likely to buy and how much they are going to spend.
- Ensure a means of communication with customers.
- Purchase and prepare raw materials to present standards in accordance with purchasing specifications and forecasted demand.
- Portion and cost the product in order to keep within company profitability objectives.
- Effectively control the complete operation from purchase to service on the plate.

Recipe development

Preparation prior to the practical aspects of producing new recipes includes the need to construct a method of recording accurate details of ingredients, their cost, quality and availability. Time needed for preparation, production and yield must be recorded. Space should be available to record comments for several attempts. Evaluation sheets or a process to enable opinions from tasting panels or people consulted should be made available, and should cover flavour, colour, texture, presentation and so on.

Research

Before developing new ideas, it is essential to have a basic foundation on which to build. Many new ideas are triggered by researching others' products or ideas. It is particularly worthwhile to keep abreast with what is happening in the industry – perhaps by looking at trade magazines (such as the *Caterer and Hotelkeeper*), watching TV programmes (for example, *MasterChef*), listening to radio programmes, or visiting other establishments, catering exhibitions, lectures or demonstrations.

Information sources for recipes are available everywhere. Every kind of establishment, from the local Chinese restaurant to a five-star hotel, can present innovative ideas. It is of particular value if it is possible to travel abroad, as well as being alert to new dishes in the UK.

If you have new developments in mind, it is necessary to pass your proposals to both senior management (who will be responsible for their implementation) and fellow members of the kitchen brigade, for their constructive comments. If possible, put your ideas to the test with respected members of the catering profession. Your proposals should include estimated food costings, time taken to produce, labour costs, equipment and facilities needed and details of staff training if required. Knowledge of the establishment's organisation is important so that the right person or people are involved.

Suppliers

It is important to take into account whether suppliers can produce the required ingredients in the right quantity, at the right price and desired quality. For more information on suppliers, see Chapter 3.

Quality of materials

The highest possible standards of ingredients should be used so that a true and valid result is available for assessment of the recipe. For more information on choosing commodities, see Chapter 3.

Staff abilities

Before implementing new recipes and menus, the standard of staff members' skills should be assessed to ensure they have the capacity to cope with innovation. You should also assess their cooperation in putting new ideas into practice, and give encouragement when the outcome is successful. Sufficient, able and willing staff, both in the kitchen and the restaurant, are necessary to achieve customer satisfaction with any menu.

Should the new dish or dishes require skills that are unfamiliar to some staff, then the workload of individuals may need to be changed while the relevant staff are trained in the appropriate skill.

Clear written instructions may need to be provided; this means the sequence in which the ingredients are to be used, with the appropriate amount (e.g. for 10 portions or 50 portions). This is to be followed by the instructions in the order that the recipe is to be followed, so that it is logical.

Should the new recipe be for a food service operation that involves preparation, cooking and presentation before the customer, then attention needs to be paid to the skills of the chef. Extra training, not only culinary skills, but customer handling skills may be needed and particular attention should be paid to hygiene. These factors need to be observed at the development stage so that customer satisfaction is guaranteed as soon as the new recipe is implemented.

Development of the dish

Having tested and arrived at the finished recipe, staff may need to practise production and presentation of the dish. This may include both small and large quantities, depending on the establishment. In all cases, careful recording of all aspects of the operation can help in the smooth-running of the exercise – in particular, basic work study should be observed.

Having validated the recipe, checked on a reliable supplier and ensured the capability of the staff, it is important that all concerned know when the dishes will be included on the menu. Storekeepers, kitchen staff and serving staff need to be briefed, as do any other departments involved, as to the time and date of implementation.

Of particular importance is how the customer sees the dish. When it is received, it needs to appeal to the senses of sight and smell, even before taste. For this reason, consideration needs to be given to presentation early in the development of the idea – what dish will be used, what will accompany it, are any particular skills needed to serve it? Foods in some establishments are prepared and cooked in front of the customer; some require the dish to be cooked fresh while the customer waits. Details of presentation must be recorded and, where possible, a test carried out in the actual situation.

The introduction of new dishes will perhaps affect the existing style of operations. If dishes are prepared in front of the customer, for example, in a department store, the new dish may require more time in preparation than others on the menu, which may cause a bottleneck. The introduction of a salad bar or sweet trolley to include new dishes can affect the service of the usual dishes. If the clientele require, say, vegetarian dishes, or people of certain cultural or religious groups have special needs, then adaptations may be necessary to accommodate this in the existing set-up.

Equipment and facilities

New recipes can affect the use of existing equipment by overloading it at peak times. The capacity of items such as pastry ovens, deep-fat fryers, salamanders, and so on, may already be fully used. New items can affect the production of the current menu; this fact should be borne in mind so that service is not impaired. When developing new menus, you must be aware of any shortcomings or deficiencies in equipment and may be wary of offering dishes that are difficult to produce. Certain items of equipment should not be overloaded by the menu requirements.

Other considerations

Finally, consider the following points:
- the elimination of waste
- the control of materials and ingredients
- the careful use of energy
- the wise use of time.

Ensure that a record is kept so that no resources are misused. Failure to control and monitor resources can be expensive in terms of time, materials and effort, and can be very wasteful.

Evaluating the product

- Adequate time needs to be allowed to test and develop any new recipe, to train staff, to appraise comments and modify the recipe if necessary.
- Constructive comments should be sought from staff and, in particular, any problems should be discussed. The results of trial runs should be conveyed to senior management and any problems that have been identified should be resolved.
- Staff (particularly serving staff) must be briefed on the composition of the dish, as well as being told when it will be included on the menu. They need to be asked if there are any problems; if required, this could be in written form.
- Senior personnel need to be asked for feedback (either orally or in writing) on the implementation of the new items. This might involve a tasting panel.
- In addition to obtaining feedback from staff, it is just as important, if not more so, to obtain comments from the customer or consumer.

Specific considerations when developing a new food product

When considering any development, it is necessary to take into account any problems and issues that may affect the outcome. For example:
- Keeping up to date on consumer choice and trends.
- Certain groups of people have restrictions on their eating habits (religious or cultural) that must be observed when producing new recipes for them.
- Extra care needs to be taken when introducing new recipes to patients in hospitals and nursing homes, and in the provision of meals in schools and residential establishments, to ensure the nutritional content is suitable.

- A balanced diet is important for health, providing the right amount and type of nutrients required to maintain a healthy lifestyle. When developing menus, consider minimising the fat, sugar and salt content of dishes.
- Some customers require special diets for health reasons. For example, those following a low-cholesterol diets, or people with allergies and intolerances or medical conditions such as diabetes.

For more information on factors influencing people's choice of food, see Chapter 3. For more information on healthy eating, dietary requirements, allergies and intolerances, see Chapter 18.

Menu design

The menu is the prime selling tool of a foodservice operation and therefore it should be written to inform and sell. The function of a menu is to inform potential customers what dishes are available and, as appropriate, the number of courses, the choice on the courses and the price. The wording should make clear to the customer what to expect. It may be used to promote specific items such as an ingredient in season, children's menus, reductions for senior citizens, or what is served at particular times etc.

If the menu is printed, the type should be clear and of readable size. If handwritten, the script should be legible so as to create a good impression. Mixing typefaces can be used to achieve emphasis but if overdone is likely to be unattractive to the eye. Emphasis may be achieved by using boxes on the menu; the menu paper and colour of the print can be carefully contrasted to make certain dishes stand out.

Menu information may appear on boards, computer screens or hand-held electronic devices, as well as printed menus.

Being able to write interesting descriptive copy is a skill – carefully devised descriptions can help to promote an individual dish, the menu generally and in turn the establishment. Factors to consider include:
- Items or groups of items should bear names people recognise and understand. Additional descriptive copy may be necessary if the name does not adequately describe the dish.
- Descriptions should describe the item realistically and not mislead the customer as this has legal implications. Care should be taken therefore in the use of terms such as fresh, British or organic.
- Some menus can be built around general descriptive copy featuring the history of the establishment or the local area.

Alternatively, copy can be based on a speciality dish, which has significant cultural importance to the area or establishment. This may include featuring the person responsible for creating and preparing the dish – especially if the chef is reasonably well known, either locally or nationally. The chef may also have had their recipes featured in the press or have appeared on radio or television – this too may be included in the menu to create further interest.

Menus are expensive to produce, but when they are attractive and fulfil the function of informing the customer, they may enhance the reputation of the establishment and increase custom.

Types of menu

Menus cover breakfast, lunch, afternoon tea, supper and dinner. There is a range of menu types: speciality menus, fixed price menus, menu du jour (menu of the day).

- **Table d'hôte** – a set menu at a set price, also known as a fixed-price menu. These menus often have a limited choice with a vegetarian option.
- **À la carte** – a menu with all the dishes individually priced. Customers compile their own menu from the dishes offered. Dishes are normally cooked and finished to order.
- **Tasting menus** – table d'hôte menus compiled by the chef with a variety of dishes and wines to match. There can be from five courses to many more. These menus are the speciality of the chef and display their creativity and innovation.
- **Function and banquet menus** – table d'hôte menus. They are well structured, costed and priced according

to the style of the establishment and the needs of the customer. Banquet dishes should not be over-complicated and require ease of assembly.

- **Cyclical menus** – menus compiled to cover a given period of time, e.g. one month or three months. They consist of a number of set menus for a particular establishment, e.g. an industrial restaurant, cafeteria, canteen, director's dining room, or hospital or college refectory. At the end of each period, the menus can be used over again. The length of the cycle is determined by management policy, by the time of year and by different foods available. These menus need to be monitored carefully to take account of changes in customer requirements and any variations in weather conditions which are likely to affect demand for certain dishes. If cyclical menus are designed to remain in operation for long periods of time, then they must be carefully compiled in order that they do not have to be changed too drastically during operation if, for instance, stock availability changes.
- **Planned and predesigned menus** – often found in banquet or function operations. Before selecting dishes, the food service operator is able to consider what the customer likes and the effect of these dishes upon the menu as a whole.
- **Company-wide menus** – the same menu made available across a chain of restaurants nationally and possibly internationally. These menus are planned centrally and the dishes may be prepared in central production units before being sent to the individual outlets.

Tables 17.1 and 17.2 list the advantages and disadvantages of cyclical and planned/pre-designed menus.

Table 17.1

Advantages of cyclical menus	Disadvantages
• They save time by removing the daily or weekly task of compiling menus, although they may require slight alterations for the next period. • When used in association with cook freeze operations, it is possible to produce the entire number of portions of each item to last the whole cycle, having determined that the standardised recipes are correct. • They give greater efficiency in time and labour. • They can cut down on the number of commodities held in stock and can assist in planning storage requirements.	• When used in establishments with a captive clientele, then the cycle has to be long enough so that customers do not get bored with the repetition of dishes. • The food service operator cannot easily take advantage of 'good buys' offered by suppliers on a daily or weekly basis unless such items are required for the cyclical menu.

Table 17.2

Advantages of planned and pre-designed menus	Disadvantages
• Enable the food service operator to ensure that good menu planning is practised. • The menu construction can be well balanced in terms of texture, colour, ingredients, temperature and structure. • Menus which are planned and costed in advance allow banqueting managers to quote prices instantly to a customer. • Menus can be planned, taking into account the availability of kitchen and service equipment and the capability of the service staff, without placing unnecessary strain on any of these. • The quality of food is likely to be higher if kitchen staff are preparing dishes they are familiar with and have prepared a number of times before.	• May be too limited to appeal to a wide range of customers. • They may reduce job satisfaction for staff who have to prepare and serve the same menu repetitively. • They may limit the chef's creativity and originality.

New technology and modern equipment

Sous vide

The term *sous vide* (pronounced *soo–veed*) is a French term, meaning 'under vacuum' and is a culinary technique in which food is immersed in a water bath or into steam and cooked at a very precise, consistent temperature.

This cooking technique typically involves cooking food for longer periods of time at a lower temperature. The precise temperature and time control allows you to cook food to perfection, while the forgiving nature of this cooking method also eliminates concerns about overcooking.

The sous vide technique has been used in kitchens since the mid-1970s. However, it has emerged to prominence again in the past few years as more kitchens consider this process to help get the best from some of the less tender cuts, as well as part of the mise-en-place to help with finishing and presenting more complex dishes within time expectations for customers in the restaurant. This has added excitement both for the customer and staff in the kitchen. This unique way of cooking can yield different and better results if used correctly and with the right ingredients. Foods cooked sous vide develop flavours and textures that simply cannot be duplicated using any other traditional cooking method.

The benefits to cooking sous vide, if used correctly, can include:

- Easy and fool-proof – perfect results, every time.
- Gourmet taste – capture the full, true flavour of foods.
- Hands-off, time-saving meal preparation – just set it and walk away.

- Added nutritional value – natural juices and nutrients are retained in the vacuum seal bag.
- Saves money – tenderises inexpensive cuts.

The science behind sous vide cooking

Sous vide cooking relies on the superior ability of water to transfer heat to food even though it is enclosed in a sealed bag. When cooking in a traditional oven, the temperature must be set much higher than the desired cooked temperature of the food, to heat the air around the food, as well as the tray the food is in, which could be around 10 to 20°C higher than is required for the food to ensure it is cooked. This may mean the outer area will be well done before the interior reaches the desired degree of cooking even in a short space of time. A few minutes too long in this super-hot environment and the food may be overcooked and tough.

With the sous vide method, because water transfers heat to and through vacuum-sealed food, it is about 10 times more efficient; the food can cook gently and precisely at the desired serving temperature, without ever exceeding it.

Food safety is a function of both time and temperature; a temperature usually considered insufficient to render food safe may be perfectly safe if maintained for long enough. Some sous vide recipes such as fish are cooked below 55°C. However, pasteurisation of food to be eaten by people with compromised immunity is highly desirable.

Health and safety ⚠️

Pregnant women eating food cooked sous vide may expose themselves and/or their foetus to risk and thus may choose to avoid unpasteurised recipes.

Generally speaking, food that is heated and served within four hours is considered safe, but meat that is cooked for longer to tenderise must reach a temperature of at least 55°C within four hours and then be kept there for sufficient time, in order to pasteurise the meat.

Table 17.3 Limits for sous vide cookery

Food item	Thickness	Cooking temperature	Minimum time	Maximum time
Sirloin beef	25mm	56.5°C	1 hour	4 hours
Lamb noisette	25mm	56.5°C	1 hour	4 hours
Pork chop	50mm	56.5°C	2 hours	4–6 hours
Leg lamb	70mm	56.5°C	10 hours	24 hours
Chicken, boneless	25mm	63.5°C	1 hour	2–4 hours

The current Food Standards Agency guidance in relation to lower cooking temperatures is:
- 60°C for a minimum of 45 minutes
- 65°C for a minimum of 10 minutes
- 70°C for a minimum of 2 minutes.

There are two approaches to cooking sous vide within the modern kitchen, depending on the commodity and dish requirements. Each has its own characteristics.

Indirect cooking is a longer process and involves a cooling and storage stage part the way through the chain:

Temperature guides

The links between time and temperature are really important in relation to sous vide cookery.

Mise-en-place → Vacuum seal → Indirect cooking → Rapid cooling (>3°C in 90mins) → Store → Regeneration (cooking) → Assembly → Serving

Direct cooking is used when the dish is for immediate consumption:

Mise-en-place → Vacuum seal → Direct cooking → Assembly → Service

Note that it is usual for many dishes to be seared in a hot pan to add colour and texture before serving.

The guidelines in Table 17.4 apply to the timing and chilling of cooked food before refrigerating or freezing in relation to sous vide cookery to support cook chill/freeze systems.

Table 17.4

Pouches of seasoned uncooked food	Cook immediately	Chill to >3°C and refrigerate up to 2 days	Freeze up to 2 weeks
Pouches of cooked food	Serve immediately	Quick chill >3°C and refrigerate up to 2 days	Quick chill and/or freeze up to 1 year for meats, and up to a few weeks for vegetables

Equipment
- **A water bath or sous vide water oven** – depending on the size of the kitchen, more than one may be used. They come in a variety of sizes, some of which are more portable than others, with options for pre-programming and settings for specific recipes.
- **Thermal circulators** could be used – this is a heating element that can be attached to gastronorm trays and add further flexibility to water bath cooking.
- **Vacuum packaging** and sealing pouches for the process – the food is placed in a pouch and placed on the plate; the lid is then closed, which creates a hermetically sealed chamber from which all or part of the atmospheric air is extracted, depending on the settings and type of vacuum system that is being used. The bag is then sealed at the end of the process. Due to the extraction or reduction of air in the bag, it adds to the preservation of the quality of the ingredients sealed. Most restaurants use a machine with a clear lid and suction to create the vacuum.

Depending on your schedule and the dishes you are preparing, you may want to just season and seal uncooked food in their pouches ahead of time and cook it later, or you can also pre-cook the food ahead of time and then chill before storing to be reheated later.

Health and safety

As water is used to operate the sous vide water oven, chefs need to be aware of safety considerations around electric points and filling or splashing water on power points.

There are a variety of machines and sizes available to seal pouches:

- Hand-held sealer – this will expel the air and is used to seal pouches. It acts like a zip-lock element to a bag.
- Table-top sealer – has limited options; removes air and seals the width of bags to a certain size.
- Multivac packaging machine – offers flexibility in terms of volume and pressure for removal and sealing of the pouches.
- Pouches – the bags themselves come in a variety of sizes and shapes, as well as density, depending on the use they are intended for.

Health and safety !

Where raw food for storage or cooking, and cooked food, are going to be vacuum-packed, it may be a requirement to have more than one vacuum packer or water bath within the kitchen.

Professional tips: sous vide

- High standards of kitchen and personal hygiene must be observed.
- Prime quality ingredients should be used.
- Adhere to the requirements for food safety in relation to minimising any potential risk of cross-contamination etc.

Thermal blenders

A thermal blender is a combination piece of equipment that can chop, blend, cook, chill and freeze depending on the setting and ingredients used. Many chefs now use these to support the production of smooth creamy purées, sorbets and fruit coulis, to name just a few items.

It features one bowl with the option of using a blade or butterfly whisk to make the dish.

Pacojet

The Pacojet is designed to 'micro-purée' frozen foods, creating mousses, ice cream, sorbets, soups and sauces. The process is designed to maintain the flavours, colours and nutrients of the food.

Food – for example, an ice cream mixture – is frozen in the Pacojet container. The spinning blade then shaves layers off the top of the food. The product has a very smooth, creamy texture, and unused mixture in the bottom of the beaker remains frozen and can be kept for later use.

An individual portion of food can be prepared in 20 seconds, and ice creams and sorbets are produced at the serving temperature.

Molecular gastronomy

Molecular gastronomy is associated with innovative modern cuisine. Chefs use a combination of unusual tastes, textures and sometimes theatrical twists to give the eating experience a new multi-sensory dimension, along with the aid of high-tech equipment and a handful of chemicals, most of which have been used with manufacturing of food products for a number of years, but have become more widely available to chefs.

A few techniques explored

Techniques that chefs start off with when introducing these ideas include:

- **Foams** – There are a few different ways to achieve froths and foams. The easiest is using a hand blender, held just under the surface of the liquid. As the foam appears, skim it off and add to your dish – this works well with creamy or buttery sauces, but the bubbles won't last long. Another method is to use a cream whipper that adds gas and put your creamy/buttery sauce through that. To give your foam a bit more stability and body, or to foam thinner liquids, stocks and juices, you can add a gelling agent such as agar agar, or thickener like lecithin before using either technique.
- **Spherification** – This creates spheres or balls, the size of caviar or larger, from almost any food substance. The process involves mixing sodium alginate with any liquid, then dripping the mixture into a calcium salt and water solution. Scooped out quickly enough, the drops should be jellied on the outside and still liquid in the middle as the calcium solution will set the sodium alginate gel. Spheres made from fruit juice can be added to desserts for decoration. Alternatively you could make balls of consommé or vegetable juice.
- **Fizz** – Many chefs will know how to make honeycomb by adding bicarbonate of soda to a light caramel, but if you mix bicarbonate of soda with any form of acid and then add water, it will fizz. You can make your own by mixing a little bicarb, citric acid and icing sugar, then dust it onto toffees, boiled sweets, or even on to the surface of fruits (only if the skins are really dry though) and get tongues tingling.

Take it further

Further information on practical gastronomy can be found in *The Theory of Hospitality and Catering for Levels 3 and 4* by David Foskett and Patricia Paskins, Chapter 16, 'Understanding molecular gastronomy and product development'.

Molecular gastronomy kits

There are kits on the market which contain all the chemicals you'll need to get started with techniques such as spherification, fizz and jelly making. They usually contain some basic equipment too such as pipettes, tubing and a recipe book or leaflet. It's worth investing in a set of precision scales too, as you'll be working with very small quantities.

There are a number of other natural and chemical ingredients that can stabilise, gel or set and hold food.

The traditional gelling and thickening agents used within kitchens were gelatine (animal protein based), flour or vegetable starch. Today more chefs are experimenting with hydrocolloids which hydrate in water, forming associations with water molecules and restricting their movement, which means they form a more structured or holding shape as in forms, gels etc. Some of the more common agents are shown in Table 17.5.

Table 17.5

Agent	Source or base	Use
Alginates	Seaweed base	Dilutes while cold with strong agitation
Citrus (sodium citrate)	Natural	Highly water soluble
Xanthan gum	Bacterial fermentation powder	Thickening agent, used hot or cold, used in gluten-free baking
Agar agar	Natural seaweed base	Gels and holds at high temperatures, often used in place of gelatine for vegetarians
Kappa or iota carragenan	Natural	Used to coat food items and gel forming at lower concentrations
Lecithin	Natural soya base	Emulsification for foams, used in place of eggs for some sauces

It is important to understand what each of these agents can do to best use them for the presentation and eating quality of the dish.

Professional tip

Remember to ensure that agents are fully dispersed and fully agitate with a hand blender when adding powders.

1 Water-bath smoked goose breast

Brine

Variations can be made according to taste.

Ingredient	For 2 breasts
Water	1 litre
Salt	100g
Sugar	100g
Peppercorns	10g
Juniper berries	8–10
Garlic cloves, crushed	4
Bay leaves	2
Noilly Prat or dry sherry	100ml
Rice vinegar	30ml

1 Warm the ingredients of the brine and allow to infuse for five minutes; cool completely.

2 Immerse goose breasts that have been pre-trimmed and soak for 2–3hrs.

3 Dry well and keep in a refrigerator until surface is fully dry (1–2hrs).

4 Place over pre-prepared blend of Earl Grey tea and wood chips (damp).

5 Smoke very gently for 20–30 minutes; keep the temperature as low as possible so that the breasts are not cooked but just slightly firm.

6 Place in a vacuum pouch and poach for 25 minutes at 51°C.

7 Chill well, colour breast fat with a blow torch then brush with honey and re-flare briefly; do not burn.

8 Slice the meat thinly and serve with a rocket salad and a poached fruit (experiment).

The breast is sufficiently large so as to obtain 8–10 portions per breast, making it more of a viable proposition for the caterer, depending on the style and course that the dish is applicable to.

Energy	Cal	Fat	Sat fat	Carb	Sugar	Protein	Fibre
146 kJ	35 kcal	0.1g	0.0g	7.0g	6.7g	0.2g	0.1g

2 Slow-cooked boiled eggs

Use 1 egg per person (you can make as many as will fit into the machine).

1 Fill and preheat the water bath to 62.5°C.

2 Put the eggs in their shells (not in cooking pouches) directly onto the perforated grill in the water bath and cook for 45 to 60 minutes.

3 Crack the eggs and peel into a bowl.

4 Pané the peeled, whole eggs. Deep fry at 180°C until golden brown. Serve with dressed leaves and brioche toast, or with asparagus.

Energy	Cal	Fat	Sat fat	Carb	Sugar	Protein	Fibre
314 kJ	76 kcal	5.6g	1.6g	0.0g	0.0g	6.3g	0.0g

3 Sous vide minted baby heirloom carrots

1 Fill and preheat the water bath to 84°C.
2 Put carrots into a small cooking pouch.
3 Add a drizzle of olive oil and a sprinkle of salt and pepper to the pouch and vacuum seal.
4 Submerge the pouch in the water bath and cook for 30 minutes.
5 Remove carrots from pouch, sprinkle with mint, and serve.

Ingredient	2 to 4 portions
Baby heirloom carrots, washed, trimmed, and sliced into coins	1 bunch
Olive oil	1 tsp (5ml)
Salt and pepper to taste	
Fresh mint leaves, minced	

Energy	Cal	Fat	Sat fat	Carb	Sugar	Protein	Fibre
217 kJ	52 kcal	2.0g	0.3g	7.3g	5.6g	1.7g	3.2g

Video: sous vide cooking
http://bit.ly/1fmtmNk

Test yourself

1 What are the six elements that influence product or dish development?
2 List three advantages and three disadvantages of cyclical menus.
3 How would you evaluate that a new product or dish is progressing or meeting the brief?
4 What are the health and safety risks in relation to using portable water baths?
5 What are the two main points that should be considered when using sous vide cooking?

18 Healthier dishes and special dietary requirements

This chapter covers the following **NVQ** units:
→ Produce healthier dishes.

Healthier dishes in the **VRQ** Diploma are incorporated and assessed in a range of units. Content in this chapter will be relevant to the following units:
→ Advanced skills and techniques in producing vegetable and vegetarian dishes.
→ Advanced skills and techniques in producing meat dishes.
→ Advanced skills and techniques in producing poultry and game dishes.
→ Advanced skills and techniques in producing fish and shellfish dishes.
→ Produce fermented dough and batter products.
→ Produce petits fours.
→ Produce paste products.
→ Produce hot, cold and frozen desserts.
→ Produce biscuits, cake and sponges.

Introduction
The aim of this chapter is to enable you to develop the necessary advanced skills, knowledge and understanding of the principles in preparing and cooking dishes to meet customer specifications and requirements in relation to healthy eating and special dietary requirements. The emphasis in this chapter is to develop precision, speed and control in existing skills and develop more refined and advanced techniques in cooking and finishing dishes to meet individual needs.

Learning outcomes
By the end of this chapter you should be able to:
→ Identify suitable commodities and recipes to meet dietary requirements.
→ Explain how to select the correct type, quality and quantity of products and ingredients to meet recipe, dish and customer requirements.
→ Understand how to cook dishes to maintain their maximum nutritional value.
→ Demonstrate a range of techniques using the correct tools and equipment appropriate to the different cooking methods.
→ Describe how to adapt recipes and cooking techniques to maximise nutritional value and healthy options in dishes and professional cookery.
→ Demonstrate professional ability in minimising faults in complex dishes.

Recipes included in this chapter

No	Recipe	Page
1	Gluten-free sultana brioche loaf	566
2	Basic ciabatta (gluten- and egg-free)	567
3	Gluten-free pastry	567
4	Dairy-free ice cream	568
You may also be interested in the following recipes in other chapters:		
	Lemon curd flourless soufflé	Ch. 14
	Flourless chocolate sponge	Ch. 15

Cooking for health

Professional chefs and caterers have to accept that they have a responsibility in helping customers achieve a balanced diet within the parameters of their eating and eating-out experience. Where the customer base takes most of its meals within a served kitchen, for example, in a care home, hospital, boarding school or even the armed services, the chef/caterer has the opportunity through planning menu cycles that can be viewed over a period of time to meet the targets of healthy eating and help customers/diners eat a healthy diet.

When the customer base is transient, such as in restaurants, pubs, hotels, bars, etc., it is normal for the menu to offer varied choice and provide information for customers to enable them to make an informed decision.

Healthy eating by stealth has more chance of success than any attempt to try and push customers into better lifestyle decisions. For example, in relation to eating out for what could be a celebration or special occasion, subtle changes can be introduced that will help to provide dishes that meet the government guidelines in terms of healthy eating without losing some of the notion of special occasion.

Nutrition and health

Eating habits have changed and lifestyles continue to change with the availability of instant meals, more grazing-style approaches to eating and an increase in the number of people being more inactive earlier in life. This has brought and continues to add to potential health problems. In particular, obesity and the added health risks that this can cause. Obesity has been clearly linked to potential diabetes, coronary heart disease and potential liver disorders, as well as possible links to cancer related to the foods we eat.

To tackle these potential health problems, people need to change their patterns of eating. The messages are simple:
- Base meals on starchy foods such as rice, bread, pasta (ideally wholegrain varieties) and potatoes.
- Eat lots of fresh fruit and vegetables (at least five portions a day).
- Eat moderate amounts of meat, fish and alternatives (including a portion of oily fish each week).
- Cut down on saturated fats, sugar and salt within the diet.

Government guidelines for healthy eating

There are clear quantitative nutritional targets for the UK population, which have been translated into daily guidance for both men and women:
- Reduce total fat intake – this is suggested as no more than 95g for men or 70g for women.
- Reduce saturated fat intake – suggested as no more than 30g for men or 20g for women.
- Reduce sugars that are not found naturally in foods or milk – suggested as no more than 60g.
- Reduce salt intake to no more than 6g.
- Increase intake of starchy carbohydrate to around 37% of energy-based food.
- Increase fibre intake to around 18g.
- Eat at least five portions of fruit and vegetables (mix the colours too).
- Eat two portions of fish per week, including one oily type.

Translating these nutritional targets into food on the plate, or what is offered on the menu, is where chefs are vitally important. They need to have the skills and knowledge to make healthy eating a positive experience for customers. Some of the best cuisines of the world are based on these guidelines. Dishes and meals can be built around lots of starchy foods with generous helpings of a wide range of vegetables, salads and fruit, adding relatively small amounts of lower-fat meats plus an abundance of fish dishes, all made with unsaturated oils such as sunflower, olive, rapeseed or sesame. Many recipes and cooking styles that are found in Italy, the Eastern Mediterranean, China, India and Thailand echo these principles.

The concept of balance

Healthy eating is about balance. Chefs can help people achieve this balance by threading the principles above through their practice in dish development and menu planning, taking into account the four areas of balance: energy, plate, meal and menu.

Energy balance

Energy balance is one of the keys to healthy eating which is about balancing the energy input through the foods we eat with the energy output we use. In simple terms, too much energy consumed leads to obesity. Energy is measured in calories (or joules – the metric version). As an example, an average 24-year-old chef needs around 2,550 kcals per day to keep his/her body working and provide the energy for the physical activity of working in a kitchen or going to the gym. Rates of calorie expenditure vary between activities and individual make up – as a rule the harder the body works, the higher the rate of energy expenditure.

Chefs should be able to meet the needs of customers who want to control their weight but still like to eat out. Chefs can make significant calorie reductions to dishes by simply making a few small changes that drive down fat levels. For example:
- Trimming the fat from meat.
- Reducing the amount of cream, butter and oils in sauces.
- Using semi-skimmed milk in place of whole milk.
- Dry-frying to seal before braising.

Balancing recipes

There are countless steps that chefs can take to drive down the levels of fat, salt and sugar while at the same time increasing the starch, fibre, fruit and vegetable content of their recipes. This balancing process may involve:
- Changing ingredients – for example, swapping full-fat crème fraiche for a half-fat version.
- Adjusting proportions with recipes – for example, increasing the amount of rice and fish in relation to meat in a paella.
- Changing the cooking method – for example, baking samosas instead of frying, or grilling in place of frying.

The example shown in Table 18.1 shows how a traditional recipe for sole mornay can be modified to make it healthier. The modified version ends up much lower in fat, especially saturated fat.

Table 18.1 Creating a healthier version of sole mornay

Traditional recipe	Modified recipe
Béchamel sauce made with butter and whole milk	Béchamel sauce made with polyunsaturated margarine and semi-skimmed milk
Sauce finished with egg yolks and cream	Fromage frais used to add texture to sauce
Large quantities of Gruyère cheese used for flavour	Small quantities of Parmesan used for flavour
28.0g fat per portion, of which 15.7g is saturated	11.8g fat per portion, of which 3.5g is saturated: a reduction of more than 12g
420 kcals per portion	272 kcals per portion: a reduction of almost 150 kcal

Balancing plates

This is about the balance of each plate of food that is served. The idea is to:

- Boost the amount of starchy food, which can be done in lots of different ways – for example, adding bread rolls or adjusting portion size.
- Increasing the content of vegetables – add side salads, garnishes or relishes; increase vegetables in recipes in proportion to other ingredients.
- Maximise fruit content in desserts – use a fruit coulis in place of cream.
- Consider offering different serving sizes for some dishes so that they can be taken as starters or main courses within the eating experience.

The example in Table 18.2 shows how the balance of a main course changes depending on the recipes used and the accompaniments chosen.

Table 18.2 Creating a healthier main course

Higher-fat main course	Lower-fat main course
Traditional sole mornay recipe	Modified sole mornay recipe (see Table 18.1)
Sauté potatoes	New potatoes
Grilled mushrooms	Broccoli and carrots
45.3g fat per portion, of which 18.0g is saturated	12.3g fat per portion, of which 3.5g is saturated
688 kcals	380 kcals

Balancing complete meals

Chefs can help customers balance individual courses so meals consumed are healthier. For example, higher-fat first courses (like deep-fried Camembert or avocado stuffed with cream cheese and walnuts) can be balanced with lower-fat main course (e.g. steamed sole with garlic, spring onion and ginger) and again, lower-fat desserts (fruit sorbets, or strawberry pavlova).

The example in Table 18.3 shows how consistently lower-fat choices across a menu, together with small changes to a traditional recipe, can improve the health profile of a complete meal.

Table 18.3 Health profile of a complete meal

Higher-fat meal	Lower-fat meal
Terrine of duck and chicken	Terrine of chicken and vegetables
Traditional sole mornay recipe with sauté potatoes and grilled mushrooms	Modified sole mornay recipe (see Table 18.1) with new potatoes, broccoli and carrots
Sticky toffee pudding with butterscotch sauce	Pears in red wine
128.0g fat per meal, of which 64.2g is saturated	29.8g fat per meal, of which 12.9g is saturated
1995 kcals	769 kcals

Balancing menus

In terms of healthy eating, chefs often have a delicate pathway to tread. Some customers will want to indulge themselves and forget about the fat and calories, while others will be consistently looking for healthier choices and ways to control their fat and calorie intake. The demand for healthier choices will be particularly high in everyday eating environments like the workplace, schools, restaurants or venues that serve business lunches. In addition, the food culture in the UK has undergone a revolution in terms of the emphasis on the highest quality, traceability and locality of ingredients used in creating dishes that reflect the principles of healthy eating. Following these rules, creative chefs can help people understand that healthy eating does not necessarily have to be brown and boring.

The demand for healthier options has never been higher and therefore chefs need to consider including the following within menus:

- A variety of dishes, including oily fish such as salmon, tuna or trout.
- A wide range of exciting vegetable dishes, as part of dishes or as dishes in their own right.

- Pasta dishes or including bread and grains in dish garnishes, making pizza with a thicker wholemeal dough.
- Desserts based on or including fruit.

In this way chefs can contribute to people achieving a dietary balance required over time as depicted in the eatwell plate. This illustration shows the proportions of different food groups that make up a healthy eating pattern.

The eatwell plate

Use the eatwell plate to help you get the balance right. It shows how much of what you eat should come from each food group.

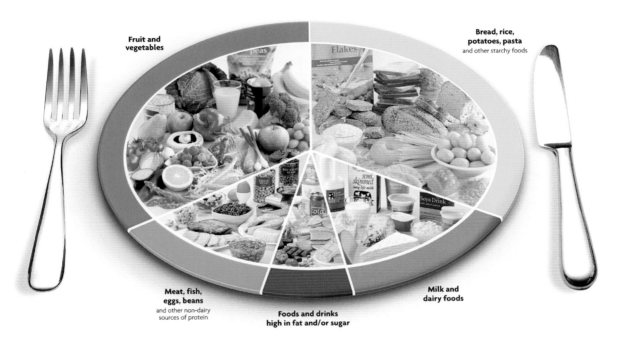

▲ The eatwell plate

Dietary requirements and intolerances

Estimates by the British Allergy Foundation and the Institute for Food Research put the proportion of the population with an allergy to at least one food at around 2 per cent, with the number growing by more than 5 per cent per year. There is further evidence that up to 30% of people could be affected by allergies which may be linked to food items. The Food Standards Agency updated food labelling, but has recently announced further major changes to food labelling legislation that will see certain allergy information become mandatory. Current food labelling laws for the food service industry will change from the end of 2014, meaning all operators or suppliers offering food items on their menus for the end consumer will need to provide allergy information, which means that specific food allergens including gluten and peanuts must be highlighted. Clear information about the food allergens contained on menu items will need to be highlighted on the menu.

Food intolerances

Food intolerances can be described as causing an adverse reaction to the food. These intolerances fall into three categories:

1 Intolerance to certain foods that cause a reaction – i.e. rashes, headaches.
2 Inability to digest certain foods – e.g. lactose where there is not enough of the correct enzyme to digest the lactose in the body.
3 Intolerance to certain chemicals – e.g. artificial colours, flavourings.

Food allergies

Food allergies occur when the body's immune system sees some particular harmless food as harmful and it therefore causes an allergic reaction. The allergic reaction that some people have to certain foods can sometimes be fatal.

Foods that may cause an allergic reaction in a very small number of people include milk and dairy products, fish, shellfish, eggs and nuts (particularly peanuts, but also cashew, pecan, Brazil and walnuts). Peanuts are often commonly used in Bombay mix, peanut butter, satay sauce, nut-coated cereals, groundnut/arachide oils, chopped nuts in vegetarian dishes and in some salads.

The Anaphylaxis Campaign has warned caterers to be alert for foods containing flour made from lupin seeds as this can cause an allergic reaction similar to nuts. Lupin flour is widely used as an ingredient in France, Holland and Italy because of its nutty flavour, attractive yellow colour and because it is GM free.

Take it further

For more information on the Anaphylaxis Campaign, visit www.anaphylaxis.org.uk.

There are 14 allergens that food businesses and venues should be aware of as ingredients or parts of products served which will need to be declared to the customer as part of labelling requirements:

1 Gluten-containing cereals
2 Crustaceans
3 Molluscs
4 Fish
5 Peanuts
6 Lupin
7 Tree nuts (such as walnut, hazelnut, almond etc.)
8 Soya
9 Eggs
10 Milk
11 Celery
12 Mustard
13 Sesame
14 Sulphur dioxide

Dietary requirements

Catering for dietary requirements for individual customers is often requested. This is based on a number of individual needs, which include:

- Religious or cultural, where specific commodities are omitted from the diet.
- Moral or belief, which is often a lifestyle choice (for example, vegetarianism).
- Therapeutic, where specific requirements are needed to balance the diet to aid health, as in controlling diabetes.

It is becoming increasingly common for people to make choices in relation to lifestyle and eating to maximise potential benefits, this includes non-vegetarians taking the vegetarian option from the menu or making a decision to reduce the amount of wheat flour products they eat.

Table 18.4 Special diets for those with food allergies and intolerances

Type of diet	Problems	Foods to avoid	Permitted/encouraged foods
Coeliac	An allergy to gluten causes severe inflammation of the gastro-intestinal tract, pain and diarrhoea. Inability to absorb nutrients causes malnutrition.	All products made from wheat, barley or rye, including bread. Always check the label on all commercial products.	Potatoes and rice Cornflour or flour made from potatoes or rice Fresh fruit and vegetables

Type of diet	Problems	Foods to avoid	Permitted/encouraged foods
Food allergies	Allergies cause severe and rapid reactions to a particular food, which can be fatal. Any food can be an allergen, but common allergies are to peanuts, sesame seeds, cashews, pecans, walnuts, hazelnuts, milk, fish, shellfish and eggs.	For a peanut allergy, avoid peanuts, chopped nuts, groundnut oil, satay sauce, arachide oil, peanut products and other nut products.	
Low cholesterol	High levels of cholesterol in the blood are associated with an increased risk of cardiovascular disease.	Liver, kidney, fatty meat, bacon, ham and paté. Egg yolks, cream, full-fat milk and yoghurt, cheese. Fried foods, pastry, biscuits, cakes and salad dressings.	Lean meat and fish, grilled or poached Low-fat milk and yoghurt Porridge, muesli, fresh fruit and vegetables
Diabetic	The body of a person with diabetes is unable to control the level of glucose in the blood. This can lead to comas and long-term problems such as increased risk of cardiovascular disease, blindness and kidney problems.	As for a low-cholesterol diet, plus all dishes that are high in sugar.	As for a low-cholesterol diet, plus wholemeal bread, pasta, rice, potatoes and pulses.

What can caterers do?

Chefs and catering staff are usually happy to cater for customers with food allergies, especially if they have prior notice. Some prefer to make up the meal at the start of their shift before other foods contaminate the kitchen and cover it ready to reheat later.

As a caterer you can:
- Be receptive to people with a food allergy asking questions about the food on offer.
- Ask the person how allergic they are and whether traces of the food from cross-contamination could be a problem.
- Let the allergic person or carer check food labels and speak to the chef themselves.
- Give the person dignity and respect – some people have a poor tolerance of people with food allergies – usually due to poor understanding about them, so learn all you can about food allergy.
- Do not offer to cater for the person if you are unable to do so safely.
- Train serving staff in pertinent food allergy issues.
- Ask suppliers to provide accurate written details about all ingredients.

- Avoid the indiscriminate use of nuts, e.g. powdered nuts as a garnish, unless this is an essential part of the recipe.
- If a dish is meant to contain nuts, why not make sure this is reflected in the name: e.g. nut & carrot salad?
- If possible, keep certain preparation areas designated as nut-free areas or food allergy preparation areas that every member of staff is aware of.
- Put up a prominent sign or a note on the menu encouraging people with allergies to question staff. For example, this could state: 'Some of our dishes contain nuts. If you are allergic to nuts, please, ask the waiter to suggest a nut-free meal'.
- Try to ensure that where a dish contains certain allergens that this is indicated in some way on the menu. Some restaurants adopt a specific symbol.
- Organise a training session on allergies for your staff. Make sure that all new staff members (including part-time and casual staff) are aware of the implications of severe allergies and how to cater for them safely by reducing the risks as much as possible.

Recipes

Healthy options are provided in a range of recipes in other chapters of this book. The following recipes are for those with allergies or intolerances.

1 Gluten-free sultana brioche loaf

Ingredient	1 loaf
Gluten-free flour mix	325g
Xanthan gum	1 tsp
Salt	1 tsp
Caster sugar	25g
Dried yeast	1 × 7g packet
Butter	200g
Warm milk	125ml
Warm water	75ml
Large egg	1
Sultanas	75g

Energy	Cal	Fat	Sat fat	Carb	Sugar	Protein	Fibre
1284 kJ	307 kcal	18.0g	11.0g	33.9g	9.1g	4.7g	1.8g

1 Put the dry ingredients into a bowl. Lightly rub in the butter (you don't need to get this to breadcrumb stage).

2 Warm the milk and mix with the water and beaten egg.

3 Make a bay in the flour mix, add sultanas into the bay and pour in the liquid. Fold the ingredients into the liquid, mix together lightly; it will be slightly lumpy.

4 Spoon the mixture into a loaf tin, 23 × 13 × 7cm in size, cover with cling film to flatten, and prove in warm place for one hour.

5 Bake in oven at 200°C for 25–30 minutes until well risen and dark golden. Remove from oven.

2 Basic ciabatta (gluten- and egg-free)

1 Preheat the oven to 200°C and lightly grease a loaf tin with the sunflower oil.

2 Place the flour, baking powder, xanthan gum, egg (or egg replacer) and sea salt into a food processor.

3 Pour over most of the water, leaving about 25ml behind, and blitz until you have a smooth and runny dough. Add a little more water at this point if you think the mixture looks too firm.

4 Tip the mixture into the loaf tin, level the top with the back of a spoon and allow to prove for 15–20 minutes.

5 Bake in the oven for 35 minutes or until golden and crisp on top.

6 Remove from the oven, turn out onto a wire rack and leave to cool for a few minutes. Serve while warm or leave to cool completely before slicing.

Ingredient	Quantity
Sunflower oil	1 tbsp
Gluten-free plain flour	225g
Baking powder	4 tsp
Xanthan gum	¼ tsp
Egg	1
OR: 1 heaped tsp Orgran egg replacer whisked together with 2 tbsp water and 1 tbsp rice milk	
Sea salt flakes	½ tsp
Warm water	300–325ml

Professional tip

Gluten-free breads, pastries and cakes may need more liquid than stated in the recipe. This depends upon the mix and quantity of gluten-free flours used.

Fault

This ciabatta was made without enough water. Weigh/measure the ingredients very carefully before mixing.

3 Gluten-free pastry

Ingredient	Makes 1 large tart case
Gluten-free flour	310g
Xanthan gum	10g
Butter or margarine	125g
Salt	Pinch
Egg yolk	1
Cider vinegar	5ml
Iced water	45ml

1 Whisk the water, egg yolk and vinegar together and set aside.

2 Sieve the flour and xanthan gum.

3 Rub in the fat, add the liquid a little at a time and mix with a fork until a soft dough forms.

4 Place in a plastic bag and chill for about 1 hour to rest before rolling out.

4 Dairy-free ice cream

Ingredient	Makes 1 litre
Soya milk powder	300g
Soya milk	750mls
Water	150mls
Sugar	300g
Vanilla	5mls
Apple cider vinegar	5mls

1 In a blender, combine the soya milk powder, water and soya milk until well blended.

2 Combine the soya milk mixture with the sugar, vanilla and vinegar in a small saucepan over medium-low heat. Stirring constantly, cook until the mixture is thick and syrupy in consistency.

3 Place in a container in the freezer uncovered for 1 hour.

4 Remove the pan from the freezer and scrape the ice cream into a blender and blend on high for 30 seconds, or until mixture is creamy.

5 Place the mixture back in the pan and back into the freezer. Repeat this procedure 3 more times at 30-minute intervals, allowing the ice cream to chill in the freezer covered for 1 hour after the last blend before serving.

6 Serve cold, adding your choice of toppings if desired.

Variation
You could use an ice cream machine or Pacojet.

Test yourself

1 What are the four reductions that should be in a balanced diet?

2 List four things as a chef you could do to avoid contamination for a customer with a nut allergy.

3 What is the difference between an intolerance and an allergy in relation to food?

4 What are the four concepts of balance that can help with healthy eating?

5 List three reasons why a customer may request a specific dietary requirement.

6 Give two examples of adjustments that could be made to a dish to meet the healthy eating guidelines.

7 As head chef, what three things could you do to ensure your kitchen team can provide food for customers requiring a coeliac diet?

Glossary

00 flour – an Italian grading of flour that is used to make pasta.

À la carte – a menu with all the dishes individually priced. Customers compile their own menu from the dishes offered. Dishes are normally cooked and finished to order.

Activated Dough Development (ADD) – used in industrial manufacture of bread. It involves high-speed milling, uses more yeast and an 'accelerator' (often ascorbic acid) to speed up the mixing and fermentation process.

Additives – chemicals (both synthetic and natural) are used to give various functional properties to foods.

Agar agar – a gelatinous marine algae. It dissolves very easily and in, addition to gelling, adds elasticity and resists heat.

Allergenic hazard – may come from nuts, wheat products, dairy products, shellfish, mushrooms, soft fruits and a range of other foods and products.

Amuse bouche – single, bite-sized hors d'oeuvres. They are served for free and according to the chef's selection.

Anaphylactic shock – the immune system overreacts to a usually harmless substances. Symptoms include faintness, skin irritation and swelling, which can lead to breathing difficulties and loss of consciousness.

Aquaculture – fish farming.

Aseptic and modified atmosphere packaging (MAP) – hermetically sealed foods that have been surrounded by gases that slow down the deterioration of food, then packaged airtight by a commercial sealing process. be added.

Ballotine – a boned-out, stuffed leg of poultry, usually chicken, duck or turkey.

Barding – laying fat, or bacon, over meat to keep it moist during cooking.

Beurre blanc – sauce made from better, vinegar and water, with chopped shallots.

Beurre fondu –melted butter, which may be added to a little white wine.

Beurre manié – also known as raw roux; this means mixing equal quantities of butter and flour.

Beurre noir – black butter, not exactly burnt, but cooked to noisette stage to which is added malt vinegar.

Beurre noisette – nut brown butter.

Biga/poolish – a soft ferment made with yeast, water and flour to boost yeast activity and encourage the dough to ripen.

Binary fission – when pathogenic bacteria are given the right conditions, they are able to multiply by dividing into two.

Biscuit glacé – a light, solid-type ice cream that can be produced without the aid of a sorbetière. The aeration is achieved from a sabayon made from egg yolks and stock syrup whisked over a bain marie of hot water. Once cold, whipped cream, Italian meringue and flavourings are folded through.

Bisque – a very rich soup with a creamy consistency.

Bivalve – mollusc with a double-hinged shell.

Blanching – this helps to preserve colour, especially in green vegetables.

Bombe glacée – frozen dessert made with a pâte à bombe mixture and frozen in a dome-shaped mould.

Bouillabaisse – a Mediterranean fish soup/stew, made of multiple types of seafood, olive oil, water, and seasonings.

Brochettes –food cooked in and sometimes served on skewers or brochettes.

Broth – an unpassed soup containing vegetables and sometimes meat or fish.

Bulk fermentation time (BFT) –the period of time after the dough is made and before it is scaled and shaped.

Canning – foods are processed at high temperatures sealed in the can.

Cephalopods – for example squid and octopus. They are technically molluscs, but look very different and tend to be cooked differently.

Chemical hazards –from various chemicals such as kitchen cleaning materials, disinfectants, insecticides, rodenticides, degreasers and agricultural chemicals.

Chowder – a North American soup, usually with a seafood base.

Cleaning – the effective removal of dirt, grease, debris and food particles usually done by using hot water and detergent.

Coeliac disease – an allergy to gluten.

Confiserie variée –other petits fours confections that do not come under the glacé or sec categories (e.g. truffles, fudges, caramel mou and pralines).

Consommé – a clear, unthickened soup, with an intense flavour derived from meat or fish bones and a good stock, clarified by a process of careful straining.

Contamination – when items get into food that should not be there.

Control of Substances Hazardous to Health 2002 (COSHH) – under these regulations, risk assessments must be completed by employers of all hazardous chemicals and substances that employees may be exposed to at work, their safe use and disposal.

Convalescent carriers – those recovering from an illness but still carry the bacteria and can pass these on to the food they handle.

Core temperature – the temperature right at the centre of the food, in the thickest part. For most foods to be safe, it is necessary for foods to remain at the core temperature for two minutes or more. Core temperature is usually checked with a disinfected food probe.

Corrective actions (CAs) – the actions that must be taken by the food handler where a CCP is identified to ensure the safe production of food.

Couverture – means 'covering' or 'coating' in English. It is chocolate with a very high percentage of cocoa butter (at least 30 per cent) and is used to flavour patisserie products such as ice creams, mousses, ganaches and soufflés.

Creaming – Used to make sweet pastry. Fat is creamed with sugar, followed by egg before the flour is added.

Crème anglaise –used as a base for desserts such as bavarois, ice cream and oeufs à la neige.

Crème chantilly – sweetened cream lightly whipped to piping consistency and flavoured with vanilla.

Crème fouettée – this is lightly whipped cream with no flavourings or sweetener, and is used to enrich and aerate certain pastry products such as mousses, parfaits and cold soufflés.

Crémeux – literally means creamy/smooth and is made from a crème anglaise base emulsified onto chocolate, which is whipped when cold to a smooth, shiny chocolate cream used for filling pastry products.

Critical Control Points (CCPs) – stages where it is essential for intervention to be taken to deal with the risks.

Critical limits (CLs) – maximum limits within a process that have been set by management in the HACCP analysis.

Cross-contamination – when pathogenic bacteria (or other contaminants) are transferred from one place to another.

Crustaceans – have tough protective outer shells and have flexible joints to allow for quick movement.

Curing – preserving through the use of salt and drying; sugar, spices or nitrates may also.

Cyclical menu – a menu compiled to cover a given period of time, e.g. one month or three months. At the end of each period, the menus can be used over again.

Danger zone – temperatures between 5°C and 63°C. These are the temperature at which pathogenic bacteria may start to multiply. Food should be kept out of these temperatures as much as possible.

Darnes – steaks cut through the bone of a round fish.

Demi-glace – a brown sauce produced from reducing brown stock and lightly thickening the stock with arrowroot.

Demi-vegetarians – this group usually chooses to exclude red meat, though they may eat it occasionally. White poultry and fish are generally acceptable.

Detergent – removes grease and dirt and holds them in suspension in water. It may be in the form of liquid, powder, gel or foam. Detergent will not kill pathogens.

Disinfectant – destroys bacteria.

Disinfection – destroys pathogens and other organisms and brings them down to a safe level. Disinfection can be achieved with chemical disinfectants, use of very hot water or by directed steam as in a steam gun.

Drawing – the removal of the entrails from the inside of poultry and game birds.

Dry curing – curing ingredients are rubbed into the meat.

Drying – removing a food's water, which inhibits the growth of micro-organisms.

Due diligence – when a person or organisation who may be subject to legal proceedings can establish a defence to show that they have taken 'all reasonable precautions and exercised due diligence' to avoid committing an offence.

Edible gelatine – a stabiliser extracted from animals' bones.

Emulsion – a mixture of oil and water.

Endotoxins – toxins produced by bacteria as they die.

Enriched dough – doughs enriched with eggs and butter.

Enterotoxins – toxins which affect the intestines.

Environmental health officer/environmental health practitioner – enforcement officers who carry out inspections of premises and have the power to serve legal notices and seize equipment.

Espuma – translates from Spanish to 'foam' or 'bubbles'. Created using a classic cream whipper.

Etuvé – a method of cooking in which food is sweated in butter or vegetable oil, covered, without colouring.

Exotoxins – toxins produced by some bacteria as they multiply in food.

Extraction rate –the extent to which the bran and the germ are extracted from the flour.

Ferment and dough – yeast is mixed with the liquid and added to a well in the flour, before flour is mixed in and sprinkled over to cover. The batter then ferments and 'erupts' through the flour crust. Often used when making savarin and blinis.

Fermentation –the chemical process of breaking down a complicated substance into simpler parts, usually with the help of bacteria, yeasts or fungi.

First in, first out (FIFO) – a stock rotation method in which foods already in storage are used before new deliveries.

Fondant – a white, opaque, solid mass made by boiling sugar, water and glucose to the soft ball stage).

Food allergy – occurs when the body's immune system sees harmless food as harmful and it therefore causes an allergic reaction.

Food intolerance – an adverse reaction to food.

Food miles – the distance that food travels from producer to consumer.

Food poisoning – an acute intestinal illness caused by eating foods contaminated with pathogenic bacteria and/or their toxins. Food poisoning may also be caused by eating poisonous fish or plants, chemicals or metals.

Food-borne illness – an illness caused by pathogenic bacteria and/or their toxins and also viruses. Pathogens do not multiply in the food, they just need to get into the intestine where they invade the cells and start to multiply. They may be transmitted person to person, in water or airborne, as well as through food.

Fruit désguisés – a petits fours glacé made by filling small fruits with coloured marzipan flavoured with kirsch, pistachio compound, praline, coffee extract and other suitable products, then dipped in boiled sugar cooked to hard-crack temperature.

Gastronomy – the study of how food influences habits, and the relationship between culture and food. It includes the study of how sociology, history, economics, geography, anthropology, marketing, science and technology impact on eating and drinking.

Gazpacho – a Spanish tomato soup served ice cold.

Goujons – skinned fillets, usually from a flat fish, cut diagonally into strips.

Granita – similar to a water ice, less sweet than a sorbet and contains no meringue. It has a slightly granular texture normally achieved by pouring into a shallow tray, freezing, then scraping the surface into frozen granules and serving in frozen glasses.

Hazard – anything that could possibly cause harm, such as chemicals, electricity, working with machinery.

Hazard Analysis and Critical Control Points (HACCP) – a food safety management food safety management system that looks at identifying the critical points or stages in any process and identifying hazards that could occur.

Healthy carriers– those who show no signs of illness but can still contaminate food.

Homogenisation – forcing milk or cream through very fine mesh under high pressure. This breaks up fat globules and distributes them evenly, which prevents the milk or cream from separating with a thicker cream layer at the top.

Hors d'oeuvres – 'outside of the main meal'. Intended to introduce the meal, create interest and stimulate the taste buds.

Insecticide – chemical product to kill insects.

Irradiation – foods are treated with low doses of gamma rays, x-rays or electrons. The energy absorbed by the food causes the formation of short-lived molecules known as free radicals, which kill bacteria that cause food poisoning.

Jus-lié – a brown sauce, which is traditionally produced from brown veal stock, flavoured with tomatoes and mushrooms and thickened with arrowroot.

Knocking back – expels old gas and bring the yeast back into contact with the dough after it is proved.

Lacto vegetarian – eats milk and cheese, but not eggs, whey or anything that is produced as a result of an animal being slaughtered.

Laminated dough – a simple or enriched dough layered with butter.

Lamination – used to make puff pastry. A series of alternating layers of a flour-based paste and a fat of the same texture are made using a series of either single or double turns during the preparation of the paste.

Larding – inserting strips of fat into the lean meat to keep it moist during cooking.

Liaison – a thickening agent of egg yolks and cream.

Maillard reaction – the effect of baking that causes browning.

Metamyoglobin – created when myoglobin is oxidised. Changes the colour of meat to dark red or brown.

Microbial hazard – a micro-organisms such as pathogens causing food poisoning or food-borne illness; also spores and toxins, moulds, viruses and parasites.

Molluscs – have either an external hinged double shell (e.g. mussels and scallops), or a single univalve spiral shell (e.g., winkles and whelks). They may also have soft bodies such as squid and octopus.

Monter au beurre – a butter-thickened sauce in which small pieces of butter are mixed through at the last moment before serving.

Mycotoxins – toxins produced by some moulds.

Myoglobin – pigment in the tissues of mammals which gives meat its bright red colour.

Neurotoxins – toxins which affect the nervous systems.

Ohmic heating – an electric current is passed through food, generating enough heat to destroy micro-organisms.

Organic farming –farms to restrict the use of pesticides, avoiding the use of chemical fertilisers.

Organoleptic assessment – using the senses to evaluate food.

Ovo-lacto vegetarians – eat eggs but otherwise are the same as lacto vegetarians.

Ozonation – ozone is an oxidising agent. It is an effective disinfectant and sanitiser for many food products.

Parfait glacé – parfait deposited into a ring mould lined with jaconde sponge, then frozen.

Pascalisation – utilises ultra-high pressures to inhibit the chemical processes of food deterioration.

Pasteurisation – kills off pathogenic bacteria, yeasts and moulds. Liquids are heated to 63°C for 30 minutes or 72°C for 15 seconds.

Pâte à bombe – this is made by whipping egg yolks until aerated, slowly pouring on sugar cooked to 121°C and whisking until cold: it is used in the base of chocolate mousses, parfait glacé, bombe glacée and fruit gratins.

Pathogenic bacteria – bacteria with the capacity to cause disease and reaction.

Paupiettes – fillets of fish spread with a suitable stuffing and rolled.

Pectin – a commonly used gelling substance from citrus peel.

Pesticides – chemical products to kill specific pests.

Petits fours glacés – petits fours with a glazed finish on the outer surface which can be achieved by using fondant, boiled sugar or chocolate.

Petits fours secs – 'dry' confections requiring further embellishment once baked to give them a decorative finish.

Physical hazard – comes from objects such as machine parts or broken machinery, paperclips, fingernails, hair, insects, packaging materials, coins, buttons, blue plasters.

Pickling – using vinegar to preserve foods.

Pin boning – removing the bones on larger fish that go into the muscles cross-wise.

Plaited (en tresse) – thin strips are cut down the length of a fillet of fish then three strips are plaited together and neatly arranged before cooking.

Potage – a French term referring to a thick soup.

Protein complementing – the process of combining various plant proteins in one dish or meal to provide the equivalent amino acid profile of animal protein.

Proving – allows yeast to ferment and increase in size, developing texture and flavour.

Retardation – process of holding dough at temperatures between 2 and 4°C to stop yeast activity. This allows the dough to be made and held, then baked as needed.

Risk – the chances, high or low, that someone could be harmed by the hazard.

Risk assessment – examination of what in the workplace could cause harm or injury to people, so that safeguards and precautions can be put in place to prevent harm.

Roux – equal quantities of butter, flour or vegetable oil in flour.

Sabayon – a mixture of whole egg or egg yolk with caster sugar, whisked over a pan of simmering water to form a cooked, aerated mass.

Safer Food, Better Business – a food safety management system launched by the Food Standards Agency and based on the principles of HACCP but in an easy-to-understand format, with pre-printed pages and charts to enter the relevant information.

Sanitiser – cleans and disinfects and usually comes in spray form.

Scores on the Doors – a strategy introduced by the Food Standards Agency to raise food safety standards and help reduce the incidence of food poisoning. A star rating for food safety (ranging from 0 to 5 stars) is awarded.

Smoking – adding colour and flavour to food using smoke.

Sorbet – alight, slightly less sweet water ice, to which Italian meringue is added during churning, after the base mixture has partially frozen.

Soubise – a purée or sauce made from onions.

Sous vide – a method of cooking and processing by which food is cooked in a vacuum pouch under vacuum at low temperatures, steamed in a water bath and reheated.

Sponge and dough – used for making enriched doughs; a batter or 'sponge' is made with the yeast, liquid, and some of the flour to enable the yeast to start working before the other ingredients are added.

Sterilisation – the elimination of all micro-organisms through extended heating at high temperatures.

Steriliser – this can be chemical or through the action of extreme heat and will kill all living micro-organisms.

Straight dough – simplest method used when making simple bread doughs. All the ingredients are added at once and mixed together.

Sublimation – the process in which a solid changes directly to a vapour without passing through a liquid phase.

Suprême – the wing and half the breast of a chicken with the trimmed wing bone attached.

Sustainability – ensuring that food is purchased, consumed and prepared with as little impact on the environment as possible, for a fair price, and which makes a positive contribution to the local economy.

Table d'hôte – a set menu at a set price, also known as a fixed-price menu.

Tempering – chocolate is taken through different temperatures to stabilise the fats in the chocolate to give a crisp, glossy finish when dry. It allows for chocolate to be manipulated and combined with other ingredients for moulding.

Tronçon – steaks cut through the bone of a large flat fish.

Trussing – tying meat with sting so that it cooks evenly.

Turning – shaping vegetables for cookery and presentation.

Ultra-heat treatment (UHT) – preservation method in which milk is heated to between 140°C and 150°C for 2 seconds, then cooled quickly.

Univalve – mollusc with a single shell.

Vegan – someone who avoids all animal products and by-products.

Vegetable gums – e.g. alginin, guar gum, locust bean gum and xanthan gum. Increase the viscosity of a liquid.

Vegetable starches – e.g. cornflour, arrowroot, fecule and rice flour. Gelatinise and thicken sauces.

Vegetative reproduction – the ability, in favourable conditions, of living bacteria to thrive and multiply by splitting in half.

Velouté – a velvety French sauce made with equal quantities of butter and flour.

Vichyssoise – a simple, flavourful, puréed potato and leek soup, thickened with the potato itself.

Water ice – frozen desserts made from simple syrup flavoured with a fruit purée and lemon juice.

Index of recipes

This index lists every recipe in the book, grouped by major commodity and by type of dish. There is a full topic index at the back of the book.

Index

CARDIFF AND VALE COLLEGE